TELECOMMUNICATION TRANSMISSION HANDBOOK

TELECOMMUNICATION TRANSMISSION HANDBOOK

SECOND EDITION

ROGER L. FREEMAN

A WILEY-INTERSCIENCE PUBLICATION

JOHN WILEY & SONS, New York • Chichester • Brisbane • Toronto • Singapore

Library of Congress Cataloging in Publication Data:

Freeman, Roger L.
    Telecommunication transmission handbook.

    "A Wiley-Interscience publication."
    Includes index.
    1. Telecommunication. I. Title.

TK5101.F66   1981   621.38   81-7499
ISBN 0-471-08029-2   AACR2

Printed in the United States of America

10 9 8 7 6 5 4 3 2

FOR

CRISTI, BOBBY, and ROSI

PREFACE

In writing this book I have sought to reach a varied readership. It is addressed to telecommunication engineering students and to transmission engineers specializing in one discipline, such as data transmission, who wish to have an appreciation of other disciplines, such as radio, telephony, or video. The book should also serve as an introduction to the problems of transmission engineering for the nonengineer. This latter group encompasses telecommunication managers and other corporate and military staff responsible for communications, as well as technicians.

To carry out this aim I have made every effort to provide explanation. It has been my intention to express, not to impress. Sufficient material and background are given so that at least a first-cut engineering effort can be made using this text as the only reference; the basic aim is to explain how to do it and why.

It must be kept in mind that a book of this size cannot treat each subdiscipline exhaustively. If more depth is desired in a specific area, a bibliography is provided at the end of each chapter to assist the reader in the search for more information as well as to document the sources used in the preparation of this work.

Engineering today is highly stratified in disciplines; and this is particularly true in the broad field of telecommunications. We have traffic engineers and radiomen, and radio itself is broken down into earth stations, radiolinks, high frequency, tropospheric scatter, and so forth. We have switching engineers, experts in signaling, and telephone transmission engineers who may or may not know anything about carrier, PCM, or the like. There are experts in outside plants, plant extension, plant operations, and data communications. We lack guidance in how to take a grouping of these fields and work them into an operational system.

The term telecommunications has grown in acceptance and usage over the last 10 years, yet it may be ambiguous. For this text the term is defined as a service that permits people or machines to communicate at a distance.

My primary concern in the preparation of this text was to emphasize the systems approach. So that we do not get bogged down in semantics, systems as used here always means the interworking of one discipline with another to reach a definite, practical end product which serves the needs of a specific user. Only point-to-point telecommunication systems are discussed in order to keep the field of discussion within manageable proportions. Television signals are treated only in that area of special system considerations necessary for the transport of a video signal on a point-to-point basis. Yet I have always kept the end user in mind. For example, how will the change of several decibels of signal-to-noise ratio at a radiolink repeater site change the picture reception for a video viewer? It should be noted that I have purposely avoided involvement with broadcast, CATV, air–ground, telemetry, and marine transmssion systems. This allows more emphasis to be placed on the basic aim of the work.

To do justice to the book, the typical reader should have a good background in algebra, trigonometry, logarithms, statistics, and electricity. They should also have some knowledge of electrical communications, particularly modulation.

The chapter order of the book is purposeful. I tried to avoid jumping into a complex concept without having introduced it previously. The readers will have a grasp of what "noise" is before having to tangle with radio, telephony before multiplex, and multiplex before broadband systems.

The object of the first chapter is to provide a leveling process. For many, admittedly, it is a review; for me it provides a useful tool to which I can refer further on in the book.

Where desirable I have referred to the appropriate CCITT (International Consultive Committee for Telephone and Telegraph), CCIR (International Consultive Committee for Radio), EIA (Electronic Industries Association), ANSI (American National Standards Institute), and U.S. military standards. Appendix A provides a guide to CCITT and CCIR recommendations and, in particular, the G. Recommendations of CCITT. Appendix B is a short glossary of acronyms, abbreviations, and some of the more difficult terms used in the text.

ROGER L. FREEMAN

Sudbury, Massachusetts
August 1981

ACKNOWLEDGMENTS

I am deeply indebted to a large group of friends and co-workers who have reviewed and suggested changes and improvements to both the first edition and the new material of the second edition. One who has given me a great deal of help and encouragement in the area of millimeter wave transmission and digital radio is Dr. James Mullen of Raytheor's R&D laboratory in Waltham, Massachusetts. John Ballard, preseident of Technology for Communications International (TCI), provided invaluable data to update the chapter on high-frequency radio. Robert W. Cullen at Raytheon's Telecommunication Operation (RDS), Norwood, Massachusetts, and John M. Rieger patiently reviewed the chapters on fiber optics and digital radio. I am appreciative of Kenneth MacDowell's suggestions on noise analysis in the fiber optics chapter and Rob Lucas's (Exxon Optical Information Systems) critique of the same chapter, as well as of George Stafford, retired vice president of Alden Electronics and Impulse Recording Corporation for his input, and the Chief Engineer of Alden Electronics, Frederick W. Simpkins, for his review and comments on the facsimile chapter. I also wish to thank Donald J. Marsh, vice president of Continental Telephone Company, and Professor Enric Vilar of Portsmouth Polytechnic University (U.K.) for their suggestions and comments, some of which I incorporated into the text and some of which I did not. Dr. G. S. Takhar of Raytheon's Communication System Directorate suggested a number of excellent improvements on channel coding in Chapter 15. As usual, my wife Paquita was patient through the difficult periods of deadlines and reviews.

R. L. F.

CONTENTS

TELECOMMUNICATION TRANSMISSION HANDBOOK

1 | INTRODUCTORY CONCEPTS

1.1 THE TRANSMISSION PROBLEM

The word transmission is often misunderstood as we try to compartmentalize tele-communications into neatly separated disciplines. A transmission engineer in telecommunications must develop a signal from a source and deliver it to the sink to the satisfaction of a customer. In a broad sense we may substitute the words transmitter for source and receiver for sink. Of major concern are those phenomena, conditions, and factors that distort or otherwise make the signal at the sink such that the customer is dissatisfied. To understand the problems of transmission, we must do away with some of the compartmentalization. Besides switching and signaling, transmission engineers must have some knowledge of maps, civil engineering, and power. A familiarity with basic traffic engineering concepts, such as busy hour, activity factor, and so forth, is also helpful.

Transmission system engineering deals with the production, transport, and delivery of a quality signal from source to sink. The following chapters describe methods of carrying out this objective.

1.2 A SIMPLIFIED TRANSMISSION SYSTEM

The simple drawing illustrates a transmission system. The source may be a telephone mouthpiece (transmitter) and the sink may be the telephone earpiece (receiver). The source converts the human intelligence, such as voice, data information, or video, into an electrical equivalent or electrical signal. The sink accepts the electrical signal and reconverts it to an approximation of the original human intelligence. The source and sink are electrical transducers. In the case of printer telegraphy, the source may be a keyboard, where each key, when depressed, transmits to the sink distinct electrical impulses. The sink in this case may be a teleprinter, which converts

1

each impulse grouping back to the intended character keyed or depressed at the source.

The transmission media can be represented as a network:

or as a series of electrical networks:

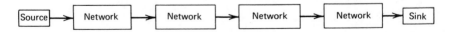

Some networks show a gain in level, others a loss. We must be prepared to discuss these gains and losses as well as electrical signal levels and the level of disturbing effects such as noise, crosstalk, or distortion. To do this we must have a firm and solid knowledge of the decibel and related measurement units.

1.3   THE DECIBEL

The decibel is a unit that describes a ratio. It is a logarithm with a base of 10. Consider first a power ratio. The number of decibels (dB) = 10 $\log_{10}$ (the power ratio).

Let us look at the following network:

The input is 1 W and its output 2 W, in the power domain. Therefore we can say the network has a 3-dB gain approximately. In this case

$$\text{Gain (dB)} = 10 \log \frac{\text{output}}{\text{input}} = 10 \log \frac{2}{1} = 10(0.3013) = 3.0103 \text{ dB}$$

or approximately, a 3-dB gain.

Now let us look at another network:

In this case there is a loss of 30 dB:

$$\text{Loss (dB)} = 10 \log \frac{\text{input}}{\text{output}} = 10 \log \frac{1000}{1} = 30 \text{ dB}$$

Or in general we can state

$$\text{Power (dB)} = 10 \log \frac{P_2}{P_1} \qquad (1.1)$$

where P_1 = lower power level, P_2 = higher power level.

A network with an input of 5 W and an output of 10 W is said to have a 3-dB gain:

$$\text{Gain (dB)} = 10 \log \tfrac{10}{5} = 10 \log \tfrac{2}{1} = 10\,(0.30103)$$

$$= 3.0103 \text{ dB} \simeq 3 \text{ dB}$$

This is a good figure to remember. Doubling the power means a 3-dB gain; likewise, halving the power means a 3-dB loss.

Consider another example, a network with a 13-dB gain:

$$\text{Gain (dB)} = 10 \log \frac{P_2}{P_1} = 10 \log \frac{P_2}{0.1} = 13 \text{ dB}$$

Then

$$P_2 \simeq 2 \text{ W}$$

Table 1.1 may be helpful. All values in the power ratio column are $X/1$, or compared to 1.

It is useful to be able to work with decibels without pencil and paper. Relationships of 10 and 3 have been reviewed. Now consider the following:

Table 1.1

Power Ratio		dB	Power Ratio	dB
10^1	(10)	+10	10^{-1} (1/10)	−10
10^2	(100)	+20	10^{-2} (1/100)	−20
10^3	(1,000)	+30	10^{-3} (1/1000)	−30
10^4	(10,000)	+40	10^{-4} (1/10,000)	−40
10^5	(100,000)	+50	10^{-5} (1/100,000)	−50
10^6	(1,000,000)	+60	10^{-6} (1/1,000,000)	−60

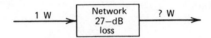

What is the power output of this network? To do this without pencil and paper, we would proceed as follows. Suppose that the network attenuated the signal 30 dB. Then the output would be 1/1000 of the input, or 1 mW, 27 dB is 3 dB less than 30 dB. Thus the output would be twice 1 mW, or 2 mW. It really is quite simple. If we have multiples of 10, as in the previous table, or 3 up or 3 down from these multiples, we can work it out in our heads, without pencil and paper.

Look at this next example:

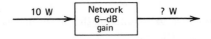

Working it out with pencil and paper, we see that the output is approximately 40 W. Here we have a multiple of 4. A 6-dB gain represents approximately a fourfold power gain. Likewise, a 6-dB loss would represent approximately one-fourth the power output. Now we should be able to work out many combinations without resorting to pencil and paper.

Consider a network with a 33-dB gain with an input level of 0.15 W. What would be the output? 30 dB represents multiplying the input power by 1000, and 3 additional decibels doubles it. In this case the input power is multiplied by 2000. Thus the answer is 0.15 × 2000 = 300 W.

The table below may further assist the reader regarding the use of decibels as power ratios.

	Approximate Power Ratio	
Decibels	Losses	Gains
1	0.8	1.25
2	0.63	1.6
3	0.5	2.0
4	0.4	2.5
5	0.32	3.2
6	0.25	4.0
7	0.2	5.0
8	0.16	6.3
9	0.125	8.0
10	0.1	10.0

The voltage and current ratios, owing to the squared term, result in twice the decibel levels as for power. Remember

$$\text{Power} = I^2 R = \frac{E^2}{R}$$

Thus

$$dB \text{ (voltage)} = 20 \log \frac{E_2}{E_1}$$

or

$$dB \text{ (current)} = 20 \log \frac{I_2}{I_1}$$

when relating current and voltage ratios. E_2 = higher voltage, E_1 = lower voltage; I_2 = higher current, I_1 = lower current.

When using the current and voltage relationships shown above, keep in mind that they must be compared against like impedances. For instance, E_2 may not be taken at a point of 600-ohm (Ω) impedance and E_1 at a point of 900 Ω.

Example 1. How many decibels correspond to a voltage ratio of 100?

$$dB = 20 \log \frac{E_2}{E_1}$$

when $E_2/E_1 = 100$,

$$dB = 20 \log 100 = 40 \text{ dB}$$

Example 2. If an amplifier has a 30-dB gain, what voltage ratio does the gain represent? Assume equal impedances at input and output of the amplifier.

$$30 = 20 \log \frac{E_2}{E_1}$$

$$\frac{E_2}{E_1} = 31.6$$

Thus the ratio is $31.6:1$.

1.4 BASIC DERIVED DECIBEL UNITS

1.4.1 The dBm

Up to now all reference to decibels has been made in terms of ratios or relative units. We *cannot* say the output of an amplifier is 33 dB. We can say that an amplifier has a gain of 33 dB or that a certain attenuator has a 6-dB loss. These figures or units give no idea whatsoever of the absolute level. Several derived decibel units do.

Perhaps dBm is the most common of these. By definition dBm is a power level related to 1 mW. A most important relationship to remember is: 0 dBm = 1 mW. The formula may then be written:

$$\text{Power (dBm)} = 10 \log \frac{\text{power (mW)}}{1 \text{ mW}}$$

Example 1. An amplifier has an output of 20 W. What is its output in dBm?

$$\text{Power (dBm)} = 10 \log \frac{20 \text{ W}}{1 \text{ mW}}$$

$$= 10 \log \frac{20 \times 10^3 \text{ mW}}{1 \text{ mW}} \simeq +43 \text{ dBm}$$

(The plus sign indicates that the quantity is above the level of reference, 0 dBm.)

Example 2. The input to a network is 0.0004 W. What is the input to dBm?

$$\text{Power (dBm)} = 10 \log \frac{0.0004 \text{ W}}{1 \text{ mW}}$$

$$= 10 \log 4 \times 10^{-1} \text{ mW} \simeq -4 \text{ dBm}$$

(The minus sign in this case tells us that the level is below reference, 0 dBm or 1 mW.)

1.4.2 The dBW

The dBW is used extensively in microwave applications. It is an absolute decibel unit and may be defined as decibels referred to 1 W:

$$\text{Power level (dBW)} = 10 \log \frac{\text{power (W)}}{1W} \qquad (1.2)$$

Remember the following relationships:

$$+30 \text{ dBm} = 0 \text{ dBW} \qquad (1.3)$$

$$-30 \text{ dBW} = 0 \text{ dBm} \qquad (1.4)$$

Consider this network:

Its output level in dBW is +20 dBW. Remember that the gain of the network is 20 dB or 100. This output is 100 W or +20 dBW.

Table 1.2

dBm	dBW	Watts	dBm	dBW	Milliwatts
+66	+36	4000	+30	0	1000
+63	+33	2000	+27	− 3	500
+60	+30	1000	+23	− 7	200
+57	+27	500	+20	−10	100
+50	+20	100	+17	−13	50
+47	+17	50	+13	−17	20
+43	+13	20	+10	−20	10
+40	+10	10	+ 7	−23	5
+37	+ 7	5	+ 6	−24	4
+33	+ 3	2	+ 3	−27	2
+30	0	1	0	−30	1
			− 3	−33	0.5
			− 6	−36	0.25
			− 7	−37	0.20
			−10	−40	0.1

Table 1.2, a table of equivalents, may be helpful.

1.4.3 The dBmV

The absolute decibel unit dBmV is used widely in video transmission. A voltage level may be expressed in decibels above or below 1 mV across 75 Ω, which is said to be the level in decibel-millivolts or dBmV. In other words,

$$\text{Voltage level (dBmV)} = 20 \log_{10} \frac{\text{voltage (mV)}}{1 \text{ mV}} \qquad (1.5)$$

Table 1.3

RMS Voltage across 75 Ω	dBmV
10 V	+80
2 V	+66
1 V	+60
10 mV	+20
2 mV	+6
1 mV	0
500 μV	−6
316 μV	−10
200 μV	−14
100 μV	−20
10 μV	−40
1 μV	−60

when the voltage is measured at the *75-Ω impedance level*. Simplified,

$$dBmV = 20 \log_{10} \text{(voltage in millivolts at 75-Ω impedance)} \qquad (1.6)$$

Table 1.3 (on preceding page) may prove helpful.

1.5 THE NEPER

A transmission unit used in a number of northern European countries as an alternate
to the decibel is the neper (Np). To convert decibels to nepers, multiply the number
of decibels by 0.1151. To convert nepers to decibels, multiply the number of
nepers by 0.686. Mathematically,

$$Np = \frac{1}{2} \log_e \frac{P_2}{P_1} \qquad (1.7)$$

where P_2, P_1 = higher and lower powers, respectively, and $e = 2.718$, the base of
the natural or Naperian logarithm. A common derived unit is the decineper $(dNp)_e$.
A decineper is one-tenth of a neper.

1.6 ADDITION OF POWER LEVELS IN dB
(dBm/dBW OR SIMILAR ABSOLUTE LOGARITHMIC UNITS)

Adding decibels corresponds to the multiplying of power ratios. Care must be taken
when adding or subtracting absolute decibel units such as dBm, dBW, and some
noise units. Consider the combining network below, which is theoretically lossless:

What is the resultant output? Answer: +33 dBm.

Figure 1.1 is a curve for directly determining the level in absolute decibel units
corresponding to the sum or difference of two levels, the values of which are known
in terms of decibels with respect to some reference.

As an example, let us add two power levels, 10 dBm and 6.0 dBm. Take the difference between them, 4 dB. Spot this value on the horizontal scale (the abscissa)
on the curve. Project the point upward to where it intersects the "addition" curve
(the upper curve). Take the corresponding number to the right and add it to the
larger level. Thus

$$10 \text{ dBm} + 1.45 \text{ dB(m)} = 11.45 \text{ dBm}$$

Suppose we subtract the 6.0 dBm signal from the 10 dBm signal. Again the difference is 4 dB. Spot this value on the horizontal scale as before. Project the point
upward to where it meets the "subtraction" curve (the lower curve). Take the

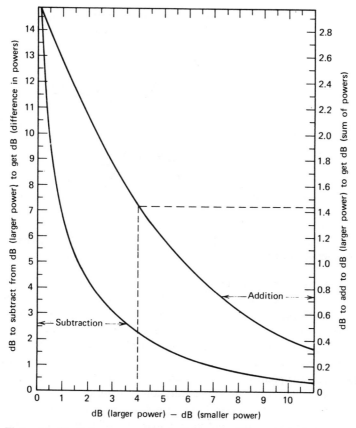

Figure 1.1 Decibels corresponding to the sum or difference of two levels.

corresponding number and subtract it from the larger level. Thus

$$10 \text{ dBm} - 2.3 \text{ dB(m)} = 7.7 \text{ dBm}$$

When it is necessary to add equal absolute levels expressed in decibels, add 10 log (the number of equal powers) to the level value. For example, add four signals of +10 dBm each. Thus

$$10 \text{ dBm} + 10 \log 4 = 10 \text{ dB(m)} + 6 \text{ dB} = +16 \text{ dBm}$$

When there are more than two levels to be added and they are not of equal value, proceed as follows. Pair them and sum the pairs, using Figure 1.1. Sum the resultants of the pairs in the same manner until one single resultant is obtained.

1.7 NORMAL DISTRIBUTION—STANDARD DEVIATION

A normal or Gaussian distribution is a binomial distribution where n, the number of points plotted (number of events), approaches infinity (Ref. 11, pp. 121, 122,

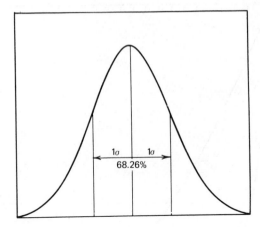

Figure 1.2 A normal distribution curve showing one standard deviation measured off either side of the arithmetic mean.

and 143). A distribution is an arrangement of data. A frequency distribution is an arrangement of numerical data according to size and magnitude. The normal distribution curve (Figure 1.2) is a symmetrical distribution. A nonsymmetrical frequency distribution curve is one in which the distributions extend further in one direction than in the other. This type of distortion is called skew. The peak of the normal distribution curve is called the point of central tendency, and its measure is its average. This is the point where the group of values tends to cluster.

The dispersion is the variation, scatteration of data, or the lack of tendency to congregate (Ref. 17). The range is the simplest measure of dispersion and is the difference between maximum and minimum values of a series. The mean deviation is another measure of dispersion. In a frequency distribution, ignoring signs, it is the average distance of items from a measure of the central tendency.

The standard deviation is the root mean square (rms) of the deviations from the arithmetic mean and is expressed by the small Greek letter sigma:

$$\sigma = \sqrt{\frac{\Sigma(X'^2)}{N}} \tag{1.8}$$

where X' = deviations from the arithmetic mean $(X - \overline{X})$, and N = total number of items.

The following expressions are useful when working with standard deviations: they refer to a "normal" distribution:

- The mean deviation = 0.7979σ.
- Measure off on both sides of the arithmetic mean one standard deviation; then 68.26% of all samples will be included within the limits.
- For two standard deviations measured off, 95.46% of all values will be included.
- For three standard deviations, 99.73% will be included.

These last three items relate to *exact* normal distributions. In cases where the distribution has moderate skew, approximate values are used, such as 68% for 1σ, 95% for 2σ, and so on.

1.8   THE SIMPLE TELEPHONE CONNECTION

Two people may speak to one another over a distance by connecting two telephone subsets together with a pair of wires and a common microphone battery supply. As the wires are extended (i.e., the distance between the talkers is increased), the speech power level decreases until at some point, depending on the distance, the diameter of the wire, and the mutual capacitance between each wire in the pair, communication becomes unacceptable. For example, in the early days of telephony in the United States it was noted that a telephone connection including as much as 30 mi (48 km) of 19-gauge nonloaded cable was at about the limit of useful transmission.

Suppose that several people want to join the network. We could add them in parallel (bridge them together). As each is added, however, the efficiency decreases because we have added subsets in parallel and, as a result, the impedance match between subset and line deteriorates. Besides, each party can overhear what is being said between any two others. Lack of privacy may be a distinct disadvantage at times.

This can be solved by using a switch so that the distant telephone may be selected. Now a signaling system must be developed so that the switch can connect the caller to the distant telephone. A system of monitoring or supervision will also be required so that on-hook and off-hook conditions may be known by the switch as well as to permit line seizure by a subscriber.

Now extend the system again, allowing several switches to be used interconnected by trunks (junctions). Because of the extension of the two-wire system without amplifiers, a reduced signal level at many of the subscribers' telephone subsets may be experienced. Now we start to reach into the transmission problem. A satisfactory signal is not being delivered to some subscribers owing to line losses because of excessive wire line lengths. Remember that line loss increases with length.

Before delving into methods of improving subscriber signal level and satisfactory signal-to-noise ratio, we must deal with basic voice channel criteria. In other words, just what are we up against? Consider also that we may want to use these telephone facilities for other types of communication such as telegraph, data, facsimile, and video transmission. The voice channel (telephone channel) criteria covered below are aimed essentially at speech transmission. However, many parameters affecting speech most certainly have bearing on the transmission of other types of signals, and other specialized criteria are peculiar to these other types of transmission. These are treated in depth in later chapters, where they become more meaningful. Where possible, cross reference is made.

Before going on, refer to the simplified sketch (Figure 1.3) of a basic telephone

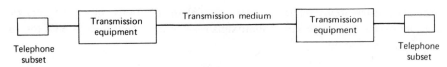

Figure 1.3 A simplified telephone transmission system.

connection. The sketch contains all the basic elements that will deteriorate the signal from source to sink. The medium may be wire, optical fiber, radio, or combinations of the three. Other transmission equipment may be used to enhance the medium by extending or expanding it. This equipment might consist of amplifiers, multiplex devices, and other signal processors such as compandors, voice terminals, and so forth.

1.9 THE PRACTICAL TRANSMISSION OF SPEECH

The telephone channel, hereafter called the voice channel, may be described technically using the following parameters:

- Nominal bandwidth
- Attenuation distortion (frequency response)
- Phase shift or envelope delay distortion
- Noise and signal-to-noise ratio
- Level

Return loss, singing, stability, echo, reference equivalent, and some other parameters deal more with the voice channel in a network and are discussed at length when we look at a transmission network later.

1.9.1 Bandwidth

The range between the lowest and highest frequencies used for a particular purpose may be defined as bandwidth. For our purposes we should consider bandwidth as those frequencies within which a performance characteristic of a device is above certain specified limits. For filters, attenuators, and amplifiers, these limits are generally taken to be where a signal will fall 3 dB below the average level in the passband or below the level at a reference frequency (1000 Hz in the United States and Canada, 800 Hz in Europe). These 3-dB points are by definition half-power points. The nominal bandwidth of a voice channel is often 4 kHz. The actual usable bandwidth is more on the order of 3.1 kHz for most modern carrier equipment. The bandwidth of a television transmitter may be defined as 6 MHz and that of a synchronous earth satellite repeater, 500 MHz.

1.9.2 Speech Transmission—The Human Factor

Frequency components of speech may be found between 20 Hz and 20 kHz. The frequency response of the ear (i.e., how it reacts to different frequencies) is a

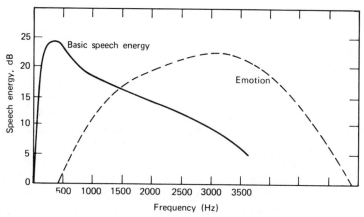

Figure 1.4 Energy and emotion distribution in speech. (Source: *Bell Sys. Tech. J.*, July 1931).

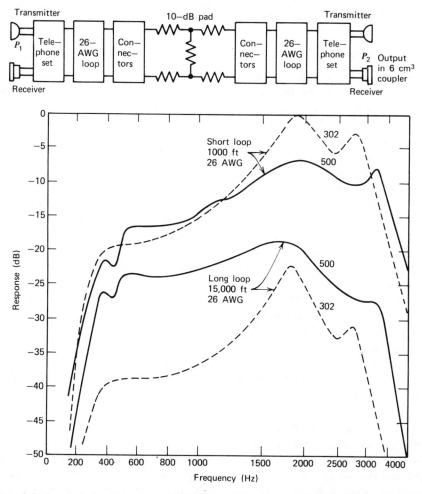

Figure 1.5 Comparison of overall response. (Source: W. F. Tuffnell, "500-Type Telephone Set," *Bell Lab. Rec.*, 29, 414–418, Sept. 1951; copyright © 1951 by Bell Telephone Laboratories.

nonlinear function between 30 Hz and 30 kHz; however, the major intelligence and energy content exists in a much narrower band. For energy distribution see Figure 1.4. The emotional content, which transfers intelligence, is carried in a band which lies above the main energy portion. Tests have shown that low frequencies up to 600-700 Hz add very little to the intelligibility of a signal to the human ear, but in this very band much of the voice energy is transferred (solid line in Figure 1.4). The dashed line in Figure 1.4 shows the portion of the frequency band that carries emotion. From this it can be seen that for economical transfer of speech intelligence, a band much narrower than 20 Hz to 20 kHz is necessary. In fact the standard bandwidth of a voice channel is 300-3400 Hz (CCITT Recs. G.132 and G.151A). As is shown later, this bandwidth is a compromise between what telephone subscribers demand (Figure 1.4) and what can be provided to them economically. However, many telephone subsets have a response range no greater than approximately 500-3000 Hz. This is shown in Figure 1.5, where the response of the more modern Bell System 500 telephone set is compared to the older 302 set.

1.9.3 Attenuation Distortion

A signal transmitted over a voice channel suffers various forms of distortion. That is, the output signal from the channel is distorted in some manner such that it is not an exact replica of the input. One form of distortion is called attenuation distortion and is the result of less than perfect amplitude-frequency response. If attenuation distortion is to be avoided, all frequencies within the passband should be subjected to the same loss (or gain). On typical wire systems higher frequencies in the passband are attenuated more than lower ones. In carrier equipment the filters used tend to attenuate frequencies around band center the least, and attenuation increases as the band edges are approached. Figure 1.6 is a good example of this. The cross-hatched areas in the figure express the specified limits of attenuation distortion, and the solid line shows measured distortion on typical carrier equipment for the channel band (see Chapter 3). It should be remembered that any practical communication channel will suffer some form of attenuation distortion.

Attenuation distortion across the voice channel is measured compared to a reference frequency. CCITT specifies the reference frequency at 800 Hz. However, 1000 Hz is used more commonly in North America (see Figure 1.6).

For example, one requirement may state that between 600 and 2800 Hz the level will vary by not more than -1, +2 dB, where the plus sign means more loss and the minus sign means less loss. Thus if a signal at -10 dBm is placed at the input of the channel, we would expect -10 dBm at the output at 800 Hz (if there was no overall loss or gain), but at other frequencies we could expect a variation between -1 and +2 dB. For instance, we might measure the level at the output at 2500 Hz at -11.9 dBm and at 1000 Hz at -9 dBm.

CCITT recommendations for attenuation distortion may be found in volume III, Recs. G.132 and G.151A. Figure 2.16 is taken from Rec. G.132 and shows permissible variation of attenuation between 300 and 3400 Hz. Often the requirement is stated as a slope in decibels. The slope is the maximum excursion that levels may vary in a band of interest about a certain frequency. A slope of 5 dB may be a curve

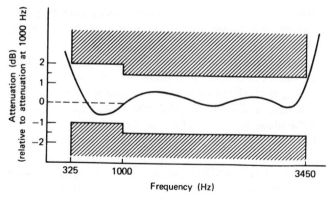

Figure 1.6 Typical attenuation distortion curve for a voice channel. Note that the reference frequency in this case is 1000 Hz.

with an excursion from −0.5 to +4.5 dB, 3 to +2 dB, and so forth. As links in a system are added in tandem, to maintain a fixed attenuation distortion across the system, the slope requirement for each link becomes more severe.

1.9.4 Phase Shift and Envelope Delay Distortion

One may look at a voice channel or any bandpass as a bandpass filter. A signal takes a finite time to pass through the filter. This time is a function of the velocity of

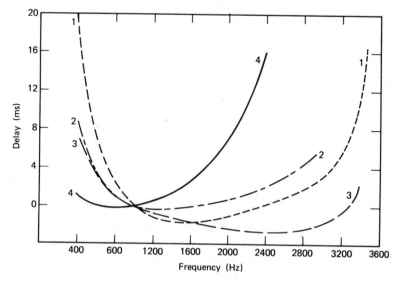

Figure 1.7 Comparison of envelope delay in some typical voice channels. Curves 1 and 3 represent the delay in several thousand miles of a toll-quality carrier system. Curve 2 shows the delay produced by 100 mi of loaded cable. Curve 4 shows the delay in 200 mi of heavily loaded cable. (Courtesy GTE Lenkurt Inc., San Carlos, CA.)

propagation. The velocity of propagation tends to vary with frequency, increasing toward band center and decreasing toward band edge, usually in the form of a parabola (see Figure 1.7).

The finite time it takes a signal to pass through the total extension of a voice channel or any other network is called delay. Absolute delay is the delay a signal experiences passing through the channel at a reference frequency. But we see that the propagation time is different for different frequencies. This is equivalent to phase shift. If the phase shift changes uniformly with frequency, the output signal will be a perfect replica of the input and there will be no distortion, whereas if the phase shift is nonlinear with respect to frequency, the output signal is distorted (i.e., it is not a perfect replica of the input). Delay distortion, or phase distortion as it is often called, is usually expressed in milliseconds or microseconds about a reference frequency.

We can relate phase distortion to phase delay. If the phase shift characteristic is known, the phase delay T_p at any frequency β_1 can be calculated as follows:

$$T_p = \frac{\beta_1 \ (\text{rad})}{\omega_1 \ (\text{rad/s})} \tag{1.9}$$

The difference between phase delays at two frequencies in a band of interest is called delay distortion T_d and can be expressed as

$$T_d = \frac{\beta_2}{\omega_2} - \frac{\beta_1}{\omega_1} \tag{1.10}$$

where β_2/ω_2 and β_1/ω_1 are the phase delays at ω_2 and ω_1, frequency being expressed as angular frequency ($2\pi f$ rad).

In essence, therefore, we are dealing with the phase linearity of a circuit. The resulting phase distortion is best measured by a parameter called envelope delay distortion (EDD). Mathematically enveloped delay is the derivative of the phase shift with respect to frequency and expresses the instantaneous slope of the phase shift characteristic. Envelope delay T_e can be stated as follows:

$$T_e = \frac{d\beta}{d\omega} \tag{1.11}$$

and the expression is valid for very small bandwidths, often referred to as apertures. For instance, the U.S. Bell System uses $166\frac{2}{3}$ Hz and CCITT, $83\frac{1}{3}$ Hz as standard apertures in measurement equipment.

The measurement of envelope delay is useful in television and facsimile transmission systems and is used in data transmission to determine intersymbol interference, a major limitation to maximum bit rate over a transmission channel. EDD is discussed in greater detail in Chapter 8.

In commercial telephony the high-frequency harmonic components, produced by the discontinuous nature of speech sounds, arrive later than the fundamental components and produce sounds which may be annoying, but do not appreciably reduce intelligibility. With present handset characteristics the evidence is that the human ear is not very sensitive to phase distortions that develop in the circuit. Although a phase delay of 12 ms between the band limits is noticeable, the transmission in commercial telephone systems often contains distortions greatly in excess of this minimum.

Owing to the larger amount of delay distortion in a telephone channel, as measured in its band and relative to a point of minimum delay, the usefulness of the entire telephone channel between its 3-dB cutoff points is severely restricted for the transmission of other than voice signals (e.g., data—see Chapter 8).

For the transmission of information which is sensitive to delay distortion, such as medium-speed digital signals, it is necessary to restrict occupancy to that part of the telephone channel in which the delay distortion can be tolerated or equalized at reasonable cost.

Applicable CCITT recommendations are Recs. G.114 and G.133.

1.9.5 Level

GENERAL

In most systems when we refer to level, we refer to a power level which may well be in dBm, dBW, or other power units. One notable exception is video, which uses voltage, usually in dBmV.

Level is an important system parameter. If levels are maintained at too high a point, amplifiers become overloaded, with resulting increases in intermodulation products or crosstalk. If levels are too low, customer satisfaction may suffer.

REFERENCE LEVEL POINTS

System levels usually are taken from a level chart or reference system drawing made by a planning group or as part of an engineered job. On the chart a 0 TLP (test level point) is established. A TLP is the location in a circuit or system at which a specified test tone level is expected during alignment. A 0 TLP is the point at which the test tone level should be 0 dBm.

From the 0 TLP other points may be shown using the unit dBr (dB reference). A minus sign shows that the level is so many decibels below reference, and a positive sign that the level is so many decibels above reference. The unit dBm0 is an absolute unit of power in dBm referred to the 0 TLP. dBm can be related to dBr and dBm0 by the following formula:

$$dBm = dBm0 + dBr \tag{1.12}$$

For instance, a value of -32 dBm at a -22-dBr point corresponds to a referenced level of -10 dBm0. A -10-dBm0 signal introduced at the 0-dBr point (0 TLP) has an absolute value of signal level of -10 dBm.

In North American practice the 0 TLP was originally defined at the transmission jack of a toll (long-distance) switchboard. Many technical changes, of course, have occurred since the days of manual switchboards. Nevertheless it was deemed desirable to maintain the 0 TLP concept. As a result, the outgoing side of a switch to which an intertoll trunk is connected (see Figure 2.9) is designated a -2-dB TLP, and the outgoing side of the switch at which a local area trunk is terminated is defined as 0 TLP (Ref. 18).

To quote from Ref. 18 in part:

In the layout of four-wire trunks, a patch bay, called the four-wire patch bay, is usually provided to facilitate test, maintenance, and circuit rearrangements between trunks and the switching machine terminations. TLPs at these four-wire patch bays have been standardized for all four-wire trunks. On the transmitting side the TLP is -16 dB, and on the receiving side the TLP is +7 dB. Thus a four-wire trunk, whether derived from voice frequency or from carrier facilities, must be designed to have 23-dB gain between four-wire patch bays. These standard TLPs are necessary to permit flexible telephone plant administration.

In four-wire circuits, the TLP concept is easily understood and applied because each transmission path has only one direction of transmission. In two-wire circuits, however, confusion or ambiguity may be introduced by the fact that a single point may be properly designated as two different TLPs, each depending on the assumed direction of transmission.

Refer to Section 2.6.3 for a discussion of two-wire and four-wire transmission and to CCITT Rec. G.141 for transmission reference point, and to Figure 2.7 for a definition of virtual switching points a-t-b.

THE VOLUME UNIT (VU)

One measure of level is the volume unit (VU). Such a unit is used to measure the power level (volume) of program channels (broadcast) and certain other types of speech or music. VU meters are usually kept on line to measure volume levels of program or speech material being transmitted. If a simple dB meter or voltmeter is bridged across the circuit to monitor the program volume level, the indicating needle tries to follow every fluctuation of speech or program power and is difficult to read; besides, the reading will have no real meaning. To further complicate matters, different meters will probably read differently because of differences in their damping and ballistic characteristics.

The indicating instrument used in VU meters is a dc milliammeter having a slow response time and damping slightly less than critical. If a steady sine wave is suddenly impressed on the VU meter, the pointer or needle will move to within 90% of the steady-state value in 0.3 s and overswing the steady-state value by no more than 1.5%.

The standard volume indicator (U.S.), which includes the meter and an associated attenuator, is calibrated to read 0 VU when connected across a 600-Ω circuit (voice pair) carrying a 1-mW sine wave power at any frequency between 35 and 10,000 Hz. For complex waves such as music and speech, a VU meter will read some value between average and peak of the complex wave. The reader must remember that there is no simple relationship between the volume measured in VUs and the power of a complex wave. It can be said, however, that for a continuous sine wave signal across 600 Ω, 0 dBm = 0 VU by definition, or that the readings in dBm and VU are the same for continuous simple sine waves in the voice frequency (VF) range. For a complex signal subtract 1.4 from the VU reading, and the result will be approximate talker power in dBm.

Talker volumes, or levels of a talker at the telephone subset, vary over wide limits for both long-term average power and peak power. Based on comprehensive tests by Holbrook and Dixon (Ref. 1), "mean talker" average power varies between −10 and −15 VU, with a mean of −13 VU.

1.9.6 Noise

GENERAL

"Noise, in its broadest definition, consists of any undesired signal in a communication circuit" (Ref. 2). The subject of noise and its reduction is probably the most important that a transmission engineer must face. It is noise that is the major limiting factor in telecommunication system performance. Noise may be divided into four categories:

1. Thermal noise (Johnson noise)
2. Intermodulation noise
3. Crosstalk
4. Impulse noise

THERMAL NOISE

Thermal noise is the noise occurring in all transmission media and in all communication equipment arising from random electron motion. It is characterized by a uniform distribution of energy over the frequency spectrum and a normal (Gaussian) distribution of levels.

Every equipment element and the transmission medium contribute thermal noise to a communication system, provided the temperature of that element of medium is above absolute zero. Thermal noise is the factor that sets the lower limit for the sensitivity of a receiving system. Often this noise is expressed as a temperature referred to absolute zero (kelvin).

Thermal noise is a general expression referring to noise based on thermal agitations. The term "white noise" refers to the average uniform spectral distribution of energy with respect to frequency. Thermal noise is directly proportional to bandwidth and temperature. The amount of thermal noise to be found in 1 Hz of bandwidth in an actual device is

$$P_n = kT \text{ (W/Hz)} \tag{1.13}$$

where k = Boltzmann's constant = 1.3803 (10^{-23}) J/K
$\qquad$ T = absolute temperature (K) of thermal noise

At room temperature, $T = 17°C$ or 290K,

$$P_n = 4.00 \ (10^{-21}) \text{ W/Hz of bandwidth}$$

$$= -204 \text{ dBW/Hz of bandwidth}$$

$$= -174 \text{ dBm/Hz of bandwidth}$$

For a band-limited system (i.e., a system with a specific bandwidth),

$$P_n = kTB \text{ (W)} \tag{1.14}$$

where B refers to what is called bandwidth (Hz). At 0 K,

$$P_n = -228.6 \text{ dBW/Hz of bandwidth} \tag{1.15}$$

and for a band-limited system,

$$P_n = -228.6 \text{ dBW} + 10 \log T + 10 \log B \tag{1.16}$$

Example 1. Given receiver with an effective noise temperature of 100 K and a 10-MHz bandwidth, what thermal noise level may we expect at its output?

$$P_n = -228.6 \text{ dBW} + 10 \log 1 \times 10^2 + 10 \log 1 \times 10^7$$

$$= -228.6 + 20 + 70$$

$$= -138.6 \text{ dBW}$$

Example 2. Given an amplifier with an effective noise temperature of 10,000 K and a 10-MHz bandwidth, what thermal noise level may we expect at its output?

$$P_n = -228.6 \text{ dBW} + 10 \log 1 \times 10^4 + 10 \log 1 \times 10^7$$

$$= -228.6 + 40 + 70$$

$$= -118.6 \text{ dBW}$$

From the examples it can be seen that there is little direct relationship between physical temperature and effective noise temperature.

INTERMODULATION NOISE

Intermodulation noise is the result of the presence of intermodulation products. Let us pass two signals with frequencies F_1 and F_2 through the nonlinear device or medium. Intermodulation products will result which are spurious frequencies. These frequencies may be present either inside or outside the band of interest for the device. Intermodulation products may be produced from harmonics of the signals in question, either as products between harmonics or as one or the other or both signals themselves.

The products result when the two (or more) signals beat together or "mix." Look at the "mixing" possibilities when passing F_1 and F_2 through a nonlinear device. The coefficients indicate first, second, or third harmonics:

- Second-order products $F_1 \pm F_2$
- Third-order products $F_1 \pm 2F_2; 2F_1 \pm F_2$
- Fourth-order products $2F_1 \pm 2F_2; 3F_1 \pm F_2$

Devices passing multiple signals, such as multichannel radio equipment, develop intermodulation products that are so varied that they resemble white noise.

Intermodulation noise may result from a number of causes:

- Improper level setting; too high a level input to a device drives the device into its nonlinear operating region (overdrive)
- Improper alignment causing a device to function nonlinearly
- Nonlinear envelope delay

To sum up, intermodulation noise results from either a nonlinearity or a malfunction having the effect of nonlinearity. The cause of intermodulation noise is different from that of thermal noise; however, its detrimental effects and physical nature are identical to those of thermal noise, particularly in multichannel systems carrying complex signals.

CROSSTALK

Crosstalk refers to unwanted coupling between signal paths. Essentially there are three causes of crosstalk. The first is the electrical coupling between transmission media, for example, between wire pairs on a VF cable system. The second is poor control of frequency response (i.e., defective filters or poor filter design), and the third is the nonlinearity performance in analog (FDM) multiplex systems. Crosstalk has been categorized into two types:

1. *Intelligible crosstalk.* At least four words are intelligible to the listener from extraneous conversation(s) in a 7-s period (Ref. 16, 3rd ed., p. 46).

2. *Unintelligible crosstalk.* Any other form of disturbing effects of one channel upon another. Babble is one form of unintelligible crosstalk.

Intelligible crosstalk presents the greatest impairment because of its distraction to the listener. One point of view is that the distraction is caused by fear of loss of privacy. Another is that the annoyance is caused primarily by the user of the primary line consciously or unconsciously trying to understand what is being said on the secondary or interfering circuits; this would be true for any interference that is syllabic in nature.

Received crosstalk varies with the volume of the disturbing talker, the loss from the disturbing talker to the point of crosstalk, the coupling loss between the two circuits under consideration, and the loss from the point of crosstalk to the listener.

As far as this discussion is concerned, the controlling element is the coupling loss between the two circuits under consideration. Talker volume or level is covered in Section 1.9.5. The effects of crosstalk are subjective, and other factors also have to be considered when the crosstalk impairment is to be measured. Among these factors are the type of people who use the channel, the acuity of listeners, traffic patterns, and operating practices.

Crosstalk coupling loss can be measured quantitatively with precision between a given sending point on a disturbing circuit and a given receiving point on a disturbed circuit. Essentially, then, it is the simple measurement of transmission loss in decibels between the two points. Between carrier circuits crosstalk coupling is most usually flat. In other words the amount of coupling experienced at one frequency will be nearly the same for every other frequency in the voice channel. For speech pairs the coupling is predominantly capacitive, and coupling loss usually has an average slope of 6 dB/octave. See CCITT Rec. 227 and the annex to CCITT Rec. G.134 which treat crosstalk measurements.

Two other units are also commonly used to measure crosstalk. One expresses the coupling in decibels above "reference coupling" and uses dBx for the unit of measure. The dBx was invented to allow crosstalk coupling to be expressed in positive units. The reference coupling is taken as a coupling loss of 90 dB between disturbing and disturbed circuits. Thus crosstalk coupling in dBx is equal to 90 minus the coupling loss in dB. If crosstalk coupling loss is 60 dB, we have 30 dBx crosstalk coupling. Thus, by definition, 0 dBx = -90 dBm at 1000 Hz.

The second unit is the crosstalk unit (CU). When the impedances of the disturbed and disturbing circuits are the same, the number of CUs is one million times the ratio of the induced crosstalk voltage or current to the disturbing voltage or current. When the impedances are not equal,

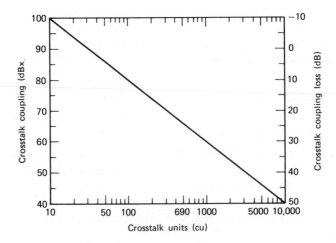

Figure 1.8 Relations between crosstalk measuring units. (Ref. 10; copyright © 1961 by American Telephone and Telegraph Company.)

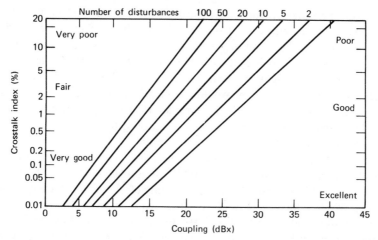

Figure 1.9 Crosstalk judgement curves. (Ref. 16; copyright © 1964 by Bell Telephone Laboratories.)

$$CU = 10^6 \sqrt{\frac{\text{crosstalk signal power}}{\text{disturbing signal power}}} \qquad (1.17)$$

Figure 1.8 relates all three units used in measuring crosstalk. CCITT recommends crosstalk criteria in Rec. G.151D (p. 4).

The percentage change of intelligible crosstalk on a circuit is defined by the crosstalk index. North American practice is to allow arbitrarily that on no more than 1% of calls will a customer hear a foreign conversation which we have defined as intelligible crosstalk. The design objective is 0.5%. The graph in Figure 1.9 may be used for guidance. It relates customer reaction to crosstalk, and crosstalk index to crosstalk coupling.

All forms of unintelligible crosstalk are covered above; its nature is very similar to intermodulation noise.

IMPULSE NOISE

This type of noise is noncontinuous, consisting of irregular pulses or noise spikes of short duration and of relatively high amplitude. Often these spikes are called "hits." Impulse noise degrades voice telephony only marginally if at all. However, it may seriously degrade the error rate on a data transmission circuit, and the subject is covered more in depth in Chapter 8.

NOISE MEASUREMENT UNITS

The interfering effect of noise on speech telephony is a function of the response of the human ear to specific frequencies in the voice channel as well as of the type of subset used.

When noise measurement units were first defined, it was decided that it would be convenient to measure the relative interfering effect of noise on the listener as a positive number. The level of a 1000-Hz tone at -90 dBm or 10^{-12} W (1 pW) was chosen by the U.S. Bell System because a tone whose level is less than -90 dBm is not ordinarily audible. Such a negative threshold meant that all noise measurements used in telephony would be greater than this number, or positive. The telephone subset then in early universal use in North America was the Western Electric 144 handset. The noise measurement unit was the dB-rn or dBrn. 0 dBrn = -90 dBm at 1000 Hz, rn standing for reference noise.

With the 144 type handset as a test receiver and with a wide distribution of "average" listeners, it was found that a 500-Hz sinusoidal signal had to have its level increased by 15 dB to have the same interfering effect on the "average" listener over the 1000-Hz reference. A 3000-Hz signal required an 18-dB increase to have the same interfering effect, 6 dB at 800 Hz, and so on. A curve showing the relative interfering effects of sinusoidal tones compared to a reference frequency is called a weighting curve. Artificial filters are made with a response resembling the weighting curve. These filters, normally used on noise measurement sets, are called weighting networks.

Subsequent to the 144 handset, the Western Electric Company developed the F1A handset, which had a considerably broader response than the older handset but was 5 dB less sensitive at 1000 Hz. The reference level for this type of handset was -85 dBm. The new weighting curve and its noise measurement weighting net-

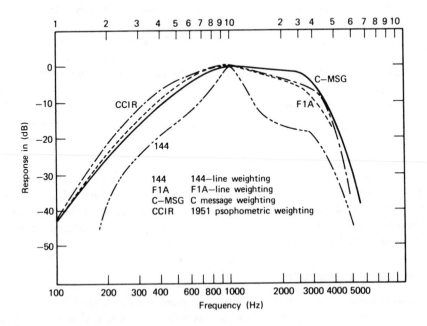

Figure 1.10 Line weightings for telephone (voice) channel noise.

Table 1.4 Conversion Chart, Psophometric, F1A, and C-Message Noise Units

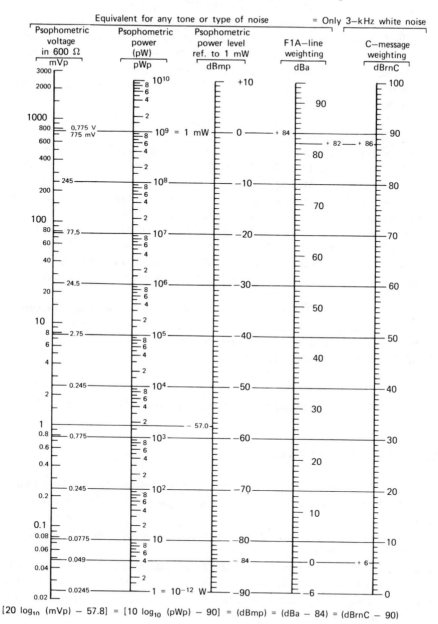

$$[20 \log_{10} (mVp) - 57.8] = [10 \log_{10} (pWp) - 90] = (dBmp) = (dBa - 84) = (dBrnC - 90)$$

work were denoted by F1A (i.e., an F1A line weighting curve, and F1A weighting network). The noise measurement unit was the dB adjusted (dBa).

A third, more sensitive handset (500 type) is now in use in North America, giving rise to the C-message line weighting curve and its companion noise measurement unit, the dBrnC. It is 3.5 dB more sensitive at the 1000-Hz reference frequency than the F1A, and 1.5 dB less sensitive than the 144 type weighting. Rather than choosing a new reference power level (-88.5 dBm), the reference power level of -90 dBm was maintained.

Figure 1.10 compares the various noise weighting curves now in use. Table 1.4 compares weighted noise units.

One important weighting curve and noise measurement unit has yet to be mentioned. The curve is the CCIR (CCITT) psophometric weighting curve. The noise measurement units associated with this curve are dBmp and pWp (dBm psophometrically weighted and picowatts psophometrically weighted, respectively). The reference frequency in this case is 800 Hz rather than 1000 Hz.

Consider now a 3-kHz band of white noise (flat, i.e., not weighted). Such a band is attenuated 8 dB when measured by a noise measurement set using a 144 weighting network, 3 dB using F1A weighting, 2.5 dB for CCIR/CCITT weighting, and 1.5 dB rounded off to 2.0 dB for C-message weighting. Table 1.4 may be used to convert from one noise measurement unit to another.

CCITT states in Rec. G.223 that, "If uniform-spectrum random noise is measured in a 3.1-kHz band with a flat attenuation frequency characteristic, the noise level must be reduced 2.5 dB to obtain a psophometric power level. For another bandwidth B, the weighting factor will be equal to

$$2.5 + 10 \log B/3.1 \text{ dB} \tag{1.18}$$

When $B = 4$ kHz, for example, this formula gives a weighting factor of 3.6 dB."

1.10 SIGNAL-TO-NOISE RATIO

The transmission system engineer deals with signal-to-noise ratio probably more frequently than with any other criterion when engineering a telecommunication system.

The signal-to-noise ratio expresses in decibels the amount by which a signal level exceeds its corresponding noise.

As we review the several types of material to be transmitted, each will require a minimum signal-to-noise ratio to satisfy the customer or to make the receiving-end instrument function within certain specified criteria. We might require the following signal-to-noise ratios with corresponding end instruments:

Voice 30 dB
Video 45 dB } based on customer satisfaction
Data 15 dB   based on a specified error rate

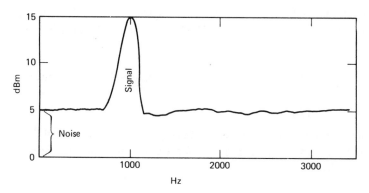

Figure 1.11   Signal-to-noise ratio S/N.

In Figure 1.11 the 1000-Hz signal has a signal-to-noise ratio S/N of 10 dB. The level of the noise is 5 dBm and the signal, 15 dBm. Thus

$$S/N_{dB} = \text{level}_{\text{signal}(dBm)} - \text{level}_{\text{noise}(dBm)} \tag{1.19}$$

1.11   THE EXPRESSION E_b/N_0

For digital transmission systems the expression E_b/N_0 is a more convenient term to qualify a received digital signal than the signal-to-noise ratio under many circumstances. E_b/N_0 expresses the received signal energy per bit per hertz of thermal noise. Thus

$$\frac{E_b}{N_0} = \frac{C}{kT \,(\text{bit rate})} \tag{1.20}$$

Where C = the receive signal level (RSL). Expressed in decibel notation,

$$\frac{E_b}{N_0} = C_{dBW} - 10 \log (\text{bit rate}) - (-228.6 \text{ dBW}) - 10 \log T_e \tag{1.21}$$

where T_e = effective noise temperature of the receiving system (see section 1.9.6).

Table 1.4   (*Continued*)

Chart basis:
  dBmp = dBa – 84
  1 mW unweighted 3-kHz white noise reads 82 dBa = 88.5 dBrnC (C-message)
  rounded off to 88.0 dBrnC. 1 mW into 600 Ω = 775 mV = 0 dBm = 10^9 pW.
Readings of noise measuring sets when calibrated on 1-mW test tone:
  F1A at 1000 Hz reads 85 dBa
  C-message at 1000 Hz reads 90 dBrn
  Psophometer at 800 Hz reads 0 dBm

Example. If the RSL for a particular digital system is -151 dBW and the receiver system effective noise temperature is 1500 K, what is E_b/N_0 for a link transmitting 2400 bps?

$$\frac{E_b}{N_0} = -151 \text{ dBW} - 10 \log 2400 - 10 \log 1500 + 228.6 \text{ dBW}$$

$$= 12 \text{ dB}$$

Depending on the modulation scheme, the type of detector used, and the coding of the transmitted signal, E_b/N_0 required for a given bit error rate (BER) may vary from 4 to 20 dB.

1.12 NOISE FIGURE

It has been established that all networks, whether passive or active, and all other forms of transmission media contribute noise to a transmission system. The noise figure is a measure of the noise produced by a practical network compared to an ideal network (i.e., one that is noiseless). For a linear system, the noise figure (NF) is expressed by

$$NF = \frac{S/N_{in}}{S/N_{out}} \tag{1.22}$$

It simply relates the signal-to-noise ratio of the output signal from the network to the signal-to-noise ratio of the input signal. From equation (1.14) the thermal noise may be expressed by the basic formula kTB, where $T = 290$ K (room temperature). As we can see, NF can be interpreted as the degradation of the signal-to-noise ratio by the network.

By letting the gain of the network G equal S_{out}/S_{in},

$$NF = \frac{N_{out}}{kTBG} \tag{1.23}$$

It should be noted that we defined the network as fully linear, so NF has not been degraded by intermodulation noise. NF more commonly is expressed in decibels, where

$$NF_{dB} = 10 \log_{10} NF \tag{1.24}$$

Example. Consider a receiver with an NF of 10 dB. Its output signal-to-noise ratio is 50 dB. What is its input equivalent signal-to-noise ratio?

$$NF_{dB} = S/N_{dB \text{ input}} - S/N_{dB \text{ output}}$$

$$10 \text{ dB} = S/N_{input} - 50 \text{ dB}$$

$$S/N_{input} = 60 \text{ dB}$$

1.13   RELATING NOISE FIGURE TO NOISE TEMPERATURE

The noise temperature of a two-port device, a receiver, for instance, is the thermal noise that that device adds to a system. If the device is connected to a noisefree source, its equivalent noise temperature

$$T_e = \frac{P_{ne}}{Gk\,df} \tag{1.25}$$

where G = gain and df = specified small band of frequencies. T_e is referred to as the effective input noise temperature of the network and is a measure of the internal noise sources of the network, and P_{ne} is the available noise power of the device (Ref. 16).

The noise temperature of a device and its NF are analytically related. Thus

$$NF = 1 + \frac{T_e}{T_0} \tag{1.26}$$

where T_0 = equivalent room temperature or 290 K.

$$T_e = T_0(NF - 1) \tag{1.27}$$

To convert NF in decibels to equivalent noise temperature T_e in kelvins, use the following formula:

$$NF_{dB} = 10 \log_{10}\left(1 + \frac{T_e}{290}\right) \tag{1.28}$$

Example 1. Consider a receiver with an equivalent noise temperature of 290 K. What is its NF?

$$NF_{dB} = 10 \log\left(1 + \frac{290}{290}\right)$$

$$NF = 10 \log 2 = 3 \text{ dB}$$

Example 2. A receiver has an NF of 10 dB. What is its equivalent noise temperature in kelvins?

$$10 \text{ dB} = 10 \log\left(1 + \frac{T_e}{290}\right)$$

$$10 = 10 \log X$$

where $X = 1 + (T_e/290)$. Thus $\log X = 1$, $X = 10$, and

$$T_e = 2900 - 290 = 2610 \text{ K}$$

Several NFs are given with their corresponding equivalent noise temperatures in Table 1.5.

Table 1.5   Noise Figure–Noise Temperature Conversion

NF_{dB}	T (K) (approx.)	NF_{dB}	T (K) (approx.)
15	8950	6	865
14	7000	5	627
13	5500	4	439
12	4300	3	289
11	3350	2.5	226
10	2610	2.0	170
9	2015	1.5	120
8	1540	1.0	75
7	1165	0.5	35.4

1.14   EFFECTIVE ISOTROPICALLY RADIATED POWER (EIRP)

A useful tool in describing antenna performance is the effective isotropically radiated power (EIRP), that is, the power in dBm or dBW over an isotropic (antenna) which is radiated. An isotropic antenna does not exist in real-life situations. It is an imaginary antenna that is used as a reference. By definition, an isotropic antenna radiates uniformly in *all* directions and therefore has a gain of 1 or 0 dB. The isotropic is a very handy tool to help radio engineers describe how common antennas function with regard to radiated power.

Perhaps an example will best assist in describing how we can use EIRP. Let an antenna have a 10-dB gain in a certain direction and connect a 10-kW transmitter to it via a very efficient transmission line, in effect, lossless (assume a VSWR of 1:1). Now we can describe the power radiated in a desired direction by converting 10 kW to dBW:

$$10 \text{ kW} = 40 \text{ dBW}$$

Add the 10-dB gain of the antenna; the EIRP in the desired direction (over an isotropic) is +50 dBW. Let us suppose that the first side lobe of the antenna was down 50 dB from the main lobe. What is the EIRP of the side lobe?

$$+50 \text{ dBW} - 50 \text{ dB} = +0 \text{ dBW} = 1 \text{ W}$$

Consider the following additional examples.

Example 1.   A radiolink transmitter has an output of 1 W; its line losses are negligible, and the antenna gain is 35 dB. What is the EIRP on the main beam?

$$1 \text{ W} = 0 \text{ dBW (or +30 dBm)}$$

$$0 \text{ dBW} + 35 \text{ dB} = +35 \text{ dBW (or +65 dBm)}$$

Example 2.   A tropospheric scatter transmitter has a power output of 2 kW, line losses are 2 dB, and its antenna gain is 43 dB. What is the EIRP on the main beam?

$$1 \text{ kW} = +30 \text{ dBW}$$

$$2 \text{ kW} = +33 \text{ dBW}$$

$$+33 \text{ dBW} - 2 \text{ dB} + 43 \text{ dB} = +74 \text{ dBW}$$

Often we hear the term ERP, which means effective radiated power. We must beware of the reference used; if it is not an isotropic but a dipole,* then all EIRP values will be somewhat less.

Note that the term dBi is sometimes used when we refer to gain over an isotropic antenna.

1.15 SOME COMMON CONVERSION FACTORS

To Convert	Into	Multiply By	Conversely, Multiply By
acres	hectares	0.4047	2.471
Btu	kilogram-calories	0.2520	3.969
°Celsius	°Fahrenheit	$9°C/5 = °F - 32$	
		$9(°C + 40)/5 = (°F + 40)$	
circular mils	square centimeters	5.067×10^{-6}	1.973×10^{5}
circular mils	square mils	0.7854	1.273
degrees (angle)	radians	1.745×10^{-2}	57.30
kilometers	feet	3281	3.048×10^{-4}
kilowatt-hours	Btu	3413	2.930×10^{-4}
liters	gallons (liq. U.S.)	0.2642	3.785
$\log_e$ or ln	$\log_{10}$	0.4343	2.303
meters	feet	3.281	0.3048
miles (nautical)	meters	1852	5.400×10^{-4}
miles (nautical)	miles (statute)	1.1508	0.8690
miles (statute)	feet	5280	1.890×10^{-4}
miles (statute)	kilometers	1.609	0.6214
nepers	decibels	8.686	0.1151
square inches	circular mils	1.273×10^{6}	7.854×10^{-7}
square millimeters	circular mils	1973	5.067×10^{-4}
Boltzmann's constant		$(1.38044 \pm 0.00007) \times 10^{-16}$ erg/deg	
velocity of light		2.998×10^{8} m/s	
		186,280 mi/s	

*A half-wave dipole has a gain of 1.64, or approximately 2.15 dB over an isotropic.

$$984 \times 10^6 \ \text{ft/s}$$

degree of longitude at the equator	68.703 statute mi or 59.661 nautical mi
1 rad	$180°/\pi = 57.2958°$
1 m	39.3701 in. = 3.28084 ft
1°	17.4533 mrad
e	2.71828

REFERENCES AND BIBLIOGRAPHY

1. B. D. Holbrook and J. T. Dixon, "Load Rating Theory for Multichannel Amplifiers," *Bell Sys. Tech. J.*, 624–644, Oct. 1939.

2. *Reference Data for Radio Engineers*, 6th ed., Howard W. Sams, Indianapolis, IN, 1977.

3. CCITT, Orange Books, Geneva, 1976, vol. III, G recommendations.

4. *DCS Engineering Installation Manual*, DCAC 330-175-1, through Change 9, U.S. Department of Defense, Washington, DC.

5. Military Standard 188C, U.S. Department of Defense, Washington, DC.

6. W. Oliver, *White Noise Loading of Multichannel Communication Systems*, Marconi Instruments, Sept. 1964.

7. N. Kramer, "Communication Needs versus Existing Facilities," lecture given at the 1964 Planning Seminar of the North Jersey Section, IEEE.

8. *Lenkurt Demodulator*, Lenkurt Electric Corp., San Carlos, CA, Dec. 1964, June 1965, Sept. 1965.

9. F. R. Connor, *Introductory Topics in Electronics and Telecommunication–Modulation*, Edward Arnold, London, 1973.

10. *Principles of Electricity Applied to Telephone and Telegraph Work*, American Telephone and Telegraph Company, New York, 1961.

11. H. Arkin and R. R. Colton, *Statistical Methods*, 5th ed., Barnes and Noble College Outline Series, New York.

12. C. E. Smith, *Applied Mathematics for Radio and Communication Engineers*, Dover, New York, 1945.

13. S. Goldman, *Information Theory*, Dover, New York, 1953.

14. H. H. Smith, "Noise Transmission Level Terms in American and International Practice," paper, ITT Communication Systems, Paramus, NJ, 1964.

15. M. M. Rosenfeld, "Noise in Aerospace Communication," *Electro-Technology*, May 1965.

16. *Transmission Systems for Communications*, 4th ed., Bell Telephone Laboratories, American Telephone and Telegraph Company, New York, 1971.

17. R. C. James (James and James), *Mathematics Dictionary*, 3rd ed., Van Nostrand, Princeton, NJ.

18. *Telecommunication Transmission Engineering*, vols. 1–3, 2nd ed., American Telephone and Telegraph Co., New York, 1977.

19. IEEE Std. 100–1977, "Dictionary of Electrical and Electronics Terms," 2nd ed., IEEE, New York, 1977.

2 | TELEPHONE TRANSMISSION

2.1 GENERAL

Section 1.8 introduced the simple telephone connection. This chapter delves into telephony and problems of telephone transmission more deeply. It exclusively treats speech transmission over wire systems. Other transmission media are treated only in the abstract so that we can consider problems in telephone networks. The subscriber loop, an important segment of the telephone network, is also covered.

2.2 THE TELEPHONE INSTRUMENT

The input–output (I/O) device that provides the human interface with the telephone network is the telephone instrument or subset. It converts sound energy into electrical energy, and vice versa. The degree of efficiency and fidelity with which it performs these functions has a vital effect upon the quality of telephone service provided. The modern telephone subset consists of a transmitter (mouthpiece), a receiver (earpiece), and an electrical network for equalization, sidetone circuitry, and devices for signaling and supervision. All these items are contained in a device that, when mass-produced, sells for about $40.

Let us discuss transmitters and receivers for a moment.

2.2.1 Transmitters

The transmitter converts acoustic energy into electric energy by means of a carbon granule transmitter. The transmitter requires a dc potential, usually on the order of 3–5 V, across its electrodes. We call this the talk battery, and in modern systems it is supplied over the line (central battery) from the switch (see Section 1.8). Current from the battery flows through the carbon granules or grains when the telephone is lifted off its cradle (off-hook). When sound impinges on the diaphragm of the transmitter, variations of air pressure are transferred to the carbon, and the resistance of the electrical path through the carbon changes in proportion to the pressure. A pulsating direct current results. The frequency response of carbon transmitters peaks between 800 and 1000 Hz.

2.2.2 Receivers

A typical receiver consists of a diaphragm of magnetic material, often soft iron alloy, placed in a steady magnetic field supplied by a permanent magnet, and a varying magnetic field, caused by the voice currents flowing through the voice coils. Such voice currents are alternating (ac) in nature and originate at the far-end telephone transmitter. These currents cause the magnetic field of the receiver to alternately increase and decrease, making the diaphragm move and respond to the variations. As a result an acoustic pressure wave is set up, reproducing, more or less exactly, the original sound wave from the distant telephone transmitter. The telephone receiver, as a converter of electrical energy to acoustic energy, has a comparatively low efficiency, on the order of 2-3%.

Sidetone is the sound of the talker's voice heard in his own receiver. The sidetone level must be controlled. When the level is high, the natural human reaction is for the talker to lower his voice. Thus by regulating the sidetone, talker levels can be regulated. If too much sidetone is fed back to the receiver, the output level of the transmitter is reduced owing to the talker lowering his or her voice, thereby reducing the level (voice volume) at the distant receiver, deteriorating performance.

2.3 THE TELEPHONE LOOP

We speak of the telephone subscriber as the user of the subset. As mentioned in Section 1.8, subscribers' telephone sets are interconnected via a switch or network of switches. Present commercial telephone service provides for transmission and reception on the same pair of wires that connect the subscriber to the local switch. Let us now define some terms.

The pair of wires connecting the subscriber to the local switch that serves him is

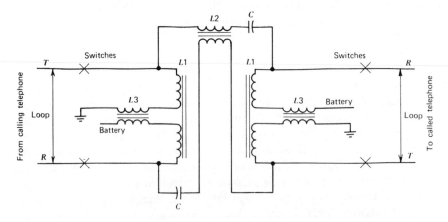

Figure 2.1 Battery feed circuit (Ref. 7). *Note.* Battery and ground are fed through inductors $L3$ and $L1$ through switch to loops. (Copyright © 1961 by Bell Telephone Laboratories.)

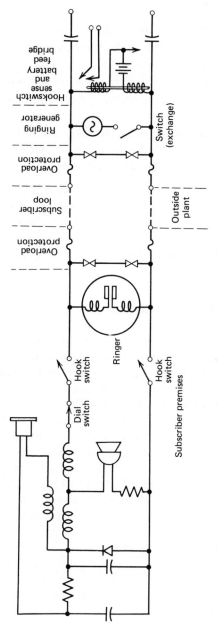

Figure 2.2 Signaling with a conventional telephone subset. Note functions of hook switch, dial, and ringer.

the *subscriber loop*. It is a dc loop in that it is a wire pair typically supplying a metallic path for the following:

1. Talk battery for the telephone transmitter.

2. An ac ringing voltage for the bell on the telephone instrument supplied from a special ringing source voltage.

3. Current to flow through the loop when the telephone instrument is taken out of its cradle, telling the switch that it requires "access" and causing line seizure at the switching center.

4. The telephone dial that, when operated, makes and breaks the dc current on the closed loop, which indicates to the switching equipment the number of the distant telephone with which communication is desired.

The typical subscriber loop is supplied battery by means of a battery feed circuit at the switch. Such a circuit is shown in Figure 2.1. One important aspect of battery feed is the line balance. The telephone battery voltage has been fairly well-standardized at -48 V.

Figure 2.2 shows the functional elements of a subscriber loop, local switch termination, and subscriber subset.

2.4 TELEPHONE LOOP LENGTH LIMITS

It is desirable from an economic viewpoint to permit subscriber loop lengths to be as long as possible. Thus the subscriber area served by a single switching center may be much larger. As a consequence, the total number of switches or telephone central offices may be reduced to a minimum. For instance, if loops were limited to 4 km in length, a switching center could serve all subscribers within a radius of something less than 4 km. If 10 km were the maximum loop length, the radius of an equivalent area that one office could cover would be extended an additional 6 km, out to a total of nearly 10 km. It is evident that to serve a large area, fewer switches (switching centers) are required for the 10-km situation than for the 4-km. The result is fewer buildings, less land to buy, fewer locations where maintenance is required, and all the benefits accruing from greater centralization, which become even more evident as subscriber density decreases, such as in rural areas.

The two basic criteria that must be considered when designing subscriber loops, and which limit their length, are the following:

- Attenuation limits (covered under what we call *transmission design*)
- Signaling limits (covered under what we call *resistance design*)

Attenuation in this case refers to loop loss in decibels (or nepers) at

- 1000 Hz in North America
- 800 Hz in Europe and many other parts of the world

As a loop is extended in length, its loss at reference frequency increases. It follows that at some point as the loop is extended, the level will be attenuated such that the subscriber cannot hear *sufficiently well.*

Likewise, as a loop is extended in length, some point is reached where signaling and/or supervision is no longer effective. This limit is a function of the *IR* drop of the line. We know that R increases as length increases. With today's modern telephone sets, the first to suffer is usually the "supervision." This is a signal sent to the switching equipment requesting "seizure" of a switch circuit and, at the same time, indicating the line is busy. "Off-hook" is a term more commonly used to describe this signal condition. When a telephone is taken "off-hook" (i.e., out of its cradle), the telephone loop is closed and current flows, closing a relay at the switch. If current flow is insufficient, the relay will not close or it will close and open intermittently (chatter) such that line seizure cannot be effected.

Signaling and supervision limits are a function of the conductivity of the cable conductor and its diameter or gauge. For this introductory discussion we can consider that the transmission limits are controlled by the same parameters.

Consider a copper conductor. The larger the conductor, the higher the conductivity, and thus the longer the loop may be for signaling purposes. Copper is expensive, so we cannot make the conductor as large as we would wish and extend subscriber loops long distances. These economic limits of loop length are discussed in detail below. First we must describe what a subscriber considers as hearing sufficiently well, which is embodied in "transmission design" (regarding subscriber loop).

2.5 THE REFERENCE EQUIVALENT

2.5.1 Definition

Hearing "sufficiently well" on a telephone connection is a subjective matter under the blanket heading of customer satisfaction. Various methods have been devised over the years to rate telephone connections regarding customer (subscriber) satisfaction. Subscriber satisfaction will be affected by the following regarding the received telephone signal:

- Level (see Section 1.9.5)
- Signal-to-noise ratio (see Section 1.10)
- Response or attenuation frequency characteristic (see Section 1.9.3)

A common rating system in use today to grade customer satisfaction is the *reference equivalent* system. This system considers only the first criterion mentioned above, namely, level. It must be emphasized that subscriber satisfaction is subjective. To measure satisfaction, the world regulative body for telecommunications, the International Telecommunication Union, devised a system of rating sufficient level to "satisfy," using the familiar decibel as the unit of measurement. It is particularly convenient in that, first disregarding the subscriber telephone subset, essentially we can add losses and gains (measured at 800 Hz) in the intervening network end to end, and determine the reference equivalent of a circuit by then adding this sum to a decibel value assigned to the subset, or to a subset plus a fixed subscriber loop length with wire gauge stated.

Let us look at how the reference equivalent system was developed, keeping in mind again that it is a subjective measurement dealing with the likes and dislikes of the "average" human being. Development took place in Europe. A standard for reference equivalent is determined using a team of qualified personnel in a laboratory. A telephone connection was established in the laboratory which was intended to be the most efficient telephone system known. The original reference system or unique master reference consisted of the following:

- A solid-back telephone transmitter
- Bell telephone receiver
- Interconnecting these, a "zero decibel loss" subscriber loop
- Connecting the loop, a manual, central battery, 22-V dc telephone exchange (switch)

The test team, to avoid ambiguity of language, used a test language which consisted of logatoms. A logatom is a one-syllable word consisting of a consonant, a vowel, and another consonant.

At present more accurate measurement methods have evolved. A more modern reference system is now available in the ITU laboratory in Geneva, Switzerland, called the NOSFER. From this master reference, field test standards are available to telephone companies, administrations, and industry to establish the reference equivalent of telephone subsets in use. These field test standards are equivalent to the NOSFER.

The NOSFER is made up of a standard telephone transmitter, receiver, and network. The reference equivalent of a subscriber's subset, together with the associated subscriber line and feeding bridge, is a quantity obtained by balancing the loudness of received speech signals and is expressed relative to the whole or a corresponding part of the NOSFER (or field) reference system.

2.5.2 Application

Essentially, as mentioned earlier, type tests are run on subscriber subsets or on the subsets plus a fixed length of subscriber loop of known characteristics. These are subjective tests carried out in a laboratory to establish the reference equivalent of a specific subset as compared to a reference standard. The microphone or transmitter and the earpiece or receiver are each rated separately and are called, respectively,

- Transmit reference equivalent (TRE)
- Receive reference equivalent (RRE)

Note. Negative values indicate that the reference equivalent is better than the laboratory standard.

Consider the following simplified telephone network:

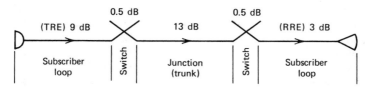

The reference equivalent for this circuit is 26 dB including a 0.5-dB loss for each switch. Junction here takes on the meaning of a circuit connecting two adjacent local or metropolitan switches.

The above circuit may be called a small transmission plan. For this discussion we can define a transmission plan as a method of assigning losses end to end on a telephone circuit. Later in this chapter we discuss why all telephone circuits that have two-wire telephone subscribers must be lossy. The reference equivalent is a handy device to rate such a plan regarding subscriber satisfaction.

When studying transmission plans or developing them, we usually consider that all sections of a circuit in a plan are symmetrical. Let us examine this. On each end of a circuit we have a subscriber loop. Thus in the plan the same loss is assigned to each loop, which may not be the case at all in real life. From the local exchange to the first long-distance exchange, called variously junctions or toll connecting trunks, a loss is assigned which is identical at each end, and so forth.

To maintain this symmetry regarding reference equivalent of telephone subsets, we use the term $(T + R)/2$. As we see from the above drawing, the TRE and the RRE of the subset have different values. We get the $(T + R)/2$ by summing the TRE and the RRE and dividing by 2. This is done to arrive at the desired symmetry. Table 2.1 gives reference equivalent data on a number of standard subscriber sets used in various parts of the world.

The ORE is the overall reference equivalent and equals the sum of the TRE, the RRE, and all the intervening losses of a telephone connection end to end with reference to 800 Hz. We would arrive at the same figure if we added twice the $(T + R)/2$ and the intervening losses at 800 Hz. On all these calculations we assume that the same telephone set is used on either end.

CCITT Rec. G.121* states that the reference equivalent from the subscriber set to an international connection should not exceed 20.8 dB (TRE) and to the subscriber set at the other end from the same point of reference RRE should not exceed 12.2 dB. (*Note.* The intervening losses already are included in these figures.)

---

*For 97% of the connections made in a country of average size.

Table 2.1 Reference Equivalents for Subscriber Sets in Various Countries

Country	Sending (dB)	Receiving (dB)
With limiting subscriber lines and exchange feeding bridges		
Australia	14[a]	6[a]
Austria	11	2.6
France	11	7
Norway	12	7
Germany	11	2
Hungary	12	3
Netherlands	17	4
United Kingdom	12	1
South Africa	9	1
Sweden	13	5
Japan	7	1
New Zealand	11	0
Spain	12	2
Finland	9.5	0.9
With no subscriber lines		
Italy	2	−5
Norway	3	−3
Sweden	3	−3
Japan	2	−1
United States (loop length 1000 ft, 83 Ω)	5	−1[b]

Source: CCITT, *Local Telephone Networks*, ITU, Geneva, July 1968, and *National Telephone Networks for the Automatic Service.*
[a] Minimum acceptable performance.
[b] Ref. 8.

By adding 12.2 and 20.8 dB, we find 33 dB to be the ORE recommended as a maximum* for an international connection. In this regard Table 2.2 should be of interest. It should also be noted that as the reference equivalent (overall, end to end) drops to about 6 dB, the subscriber begins to complain that the call is too loud.

It is noted in Table 2.2 that the 33-dB ORE discussed above is unsatisfactory for more than 10% of calls. Therefore the tendency in many telephone administrations is to reduce this figure as much as possible. In fact the long-term design objective of CCITT for traffic weighted mean values of the reference equivalent should lie in the range of 13–18 dB for international connections (CCITT Rec. G.111b). As we shall see later, this process is difficult and can prove costly.

*For 97% of the connections made in a country of average size.

Table 2.2   British Post Office Survey of Subscribers for Percentage of Unsatisfactory Calls

Overall Reference Equivalent (dB)	% of Unsatisfactory Calls
40	33.6
36	18.9
32	9.7
28	4.2
24	1.7
20	0.67
16	0.228

2.6   TELEPHONE NETWORKS

2.6.1   General

The next logical step in our discussion of telephone transmission is to consider the large-scale interconnection of telephones. As we have seen, subscribers within a reasonable distance of one another can be interconnected by wire lines and we can still expect satisfactory communication. A switch is used so that a subscriber can speak with some other discrete subscriber as he or she chooses. As we extend the network to include more subscribers and circumscribe a wider area, two technical/ economic factors must be taken into account:

1.   More than one switch must be used.

2.   Wire pair transmission losses on longer circuits must be offset by amplifiers, or the pairs must be replaced by other, more efficient means.

Let us accept item 1. The remainder of this section concentrates on item 2. The reason for the second statement becomes obvious when the salient point of Section 2.5 regarding reference equivalents is reviewed.

Example. How far can a two-wire line be run without amplifiers following the rules of reference equivalent? Allow, in this case, no more than an ORE of 33 dB, a high value for a design goal. Referring to Table 2.1 and using Spain as an example, the TRE + RRE for telephone subsets sums to 14 dB. This leaves us with a limiting loss of 19 dB (33 - 14) for the remaining network. Use 19-gauge (0.91-mm) telephone cable, typical for "long-distance" communication (Table 2.8). This cable has a loss of 0.71 dB/km. Thus the total extension of the network will be about 26 km allowing no loss for a switch.

2.6.2 Basic Considerations

What are some of the more common approaches that may be used to extend the network? We may use:

1. Coarser gauge cable (larger diameter conductors)
2. Amplifiers in the present wire pair system
3. Carrier transmission techniques (Chapters 3 and 11)
4. Radio transmission techniques (Chapters 4-7, 10, and 15)

Items 1 and 2 are used quite widely but often become unattractive from an economic point of view as the length increases. Items 3 and 4 become attractive for multichannel transmission over longer distances. Discussion of the formation of a multichannel signal is left for Chapter 3. Such a multichannel transmission technique is referred to as *carrier* transmission.

For our purposes we can consider carrier as a method of high-velocity transmission of bands of frequency above the VF region (i.e., above 4000 Hz) over wire, optical fiber, or radio.

2.6.3 Two-Wire/Four-Wire Transmission

TWO-WIRE TRANSMISSION

By its basic nature a telephone conversation requires transmission in both directions. When both directions are carried on the same wire pair, we call it two-wire transmission. The telephones in our home and office are connected to a local switching center by means of two-wire circuits. A more proper definition for transmitting and switching purposes is that when oppositely directed portions of a single telephone conversation occur over the same electrical transmission channel or path, we call this two-wire operation.

FOUR-WIRE TRANSMISSION

Carrier and radio systems require that oppositely directed portions of a single conversation occur over separate transmission channels or paths (or using mutually exclusive time periods). Thus we have two wires for the transmit path and two wires for the receive path, or a total of four wires for a full-duplex (two-way) telephone conversation. For almost all operational telephone systems, the end instrument (i.e., the telephone subset) is connected to its intervening network on a two-wire basis.*

Nearly all long-distance telephone connections traverse four-wire links. From the near-end user the connection to the long-distance network is two wire. Likewise,

---

*A notable exception is the U.S. military telephone network Autovon, where end users are connected on a four-wire basis.

Direction of transmission

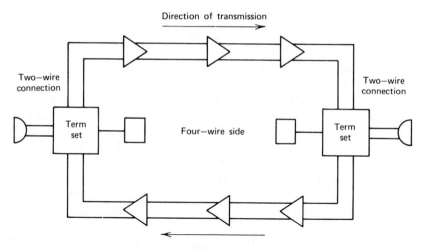

Two—wire
connection

Two—wire
connection

Term
set

Four—wire side

Term
set

Figure 2.3   Typical long-distance telephone connection.

the far-end user is also connected to the long-distance network via a two-wire link. Such a long-distance connection is shown in Figure 2.3. Schematically the four-wire interconnection is shown as if it were wire line, single channel with amplifiers. More likely it would be multichannel carrier on cable and/or multiplex on radio. However, the amplifiers in the figure serve to convey the ideas that this chapter considers.

As shown in Figure 2.3, conversion from two-wire to four-wire operation is carried out by a terminating set, more commonly referred to in the industry as a *term set*. This set contains a three-winding balanced transformer (a hybrid) or a resistive network, the latter being less common.

OPERATION OF A HYBRID

A hybrid, for telephone work (at VF), is a transformer. For a simplified description, a hybrid may be viewed as a power splitter with four sets of wire pair connections. A functional block diagram of a hybrid device is shown in Figure 2.4. Two of these wire pair connections belong to the four-wire path, which consists of a transmit pair

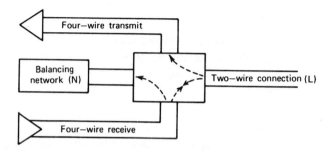

Four—wire transmit

Balancing
network (N)

Two—wire connection (L)

Four—wire receive

Figure 2.4   Operation of a hybrid transformer.

and a receive pair. The third pair is the connection to the two-wire link to the subscriber subset. The last wire pair connects the hybrid to a resistance–capacitance
balancing network which electrically balances the hybrid with the two-wire connection to the subscriber's subset over the frequency range of the balancing network. An artificial line may also be used for this purpose.

The hybrid function permits signals to pass from any pair through the transformer
to both adjacent pairs but blocks signals to the opposite pair (as shown in Figure
2.4). Signal energy entering from the four-wire side divides equally, half dissipating
into the balancing network and half going to the desired two-wire connection.
Ideally no signal energy in this path crosses over the four-wire transmit side. This is
an important point, which we take up later.

Signal energy entering from the two-wire subset connection divides equally, half
of it dissipating in the impedance of the four-wire side receive path, and half going
to the four-wire side transmit path. Here the *ideal* situation is that no energy is to
be dissipated by the balancing network (i.e., there is a perfect balance). The balancing network is supposed to display the characteristic impedance of the two-wire
line (subscriber connection) to the hybrid.

The reader notes that in the description of the hybrid, in every case, ideally half
of the signal energy entering the hybrid is used to advantage and only half is dissipated, wasted. Also keep in mind that any passive device inserted in a circuit such
as a hybrid has an insertion loss. As a rule of thumb we say that the insertion loss of
a hybrid is 0.5 dB. Thus there are two losses here of which the reader must not lose
sight:

Hybrid insertion loss	0.5 dB
Hybrid dissipation loss	3.0 dB (half power)
	3.5 dB total

As far as this chapter is concerned, any signal passing through a hybrid suffers a
3.5-dB loss. Some hybrids used on short subscriber loops purposely have higher
losses, as do special resistance type hybrids.

2.6.4 Echo, Singing, and Design Loss

GENERAL

The operation of the hybrid with its two-wire connection on one end and four-wire
connection on the other leads us to the discussion of two phenomena which, if
not properly designed for, may lead to major impairments in communication. These
impairments are echo and singing.

Echo

As the name implies, echo in telephone systems is the return of a talker's voice.
The returned voice, to be an impairment, must suffer some noticeable delay.

Thus we can say that echo is a reflection of the voice. Analogously it may be considered as that part of the voice energy that bounces off obstacles in a telephone connection. These obstacles are impedance irregularities, more properly called impedance mismatches.

Echo is a major annoyance to the telephone user. It affects the talker more than the listener. Two factors determine the degree of annoyance of echo: its loudness and how long it is delayed.

Singing

Singing is the result of sustained oscillations due to positive feedback in telephone amplifiers or amplifying circuits. Circuits that sing are unusable and promptly overload multichannel carrier equipment (FDM, see Chapter 3).

Singing may be thought of as echo that is completely out of control. This can occur at the frequency at which the circuit is resonant. Under such conditions the circuit losses at the singing frequency are so low that oscillation will continue even after the impulse that started it ceases to exist.

The primary cause of echo and singing generally can be attributed to the mismatch between the balancing network and its two-wire connection associated with the subscriber loop. It is at this point that the major impedance mismatch usually occurs and an echo path exists. To understand the cause of the mismatch, remember that we always have at least one two-wire switch between the hybrid and the subscriber. Ideally the hybrid balancing network must match each and every subscriber line to which it may be switched. Obviously the impedances of the four-wire trunks (lines) may be kept fairly uniform. However, the two-wire subscriber lines may vary over a wide range. The subscriber loop may be long or short, may or may not have inductive loading (see Section 2.8.4), and may or may not be carrier derived (see Chapter 3). The hybrid imbalance causes signal reflection or signal "return." The better the match, the more the return signal is attenuated. The amount that the return signal (or reflected signal) is attenuated is called the *return loss* and is expressed in decibels. The reader should remember that any four-wire circuit may be switched to hundreds or even thousands of different subscribers. If not, it would be a simple matter to match the four-wire circuit to its single subscriber through the hybrid. This is why the hybrid to which we refer has a compromise balancing network rather than a precision network. A compromise network is usually adjusted for a compromise in the range of impedance that is expected to be encountered on the two-wire side.

Let us consider now the problem of match. For our case the impedance match is between the balancing network N and the two-wire line L (see Figure 2.4). With this in mind,

$$\text{Return loss}_{dB} = 20 \log_{10} \frac{Z_N + Z_L}{Z_N - Z_L}$$

If the network perfectly balances the line, $Z_N = Z_L$, and the return loss would be infinite.

The return loss may also be expressed in terms of the reflection coefficient:

$$\text{Return loss}_{dB} = 20 \log_{10} \frac{1}{\text{reflection coefficient}}$$

where the reflection coefficient is the ratio of reflected signal to incident signal.

The CCITT uses the term *balance return loss* (see CCITT Rec. G.122) and classifies it as two types:

1. Balance return loss from the point of view of echo.* This is the return loss across the band of frequencies from 500 to 2500 Hz.

2. Balance return loss from the point of view of stability. This is the return loss between 0 and 4000 Hz.

The band of frequencies which is most important from the standpoint of echo for the voice channel is that between 500 and 2500 Hz. A good value for the echo return loss (ERL) for a toll telephone plant is 11 dB, with values on some connections dropping to as low as 6 dB. For the local telephone network, CCITT recommends better than 6 dB, with a standard deviation of 2.5 dB (Ref. 18).

For frequencies outside the 500–2500-Hz band, return loss values often are below the desired 11 dB. For these frequencies we are dealing with return loss from the point of view of stability. CCITT recommends that balance return loss from the point of view of stability (singing) should have a value of not less than 2 dB for all terminal conditions encountered during normal operation (CCITT Rec. G.122, p. 3). For further information the reader should consult Appendix A and CCITT Recs. G.122 and G.131.

Echo and singing may be controlled by

- Improved return loss at the term set (hybrid)
- Adding loss on the four-wire side (or on the two-wire side)
- Reducing the gain of the individual four-wire amplifiers

The annoyance of echo to a subscriber is also a function of its delay. Delay is a function of the velocity of propagation of the intervening transmission facility. A telephone signal requires considerably more time to traverse 100 km of a voice pair cable facility, particularly if it has inductive loading, than 100 km of radio facility.

Delay is expressed in one-way or round-trip propagation time measured in milliseconds. CCITT recommends that if the mean round-trip propagation time exceeds 50 ms for a particular circuit, an echo suppressor should be used. Bell System practices in North America use 45 ms as a dividing line. In other words, where the echo delay is less than that stated above, the echo will be controlled by adding loss.

An echo suppressor is an electronic device inserted in a four-wire circuit which

---

*Called echo return loss (ERL) in via net loss (North American practice, Section 2.6.5), but uses a weight distribution of level.

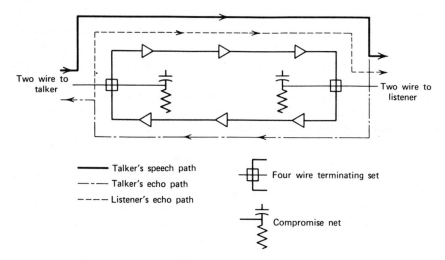

Talker's speech path
Talker's echo path
Listener's echo path

Four wire terminating set

Compromise net

Figure 2.5 Echo paths in a four-wire circuit.

effectively blocks passage of reflected signal energy. The device is voice operated with a sufficiently fast reaction time to "reverse" the direction of transmission, depending on which subscriber is talking at the moment. The blocking of reflected energy is carried out by simply inserting a high loss in the return four-wire path.

Figure 2.5 shows the echo path on a four-wire circuit.

TRANSMISSION DESIGN TO CONTROL ECHO AND SINGING

As stated previously, echo is an annoyance to the subscriber. Figure 2.6 relates echo path delay to echo path loss. The curve in Figure 2.6 is a group of points at

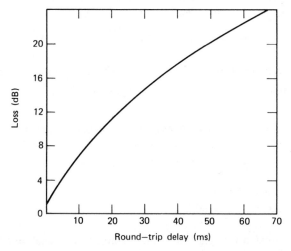

Figure 2.6 Talker echo tolerance for average telephone users.

which the average subscriber will tolerate echo as a function of its delay. Remember that the greater the return signal delay, the more annoying it is to the telephone talker (i.e., the more the echo signal must be attenuated). For instance, if the echo path delay on a particular circuit is 20 ms, an 11-dB loss must be inserted to make echo tolerable to the talker. The careful reader will note that the 11 dB designed into the circuit will increase the end-to-end reference equivalent by that amount, something quite undesirable. The effect of loss design on reference equivalents and the trade-offs available are discussed below.

To control singing all four-wire paths must have some loss. Once they go into a gain condition, and we refer here to overall circuit gain, we will have positive feedback and the amplifiers will begin to oscillate or "sing." North American practice calls for a 4-dB loss on all four-wire circuits to ensure against singing.

Almost all four-wire circuits have some form of amplifier and level control. Often such amplifiers are embodied in the channel banks of the carrier equipment. For a discussion of carrier equipment, see Chapter 3.

AN INTRODUCTION TO TRANSMISSION LOSS PLANNING

One major aspect of transmission system design for a telephone network is to establish a transmission loss plan. Such a plan, when implemented, is formulated to accomplish three goals:

- Control singing
- Keep echo levels within limits tolerable to the subscriber
- Provide an acceptable overall reference equivalent to the subscriber

For North America the via net loss (VNL) concept embodies the transmission plan idea. VNL is covered in Section 2.6.5.

From our discussions above we have much of the basic background necessary to develop a transmission loss plan. We know the following:

1. A certain minimum loss must be maintained in four-wire circuits to ensure against singing.

2. Up to a certain limit of round-trip delay, echo is controlled by loss.

3. It is desirable to limit these losses as much as possible to improve the reference equivalent.

National transmission plans vary considerably. Obviously the length of circuit is important as well as the velocity of propagation of the transmission media. Two approaches are available in the preparation of a loss plan:

- Variable loss plan (i.e., VNL)
- Fixed loss plan (i.e., as used in Europe)

A national transmission loss plan for a small country (i.e., small in extension) such as Belgium could be quite simple. Assume that a 4-dB loss is inserted in all four-wire circuits to prevent singing (see Figure 2.6). Here 4 dB allows for 5 ms of round-trip delay. If we assume carrier transmission for the entire length of the connection and use 105,000 mi/s for the velocity of propagation, we can then satisfy Belgium's echo problem. The velocity of propagation used comes out to 105 mi (168 km)/ms. By simple arithmetic we see that a 4-dB loss on all four-wire circuits will make echo tolerable for all circuits extending 210 mi (336 km). This is an application of the fixed loss type of transmission plan. In the case of small countries or telephone companies operating over a small geographical extension, the minimum loss inserted to control singing controls echo as well for the entire country.

Let us try another example. Assume that all four-wire connections have a 7-dB loss. Figure 2.6 indicates that 7 dB permits an 11-ms round-trip delay. Assume that the velocity of propagation is 105,000 mi/s. Remember that we deal with round-trip delay. The talker's voice goes out to the far-end hybrid and is then reflected back. This means that the signal traverses the system twice, as shown:

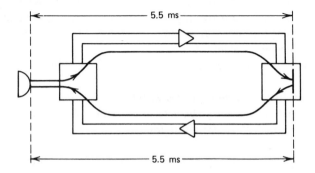

In this example the round-trip delay is 5.5 + 5.5 = 11 ms.

Thus 7 dB of loss for the velocity of propagation given allows about 578 mi of extension or, for all intents and purposes, the distance between subscribers.

It has become evident by now that we cannot continue increasing losses indefinitely to compensate for echo on longer circuits. Most telephone companies and administrations have set the 45- or 50-ms round-trip delay criterion. This sets a top figure above which echo suppressors shall be used.

One major goal of the transmission loss plan is to improve overall reference equivalents to apportion more loss to the subscriber plant so that subscriber loops can be longer, or to allow the use of less copper (i.e., smaller diameter conductors). The question is, what measures can be taken to reduce losses and still keep echo within tolerable limits? One obvious target is to improve return losses at the hydrids. If all hybrid return losses are improved, then the echo tolerance curve gets shifted. This is so because improved return losses reduce the intensity of the echo returned to the talker. Thus the subscriber is less annoyed by the echo effect.

One way of improving return loss is to make all two-wire lines out of hybrid look alike, that is, have the same impedance. The switch at the other end of the hybrid (i.e., on the two-wire side) connects two-wire loops of varying lengths causing the resulting impedances to vary greatly. One approach is to extend four-wire transmission to the local office such that each hybrid can be better balanced. This is being carried out with success in Japan. The U.S. Department of Defense has its Autovon (automatic voice network) in which every subscriber line is operated on a four-wire basis. Two-wire subscribers connect through the system on a private automatic branch exchange (PABX).

Let us return to standard telephone networks using two-wire switches in the subscriber area. Suppose that the balance return loss could be improved to 27 dB. Thus the minimum loss to assure against singing could be reduced to 0.4 dB. Suppose that we distributed this loss across four four-wire circuits in tandem. Thus each four-wire circuit would be assigned a 0.1-dB loss. If we have gain in the network, singing will result. The safety factor between loss and gain is 0.4 dB. The loss in each circuit or link is maintained by amplifiers. It is difficult to adjust the gain of an amplifier to 0.1 dB, much less keep it there over long periods, even with good automatic regulation. *Stability* or gain stability is the term used to describe how well a circuit can maintain a desired level. Of course in this case we refer to a testtone level. In the example above it would take only one amplifier to shift 0.4 dB, two to shift in the positive direction 0.2 dB, and so forth. The importance of stability, then, becomes evident.

The stability of a telephone connection depends on

- The variation of transmission level with time
- The attenuation-frequency characteristics of the links in tandem
- The distribution of balance return loss

Each of these criteria becomes magnified when circuits are switched in tandem. To handle the problem properly we must talk about statistical methods and standard distributions.

Returning to the criteria above, in the case of the first two items we refer to the tandeming of four-wire circuits. The last item refers to switching subscriber loops/ hybrid combinations that will give a poorer return loss than the 11 dB stated above. Return losses on some connections can drop to 3 dB or less.

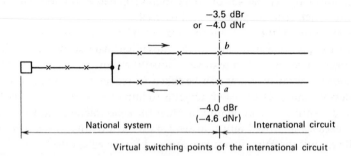

Virtual switching points of the international circuit

Figure 2.7 Definition of points *a–t–b* (CCITT Rec. G.122); × indicates a switch. (Courtesy ITU–CCITT.)

CCITT Recs. G.122, G.131, and G.151C treat stability. In essence the loss through points a-t-b in Figure 2.7 shall have a value not less than $(6 + N)$ dB, where N is the number of four-wire circuits in the national chain. Thus the minimum loss is stated (CCITT Rec. G.122). Rec. G.131 is quoted in part below:

The standard deviation of transmission loss among international circuits routed in groups equipped with automatic regulation is 1 dB . . . . This accords with . . . that the tests . . . indicate that this target is being approached in that 1.1 dB was the standard deviation of the recorded data . . . .

CCITT Rec. G.131 continues:

It is also evident that those national networks which can exhibit no better stability balance return loss than 3 dB, 1.5 dB standard deviation, are unlikely to seriously jeopardize the stability of international connections as far as oscillation is concerned. However, the near-singing (rain barrel effect) distortion and echo effects that may result give no grounds for complacency in this matter.

Stability requirements in regard to North American practice are embodied in the VNL concept discussed in the next section.

2.6.5 Via Net Loss (VNL)

VNL is a concept or method of transmission planning which permits a relatively close approach to an overall zero transmission loss in the telephone network (lowest practicably attainable) and maintains singing and echo within specified limits. The two criteria that follow are basic to VNL design:

1. The customer-to-customer talker echo shall be satisfactorily low on more than 99% of all telephone connections that encounter the maximum delay likely to be experienced.

2. The total amount of overall loss is distributed throughout the trunk segments of the connection by the allocation of loss to the echo characteristics of each segment.

One important concept in the development of the discussion of VNL is ERL (see Section 2.6.4). For this discussion we consider ERL as a single-valued weighted figure of return losses in the frequency band of 500–2500 Hz. ERL differs from return loss in that it takes into account a weighted distribution of level versus frequency in order to simulate the nonlinear characteristics of the transmitter and receiver of the telephone instrument. By using ERL measurements it is possible to arrive at a basic factor for the development of the VNL formula. This design factor states that the average return loss at Class 5 offices (local offices) is 11 dB with a standard deviation of 3 dB. Considering then a standard distribution curve and the

1, 2, or 3σ points on the curve, we could therefore expect practically all measurements of ERL to fall between 2 and 20 dB at Class 5 offices (local offices). VNL also considers that reflection occurs at the far end in relation to the talker where the toll connecting trunks are switched to the intertoll trunks (see Section 2.7).

The next concept in the development of the VNL discussion is that of overall connection loss (OCL), which is the value of one-way trunk loss between two end (local) offices (not subscribers).

Consider that

$$\text{Echo path loss} = 2 \times \text{trunk loss (one way)} + \text{return loss (hybrid)}$$

Now let us consider the average tolerance for a particular echo path loss. The average echo tolerance is taken from the curve in Figure 2.6. Therefore,

$$\text{OCL} = \frac{\text{average echo tolerance (loss)} - \text{return loss}}{2}$$

Return loss in this case is the average echo return loss that must be maintained at the distant Class 5 office—the 11 dB given above.

An important variability factor has not been considered in the formula, namely, trunk stability. This factor defines how close assigned levels are maintained on a trunk. VNL practice dictates trunk stability to be maintained with a normal distribution of levels and a standard deviation of 1 dB in each direction. For a round-trip echo path the deviation is taken as 2 dB. This variability applies to each trunk in a tandem connection. If there are three trunks in tandem, this deviation must be applied to each of them.

The reader will recall that the service requirement in VNL practice is satisfactory echo performance for 99% of all connections. This may be considered as a cumulative distribution or +2.33 standard deviations summing from negative infinity toward the positive direction.

The OCL formula may now be rewritten:

$$\text{OCL} = \frac{\text{average echo tolerance} - \text{average return loss} + 2.33\sigma'}{2}$$

where σ' = composite standard deviation of all functions, namely,

$$\sigma' = \sqrt{\sigma_t^2 + \sigma_{r1}^2 + N\sigma_1^2}$$

with σ_t = standard deviation of the distribution of echo tolerance among a large group of observers, 2.5 dB

σ_{r1} = standard deviation of the distribution of return loss, 3 dB

σ_1 = standard deviation of the distribution of the variability of trunk loss for a round-trip echo path, given as 2 dB

N = number of trunks switched in tandem to form a connection Class 5 office to Class 5 office

Consider now several trunks in tandem. Then it can be calculated that at just about any given echo path delay, the OCL increases approximately 0.4 dB for each trunk added. With this simplification, once we have the OCL for one trunk, all that is needed to compute the OCL for additional trunks is to add 0.4 dB times the number of trunks added in tandem. This loss may be regarded as an additional constant needed to compensate for variations in trunk loss in the VNL formula.

Figure 2.8 relates echo path delay (round trip) to OCL (for one trunk, then for a second trunk in tandem and for four and six trunks in tandem). The straight-line curve has been simplified, yet the approximation is sufficient for engineering VNL circuits. On examining the straight-line curve in Figure 2.8 it will be noted that that curve cuts the Y axis at 4.4 dB where round-trip delay is 0. This 4.4 dB is made up of two elements, namely, that all trunks have a minimum of 4 dB to control singing and 0.4 dB protection against negative variation of trunk loss.

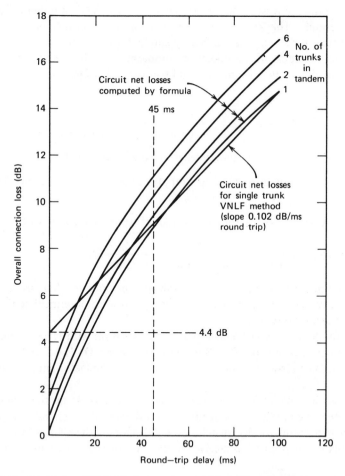

Figure 2.8 Approximate relationship between round-trip echo delay and overall connection loss (OCL).

Another important point to be defined on the linear curve in Figure 2.8 is that of a round-trip delay of 45 ms, which corresponds to an OCL of 9.3 dB. Empirically it has been determined that for delays greater than 45 ms, echo suppressors must be used.

From the linear curve in the figure the following formula for OCL may be derived:

$$OCL = (0.102) \text{ (path delay in ms)}$$

$$+ (0.4 \text{ dB}) \text{ (number of trunks in tandem)} + 4 \text{ dB}$$

A word now about the last term in the OCL equation, a constant (4 dB): usually the 4 dB is applied to the extremity of each trunk network, namely, to the toll-connecting trunks, 2 dB to each.*

OCL deals with the losses of an entire network consisting of trunks in tandem. VNL deals with the losses assigned to one trunk. The VNL formula follows from the OCL formula. The key here is the round-trip delay on the trunk in question. The delay time for a transmission facility employing only one particular medium is equal to the reciprocal of the velocity of propagation of the medium multiplied by the length of the trunk. To obtain round-trip time, this figure must be multiplied by 2. Thus

$$VNL = 0.102 \times 2 \times \left(\frac{1}{\text{velocity of propagation}} \right)$$

$$\times \text{ (one-way length of the trunk)} + 0.4 \text{ dB}$$

Often another term is introduced to simplify the equation, the via net loss factor (VNLF):

$$VNL = VNLF \times \text{ (one-way length of trunk in miles)} + 0.4 \text{ dB}$$

$$VNLF = \frac{2 \times 0.102}{\text{velocity of propagation of medium}} \text{ (dB/mi)}$$

It should be noted that the velocity of propagation of the medium used here must be modified by such things as delays caused by repeaters, intermediate modulation points, and facility terminals.

VNLFs for loaded two-wire facilities are 0.02 dB/mi with H-88 loading on 19-gauge wire and increase to 0.04 dB/mi on B-88 loading. On H-44 facilities they are 0.06 dB/mi. On four-wire carrier and radio facilities the factor improves to 0.0015 dB/mi.

For connections with round-trip delay times in excess of 45 ms the standard VNL approach must be modified. As mentioned previously, these circuits use echo suppressors which automatically switch about 50 dB into the echo return path. The switch actuates when speech is received in the "return" path, switching the pad into

---

Note. This is now shown as 2.5 dB (Figure 2.9). Where previously a local switching loss of 0.5 dB was lumped with subscriber loop loss, it is now added to the toll-connecting trunks.

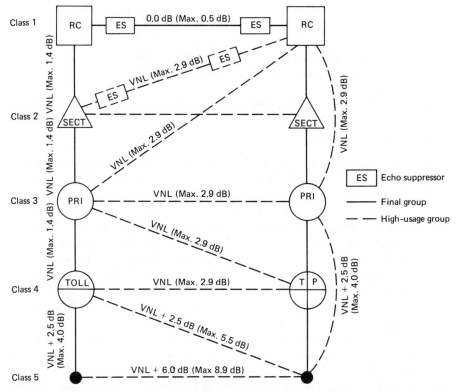

Figure 2.9 Trunk losses with VNL design. RC = regional center; SECT = sectional center; PRI = primary center; TOLL = toll center; TP = tool point. (Copyright © 1977 by American Telephone and Telegraph Company.)

the "go" path. Thus whenever delay exceeds 45 ms, echo suppressors are used. In North America they are used on interregional high-usage intertoll trunks (Figure 2.9) and on interregional toll-connecting and end office (local exchange) toll trunks more than 1850 mi long.

When dealing with VNL in the toll network (long-distance network), trunk losses are stated in terms of inserted connection loss (ICL), which is defined as the 1000-Hz loss inserted by switching the trunk into actual operating connection. The losses shown in Figure 2.9 are all ICL.

In the Bell System of North America nearly all local exchanges (Class 5) use an impedance of 900 Ω, whereas toll (long-distance) exchanges use 600 Ω. However, tandem exchanges using No. 5 crossbar switches use the 900-Ω impedance value. These values are important for terminal and through balance to meet return loss objectives.

In summary, in VNL design we have three types of losses that may be assigned to a trunk:

Type	Loss
Toll-connecting trunk	VNL + 2.5 dB
Intertoll trunk (no echo suppressor)	VNL
Intertoll trunk (with echo suppressor)	0 dB

Note. See discussion on echo suppressors for an explanation of the 0-dB figure.

VNL PENALTY FACTORS

Must of the toll network using the VNL concept utilizes two-wire switches even though the network is considered four wire. At each point where the network has a two-wire to four-wire transformation, another source of echo occurs. Again the amount of echo is a function of the ERL at each point of transformation. If the return loss at each of these intermediate points, often called *through return loss*, is high enough, the point is transparent regarding echo. Echo, then, can be considered only as a function of the return loss at the terminating points of transformation, often called *terminating return loss*. If a two-wire toll switch has an ERL of 27 dB on at least 50% of the through connections, it meets through return loss objectives, and as far as VNL design is concerned, it may be considered transparent regarding echo.

However, a number of two-wire toll exchanges do not meet this minimum criterion. Therefore a penalty is assigned to each in the VNL design. This penalty is called the *B* factor, which is the amount of loss that must be added to the VNL value of each trunk (incoming as well as outgoing) to compensate for the excessive amount of echo and singing current reflection which this office will create in an intertoll connection. The following table provides median office ERL and the corresponding *B* factors:

Median Office ERL (dB)	*B* Factor (dB)
27	0
21	0.3
18	0.6
16	0.9
15	1.2
14	1.5

FIXED-LOSS PLAN

VNL is a variable-loss plan for the North American network. It is one valid approach for a minimum-loss network to reasonably meet subscriber received volume objectives on the one hand and echo performance on the other. The North American network is now in a stage of evolvement from analog to digital, and a fixed-loss plan is more appropriate once the network is all digital.

The fixed-loss plan being introduced in North America by the Bell System is based on a 0-dB loss on all-digital intertoll trunks, that is, on digital transmission facilities interconnecting digital toll exchanges. The plan specifies a 6-dB trunk loss between local exchanges (Class 5 offices) regardless of the connection milage. This 6-dB loss provides an acceptable compromise between loss and echo performance over a wide range of connection lengths and loop losses found in this environment. Toll-connecting trunks are assigned a 3-dB loss under the plan.

It will take many years for the present predominantly analog network to be converted to a fully digital network. In the meantime the network will be a mix of analog and digital facilities which will increasingly trend toward the digital. For the combined analog–digital network of the interim period, the following principal characteristics and constraints will be adhered to:

1. The expected measured loss and inserted connection loss of each trunk must be the same for both directions of transmission.

2. The -2-dB TLP at the outgoing side of analog toll switches and the 0-dB TLP at the Class 5 (local) exchanges are retained, and a -3-dB TLP is established for digital toll exchanges.

3. The -16-dB and $+7$-dB TLPs at carrier system input and output are retained.

4. Existing test and lineup procedures for digital channel banks are retained.

5. Combination intertoll trunks, namely, those terminating in digital terminals at a digital (i.e., No. 4 ESS) switching machine at one end and in D-type channel banks at an analog switch at the other end, are designed to have a 1-dB ICL.

6. Analog intertoll trunks are designed according to the VNL plan.

2.7 NOTES ON NETWORK HIERARCHY

Telephone networks require some form of hierarchy regarding switches to route traffic effectively and economically. Earlier in the chapter we discussed the subscriber as being connected to a local office, a switch. Once the subscriber lifts a telephone "off-hook," he seizes a line at the switch, gets a dial tone in return, and by dialing gives himself access to all other subscribers in the common user network to which he is connected. Figure 2.9 shows the hierarchy of the North American network. Figure 2.10 shows the CCITT hierarchy.

Networks have evolved and so developed in accordance with the principal requirements of traffic volume, geographic coverage, and community of interest of the population as spread across that geographic area. A simple network is one where all switches are mutually interconnected. This, by definition, is a mesh connection, and is a common form of interconnection in many European metropolitan areas where traffic volumes between all switches are high. In today's complex society, where the demand for telephone service is very high with resulting higher traffic volumes which are nonuniform, such a simple configuration is inefficient and cannot be supported economically.

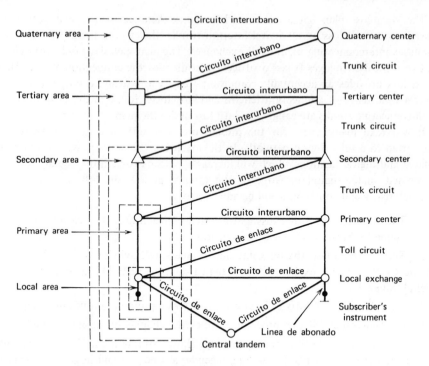

Figure 2.10 CCITT routing plan (network hierarchy). Linea de abonado = subscriber line; Central tandem = tandem switch; Circuito de enlace = junction (trunk); Circuito interurbano = long-distance (trunk) circuit. (Courtesy ITU–CCITT.)

A natural development from the simple network is the star network. A simple star configured network is shown below:

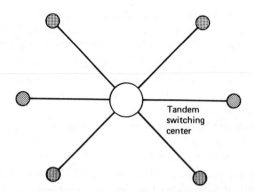

A star network allows efficient utilization of interconnecting facilities when traffic between switches is relatively small. This is done by connecting each switch to a central point where tandem switching may be carried out to interconnect outlying

Table 2.3 Hierarchy Nomenclature

CCITT	North American
Quaternary center (highest ranking)	Class 1 office
Tertiary center	Class 2 office
Secondary center	Class 3 office
Primary center	Class 4 office
Local exchange	Class 5 office

switches by trunk groups. A trunk group is a grouping of telephone circuits representing a concentration of traffic.

As a network grows in extension and in traffic volume carried, successive stages of concentration are required to switch relatively small traffic volumes greater distances. The hierarchical network naturally evolves from a series of simple star configurations. Figures 2.9 and 2.10 are good examples. Both the North American and the CCITT networks comply with the following criteria:

- Any switch, anywhere, can be connected with any other switch with no more than 12 links connected in tandem for the worst connection (CCITT Rec. E.171, p. 3.) (international).
- It is possible for calls to bypass one or more intermediate exchanges on the final route path.
- Direct trunk groups available between switching centers on the final route may be employed to advance a call toward its destination.

Table 2.3 compares the CCITT hierarchy switching/routing nomenclature (Figure 2.10) with that of North America (Figure 2.9). For more information on telephone and data network design, consult (Ref. 22).

2.8 DESIGN OF SUBSCRIBER LOOP

2.8.1 Introduction

The subscriber loop connects a subscriber telephone subset with a local switching center (Class 5 office in Figure 2.9). A subscriber loop in nearly all cases is two wire with simultaneous transmission in both directions. The simplified drawing below will help to illustrate the problem:

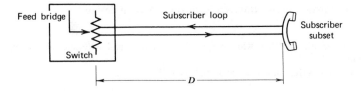

Distance D, the loop length, is most important. By Section 2.4 D must be limited in length owing to (1) attenuation of the voice signal and (2) dc resistance for signaling.

The attenuation is taken from the national transmission plan covered in Section 2.6.4. For our discussion we shall assign 6 dB as the loop attenuation limit (referred to 800 Hz). For the loop resistance limit, many crossbar switches will accept up to 1300 Ω.* From this figure we subtract 50 Ω, the nominal resistance for the telephone subset in series with the loop, leaving us with a 1250-Ω limit for the wire pair (bridge resistance disregarded). Therefore in the paragraphs that follow we shall use the following figures:

6 dB (attenuation limit for a loop)†

1250 Ω (resistance limit)

2.8.2 Basic Resistance Design

To calculate the dc loop resistance for copper conductors, the following formula is applicable:

$$R_{dc} = \frac{0.1095}{d^2}$$

where R_{dc} = loop resistance, Ω/mi, and d = diameter of the conductor, in.

If we want a 10-mi loop and allow 125 Ω/mi of loop (for the 1250-Ω limit), what diameter of copper wire would we need?

$$125 = \frac{0.1095}{d^2}$$

$$d^2 = \frac{0.1095}{125} = 0.0089$$

$$d = 0.03 \text{ in. or } 0.76 \text{ mm (rounded off to 0.80 mm)}$$

Using Table 2.4, we can compute maximum loop lengths for 1250-Ω signaling resistance. As an example, for a 26-gauge loop:

$$\frac{1250}{83.5} = 14.97 \text{ kft} \quad \text{or} \quad 14,970 \text{ ft}$$

This, then, is the signaling limit and not the loss (attenuation) limit, or what some call the "transmission limit" referred to in transmission design.

---

*Many semielectronic switches will accept 1800-Ω loops, and with special line equipment, 2400-Ω loops.
†In the United States this value may be as high as 9 dB.

Table 2.4

Gauge of Conductor	$\Omega/1000$ ft of loop	Ω/mi of loop	Ω/km of loop
26	83.5	440	237
24	51.9	274	148
22	32.4	171	92.4
19	16.1	85	45.9

Table 2.5   American Wire Gauge (B & S) Versus Wire Diameter and Resistance

American Wire Gauge	Diameter (mm)	Resistance (Ω/km)[a] at 20°C
11	2.305	4.134
12	2.053	5.210
13	1.828	6.571
14	1.628	8.284
15	1.450	10.45
16	1.291	13.18
17	1.150	16.61
18	1.024	20.95
19	0.9116	26.39
20	0.8118	33.30
21	0.7229	41.99
22	0.6439	52.95
23	0.5733	66.80
24	0.5105	84.22
25	0.4547	106.20
26	0.4049	133.9
27	0.3607	168.9
28	0.3211	212.9
29	0.2859	268.6
30	0.2547	338.6
31	0.2268	426.8
32	0.2019	538.4

[a] These figures must be doubled for loop/km. Remember it has a "go" and "return" path.

To assist relating American Wire Gauge (AWG) to cable diameter in millimeters, Table 2.5 is presented.

Another guideline in the design of subscriber loops is the minimum loop current off-hook for effective subset operation. For instance, the Bell System 500-type subset requires at least 23 mA for efficient operation.

2.8.3 Basic Transmission Design

The second design consideration mentioned above was attenuation or loss. The attenuation of a wire pair used on a subscriber loop varies with frequency, resistance, inductance, capacitance, and leakage conductance. Resistance of the line will depend on the temperature. For open-wire lines attenuation may vary ±12% between winter and summer conditions. For buried cable, with which we are more concerned, loss variations due to temperature are much less.

Table 2.6 gives losses of some common subscriber cable per 1000 ft. If we are limited to a 6-dB (loss) subscriber loop, then by simple division, we can derive the maximum loop length permissible for transmission design considerations for the wire gauges shown:

$$26 \quad \frac{6}{0.51} = 11.7 \text{ kft}$$

$$24 \quad \frac{6}{0.41} = 14.6 \text{ kft}$$

$$22 \quad \frac{6}{0.32} = 19.0 \text{ kft}$$

$$19 \quad \frac{6}{0.21} = 28.5 \text{ kft}$$

$$16 \quad \frac{6}{0.14} = 42.8 \text{ kft}$$

Table 2.6 Loss/1000 ft of Subscriber Cable[a]

Cable Gauge	Loss/1000 ft (dB)
26	0.51
24	0.41
22	0.32
19	0.21
16	0.14

[a] Cable is low capacitance type (i.e., under 0.075 nF/mi).

2.8.4 Loading

In many situations it is desirable to extend subscriber loop lengths beyond the limits described in Section 2.8.3. Common methods to attain longer loops without exceeding loss limits are the following:

1. Increase conductor diameter
2. Use amplifiers and/or loop extenders*
3. Use inductive loading

Loading tends to reduce the transmission loss on subscriber loops and other types of voice pairs at the expense of good attenuation–frequency response beyond 3000–3400 Hz. Loading a particular voice pair loop consists of inserting series inductances (loading coils) into the loop at fixed intervals. Adding load coils tends to

- Decrease the velocity of propagation
- Increase impedance

Loaded cables are coded according to the spacing of the load coils. The standard code for load coils regarding spacing is shown in Table 2.7. Loaded cables typically are designated 19-H-44, 24-B-88, and so forth. The first number indicates the wire gauge, the letter is taken from Table 2.7 and is indicative of the spacing, and the third item is the inductance of the coil in millihenrys (mH). 19-H-66 is a cable commonly used for long-distance operation in Europe. Thus the cable has 19-gauge voice pairs loaded at 1830-m intervals with coils of 66-mH inductance. The most commonly used spacings are B, D, and H.

*A loop extender is a device that increases the battery voltage on a loop, extending its signaling range. It may also contain an amplifier, thereby extending the transmission loss limits.

Table 2.7 Code for Load Coil Spacing

Code Letter	Spacing (ft)	Spacing (m)
A	700	213.5
B	3000	915
C	929	283.3
D	4500	1372.5
E	5575	1700.4
F	2787	850
H	6000	1830
X	680	207.4
Y	2130	649.6

Table 2.8 Some Properties of Cable Conductors

Diameter (mm)	AWG No.	Mutual Capacitance (nF/km)	Type of Loading	Loop Resistance (Ω/km)	Attenuation at 1000 Hz (dB/km)
0.32	28	40	None	433	2.03
		50	None		2.27
0.40		40	None	277	1.62
		50	H-66		1.42
		50	H-88		1.24
0.405	26	40	None	270	1.61
		50	None		1.79
		40	H-66	273	1.25
		50	H-66		1.39
		40	H-88	274	1.09
		50	H-88		1.21
0.50		40	None	177	1.30
		50	H-66	180	0.92
		50	H-88	181	0.80
0.511	24	40	None	170	1.27
		50	None		1.42
		40	H-66	173	0.79
		50	H-66		0.88
		40	H-88	174	0.69
		50	H-88		0.77
0.60		40	None	123	1.08
		50	None		1.21
		40	H-66	126	0.58
		50	H-88	127	0.56
0.644	22	40	None	107	1.01
		50	None		1.12
		40	H-66	110	0.50
		50	H-66		0.56
		40	H-88	111	0.44
0.70		40	None	90	0.92
		50	H-66		0.48
		40	H-88	94	0.37
0.80		40	None	69	0.81
		50	H-66	72	0.38
		40	H-88	73	0.29
0.90		40	None	55	0.72
0.91	19	40	None	53	0.71
		50	None		0.79
		40	H-44	55	0.31
		50	H-66	56	0.29
		50	H-88	57	0.26

Source: ITT, *Telecommunication Planning Documents—Outside Plant.*

Table 2.8 will be useful to calculate the attenuation of loaded loops for a given length. For example, for 19-H-88 (last entry in table) cable, the attenuation per kilometer is 0.26 dB (0.42 dB/statute mi). Thus for our 6-dB loop loss limit, we have 6/0.26, limiting the loop to 23 km in length (14.3 statute mi).

When determining signaling limits in loop design, add about 15 Ω per load coil as series resistors.

The tendency in many administrations is to use a new loading technique. This has been taken from "unigauge design" discussed in the next section. With this technique no loading is required on any loop less than 5000 m long (15,000 ft). For loops longer than 5000 m, loading starts at the 4200-m point, and load coils are installed at 1830-m intervals from there on. The loading intervals should not vary by more than 2%.

2.8.5 Other Approaches to Subscriber Loop Design

INTRODUCTION

Between 30 and 50% of a telephone company's investment is tied up in what is generally referred to as "outside plant." Outside plant, for this discussion, can be defined as that part of the telephone plant that takes the signal from the local switch and delivers it to the subscriber. Much of this expense is attributable to copper in the subscriber cable. Another important expense is cable installation, such as that incurred in tearing up city streets to augment present installation or install new plant. Much work today is being done to devise methods to reduce these expenses. Among these methods are unigauge design, dedicated plant, and fine gauge design. All have a direct impact on the outside plant transmission engineer.

UNIGAUGE DESIGN

Unigauge is a concept developed by the Bell System of North America to save on the expense of copper in the subscriber loop plant. This is done by reducing the gauge (diameter) of wire pairs to a minimum, retaining specific resistance and transmission limits. The description that follows is an attempt at applying unigauge to the more general case.

To start with, let us review the basic rules set down up to this point for subscriber loop design:

- Maximum loop resistance 1250 Ω (*Note*. 1300 Ω is used in North America.)
- Maximum loop loss 6 dB or a figure taken from the national transmission plan. (The Bell System objective is 8 dB.)

To these, let us add two more:

- The use of modern equalized telephone subsets [or at least on all loops longer than 10,000 ft (3000 m)]

- Cable gauges limited to AWG 19, 22, 24, and 26, while maintaining the 1250-Ω limit or whichever limit needs to be established inside which signaling can be effected.

All further argument will be based on minimizing the amount of copper used (e.g., using the smallest diameter wire possible within the limits prescribed above).

The reader should keep in mind that many telephone companies must install subscriber loops in excess of 15,000 ft (5000 m) and thus must use a conductor size (or sizes) larger than 26 gauge. The Bell System, for example, found that 20% of its subscriber loops exceeded 15,000 ft in 1964. To meet transmission and signaling

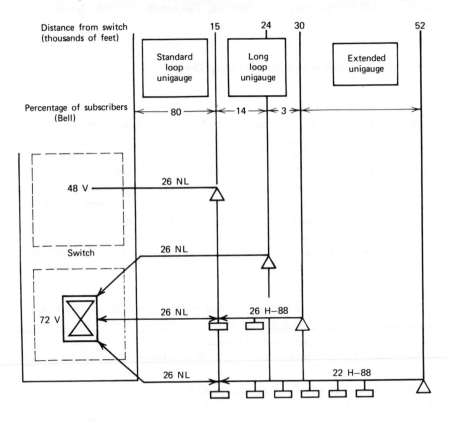

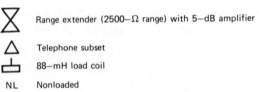

Figure 2.11 Layout of a unigauge subscriber plant.

objectives, the cost of such loops tends to mount excessively for those loops over 15,000 ft in length.

The concept of unigauge design basically is one that takes advantage of relatively inexpensive voice frequency gain devices and range extenders to permit the use of 26-gauge conductors on the greater percentage of its longer loops. Also, unigauge design makes it mandatory that a unified design approach be used. In other words, the design of the subscriber loop plant is an integrated process and not one of piece-by-piece engineering. Unigauge allows the use of 26-gauge cable on subscriber loops up to 30,000 ft in length. This results in very significant savings in copper and a general overall economy in present-day outside plant installation.

APPLICATION OF UNIGAUGE DESIGN

A typical layout of a subscriber plant based on unigauge design, taken from the Bell System, is shown in Figure 2.11 (Ref. 10). In the figure it will be seen that subscribers within 15,000 ft of the switch are connected over loops made up of 26-gauge nonloaded cable with standard 48-V battery. Their connection at the switch is conventional. It is also seen that 80% of Bell System subscribers are within this radius. Loops 15,000–30,000 ft long are called unigauge loops. Subscribers in the range of 15,000–24,000 ft from the switch are connected by 26-gauge nonloaded cable as well, but require a range extender to provide sufficient voltage for signaling and supervision. In the drawing a 72-V range extender is shown equipped with an amplifier that gives a midband gain of 5 dB. The output of the amplifier "emphasizes" the higher frequencies. This offsets that additional loss suffered at the higher frequencies of the voice channel on the long nonloaded loops. To extend the loops to the full 30,000 ft, the Bell System adds 88 mH loading coils at the 15,000- and 21,000-ft points.

For long loops (more than 15,000 ft long) the range extender–amplifier combinations are not connected on a line-for-line basis. It is standard practice to equip four or five subscriber loops with only one range extender–amplifier used on a shared basis. When the subscriber goes off-hook on a long unigauge loop, a line is seized, and the range extender is switched in. This concentration is another point in favor of unigauge because of the economics involved. It should be noted that the long 15,000-ft nonloaded sections, into which the switch faces, provide a fairly uniform impedance for all conditions when an active amplifier is switched in. This is a positive factor with regard to stability.

Loops more than 30,000 ft long may also use the unigauge principle and often are referred to as *extended unigauge*. Such a loop is equipped with 26-gauge nonloaded cable from the switch out to the 15,000-ft point. Beyond 15,000 ft 22-gauge cable is used with H-88 loading. As with all loops more than 15,000 ft long, a range extender–amplifier is switched in when the loop is in use. The loop length limit for this combination is 52,000 ft. Loops more than 52,000 ft long may also be installed by using a gauge of diameter larger than 22.

It should be noted that the Bell System replaces its line relays with others that are sensitive up to 2500 Ω of loop resistance with 48-V battery. Such a modification to the switch is done on long loops only. The 72 V supplied by the range extender is for pulsing, and the ringing voltage (to ring the distant telephone) is superimposed on the line with the subscriber's subset being in an on-hook condition.

Besides the savings in the expenditure on copper, unigauge displays some small improvements in transmission characteristics over older design methods of subscriber loops:

- Unigauge has a slightly lower average loss (when we look at a statistical distribution of subscribers).
- There is 15-dB average return loss on the switch side of an amplifier, compared with an average of 11 dB for older design methods.

These two points echo the Bell System experience with unigauge design (Refs. 10 and 21).

DEDICATED PLANT

The dedicated plant assigns wire pairs to subscribers or would-be subscriber locations. In the past bridged taps were used so that subscriber loops were more versatile regarding assignment. The idea of the bridged tap is shown below:

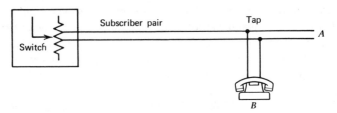

In this case station *A* is not connected; it is not in use but available in case station *B* disconnects telephone service.

The primary impact of the dedicated plant is that it eliminates the use of bridged taps. Bridged taps deteriorate the quality of transmission by notably increasing the capacitance of a loop.

OTHER LOOP DESIGN TECHNIQUES

Fine gauge and minigauge techniques essentially are refinements of the unigauge concept. In each case the principal object is to reduce the amount of copper. Obviously, one method is to use still smaller gauge pairs on shorter loops. Consideration has been given to the use of gauges as small as 32. Another approach is to use aluminum as the conductor. When aluminum is used, a handy rule of thumb to follow is that ohmic and attenuation losses of aluminum may be equated to copper in that aluminum wire always is the next "standard gauge" larger than its copper counterpart if copper were to be used. Some examples are listed as follows:

Copper	Aluminum
19	17
22	20
24	22
26	24

Aluminum has some drawbacks as well. The major ones are summed up as follows:

- Not to be used on the first 500 yd of cable where the cable has a large diameter (i.e., more loops before branching).
- More difficult to splice than copper.
- More brittle.
- Because the equivalent conductor is larger than its copper counterpart, an equivalent aluminum cable with the same conductivity–loss characteristics as copper will have a smaller pair count in the same sheath.

2.9   DESIGN OF LOCAL AREA TRUNKS (JUNCTIONS)

Exchanges in a common local area often are connected on a full mesh basis (see Section 2.7 for definition of mesh connection). Depending on distance and certain other economic factors, these trunk circuits use VF transmission over cable. In view of the relatively small number of these trunk circuits when compared to the number of subscriber lines,* it is generally economically desirable to minimize attenuation in this portion of the network.

One approach used by some telephone companies (administrations) is to allot one-third of the total end-to-end reference equivalent to each subscriber's loop and one-third to the trunk network. Figure 2.12 illustrates this concept. For instance, if the transmission plan called for a 24-dB ORE, then one-third of 24 dB, or 8 dB, would be assigned to the trunk plant. Of this we may assign 4 dB to the four-wire portion or toll segment of the network, leaving 4 dB for local VF trunks or 2 dB at each end. The example has been highly simplified, of course.

*Due to the inherent concentration in local switches (e.g., 1 trunk for 8–25 subscribers, depending on design).

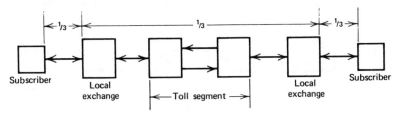

Figure 2.12   One approach to loss assignment.

Table 2.9. Inserted Connection Loss Objectives for Local Trunks (Ref. 21)

Trunk Types	Inserted Connection Loss (dB)	
	Nongain	Gain
Direct trunks[a]	0–5.0	3.0 (5.0 max.)
Tandem trunks	0–4.0	3.0 (4.0 max.)
Intertandem trunks		
Terminated at sector tandems at both ends	–	1.5

Terminated in a toll center or
    sector tandem that meets
    terminal balance objectives
    at one or both ends
	–	0.5

[a] These ICLs apply to direct trunks less than 200 mi in length. Direct trunks more than 200 mi long are designed in accordance with the VNL plan.

For the toll-connecting trunks (e.g., those trunks which connect the local network to the toll network), if a good return loss cannot be maintained on all or nearly all connections, losses on two-wire toll-connecting trunks may have to be increased to reduce possibilities of echo and singing. Sometimes the range of loss for these two-wire circuits must be extended to 5 or 6 dB. It is just these circuits that the four-wire toll network looks into directly.

The Bell System uses the design objectives given in Table 2.9 for inserted connection loss assignment for local trunks.

From this it can be seen that the approach to the design of VF trunks varies considerably from that used for subscriber loop design. Although we must ensure that signaling limits are not exceeded, almost always the transmission limit will be exceeded well before the signaling limit. The tendency to use larger diameter cable on long routes is also evident.

One major difference from the subscriber loop approach is in the loading. If loading is to be used, the first load coil is installed at distance $D/2$, where D is the normal separation distance between load points. Take the case of H loading, for instance. The distance between load points is 1830 m, but the first load coil from the exchange is placed at $D/2$ or 915 m from the exchange. Then if an exchange is bypassed, a full load section exists. This concept is illustrated in Figure 2.13.

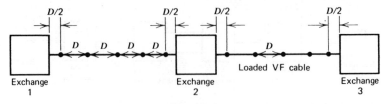

Figure 2.13 Loading of VF trunks (junctions).

Now consider this example. A loaded 500-pair VF trunk cable extends across town. A new switching center is to be installed along the route where 50 pairs are to be dropped and 50 inserted. It would be desirable to establish the new switch midway between load points. At the switch 450 circuits will bypass the office (switch). Using this $D/2$ technique, these circuits need no conditioning; they are full-load sections (i.e., $D/2 + D/2 = 1D$, a full-load section). Meanwhile the 50 circuits entering from each direction are terminated for switching and need conditioning so that each looks electrically like a full-load section. However, the physical distance from the switch out to the first load point is $D/2$ or, in the case of H loading, 915 m. To make the load distance electrically equivalent to 1830 m, line build out (LBO) is used. This is done simply by adding capacity to the line.

Suppose that the location of the new switching center was such that it was not halfway, but at some other fractional distance. For the section comprising the shorter distance, LBO is used. For the other, longer run, often a half-load coil is installed at the switching center and LBO is added to trim up the remaining electrical distance.

2.10   VF REPEATERS

VF repeaters in telephone terminology imply the use of *uni*directional amplifiers at VF on VF trunks. On a two-wire trunk two amplifiers must be used on each pair with a hybrid in and a hybrid out. A simplified block diagram is shown in Figure 2.14.

The gain of a VF repeater can be run up as high as 20 or 25 dB, and originally they were used at 50 mi intervals on 19-gauge loaded cable in the long-distance (toll) plant. Today they are seldom found on long-distance circuits, but they do have application on local trunk circuits where the gain requirements are considerably less. Trunks using VF repeaters have the repeater's gain adjusted to the equivalent loss of the circuit minus the 4-dB loss to provide the necessary singing margin. In practice a repeater is installed at each end of the trunk circuit to simplify maintenance and power feeding. Gains may be as high as 6–8 dB.

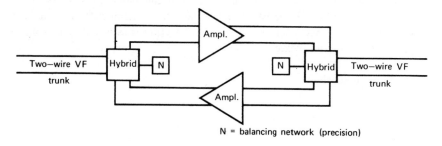

N = balancing network (precision)

Figure 2.14 Simplified block diagram of a VF repeater.

An important consideration with VF repeaters is the balance at the hybrids. Here precision balancing networks may be used instead of the compromise networks employed at the two-wire–four-wire interface (Section 2.6.3). It is common to achieve a 21-dB return loss, 27 dB is also possible, and, theoretically, 35 dB can be reached.

Another repeater commonly used on two-wire trunks is the negative-impedance repeater. This repeater can provide a gain as high as 12 dB, but 7 or 8 dB is more common in practice. The negative-impedance repeater requires an LBO at each port and is a true two-way, two-wire repeater. The repeater action is based on regenerative feedback of two amplifiers. The advantage of negative-impedance repeaters is that they are transparent to dc signaling. On the other hand, VF repeaters require a composite arrangement to pass dc signaling. This consists of a transformer bypass.

2.11 TRANSMISSION CONSIDERATIONS OF TELEPHONE SWITCHES IN THE LONG-DISTANCE NETWORK (FOUR-WIRE)

Any device placed in an analog transmission circuit tends to degrade the quality of transmission. Telephone switches, unless properly designed, well selected from a system's engineering point of view, and properly installed, may seriously degrade a transmission network. Transmission specifications of a switch must be set forth clearly when switches are to be purchased. One reference which sets forth a series of specifications for toll switches is CCITT Rec. Q.45. The reader should also consult the G.100 series of CCITT recommendations (see Appendix A).

In practice it has been found that much of the criterion specified in CCITT Rec. Q.45 is rather loose. Table 2.10 compares Q.45 recommendations with a stricter criterion (based on Ref. 19).

2.12 CCITT INTERFACE

2.12.1 Introduction

To facilitate satisfactory communications between telephone subscribers in different countries, the CCITT has established certain transmission criteria in the form of recommendations. According to CCITT a connection is satisfactory if it meets certain criteria for the following:

- Reference equivalent
- Noise
- Echo
- Singing

The following paragraphs outline and highlight these criteria from the point of view of a telephone company or administration's international interface. This is the point where the national network meets the international system.

2.12.2 Maximum Number of Circuits in Tandem

Consulting Figure 2.15, we count six links in the portion of a connection that we call international. The interface for the country of origin is a CT3 [the CT (central transito) is ranked 3 in the hierarchy], and that for the country of destination, the

Table 2.10 Transmission Characteristics of Switches (Four-Wire Switching)

Item	1 (Q.45)[c]	2 (Ref. 19 Criterion)[c]
Loss	0.5 dB	0.5 dB
Loss, dispersion[a]	<0.2 dB	<0.2 dB
Attenuation–	300–400 Hz: -0.2/+0.5 dB	-0.1/+0.2 dB
frequency response	400–2400 Hz: -0.2/+0.3 dB	-0.1/+0.2 dB
	2400–3400 Hz: -0.2/+0.5 dB	-0.1/+0.3 dB
Impulse noise	5 in 5 min. above -35 dBm0	5 in 5 min, 12 dB above floor of random noise[d]
Noise		
Weighted	200 pWp	25 pWp
Unweighted	1000,000 pW	3000 pW
Unbalance against	300–600 Hz: 40 dB	300–3000 Hz: 55 dB
ground	600–3400 Hz: 46 dB	3000–3400 Hz: 53 dB
Crosstalk		
Between go and return paths	60 dB	65 dB } in the band
Between any two paths	70 dB	80 dB } 200–3200 Hz
Harmonic distortion	–	50 dB down with a -10-dBm signal for 2nd harmonic, 60 dB down for 3rd harmonic
Impedance variation with frequency[b] (or unbalance to ground)	200 Hz: 15 dB 300–600 Hz: 15 dB 600–3400 Hz: 20 dB	300 Hz: 18 dB 500–2500 Hz: 20 dB 3000 Hz: 18 dB 3400 Hz: 15 dB

[a]Dispersion loss is the variation in loss from calls with the highest loss to those with the lowest loss. This important parameter affects circuit stability.
[b]Expressed as return loss.
[c]Reference frequency, where required, is 800 Hz for column 1 and 1000 Hz for column 2.
[d]Ref. 23.

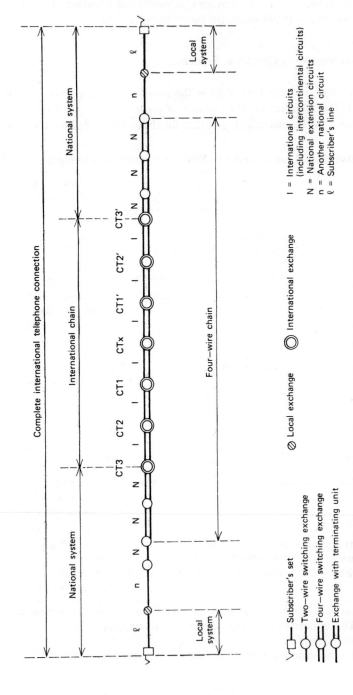

Figure 2.15 A CCITT international telephone connection with national extensions. Transit exchanges (CT) are hierarchically ranked. (Courtesy ITU–CCITT.)

74

other CT3. A call must be limited from local exchange (origin) to local exchange (destination) to have no more than 14 links (circuits) in tandem, of which four may be in the country of origin and four in the country of destination. The number of links is limited to assure maintenance of the limits of noise, stability, and reference equivalent.

The number of links in a national chain is further limited (CCITT Rec. G.101). If the average distance to a subscriber is 600 mi (1000 km) from the international interface, at most three national four-wire circuits can be connected on a four-wire basis between each other. In countries that have average distances to subscribers in excess of 600 mi, a fourth or possibly a fifth national circuit may be added on a four-wire basis (see CCITT Rec. E.171).

2.12.3 Noise

CCITT Rec. G.103 treats noise. It equates noise to the length of a circuit. This is equated at the rate of 4 pWp/km. Assume that a national connection is 2500 km long, or 2500 km to reach the international interface. Then we would expect to measure no more than 10,000 pWp at that point. CCITT Recs. G.152 and G.153 contain further information. Actual design objectives should improve on these specifications by proper choice of equipment and system layout and engineering.

Absolute noise maxima on the receive side of an international connection should not exceed 50,000 pWp referred to a zero relative level point of the first circuit in the chain (CCITT Rec. G.143). This maximum noise level is permissible if there are six international circuits in tandem.

2.12.4 Variation of Transmission Loss with Time

This important parameter effects stability. CCITT Rec. G.151C states that

> The standard deviation of the variation in transmission loss of a circuit should not exceed 1 dB . . . . The difference between the mean value and the nominal value of the transmission loss for each circuit should not exceed 0.5 dB.

2.12.5 Crosstalk (CCITT Rec. G.151D)

The near-end and far-end crosstalk (intelligible crosstalk only) measured at audio frequencies at trunk exchanges between two complete circuits in terminal service position should not numerically be less than 58 dB.

Between go and return paths of the same circuit in a four-wire long-distance (toll) exchange intelligible crosstalk should be at least 43 dB down.

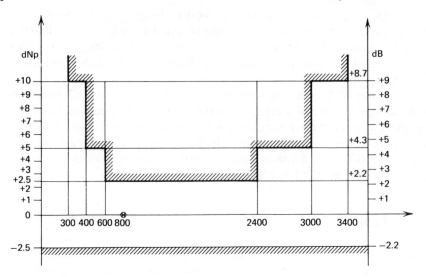

Figure 2.16 Permissible attenuation variation with respect to its value measured at 800 Hz (objective for worldwide four-wire chain of 12 circuits in terminal service). (CCITT Rec. G.132; Courtesy ITU–CCITT.)

2.12.6 Attenuation Distortion

The worst condition for attenuation distortion is shown in Figure 2.16 (taken from CCITT Rec. G.131). It assumes that the nominal 4-kHz voice channel is used straight through. The attenuation distortion as shown in the figure is a result of 12 circuits of a four-wire chain in tandem. Note that the slope from 300 to 600 Hz and from 2400 to 3400 Hz is approximately 6.5 dB. The slope for one link, therefore, would be 6.5/12 or approximately 0.5 dB. From this we can derive the attenuation distortion permissible for each link to the international interface. For further information consult Chapter 3 which gives a discussion of attenuation distortion of FDM carrier equipment back to back.

2.12.7 Reference Equivalent

CCITT Rec. G.111 states:

> For 97 percent of the connections made in a country of average size (1000 km average distance to the CT), the nominal reference equivalent between a subscriber and the 4-wire terminals of the international circuit (the international interface) . . . should not exceed 20.8 dB sending and 12.2 dB receiving.

This gives an ORE of 33 dB. It is recommended that telephone companies and administrations attempt to lower this figure as much as possible to improve subscriber satisfaction. A good target would be an ORE from a low of 6 dB to a maximum of 20 dB.

2.12.8 Propagation Time

The propagation time at 800 Hz of the national sending of receiving system should not exceed 50 ms (CCITT Rec. G.114).

2.12.9 Echo Suppressors

Echo suppressors should be used on all connections where the *round-trip* delay (propagation time) exceeds 50 ms. Echo suppressors are covered in CCITT Rec. G.161.

REFERENCES AND BIBLIOGRAPHY

1. *Reference Data for Radio Engineers*, 6th ed., Howard W. Sams, Indianapolis, IN, 1977.
2. *Outside Plant*, U.S. Army Tech. Manual TM-486-5.
3. M. A. Clement, "Transmission," *Telephony*.
4. CCITT Orange Books, Geneva, 1976, vol. III, G recommendations.
5. *National Networks for the Automatic Service*, ITU, Geneva, chap. V.
6. *Overall Communications System Planning*, vols. I–III, IEEE New Jersey Section Seminar, 1964.
7. *Transmission Systems for Communications*, 4th ed., Bell Telephone Laboratories, American Telephone and Telegraph Company, New York, 1971.
8. F. T. Andrew and R. W. Hatch, "National Telephone Network Planning in the ATT," *IEEE Trans. Comm. Technol.*, June 1971.
9. "Terminal Balance, Description and Test Methods," *Autom. Electr. Tech. Bull.*, 305–351.
10. P. A. Gresh *et al.*, "A Unigauge Design Concept for Telephone Customer Loop Plant," *IEEE Trans. Comm. Technol.*, Apr. 1968.
11. *Principles of Electricity Applied to Telephone and Telegraph Work*, American Telephone and Telegraph Co., New York, 1961.
12. *Rural Electrification Administration Telephone Engineering and Construction Manual*, Sec. 400 series.
13. *ITT Pentaconta Manual*, PC-1000 A/B/B1.
14. H. R. Huntley, "Transmission Design of Intertoll Telephone Trunks," *Bell Sys. Tech. J.*, Sept. 1953.
15. Military Standard 188C with Notice 1, U.S. Department of Defense, Washington, DC.
16. *DCS Engineering Installation Manual*, DCA C 330-175-1, through Change 9, U.S. Department of Defense, Washington, DC.
17. *Lenkurt Demodulator*, Lenkurt Electric Corp., San Carlos, CA, July 1960, Oct. 1962, Mar. 1964, Jan. 1964, Aug. 1966, June 1968, July 1966, Aug. 1973.

18. *Local Telephone Networks*, ITU–CCITT, Geneva, 1968.

19. USITA Symposium, Apr. 1970, Open Questions 18–37.

20. *Outside Plant*, Telecommunication Planning Documents, ITT Laboratories, Spain, 1973.

21. *Telecommunication Transmission Engineering*, vols. 1–3, 2nd ed., American Telephone and Telegraph Co., New York, 1977.

22. R. L. Freeman, *Telecommunication System Engineering*, Wiley Interscience, New York, 1980.

23. Bell System Technical Reference, *Transmission Parameters Affecting Voiceband Data Transmission–Description of Parameters*, Publ. 41008, American Telephone and Telegraph Co., New York, July 1964.

3 | FREQUENCY DIVISION MULTIPLEX

3.1 INTRODUCTION

Multiplex deals with the transmission of two or more signals simultaneously over a single transmission facility. Multiplexing may be accomplished either in the frequency domain or in the time domain. The first is called frequency division multiplex (FDM), and the second, time division multiplex (TDM). This chapter covers the former. The latter is discussed in Chapter 11.

Before we launch into multiplexing, keep in mind that all multiplex systems work on a four-wire basis. The transmit and receive paths are separate. Two-wire and four-wire transmission and conversion from two-wire to four-wire systems are covered in Section 2.6.3.

Carrier is a word that has become associated with FDM systems, probably owing to the fact that FDM employs "carrier" waves. Unfortunately its use has spread to TDM systems as well. As used in this text, however, carrier refers exclusively to FDM systems.

It is assumed that the reader has some background in how a single-sideband (SSB) signal is developed as well as how a carrier is suppressed in (SSBSC) systems.

A simplified diagram of an FDM link is shown in Figure 3.1.

3.2 MIXING

The heterodyning or mixing of signals of frequencies A and B is shown below. What frequencies may be found at the output of the mixer?

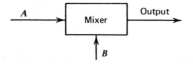

Both the original signals will be present as well as the signals representing their sum and their difference in the frequency domain. Thus at the output of the above mixer we will have present the signals of frequency $A, B, A + B$, and $A - B$. Such a mixing process is repeated many times in FDM equipment.

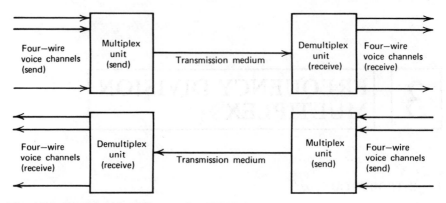

Figure 3.1 Simplified block diagram of an FDM link.

Let us now look at the boundaries of the nominal 4-kHz voice channel. These are 300 and 3400 Hz. Let us further consider these frequencies as simple tones of 300 and 3400 Hz. Now consider the mixer below and examine the possibilities at its output.

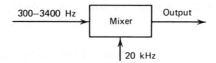

First, the output may be the sum or

$$
\begin{array}{r}
20{,}000 \text{ Hz} \\
+ \quad 300 \text{ Hz} \\
\hline
20{,}300 \text{ Hz}
\end{array}
\qquad
\begin{array}{r}
20{,}000 \text{ Hz} \\
+ \ 3{,}400 \text{ Hz} \\
\hline
23{,}400 \text{ Hz}
\end{array}
$$

A simple low-pass filter could filter out all frequencies below 20,300 Hz.

Now imagine that instead of two frequencies, we have a continuous spectrum of frequencies between 300 and 3400 Hz (i.e., we have the voice channel). We represent the spectrum as a triangle:

As a result of the mixing process (translation) we have another triangle as follows:

When we take the sum, as we did above, and filter out all other frequencies, we say we have selected the upper sideband. Thus we have a triangle facing to the right, and we call this an upright or erect sideband.

We can also take the difference, such that

$$
\begin{array}{rr}
20{,}000 \text{ Hz} & 20{,}000 \text{ Hz} \\
- \quad 300 \text{ Hz} & - \ 3{,}400 \text{ Hz} \\
\hline
19{,}700 \text{ Hz} & 16{,}000 \text{ Hz}
\end{array}
$$

and we see that in the translation (mixing process) we have had an inversion of frequencies. The higher frequencies of the voice channel become the lower frequencies of the translated spectrum, and the lower frequencies of the voice channel become the higher when the difference is taken. We represent this by a right triangle facing the other direction:

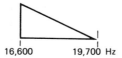

This is called an inverted sideband. To review, when we take the sum, we get an erect sideband. When we take the difference, frequencies invert and we have an inverted sideband represented by a triangle facing left.

Now let us complicate the process a little by translating three voice channels into the radio electric spectrum for simultaneous transmission on a specific medium, a pair of wire lines, for example. Let the local oscillator (mixing) frequency in each case be 20, 16, and 12 kHz. The mixing process is shown in Figure 3.2.

From Figure 3.2 the difference frequencies are selected in each case as follows:

$$
\begin{array}{lrr}
\text{For channel 1} & 20{,}000 \text{ Hz} & 20{,}000 \text{ Hz} \\
& - \quad 300 \text{ Hz} & - \ 3{,}400 \text{ Hz} \\
\hline
& 19{,}700 \text{ Hz} & 16{,}600 \text{ Hz} \\
\\
\text{For channel 2} & 16{,}000 \text{ Hz} & 16{,}000 \text{ Hz} \\
& - \quad 300 \text{ Hz} & - \ 3{,}400 \text{ Hz} \\
\hline
& 15{,}700 \text{ Hz} & 12{,}600 \text{ Hz} \\
\\
\text{For channel 3} & 12{,}000 \text{ Hz} & 12{,}000 \text{ Hz} \\
& - \quad 300 \text{ Hz} & - \ 3{,}400 \text{ Hz} \\
\hline
& 11{,}700 \text{ Hz} & 8{,}600 \text{ Hz}
\end{array}
$$

In each case the lower sidebands have been selected as mentioned above, and all frequencies above 19,700 Hz have been filtered from the output as well as the local oscillator carriers themselves. The outputs from the modulators terminate on a common bus. The common output appearing on this bus is a band of frequencies between 8.6 and 19.7 kHz containing the three voice channels which have been

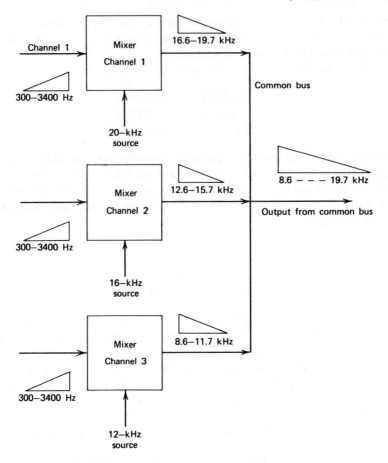

Figure 3.2   Simple FDM (transmit portion only shown).

translated in frequency. They now appear on one two-wire circuit ready for transmission. They may be represented by a single inverted triangle as shown:

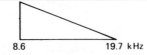

3.3   THE CCITT MODULATION PLAN

3.3.1   Introduction

A modulation plan sets forth the development of a band of frequencies called the line frequency (i.e., ready for transmission on the line or transmission medium). The modulation plan usually is a diagram showing the necessary mixing, local oscillator mixing frequencies, and the sidebands selected by means of the triangles described previously in a step-by-step process from voice channel input to line

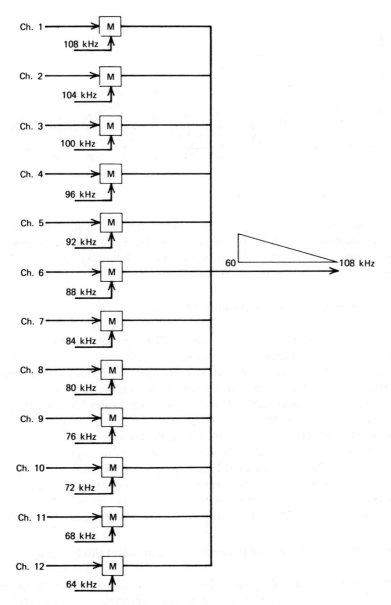

Figure 3.3 Formation of the standard CCITT group.

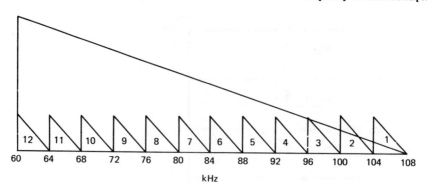

Figure 3.4 Standard CCITT group.

frequency output. The CCITT has recommended a standardized modulation plan with a common terminology. This allows large telephone networks, on both national and multinational systems, to interconnect. In the following paragraphs the reader is advised to be careful with terminology.

3.3.2 Formation of the Standard CCITT Group

The standard *group* as defined by the CCITT occupies the frequency band of 60–108 kHz and contains 12 voice channels. Each voice channel is the nominal 4-kHz channel occupying the 300–3400-Hz spectrum. The group is formed by mixing each of the 12 voice channels with a particular carrier frequency associated with the channel. Lower sidebands are then selected. Figure 3.3 shows the preferred approach to the formation of the standard CCITT group. It should be noted that in the 60–108 kHz band voice channel 1 occupies the highest frequency segment by convention, between 104 and 108 kHz. The layout of the standard group is shown in Figure 3.4. The applicable CCITT recommendation is G.232.

SSBSC modulation techniques are recommended except under special circumstances discussed later in this chapter. CCITT recommends that carrier leak be down to at least −26 dBm0 referred to a 0 relative level point (see Section 1.9.5).

3.3.3 Alternative Method of Formation of the Standard CCITT Group

The economy of filter design has caused some manufacturers to use an alternate method to form a group. This is done by an intermediate modulation step forming four pregroups. Each pregroup translates three voice channels in the intermediate modulation step. The translation process for this alternative method is shown in Figure 3.5. For each pregroup the first voice channel modulates a 12-kHz carrier, the second a 16-kHz carrier, and the third a 20-kHz carrier. The upper sidebands are selected in this case and carriers are suppressed.

The result is a subgroup occupying a band of frequencies from 12 to 24 kHz. The second modulation step is to take four of these pregroups so formed and translate

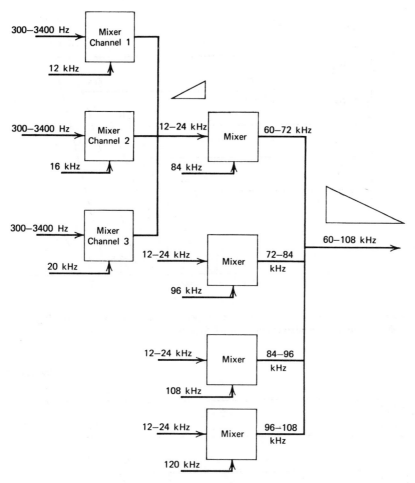

Figure 3.5 Formation of standard CCITT group by two steps of modulation (mixing).

them, each to their own frequency segment, in the band of 60–108 kHz. To achieve this the pregroups are modulated by carrier frequencies of 84, 96, 108, and 120 kHz, and the lower sidebands are selected, properly inverting the voice channels. This dual modulation process is shown in Figure 3.5.

The choice of the one-step or two-step modulation is an economic trade-off. Adding a modulation stage adds noise to the system:

3.3.4 Formation of the Standard CCITT Supergroup

A supergroup contains five standard CCITT groups, equivalent to 60 voice channels. The standard supergroup before translation occupies the frequency band of 312–552 kHz. Each of the five groups making up the supergroup is translated in fre-

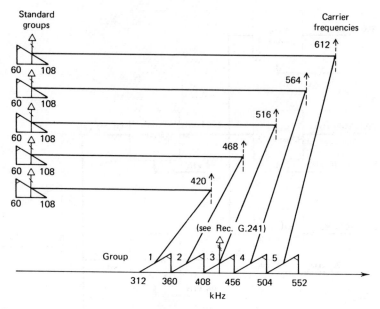

Figure 3.6 Formation of the standard CCITT supergroup (CCITT Rec. G.233). *Note.* Vertical arrows show group level regulating pilot tones (see Section 3.5). (Courtesy ITU–CCITT.)

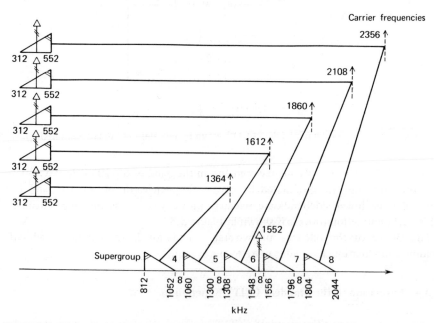

Figure 3.7 Formation of the standard CCITT mastergroup (CCITT Rec. G.233; Courtesy ITU–CCITT.)

quency to the supergroup band by mixing with the proper carrier frequencies. The carrier frequencies are 420 kHz for group 1, 468 kHz for group 2, 516 kHz for group 3, 564 kHz for group 4, and 612 kHz for group 5. In the mixing process the difference is taken (lower sidebands are selected). This translation process is shown in Figure 3.6.

3.3.5 Formation of the Standard CCITT Basic Mastergroup and Supermastergroup

The basic mastergroup contains five supergroups, 300 voice channels, and occupies the spectrum of 812–2044 kHz. It is formed by translating the five standard super-groups, each occupying the 312–552-kHz band, by a process similar to that used to form the supergroup from five standard CCITT groups. This process is shown in Figure 3.7.

The basic supermastergroup contains three mastergroups and occupies the band of 8516–12,388 kHz. The formation of the supermastergroup is shown in Figure 3.8.

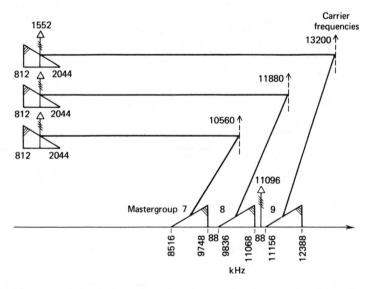

Figure 3.8 Formation of the standard CCITT supermastergroup (CCITT Rec. G.233; Courtesy ITU–CCITT.)

3.3.6 The "Line" Frequency

The band of frequencies that the multiplex applies to the line, whether the line is a radiolink, coaxial cable, wire pair or open-wire line, is called the line frequency.

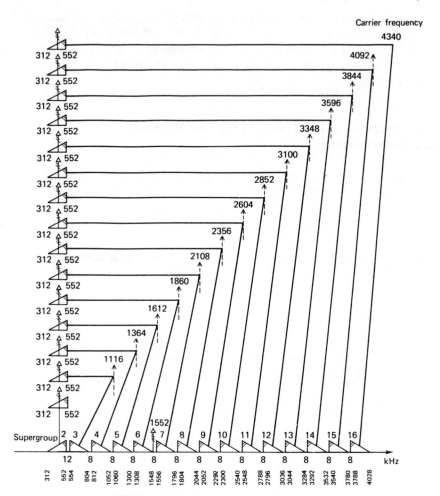

Figure 3.9 Makeup of basic CCITT 15-supergroup assembly. (Courtesy ITU–CCITT.)

Another expression often used is HF (or high frequency), not to be confused with high frequency radio, discussed in Chapter 4.

The line frequency in this case may be the direct application of a group or supergroup to the line. However, more commonly a final translation stage occurs, particularly on high density systems. Several of these line configurations are shown below.

Figure 3.9 shows the makeup of the basic 15-supergroup assembly. Figure 3.10 shows the makeup of the standard 15-supergroup assembly No. 3 as derived from the basic 15-supergroup assembly shown in Figure 3.9. Figure 3.11 shows the development of a 600-channel standard CCITT line frequency.

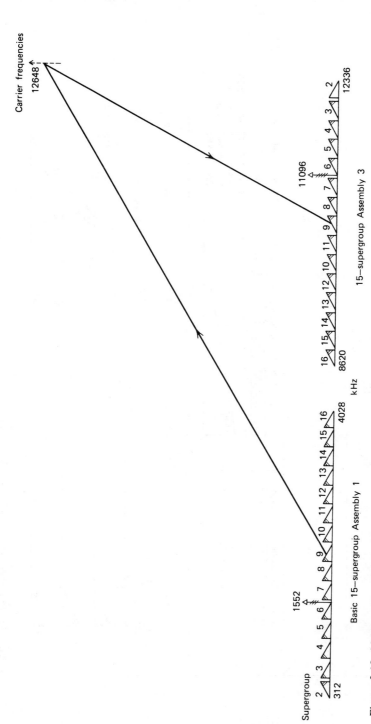

Figure 3.10 Makeup of standard 15-supergroup assembly 3 as derived from basic 15-supergroup assembly. (Courtesy ITU–CCITT.)

89

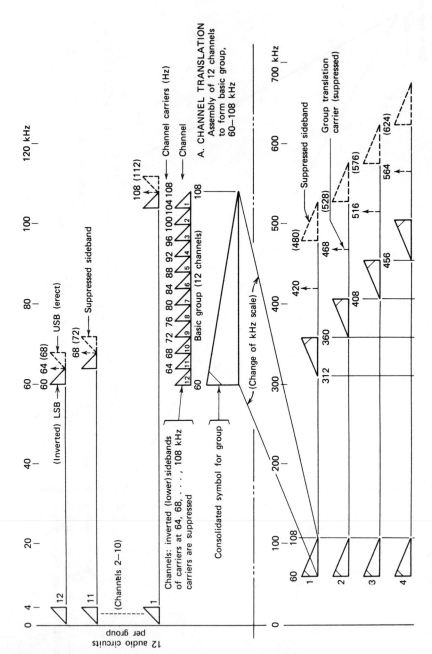

Figure 3.11 Development of a 600-channel line frequency.

90

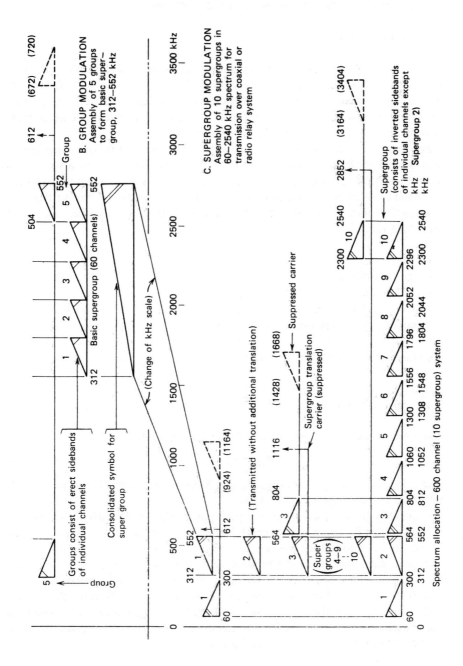

Figure 3.11 (*Continued*)

91

3.4   LOADING OF MULTICHANNEL FDM SYSTEMS

3.4.1   Introduction

Most of the FDM (carrier) equipment in use today carries speech traffic, sometimes misnamed in North America "message traffic." In this context we refer to full-duplex conversations by telephone between two "talkers." However, the reader should not lose sight of the fact that there is a marked increase in the use of these same intervening talker facilities for data transmission.

For this discussion the problem essentially boils down to that of human speech and how multiple telephone users may load a carrier system. If we load a carrier system too heavily, meaning here that the input levels are too high, intermodulation noise and crosstalk will become intolerable. If we do not load the system sufficiently, the signal-to-noise ratio will suffer. The problem is fairly complex because speech amplitude varies:

- With talker volume
- At a syllabic rate
- At an audio rate
- With varying circuit losses as different loops and trunks are switched into the same channel bank voice channel input

Also the loading of a particular system varies with the busy hour.*

3.4.2   Speech Measurement

The average power measured in dBm of a typical single talker is

$$P_{dBm} = V_{VU} - 1.4 \tag{3.1}$$

where V_{VU} = reading of a VU meter (see Section 1.9.5). In other words a 0 VU talker has an average power of -1.4 dBm.

Empirically for a typical talker the peak power is about 18.6 dB higher than the average power. Peakiness of the speech level means that the carrier equipment must be operated at a low average power to withstand voice peaks so as not to overload and cause distortion. Thus the primary concern is that of voice peaks or spurts. These can be related to an activity factor T_a. T_a is defined as that proportion of the time that the rectified speech envelope exceeds some threshold. If the threshold is about 20 dB below the average power, the activity dependence on threshold is fairly weak. We can now rewrite our equation for average talker power in dBm, relating it to the activity factor T_a as follows:

---

*The *busy hour* is a term used in traffic engineering and is defined by the CCITT as "the uninterrupted period of 60 minutes during which the average traffic flow is maximum."

$$P_{\text{dBm}} = V_{\text{VU}} + 10 \log T_{\text{a}}$$

If $T_{\text{a}} = 0.725$, the results will be the same as for equation (3.1).

Consider now adding a second talker operating on a different frequency segment on the same equipment, but independent of the first talker. Translations to separate frequency segments are described earlier in the chapter. With the second talker added the system average power will increase by 3 dB. If we have N talkers, each on a different frequency segment, the average power developed will be:

$$P_{\text{dBm}} = V_{\text{VU}} - 1.4 + 10 \log N \tag{3.2}$$

where P_{dBm} = power developed across the frequency band occupied by all the talkers.

Empirically we have found that the peakiness or peak factor of multitalkers over a multichannel analog system reaches the characteristics of random noise peaks when the number of talkers N exceeds 64. When $N = 2$, the peaking factor is 18 dB; for 10 talkers it is 16 dB, for 50 talkers 14 dB, and so forth. Above 64 talkers the peak factor is 12.5 dB.

An activity factor of 1, which we have been using above, is unrealistic. This means that someone is talking all the time on the circuit. The traditional figure for activity factor accepted by CCITT and used in North American practice is 0.25. Let us see how we reach this lower figure.

For one thing, the multichannel equipment cannot be designed for N callers and no more. If this were true, a new call would have to be initiated every time a call terminated or calls would have to be turned away for an "all trunks busy" condition. In the real life situation, particularly for automatic service, carrier equipment, like switches, must have a certain margin by being overdimensioned for busy hour service. For this overdimensioning we drop the activity factor to 0.70. Other causes reduce the figure even more. For instance, circuits are essentially inactive during call setup as well as during pauses for thinking during a conversation. The factor of 0.70 now reduces to 0.50. This latter figure is divided in half owing to the talk–listen effect. If we disregard isolated cases of "double-talking," it is obvious that on a full-duplex telephone circuit, while one end is talking, the other is listening. Thus a circuit (in one direction) is idle half the time during the "listen" period. The resulting activity factor is 0.25.

3.4.3 Overload

In Section 1.9.6, where we discussed intermodulation (IM) products, we showed that one cause of IM products is overload. One definition of overload (Ref. 4) is:

The overload point, or overload level, of a telephone transmission system is 6 dB higher than the average power in dBm0 of each of two applied sinusoids of equal amplitude and of frequencies a and b, when these input levels are so adjusted that an increase of 1 dB in both their separate levels causes an increase, at the output, of 20 dB in the intermodulation product of frequency $2a - b$.

Up to this point we have been talking about average power. Overload usually occurs when instantaneous signal peaks exceed some threshold. Consider that peak instantaneous power exceeds average power of a simple sinusoid by 3 dB. For multichannel systems the peak factor may exceed that of a sinusoid by 10 dB.

White noise is often used to simulate multitalker situations for systems with more than 64 operative channels.

3.4.4 Loading

For loading multichannel FDM systems, CCITT recommends (CCITT Rec. G.223):

It will be assumed for the calculation of intermodulation below the overload point that the multiplex signal during the busy hour can be represented by a uniform spectrum [of] random noise signal, the mean absolute power level of which, at a zero relative level point—(in dBm0)

$$P_{av} = -15 + 10 \log N \qquad (3.3)$$

$$\text{when } N \geq 240$$

and

$$P_{av} = -1 + 4 \log N \qquad (3.4)$$

$$\text{when } 12 \leq N < 240 \ldots.$$

where N = number of 4-kHz voice channels.

All logs are to the base 10. *Note.* These equations apply only to systems without preemphasis and using independent amplifiers in both directions. Preemphasis is discussed in Section 5.5.2. An activity factor of 0.25 is assumed. See Figure 5.15 and the discussion therewith. Examples of the application of these formulas are discussed in Section 5.6.3. It should also be noted that the formulas above include a small margin for loads caused by signaling tones, pilot tones, and carrier leaks.

Example 1. What is the average power of the composite signal for a 600-voice channel system using CCITT loading? N is greater than 240; thus equation (3.3) is valid.

$$P_{av} = -15 + 10 \log 600$$

$$= -15 + 10 \times 2.7782$$

$$= +12.782 \text{ dBm0}$$

Example 2. What is the average power of the composite signal for a 24-voice channel system using CCITT loading? N is less than 240, thus equation (3.4) is valid.

$$P_{av} = -1 + 4 \log 24$$

$$= -1 + 4 \times 1.3802$$

$$= +4.5208 \text{ dBm0}$$

3.4.5 Single-Channel Loading

A number of telephone administrations have attempted to standardize on -16 dBm0 for single-channel speech input to multichannel FDM equipment. With this input, peaks in speech level may reach -3 dBm0. Tests indicated that such peaks will not be exceeded more than 1% of the time. However, the conventional value of average power per voice channel allowed by the CCITT is -15 dBm0. (Refer to formula (3.3) and CCITT Rec. G.223.) This assumes a standard deviation of 5.8 dB and the traditional activity factor of 0.25. Average talker level is assumed to be at -11.5 VU. We must turn to the use of standard deviation because we are dealing with talker levels that vary with each talker, and thus with the mean or average.

3.4.6 Loading with Constant-Amplitude Signals

Speech on multichannel systems has a low duty cycle or activity factor. We established the traditional figure of 0.25. Certain other types of signals transmitted over the multichannel equipment have an activity factor of 1. This means that they are transmitted continuously, or continuously over fixed time frames. They are also characterized by constant amplitude. Examples of these types of signals follow:

- Telegraph tone or tones
- Signaling tone or tones
- Pilot tones
- Data signals (particularly FSK and PSK; see Chapter 8)

Here again, if we reduce the level too much to ensure against overload, the signal-to-noise ratio will suffer, and hence the error rate will suffer.

For typical constant-amplitude signals, traditional* transmit levels [input to the channel modulator on the carrier (FDM) equipment] are as follows:

*As taken from CCITT.

- Data: -13 dBm0
- Signaling (SF supervision), tone-on when idle: -20 dBm0
- Composite telegraph: -8.7 dBm0

For one FDM system now on the market with 75% speech loading and 25% data/telegraph loading with more than 240 voice channels, the manufacturer recommends the following:

$$P_{rms} = -11 + 10 \log N$$

using -5 dBm0 per channel for the data input levels† and -8 dBm0 for the composite telegraph level.

Table 3.1 shows some of the standard practice for data/telegraph loading on a per-channel basis. Data and telegraph should be loaded uniformly. For instance, if equipment is designed for 25% data and telegraph loading, then voice channel assignment should, whenever possible, load each group and supergroup uniformly. For instance, it is bad practice to load one group with 75% data, while another group carries no data traffic at all. Data should also be assigned to voice channels that will not be near group band edge. Avoid channels 1 and 12 on each group for the transmission of data, particularly medium- and high-speed data. It is precisely these channels that display the poorest attenuation distortion and group delay due to the sharp roll-off of group filters (see Chapter 8).

†All VF channels may be loaded at -8-dBm0 level, whether data or telegraph with this equipment, but for -5-dBm0 level data, only two channels per group may be assigned this level, the remainder voice, or the group must be "deloaded" (i.e., idle channels).

Table 3.1 Voice Channel Loading of Data/Telegraph Signals

Signal Type	CCITT	North American
High speed data	−10 dBm0 simplex −13 dbm0 duplex	−10 dBm0 switched network −8 dBm0 leased line −5 dBm0 occasionally
Medium speed Telegraph (multichannel)	—	−8 dBm0 total power
≤12 channels	−19.5 dBm0 /channel −8.7 dBm0 total	
≤18 channels	−21.25 dBm0 /channel −8.7 dBm0 total	
≤24 channels	−22.25 dBm0 /channel −8.7 dBm0 total	

Source: *Lenkurt Demodulator*, July 1968.

3.5 PILOT TONES

3.5.1 Introduction

Pilot tones in FDM carrier equipment have essentially two purposes:

- Control of level
- Frequency synchronization

Separate tones are used for each application. However, it should be noted that on a number of systems frequency synchronizing pilots are not standard design features, owing to the improved stabilities now available in master oscillators.

Secondarily, pilots are used for alarms.

3.5.2 Level Regulating Pilots

The nature of speech, particularly its varying amplitude, makes it a poor prospect as a reference for level control. Ideally simple single-sinusoid constant-amplitude signals with 100% duty cycles provide simple control information for level regulating equipment. Multiplex level regulators operate in the same manner as automatic gain control circuits on radio systems, except that their dynamic range is considerably smaller.

Modern carrier systems initiate a level regulating pilot tone on each group at the transmit end. Individual level regulating pilots are also initiated on all supergroups and mastergroups. The intent is to regulate the system level within ±0.5 dB.

Pilots are assigned frequencies that are part of the transmitted spectrum yet do not interfere with voice channel operation. They usually are assigned a frequency appearing in the guard band between voice channels or are residual carriers (i.e., partially suppressed carriers). CCITT has assigned the following as group regulation pilots:

- 84.080 kHz (at a level of −20 dBm0)
- 84.140 kHz (at a level of −25 dBm0)

The Defense Communications Agency of the U.S. Department of Defense recommends 104.08 kHz ± 1 Hz for group regulation and alarm.

For CCITT group pilots, the maximum level of interference permissible in the voice channel is −73 dBm0p. CCITT pilot filters have essentially a bandwidth at the 3-dB points of 50 Hz (refer to CCITT Rec. G.232).

Table 3.2 presents other CCITT pilot tone frequencies as well as those standard for group regulation. Respective levels are also shown. This table was taken from CCITT Rec. G.241. The operating range of level control equipment activated by

Table 3.2 Frequency and Level of CCITT Recommended Pilots

Pilot for	Frequency (kHz)	Absolute Power Level at a Zero Relative Level Point (dB) (Np)
Basic group B	84.080	−20 (−2.3)
	84.140	−25 (−2.9)
	104.080	−20 (−2.3)
Basic supergroup	411.860	−25 (−2.9)
	411.920	−20 (−2.3)
	547.920	−20 (−2.3)
Basic mastergroup	1552	−20 (−2.3)
Basic supermastergroup	11096	−20 (−2.3)
Basic 15-supergroup assembly (No. 1)	1552	−20 (−2.3)

pilot tones is usually about ±4 or 5 dB. If the incoming level of a pilot tone in the multiplex receive equipment drops outside the level-regulating range, then an alarm will be indicated (if such an alarm is included in the system design). CCITT recommends such an alarm when the incoming level varies 4 dB up or down from the nominal (CCITT Rec. G.241).

3.5.3 Frequency Synchronizing Pilots

End-to-end frequency tolerance on international circuits should be better than 2 Hz. To maintain this accuracy, carrier frequencies used in FDM equipment must be very accurate or a frequency synchronizing pilot must be used.

The basis of all carrier frequency generation for modern FDM equipment is a master frequency source. On the transmit side, called the master station, the frequency synchronizing pilot is derived from this source. It is thence transmitted to the receive side, called a slave station. The receive master oscillator is phase locked to the incoming pilot tone. Thus for any variation in the transmit master frequency source, the receive master frequency source at the other end of the link is also varied. The Defense Communication Agency recommends 96 kHz as a frequency synchronizing pilot on group 5 of supergroup 1 (DCA Circular 330-175-1); other systems use 60 kHz. The transmit level is at −16 dBm0.

CCITT Rec. G.225 does not recommend a frequency synchronizing pilot. Individual master frequency sources should have sufficient stability and accuracy to meet the following:

- Virtual channel carrier frequency, $\pm 10^{-6}$
- Group and supergroup carrier frequencies, $\pm 10^{-7}$

- Master group and supermaster group carrier frequencies,
    for 12 MHz (line frequency), $\pm 5 \times 10^{-8}$
    for 60 MHz (above 12 MHz), $\pm 10^{-8}$

3.6 FREQUENCY GENERATION

In modern FDM carrier equipment a redundant master frequency generator serves as the prime frequency source from which all carriers are derived or to which they are phase locked. Providing redundant oscillators with fail-safe circuitry gives markedly improved reliability figures.

One equipment on the market has a master frequency generator with three outputs: 4, 12, and 124 kHz. Automatic frequency synchronization is available as an option. This enables the slave terminal to stay in exact frequency synchronization with the master terminal providing drop-to-drop frequency stability (see Section 3.5.3).

The 4-kHz output of the master supply drives a harmonic generator in the channel-group carrier supply. Harmonics of the 4-kHz signal falling between 64 and 108 kHz are selected for use as channel carrier frequencies. The 12-kHz output is used in a similar manner to derive translation frequencies to form the basic CCITT super-group (420, 468, 516, 564, and 612 kHz). The 124-kHz output drives a similar harmonic generator providing the necessary carriers to translate standard super-groups to the line frequency.

These same carrier frequencies are also used on demultiplex at a slave terminal, or at the demultiplex at a master terminal if that demultiplex is not slaved to a distant terminal.

A simplified block diagram of a typical SSB suppressed carrier multiplex–demultiplex terminal is shown in Figure 3.12.

3.7 NOISE AND NOISE CALCULATIONS

3.7.1 General

Carrier equipment is the principal contributor of noise on coaxial cable systems and other metallic transmission media. On radiolinks it makes up about one quarter of the total noise. The traditional approach is to consider noise from the point of view of a hypothetical reference circuit. Two methods are possible, depending on the application. The first is the CCITT method, which is based on a 2500-km hypothetical reference circuit. The second is used by the U.S. Department of Defense in specifying communication systems. Such military systems are based on a 12,000-nautical mi reference circuit with 1000-mi links and 333-mi sections.

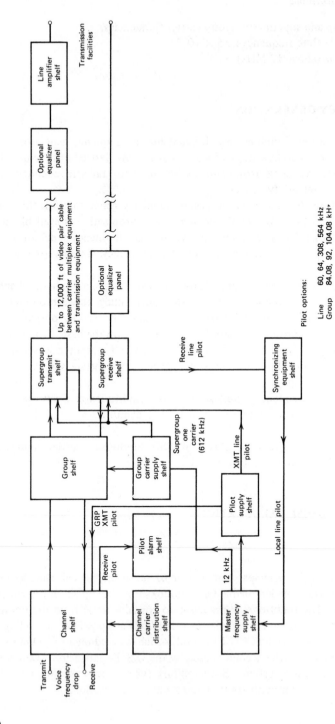

Figure 3.12 Simplified block diagram of a typical 120-channel terminal arrangement (Courtesy GTE Lenkurt Inc., San Carlos, CA.)

100

3.7.2   CCITT Approach

CCITT Rec. G.222 states:

> ... The mean psophometric power, which corresponds to the noise produced
> by all modulating (multiplex) equipment ... shall not exceed 2500 pW at a
> zero relative level point. This value of power refers to the whole of the noise
> due to various causes (thermal, intermodulation, crosstalk, power supplies,
> etc.). Its allocation between various equipments can be to a certain extent left
> to the discretion of design engineers. However, to ensure a measurement agree-
> ment in the allocation chosen by different administrations, the following values
> are given as a guide to the target values:

Equipment	Maximum Value Contributed by the Send and Receive Side Together	Assumptions About Loading		
Channel modulators	200 pW0p[a]	Adjacent channels loaded with:	−15 dBm0	(Signal corresponding to Rec. G.227)
		Other channels loaded with:	−6.4 dBm0	
Group modulators	80 pW0p	Load in group to be measured:	+3.3 dBm0	
		Load in other groups:	−3.1 dBm0 (each)	
Supergroup modulators	60 pW0p	Load in supergroup to be measured:	+6.1 dBm0	
		Load in other supergroups:	+2.3 dBm0 (each)	
Mastergroup modulators	60 pW0p	Load in each mastergroup:	+9.8 dBm0	
Supermastergroup modulators	60 pW0p	Load in each supermastergroup: +14.5 dBm0		
Basic 15-supergroup assembly modulators	60 pW0p	Load in each 15-supergroup assembly:	+14.5 dBm0	

[a]No account is taken of the values attributed to pilot frequencies and carrier leaks.

Experience has shown that often these target figures can be improved upon considerably. The CCITT notes that they purposely loosened the value for channel
modulators. This permits the use of the subgroup modulation scheme shown above
as the alternate method for forming the basic 12-channel group.

For instance, one solid-state equipment now on the market, when operated with
CCITT loading, has the following characteristics:

1 pair of channel modulators	224 pWp
1 pair of group modulators	62 pWp
1 pair of supergroup modulators	25 pWp
1 (single) line amplifier	30 pWp

If out-of-band signaling is used, the noise in a pair of channel modulators reduces
to 75 pWp. Out-of-band signaling is discussed in Section 3.13.

Using the same solid-state equipment mentioned above and increasing the loading to 75% voice, 17% telegraph tones, and 8% data, the following noise information is applicable:

1 pair of channel modulators	322 pWp
1 pair of group modulators	100 pWp
1 pair of supergroup modulators	63 pWp
1 (single) line amplifier	51 pWp

If this equipment is used on a real circuit with the heavier loading, the sum for noise for channel modulators, group modulators, and supergroup modulator pairs is 485 pWp. Thus a system would be permitted to demodulate to voice only five times over a 2500-km route (i.e., $5 \times 485 = 2425$ pWp). This leads to the use of through-group and through-supergroup techniques discussed in Section 3.9. Figure 3.13 shows a typical application of this same equipment using CCITT loading.

3.7.3 U.S. Military Approach

The following is excerpted from Mil-Std-188-100 under Multiplex Noise for Voice Bandwidth Links:

The multiplex idle channel noise and the loaded noise of a long haul or of a tactical less maneuverable FDM reference voice bandwidth link when referenced to a 0 TLP shall not exceed the values given:

	FDM Noise Allocation	
Configuration	Idle Noise (pWp0)	Loaded Noise (pWp0)
Pair, channel translation sets	10	31
Pair, group translation sets	40	50
Pair, supergroup translation sets	25	50
Through-group equipment	N/A	10
Through-supergroup equipment	25	50

3.8 OTHER CHARACTERISTICS OF CARRIER EQUIPMENT

3.8.1 Attenuation Distortion

Our interest in this discussion is centered on the attenuation distortion of the voice channel (i.e., not the group, supergroup, etc.) (see CCITT Rec. G.232A). Figure 3.14*a* shows limits of attenuation distortion (amplitude–frequency response) as set

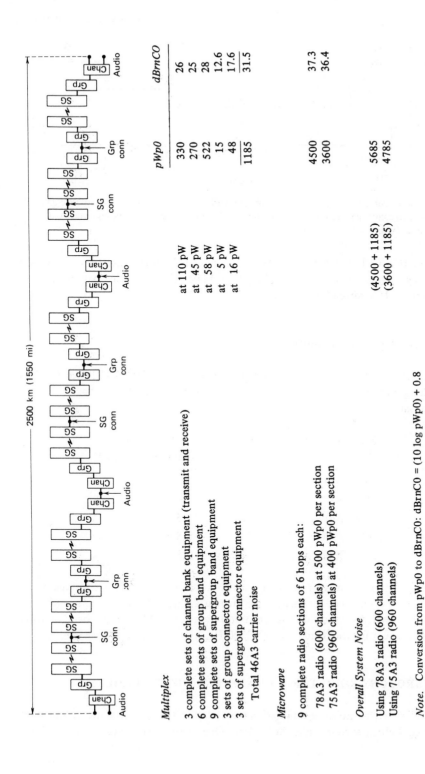

2500 km (1550 mi)

Audio — Chan | Grp | SG | SG | Grp | Grp | SG | SG | SG | SG | Grp | Chan | Chan — Audio | Grp | SG | SG | Grp | Grp | SG | SG | SG | SG | Grp | Chan | Chan — Audio | Grp | Chan | Chan — Audio

Grp conn / SG conn / Grp conn / SG conn / Grp conn

	pWp0	dBrnCO

Multiplex

	pWp0	dBrnCO
3 complete sets of channel bank equipment (transmit and receive) at 110 pW	330	26
6 complete sets of group band equipment at 45 pW	270	25
9 complete sets of supergroup band equipment at 58 pW	522	28
3 sets of group connector equipment at 5 pW	15	12.6
3 sets of supergroup connector equipment at 16 pW	48	17.6
Total 46A3 carrier noise	1185	31.5

Microwave

9 complete radio sections of 6 hops each:

	pWp0	dBrnCO
78A3 radio (600 channels) at 500 pWp0 per section	4500	37.3
75A3 radio (960 channels) at 400 pWp0 per section	3600	36.4

Overall System Noise

	pWp0	
Using 78A3 radio (600 channels)	5685	(4500 + 1185)
Using 75A3 radio (960 channels)	4785	(3600 + 1185)

Note. Conversion from pWp0 to dBrnCO: dBrnC0 = (10 log pWp0) + 0.8

Figure 3.13 Typical CCITT reference system noise calculations. (Courtesy GTE Lenkurt Inc., San Carlos, CA.)

103

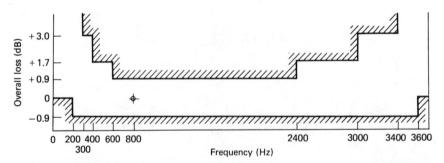

Figure 3.14a Permissible limits for the variation with frequency of the overall loss for any pair of channel transmitting and receiving equipments of one 12-channel terminal equipment. (CCITT Rec. G.232; Courtesy ITU–CCITT.)

forth in this CCITT recommendation. Figure 3.14*b* shows insertion loss versus frequency characteristic for single long-haul FDM reference voice bandwidth link according to U.S. Mil-Std-188-100.

3.8.2 Envelope Delay Distortion

The causes and effects of envelope and group delay are similar. CCITT Rec. G.232C provides guidance on group delay for a pair of channel modulators (back-to-back). Typical values are:

Frequency Band (Hz)	Group-Delay Distortion (relative to minimum delay) (ms)
400–500	5
500–600	3
600–1000	1.5
1000–2600	0.5
2600–3000	2.5

On U.S. military circuits in accordance with Mil-Std-188-100, the envelope delay distortion of a long-haul or tactical less maneuverable FDM voice bandwidth link shall not exceed, for a pair of channel translation sets, 120 μs in the band of 600–3200 Hz, except for the band of 1000–2500 Hz, where it shall not exceed 80 μs.

3.9 THROUGH-GROUP, THROUGH-SUPERGROUP TECHNIQUES

To avoid excessive noise accumulation, modulation/translation steps in long-haul carrier equipment systems must be limited. One method widely used is that of em-

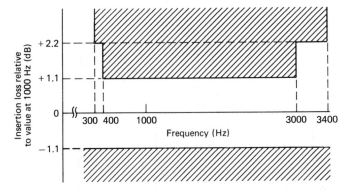

Figure 3.14b Insertion loss versus frequency characteristic for single long-haul reference voice bandwidth link. (Mil-Std-188-100).

ploying group connectors and through-supergroup devices. A simple application of supergroup connectors (through-supergroup devices) is shown in Figure 3.15. Here supergroup 1 is passed directly from point A to B, while supergroup 2 is dropped at C, a new supergroup is inserted for onward transmission to E, and so forth. At the same time supergroups 3–15 are passed directly from A to E on the same line frequency (baseband).

The expression "drop and insert" is terminology used in carrier systems to indicate that at some point a number of channels are "dropped" to voice (if you will) and an equal number are "inserted" for transmission back in the opposite direction. If channels are dropped at B from A, B necessarily must insert channels going back to A again.

Through-group and through-supergroup techniques are much more used on long trunk routes where excessive noise accumulation can be a problem. Such route plans can be very complex. However, the savings on equipment and the reduction of noise accumulation are obvious.

When through-supergroup techniques are used, the supergroup pilot may be

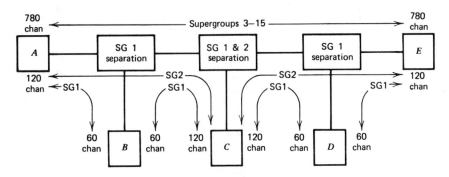

Figure 3.15 Typical drop and insert of supergroups.

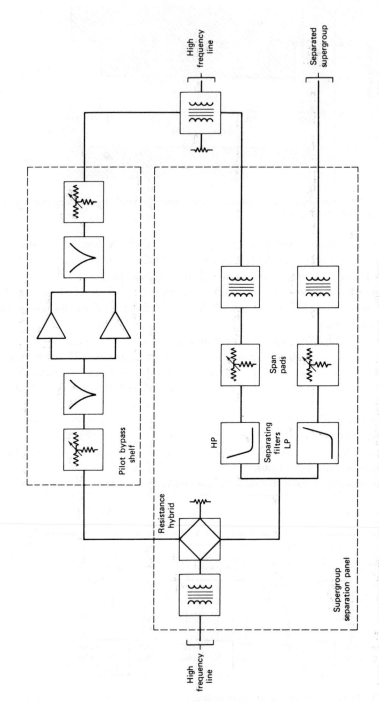

Figure 3.16 Simplified block diagram of typical supergroup drop equipment with pilot tone bypass. (Courtesy GTE Lenkurt Inc., San Carlos, CA.)

picked off and used for level regulation. Nearly all carrier equipment manufacturers include level regulators as an option on through-supergroup equipment, whereas through-group equipment does not usually have the option. Figure 3.16 is a simplified block diagram showing how a supergroup may be dropped (separated and inserted). CCITT Recs. G.242 and G.243 apply.

3.10 LINE FREQUENCY CONFIGURATIONS

3.10.1 General

When applying carrier techniques to a specific medium such as open-wire or radio-link, consideration must be given to some of the following:

- Type of medium, whether metallic, optical fiber, or radio?
- If metallic, is it pair, quad, open wire, or coaxial?
- If metallic, what are amplitude distortion, envelope delay, and loss limitations?
- If optical fiber (Chapter 14), single mode or multimode, power limited or dispersion limited?
- If radio, what are the bandwidth limitations?
- What are considerations of international regulations, CCITT recommendations?
- What are we looking into at the other end or ends?

The following paragraphs review some of the more standard line frequency configurations and their applications. The idea is to answer the questions posed above.

3.10.2 12-Channel Open-Wire Carrier

This is a line frequency configuration that permits transmission of 12 full-duplex voice channels on open-wire pole lines using SSB suppressed carrier multiplex techniques. The industry usually refers to the "go" channels as west–east and the "return" channels as east–west. The standard modulation approach is to develop the standard CCITT 12-channel basic group and through an intermediate modulation process translate the group to one of the following (CCITT terminology):

System Type	West–East (kHz)	East–West (kHz)
SOJ-A-12	36–84	92–140
SOJ-B-12	36–84	95–143
SOJ-C-12	36–84	93–141
SOJ-D-12	36–84	94–142

It should be noted that this type of carrier arrangement replaces a single voice pair on open wire with 12 full-duplex channels. The go and return segments on the same pair are separated by directional filters. Here CCITT Rec. G.311 applies with regard to modulation plans.

Baseband or line frequency repeaters have gains on the order of 43–64 dB. Loss in an open-wire section should not exceed 34 dB under worst conditions at the highest transmitted frequency. This follows the intent of CCITT Rec. G.313. The Bell System uses a similar configuration called J-carrier. Here average repeater spacing is on the order of 50 mi. Two pilot tones are used. One is for flat gain at 80 kHz for west–east and 92 kHz for east–west directions. The other pilot is called a slope pilot. Such slope pilots regulate the lossier part of the transmitted band as weather conditions change (i.e., dry to wet conditions or the reverse) as these losses increase and decrease.

3.10.3   Carrier Transmission over Nonloaded Cable Pairs (K-Carrier)

This technique increases the capacity of nonloaded cable facilities. Typical is the North American cable carrier called K-carrier. Separate cable pairs are used in each direction. Each pair carries 12 voice channels in the band of 12–60 kHz. This is the standard CCITT subgroup A, which is the standard CCITT basic group, 60–108 kHz, translated in frequency. On 19-gauge cable repeaters are required about every 17 mi. Pilots are used for automatic regulation on one of the following frequencies: 12, 28, 56, or 60 kHz.

3.10.4   Type N Carrier for Transmission over Nonloaded Cable Pairs

The N-Carrier is designed to provide 12 full-duplex voice channels on nonloaded cable pairs over distances of 20–200 mi. The modulation plan is nonstandard. The N-carrier uses double sideband emitted carrier with carrier spacings every 8 kHz. Nominal voice channel bandwidth is 250–3100 Hz. The 12 channels for one direction of transmission are contained in a band of frequencies from 44 to 140 kHz called the low band. The other direction of transmission is in a high band, 164–260 kHz. The emitted carriers serve, as pilot tones do in other systems, as level references in level regulating equipment.

A technique known as "frogging" or "frequency frogging" is used with the N-carrier whereby the frequency groups in each direction of transmission are transposed and reversed at each repeater so that all repeater outputs are always in one frequency band and all repeater inputs are always in the other. This minimizes the possibility of "interaction crosstalk" around the repeaters through paralleling VF cables. This reversal or transposition of channel groups at each repeater provides automatic self-equalization. A 304-kHz oscillator is basic to every repeater.

N3 is the latest in the N-carrier series of the Bell System of North America. It is a 24-channel system using conventional FDM carrier techniques. It finds application on wire pair routes from about 10 to over 200 mi long. Supervisory signaling is in-band, and as many as 26 N3 terminals can share a common carrier supply.

The 24-channel line frequency can be a high- or low-group signal and is composed of two identical 12-channel groups. For each of these channel groups 12 carriers spaced at 4-kHz intervals from 148 to 192 kHz are modulated by 12 individual voice channels. After modulation the upper sideband is selected in each case and the lower sideband and carrier are suppressed. Six of the 12 carriers are reinserted as pilots for repeater regulation and at receiver terminals for regulation, frequency correction, and demodulation.

Each of the identical derived 12-channel groups is modulated by a different carrier frequency to form a 24-channel configuration occupying the band 36–132 kHz. These modulating carrier frequencies are 232 and 280 kHz. The line frequency may be a high group or a low group. To form the high group the basic 24-channel configuration is modulated again by a 304-kHz carrier, the lower sideband is selected deriving a line frequency of 172–268 kHz. If the low group is desired, the basic 24-channel configuration (36–132 kHz) is placed directly on line.

The N3 repeater is similar in its modulation scheme as the other N-carrier repeaters described above, utilizing frequency frogging by modulating the incoming carrier with a 304-kHz signal inverting the incoming line signals.

3.10.5 Carrier Transmission on Star or Quad Type Cables

Standard carrier systems recommended by the CCITT for transmission over star or quad cables allow transmission of 12, 24, 36, 48, 60, or 120 full-duplex VF channels. The cables in all cases are nonloaded or deloaded. Regarding repeater sections CCITT Rec. G.321 recommends that these sections have no more than a 41-dB loss at the highest modulating frequency for systems with one, two, or three groups, and 36 dB for those with four or five groups and up to two supergroups. We refer here to transistor types repeaters. Figure 3.17 shows CCITT line frequency configurations recommended for star or quad cables.

3.10.6 Coaxial Cable Carrier Transmission Systems

Coaxial cable transmission systems using FDM carrier configurations are among the highest density transmission media in common use today. Nominal cable impedance is 75 Ω. Repeater spacing is a function of the highest modulating frequency.

A number of line frequency arrangements are recommended by the CCITT. Only several of these are discussed here. Figures 3.18 and 3.19 show several configurations for 12-MHz cables. Here repeater spacing is on the order of 3 mi (4.5 km). CCITT Rec. G.332 applies.

CCITT recommends 12,435 kHz to be used for the main-line pilot on 12-MHz

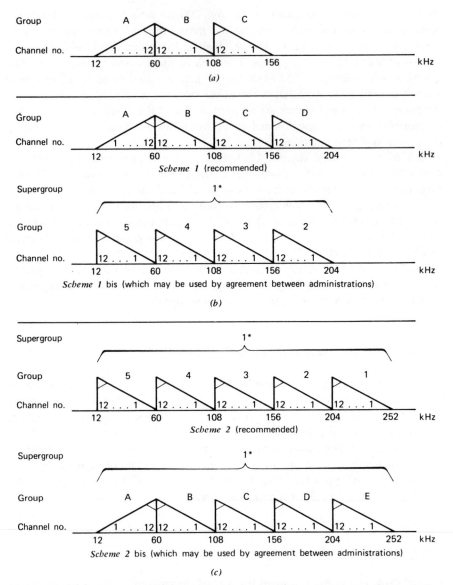

Figure 3.17 Line frequency configurations for star or quad type cables. (*a*) Systems providing one, two, or three groups. (*b*) Systems providing four groups. (*c*) Systems providing five groups. (Courtesy ITU–CCITT.)

110

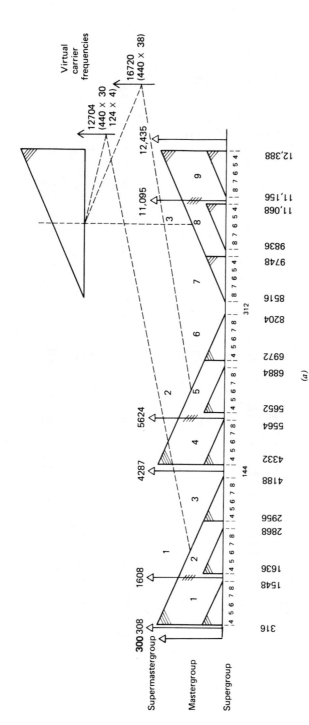

Figure 3.18 (*a*) Plan 1A frequency arrangement for 12-MHz systems. (*b*) Plan 1B frequency arrangement for 12-MHz systems. (*c*) Plan 1B frequency arrangement for 12-MHz systems, frequencies below 4287 kHz. (Courtesy ITU–CCITT.)

111

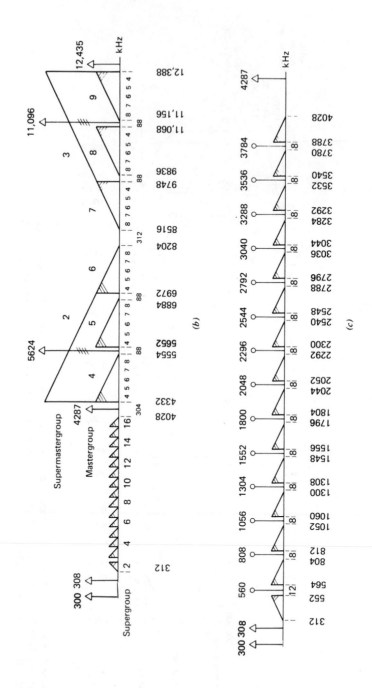

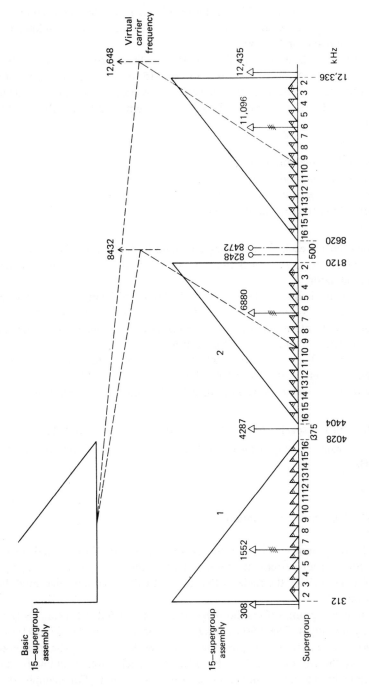

Figure 3.19 Plan 2 frequency arrangement for 12-MHz systems. (Courtesy ITU–CCITT.)

line frequency configurations. This is the main level regulating pilot, and it should maintain a frequency accuracy of $\pm 1 \times 10^{-5}$. For auxiliary-line pilots, CCITT recommends 308 and/or 4287 kHz.

The CCITT 12-MHz hypothetical reference circuit for coaxial cable systems is shown in Figure 3.20. The circuit is 2500 km (1550 mi) long, consisting of nine homogeneous sections. Such an imaginary circuit is used as guidance in real circuit design for the allocation of noise. This particular circuit has in each direction of transmission

- 3 pairs of channel modulators
- 3 pairs of group modulators
- 6 pairs of supergroup modulators
- 9 pairs of mastergroup modulators

Noise allocation for each of these may be found in Section 3.7. When referring to pairs of modulators, the intent is that a pair consists of one modulator and a companion demodulator.

Each telephone channel at the end of the 12-MHz (2500-km) coaxial cable carrier system should not exceed 10,000 pWp of noise during any period of 1 hr. 10,000 pWp is the sum of intermodulation and thermal noise. No specific recommendation regarding allocation to either type of noise is made.

Of the 10,000 pWp total noise, 2500 pWp is assigned to terminal equipment and 7500 to line equipment. Refer to CCITT Rec. G.332.

For line frequency allocations other than 12 MHz, consult the following:

2.6 MHz: repeater spacing 6 mi (9 km), line frequency 60–2540 kHz

4 MHz: repeater spacing 6 mi (9 km), 15 supergroups with line frequency 60–4028 kHz

3.10.7 The L-Carrier Configuration

L-carrier is the generic name given by the Bell System of North America to their long-haul SSB carrier system. Its development of the basic group and supergroup assemblies is essentially the same as that in the CCITT recommended modulation plan (Section 3.3).

The basic mastergroup differs, however. It consists of 600 VF channels (i.e., 10 standard supergroups). The L600 configuration occupies the 60–2788-kHz band. The U600 configuration occupies the 564–2084-kHz band. The mastergroup assemblies are shown in Figure 3.21. Bell identifies specific long-haul line frequency configurations by adding a simple number after the letter L. For example, the L3-

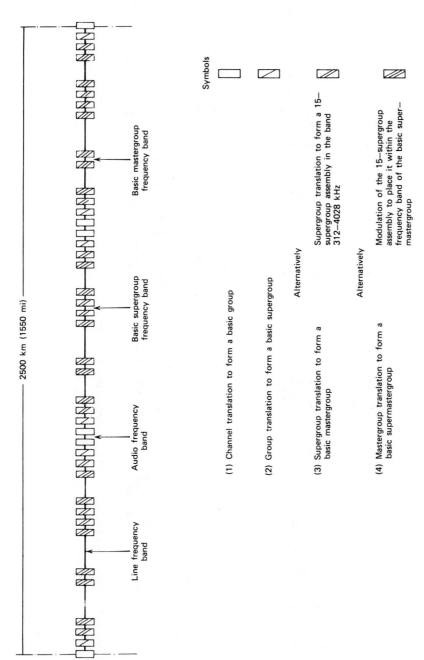

Line frequency band   Audio frequency band   Basic supergroup frequency band   Basic mastergroup frequency band

— 2500 km (1550 mi) —

Symbols

(1) Channel translation to form a basic group

(2) Group translation to form a basic supergroup

Alternatively

(3) Supergroup translation to form a basic mastergroup

Supergroup translation to form a 15—supergroup assembly in the band 312—4028 kHz

Alternatively

(4) Mastergroup translation to form a basic supermastergroup

Modulation of the 15—supergroup assembly to place it within the frequency band of the basic super—mastergroup

Figure 3.20 CCITT 12-MHz hypothetical reference circuit for coaxial cable systems. (CCITT Rec. G.322; Courtesy ITU–CCITT.)

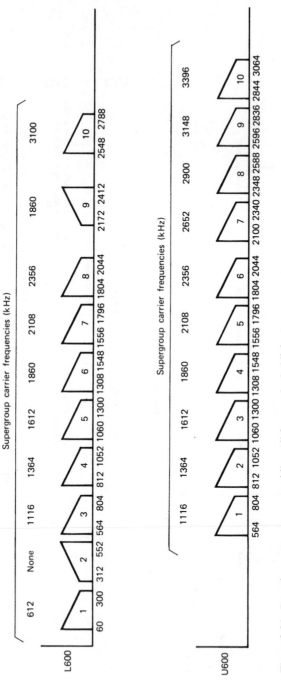

Figure 3.21 L-carrier mastergroup assemblies. All frequencies in kilohertz.

116

carrier, which is used on coaxial cable and the TH microwave,* has three master-groups plus one supergroup comprising 1860 VF channels occupying the band of 312–8284 kHz (see Figure 3.22). L4 consists of six U600 mastergroups in a 3600 VF channel configuration. L5 is discussed in Chapter 11.

Table 3.3 compares some basic L-carrier and CCITT system parameters.

A number of multiplex arrangements are available to translate and combine U600 mastergroup signal spectra for transmission over broadband coaxial cable and microwave radio systems. One such system is the MMX-1 manufactured by Western Electric Company. This derives the L1860 line configuration, originally designed for the L3-carrier, which is shown in Figure 3.22. The three-digit numbering system

*A Bell System type microwave.

Table 3.3   L-Carrier and CCITT Comparison Table

Item	ATT L-Carrier	CCITT
Level		
Group		
Transmit	-42 dBm	-37 dBm
Receive	-5 dBm	-8 dBm
Supergroup		
Transmit	-25 dBm	-35 dBm
Receive	-28 dBm	-30 dBm
Impedance		
Group	130 Ω balanced	75 Ω unbalanced
Supergroup	75 Ω unbalanced	75 Ω unbalanced
VF channel	200–3350 Hz	300–3400 Hz
Response	+1.0 to -1.0 dB	+0.9 to -3.5 dB
Channel carrier		
Levels	0 dBm	Not specified
Impedances	130 Ω balanced	Not specified
Signaling	2600 Hz in band	3825 Hz out of band
Group pilot		
Frequencies	104.08 kHz	84.08 kHz
Relative levels	-20 dBm0	-20 dBm0
Supergroup carrier		
Levels	+19.0 dBm per mod or demod	Not specified
Impedances	75 Ω unbalanced	Not specified
Supergroup pilot frequency	315.92 kHz	411.92 kHz
Relative supergroup pilot levels	-20 dBm0	-20 dBm0
Frequency synchronization	Yes, 64 kHz	Not specified
Line pilot frequency	64 kHz	60/308 kHz
Relative line pilot level	-14 dBm0	-10 dBm0
Regulation		
Group	Yes	Yes
Supergroup	Yes	Yes

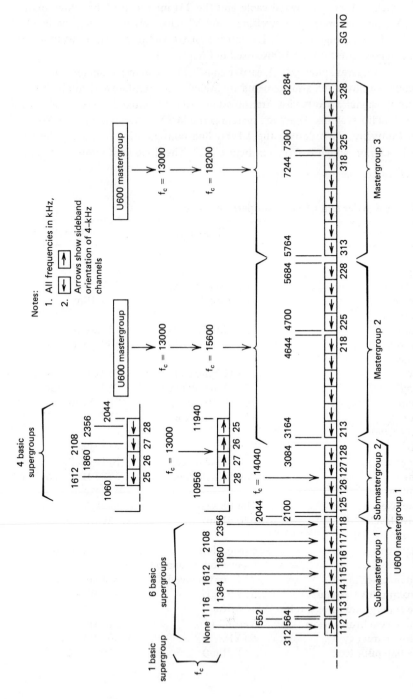

Figure 3.22 Derivation of the L1860 spectrum. (Copyright © 1977 by America1 Telephone and Telegraph Company.)

118

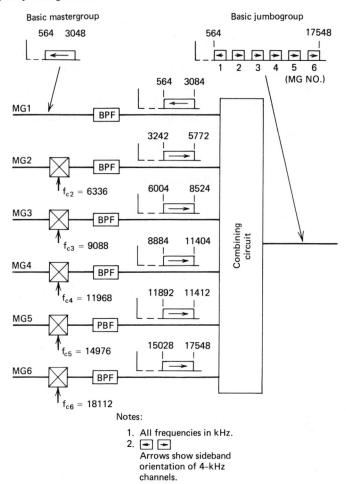

Figure 3.23 MMX-2 terminal, transmitting side. (Copyright © 1977 by American Telephone and Telegraph Company.)

shown in the figure was adopted to identify each supergroup. The first digit represents the mastergroup number and the second digit the submastergroup number. The third digit represents the supergroup number in the L600 spectrum. Elements of this numbering system are retained in later Bell System/Western Electric designs.

One of the later equipments that translate mastergroups used in North America is the MMX-2 terminal, shown in Figure 3.23, which provides a line frequency spectrum from 564 to 17,548 kHz.

3.10.8 Direct-to-Line (DTL) Multiplex

DTL FDM equipment is an economic alternative for main route spurs and other light route applications where 600 or less VF channels are required. Each individual channel unit translates the nominal 4-kHz VF channel to its proper line frequency slot, eliminating group and supergroup translations. Further, all channel units are identical and are equipped with a field-programmable channel selection capability. Line frequencies through supergroup 10 are CCITT compatible in their modulation plan. A major advantage of this equipment is that all common equipment is eliminated, each channel unit being equipped with a standard frequency generator/synthesizer. The synthesizer is adjustable in the field to set the channel unit output/input to the required line frequency slot.

3.11 SUBSCRIBER CARRIER–STATION CARRIER

In the subscriber distribution plant, when an additional subscriber line is required, it is often more economical to superimpose a carrier signal above the VF signal on a

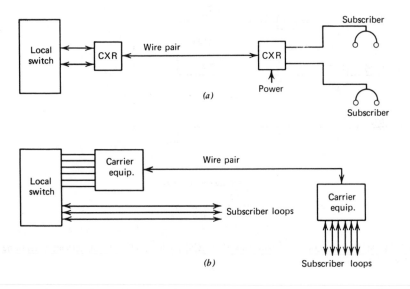

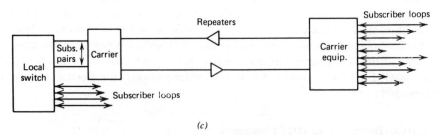

Figure 3.24 (a) Typical 1 + 1 subscriber carrier system. (b) Typical station carrier system serving six subscribers from a single wire pair. (c) Typical station carrier system using a wire pair for each direction of transmission to the remote carrier equipment.

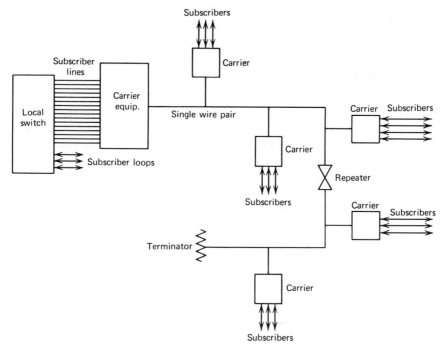

Figure 3.25 Arrangement of station carrier equipment with distributed subscribers.

particular subscriber pair. This is particularly true when a cable has reached or is near exhaustion (i.e., all the subscriber voice pairs are assigned). This form of subscriber carrier is often referred to as a 1 + 1 system and is the most commonly encountered today.

The terms *subscriber carrier* and *station carrier* often are used synonymously because both systems perform the same function in much the same way. There is a difference, however. By convention we say that subscriber carrier systems are powered locally at remote distribution points. Station carrier systems are powered at the serving local switch (central office). Figures 3.24 and 3.25 outline general applications of these types of systems.

Figure 3.24c shows one type of station carrier system. Several are available on the market today. Some allow a local switch to serve subscribers more than 30 mi (50 km) away. This particular system carries 20 voice channels on two pairs of cable to a distant distribution point. From the distribution point standard voice pair subscriber loops can be installed with loop resistance up to 1000 Ω (see Section 2.8.2). Another system provides for the addition of six subscribers to a single wire pair (Figure 3.24b). In this case transmission from the serving switch to the subscriber is in the band of 72–140 kHz. From the subscribers to the switch, frequency assignment is in the band of 8–56 kHz.

An important point when applying subscriber carrier to an existing subscriber loop is that load coils and bridged taps must be removed. The taps may act as tuned stubs and the load coils produce a sharp cutoff, usually right in the band of interest. The load points are candidate locations for line repeaters if required.

3.12 ECONOMICS OF CARRIER TRANSMISSION

We often hear that a particular brand of FDM carrier equipment costs so much per channel end. A channel end is taken to be a full-duplex voice channel, including the cost of the channel modulator–demodulator pair, its portion of the group modulation equipment, supergroup, and so forth. However, such figures are illusory. It is illusory in that much meaning is lost or hidden in what is called common equipment expense. Common equipment is that equipment that serves more than one module. For instance, a 108-kHz carrier frequency source will be used for both modulation and demodulation of voice channel 1 in each and every channel bank. A 24-channel system must amortize this frequency source for four modules. A 600-channel system amortizes the same source across 100 modules.

For channel modulation and demodulation equipment (channel banks), when using the standard CCITT modulation plan, 12 frequency sources are required. These are carriers starting at 64 kHz and spaced at 4-kHz intervals extending to 108 kHz. Let us say, for example, that the cost of these frequency sources totaled $1200 including the master frequency generator. Thus the cost of frequency sources for a 24-channel system would total $50 per channel added to the other channel bank costs. The cost of a 600-channel system would be only $2 per channel. Thus per-channel costs tend to drop as we increase the number of channels in a system.

A telecommunication planner must decide: is it more economical to use wire pairs for a particular number of voice channels to be transmitted or carrier on one wire pair? It may cost $10,000 to lay a cable for 60 voice pairs for a distance of 10 mi, $20,000 for 20 mi, and so on. At some point a decision is made that to extend the system further, we will abandon wire pairs and use carrier systems because of the economy involved.

There are many inputs to this decision-making process. Today more than ever, we usually deal with service expansion. Here wire pair systems are already in existence, such as interexchange trunks, toll-connecting trunks, and tandem office trunks. One case that arises is that the exhaustion date is approaching (the date when nearly all the cable pairs have been assigned). Here a telecommunication planner must decide: is it more economical to rip up the streets and add pair cable, or to add carrier systems to the existing plant?

One term we use in this regard is "prove-in distance." For one type of carrier system we say that on junctions (trunks) the prove-in distance is 8–12 km (5–7.5 mi). Here we mean that if a trunk route is less than 8 km in length and we wish to expand it, we add pairs. If it is more than 12 km long, our decision is to use a specific carrier system. For intermediate ranges a careful economic study is necessary to decide whether pair cable or carrier should be used.

Prove-in distances can be reduced by reducing the costs of carrier systems. It is here that double-sideband emitted carrier systems (DSBEC) show feasibility from an economic standpoint. One such DSBEC system has been discussed in Section 3.10.4 (N-carrier). Here, by reducing the cost of equipment, prove-in distance is reduced as well, usually at the expense of deteriorating some of the desirable features. For instance, DSBEC systems have increased intermodulation noise over their SSBSC counterparts. One major source of the noise is the fact that carriers are always present. More power is consumed because of the presence of the carriers, and loading problems exist.

3.13 COMPANDORS

The word compandor is derived from two words which describe its functions: compressor–expandor, to compress and to expand. A compandor does just that. It compresses a signal on one end of a circuit and expands it on the other. A simplified functional diagram and its analogy are shown in Figure 3.26.

The compressor compresses the intensity range of speech signals at the input circuit of a communication channel by imparting more gain to weak signals than to strong signals. At the far-end output of the communication circuit the expandor performs the reverse function. It restores the intensity of the signal to its original dynamic range. We cover only syllabic compandors in this discussion.

There are three advantages of compandors:

1. Tend to improve the signal-to-noise ratio on noisy speech circuits

2. Limit the dynamic power range of voice signals reducing the chances of overload of carrier systems

3. Reduce the possibility of crosstalk

A basic problem in telephony stems from the dynamic range of talker levels. This intensity range can vary 60 dB for the weakest syllables of the weakest talker to the loudest syllables of the loudest talker. The compandor brings this range down to more manageable proportions.

An important parameter of a compandor is its compression–expansion ratio, the degree to which speech energy is compressed and expanded. It is expressed by the ratio of the *input* to the *output* power (dB) in the compressor and expandor,

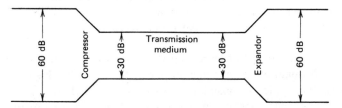

Figure 3.26 Functional analogy of a compandor.

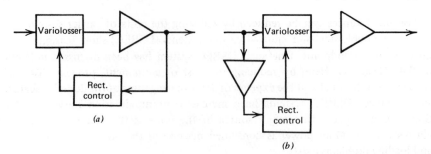

Figure 3.27   Simplified functional block diagram. (*a*) Compressor. (*b*) Expandor.

respectively. Compression ratios are always greater than 1 and expansion ratios are less than 1. The most common compression ratio is 2 (2:1). The corresponding expansion ratio is thus $\frac{1}{2}$. The meaning of a compression ratio of 2 is that the dynamic range of the speech volume has been cut in half from the input of the compressor to its output. Figure 3.27 is a simplified functional block diagram of a compressor and an expandor.

Another important criterion for a compandor is its companding range. This is the range of intensity levels a compressor can handle at its input. Usually 50–60 dB is sufficient to provide the expected signal-to-noise ratio and reduce the possibility of distortion. High-level signals appearing outside this range are limited without markedly affecting intelligibility.

But just what are the high and the low levels? Such a high or low is referred to as an "unaffected level" or focal point. CCITT Rec. G.162 defines the unaffected level as follows:

> . . . the absolute level, at a point of zero relative level on the line between the compressor and expandor of a signal at 800 Hz, which remains unchanged whether the circuit is operated with the compressor or not.

CCITT goes on to comment:

> The unaffected level should be, in principle, 0 dBm0. Nevertheless, to make allowances for the increase in mean power introduced by the compressor, and to avoid the risk of increasing the intermodulation noise and overload which might result. the unaffected level may, in some cases, be reduced as much as 5 dB. However, this reduction of unaffected level entails a diminution of improvement in signal-to-noise ratio provided by the compandor . . . . No reduction is necessary, in general, for systems with less than 60 channels.

CCITT recommends a range of level from +5 to −45 dBm0 at the compressor input and +5 to −50 dBm0 at the nominal output of the expandor. Figure 3.28 shows diagrammatically a typical compandor range of +10 to −50 dBm.

A syllabic compandor operates much in the fashion of any level control device where the output level acts as a source for controlling input to the device (see Figure

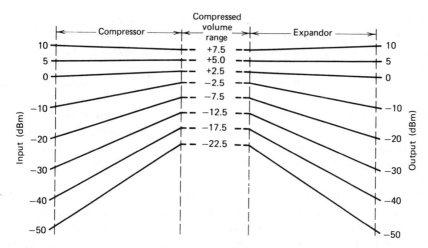

Figure 3.28 Input–output characteristics of a compandor. (Copyright © 1970 by Bell Telephone Laboratories.)

3.27). Automatic gain control (AGC) on radio receivers operates in the same manner. This brings up the third important design parameter of syllabic compandors: attack and recovery times. These are the response to suddenly applied signals such as a loud speech syllable or burst of syllables. Attack and recovery times are a function of design time constants and are adjusted by the designer to operate as a function of the speech envelope (syllabic variations) and *not* with instantaneous amplitude changes (such as used in PCM). If the operation time is too fast, wide bandwidths would be required for faithful transmission. When attack times are too slow, the system may be prone to overload.

CCITT Rec. G.162 specifies an attack time equal to or less than 5 ms and a recovery time equal to or less than 22.5 ms.

The signal-to-noise ratio advantage of a compandor varies with the multichannel loading factor of FDM equipment and thus depends on the voice level into the FDM channel modulation equipment. At best an advantage of 20 dB may be attained on low-level signals.

3.14 SIGNALING ON CARRIER SYSTEMS

In any discussion of carrier equipment, some space must be dedicated to signaling over carrier channels. For this discussion signaling may be broken down into two categories, supervisory and address. Supervisory signaling conveys information regarding on-hook and off-hook conditions. Address signaling routes the calls through the switching equipment. It is that signaling which contains the dialing information. For this discussion we are concerned with only the former, supervisory signaling.

The problem stems from the fact that it would be desirable to have continuous supervisory information being exchanged during an entire telephone conversation. This may be done by one of two methods, in-band signaling and out-of-band signaling.

In-band signaling accomplishes both supervisory and address signaling inside the operative voice band spectrum (i.e., in the band of 300–3400 Hz). The supervisory function is carried out when a call is set up and when it is terminated. Thus it is not continuous. The most common type in use today is called SF signaling. SF means single frequency and is a tone, usually at 2600 Hz. Other frequencies also may be used, and most often are selected purposely in the higher end of the voice band.

A major problem with in-band signaling is the possibility of "talk-down." Talk-down refers to the activation or deactivation of supervisory equipment by an inadvertent sequence of voice tones through normal speech usage of the channel. One approach is to use slot filters to bypass the tones as well as a time-delay protection circuit to avoid the possibility of talk-down.

With out-of-band signaling supervisory information is transmitted out of band. Here we mean above 3400 Hz (i.e., outside the speech channel).

Supervisory information is binary, either on-hook or off-hook. Some systems use tone-on to indicate on-hook and others use tone-off. One expression used in the industry is "tone-on when idle." When a circuit is idle, it is on-hook. The advantage of out-of-band signaling is that either system may be used, tone-on or tone-off when idle. There is no possibility of talk-down occurring because all supervisory information is passed out of band, away from the voice.

The most common out-of-band signaling frequencies are 3700 and 3825 Hz. 3700 Hz finds more application in North America; 3825 Hz is that recommended by CCITT (Rec. Q.21).

In the short run, out-of-band signaling is attractive from an economic and design standpoint. One drawback is that when patching is required, signaling leads have to be patched as well. In the long run, signaling equipment required may indeed make out-of-band signaling even more costly owing to the extra supervisory signaling equipment and signaling lead extensions required at each end and at each time the FDM equipment demodulates to voice. The advantage is that continuous supervision is provided, whether tone-on or tone-off, during the entire telephone conversation.

REFERENCES AND BIBLIOGRAPHY

1. *Reference Data for Radio Engineers*, 6th ed., Howard W. Sams, Indianapolis, IN, 1977.
2. *Lenkurt Demodulator*, Lenkurt Electric Corp., San Carlos, CA, Dec. 1965, May 1966, Mar. 1965, Oct. 1964, Oct. 1972, Oct. 1964, March 1970, July 1968, Sept. 1967, Sept. 1973.
3. CCITT Orange Books, Geneva, 1976, vol. III, G recommendations.
4. *Transmission Systems for Communications*, 4th ed., Bell Telephone Laboratories, American Telephone and Telegraph Company, New York, 1976.

5. Military Standard 188C, with Notice 1, U.S. Department of Defense, Washington, DC.

6. *Principles of Electricity Applied to Telephone and Telegraph Work*, American Telephone and Telegraph Co., 1961, New York, 1961.

7. D. M. Hamsher, *Communication System Engineering Handbook*, McGraw-Hill, New York, 1967.

8. Military Standard Mil-Std-188-311, U.S. Department of Defense, Washington, DC, 1971.

9. R. L. Marks *et al.*, *Some Aspects of Design for FM Line-of-Sight Microwave and Tropo-scatter Systems*, USAF Rome Air Development Center, NY, U.S. Technical Information Service, AD 617-686, Springfield, VA, Apr. 1965.

10. B. D. Holbrook and J. T. Dixon, "Load Rating Theory for Multichannel Amplifiers," *Bell Sys. Tech. J.*, vol. 18, 624-644, Oct. 1939.

11. D. Talley, *Basic Carrier Telephony*, J. R. Rider, New York, 1960 (No. 268).

12. P. F. Panter, *Communication System Design—Line-of-Sight and Troposcatter Systems*, McGraw-Hill, New York, 1972.

13. *Engineering and Equipment Considerations—46A*, Issue 1, Lenkurt Electric Corp., San Carlos, CA.

14. R. L. Freeman, *Telecommunication System Engineering*, Wiley-Interscience, New York, 1980.

15. *Telecommunication Transmission Engineering*, Vols. 1-3, 2nd ed., American Telephone and Telegraph Co., New York, 1977.

16. Military Standard Mil-Std-188-100, U.S. Department of Defense, Washington, DC, 1972.

4 | HIGH-FREQUENCY RADIO

4.1 GENERAL

Radio frequency transmission between 3 and 30 MHz by convention is called high-frequency radio, or simply HF. HF radio is in a class of its own because of certain characteristics of propagation; the phenomenon is such that many radio amateurs at certain times carry out satisfactory communication halfway around the world with 1–2 W of radiated power.

4.2 BASIC HF PROPAGATION

4.2.1 General

HF propagation is characterized by a groundwave component and a skywave component. The groundwave follows the surface of the earth and can provide useful communications up to about 400 mi (640 km) from the transmitter location, particularly over water, in the lower part of the band. It is the skywave, however, that gives HF an advantage. Transmission engineers design systems to take advantage of skywave propagation which permits reliable communication (90% path reliability) for distances up to 4000 mi (6400 km). On some links somewhat greater than 90% reliability has been reported. The same reliability has been obtained on even longer paths, except in years with low sunspot number, using oblique ionospheric sounding in parallel with the link to accurately determine optimum transmitting frequency. Ionospheric sounding is discussed in Section 4.14.

4.2.2 Skywave Transmission

The skywave transmission phenomenon of HF depends on ionospheric refraction. Transmitted radio waves hitting the ionosphere are bent or refracted. When they are bent sufficiently, the waves are returned to earth at a distant location. Often at the

128

distant location they are reflected back to the sky again, only to be returned to earth still again, even further from the transmitter.

The ionosphere is the key to HF skywave communication. Look at the ionosphere as a layered region of ionized gas above the earth. The amount of refraction varies with the degree of ionization. The degree of ionization is primarily a function of the sun's ultraviolet radiation. Depending on the intensity of the ultraviolet radiation, more than one ionized layer may form (see Figure 4.1). The existence of more than one ionized layer in the atmosphere is explained by the existence of different ultraviolet frequencies in the sun's radiation. The lower frequencies produce the upper ionospheric layers, expending all their energy at high altitude. The higher frequency ultraviolet waves penetrate the atmosphere more deeply before producing appreciable ionization. Ionization of the atmosphere may also be caused by particle radiation from sunspots, cosmic rays, and meteor activity.

For all practical purposes four layers of the ionosphere have been identified and labeled as follows:

D region. Not always present, but when it does exist, it is a daytime phenomenon. It is the lowest of the four layers. When it exists, it occupies an area between 50 and 90 km above the earth. The D region is usually highly absorptive.

E layer. A daylight phenomenon, existing between 90 and 140 km above the earth. It depends directly on the sun's ultraviolet radiation and thus it is most dense directly under the sun. The layer disappears shortly after sunset. Layer density varies with seasons owing to variations in the sun's zenith angle with seasons.

F_1 *layer.* A daylight phenomenon existing between 140 and 250 km above the earth. Its behavior is similar to that of the E layer in that it tends to follow the sun (i.e., most dense under the sun). At sunset the F_1 layer rises, merging with the next higher layer, the F_2 layer.

F_2 *layer.* This layer exists day and night between 150 and 250 km (night) and 250 and 300 km above the earth (day). During winter in the daytime, it extends from 150 to 300 km above the earth. Variations in height are due to solar heat. It is believed that the F_2 layer is also strongly influenced by the earth's magnetic field. The earth is divided into three magnetic zones representing different degrees of magnetic intensity called east, west, and intermediate. Monthly F_2 propagation predictions are made for each zone.* The north and south auroral zones are also important for F_2 propagation, particularly during high sunspot activity.

Consider these layers as mirrors or partial mirrors, depending on the amount of ionization present. Thus transmitted waves striking an ionospheric layer, particularly the F layer, may be refracted directly back to earth and received after their first

---

*Monthly propagation forecasts are made by the Central Radio Propagation Laboratory (CRPL), U.S. National Bureau of Standards.

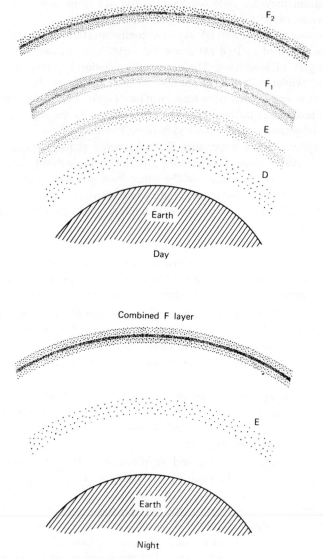

Figure 4.1 Ionized layers of the atmosphere.

hop, or they may be reflected from the earth back to the ionosphere again and repeat the process several times before reaching the distant receiver. The latter phenomenon is called multihop transmission. Single and multihop transmission are illustrated diagrammatically in Figure 4.2.

To obtain some idea of the estimated least possible number of F layer hops as related to path length, the following may be used as a guide:

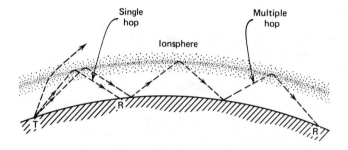

Figure 4.2 Single and multihop HF skywave transmission. T = transmitter; R = receiver.

Number of Hops	Path Length (km)
1	< 4,000
2	4,000–8,000
3	8,000–12,000

An important concept at this point in the discussion is that the higher and more strongly ionized layers refract progressively higher frequencies. At some point as the frequency is increased, the wave will not be refracted but will pierce the ionosphere and continue out into space. This point is called the *critical frequency* and varies with time of day, geographical position, season of the year, and the time position in the sunspot cycle.

HF propagation above about 8 MHz encounters what is called a *skip zone*. This is an "area of silence" or a zone of no reception extending from the outer limit of groundwave communication to the inner limit of skywave communication (first hop). The skip zone is shown graphically in Figure 4.3.

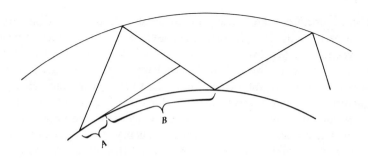

Figure 4.3 Skip zone. A = limit of groundwave communication; B = skip zone.

4.3 OPERATING FREQUENCY

One of the most important elements to the successful operation of an HF system using the skywave phenomenon is an operating frequency that will assure 100% path reliability day and night, year-round. Seldom, if ever, can this be achieved. To arrive as close as possible to this ideal, we must deal with two factors: (1) the design of an efficient HF installation,* and (2) the determination of an assigned frequency that will provide optimum communication.

The latter portion of this section deals with item 1. Our concern here is the proper determination of the operating frequency. From the point of view of an HF installation operator on long-haul point-to-point service, HF propagation varies with the time of day, from day to day, with the season of the year, and on a cyclic yearly basis. The operator will tell you that propagation variations are most notable around local sunrise and sunset. As midday approaches, conditions seem to become more stable, and higher frequencies prove more attractive. After local sunset, depending on the location of the distant end (i.e., whether it is in darkness, too), lower frequencies become more desirable. These are only generalities, which might be set forth from the point of view of a serious shortwave listener (SWL). To the communicator, the propagation problem of HF is complex. The vagaries (as it seems) of the ionosphere are beyond comprehension. Long circuits, particularly those crossing high latitude regions, are particularly troublesome.

The HF operator, to maintain good communication, is required to make frequency changes during any 24-h period on long-haul circuits. At this juncture we can make a generality. Well designed HF point-to-point installations not operating in or crossing the auroral zones, during the greater portion of the sunspot cycle will require operating frequency changes (QSYs) no more than two to four times a day.

There are three methods to determine operating frequency:

1. By experience
2. By use of CRPL predictions
3. By oblique ionospheric sounder information (sounders are covered in Section 4.14)

Many old-time operators rely on experience. First they listen to their receivers, then they judge if an operating frequency change is required. An operator may well feel that "yesterday I had to QSY at 7 PM to 13.7 MHz, so today I must do the same." Listening to his receiver will confirm or deny this belief. He hears his present operating frequency (from the distant end) start to take deep fades. He checks a new frequency on a spare receiver, listening for other identifiable signals nearby to determine conditions on the new frequency. If he finds conditions to his liking, he orders the transmitter operator to change frequency (QSY at the distant end). If not, he may check other assigned frequencies and then order a move to one of these. This is the "experience method."

*The reader should not lose sight of the interaction of item 2 on item 1.

The next approach is to carefully study the CRPL predictions issued by the U.S. National Bureau of Standards. These are published monthly, three months in advance of their effective dates.

We choose an optimum operating frequency to minimize communication outrages—to keep the circuit operating at optimum. On skywave HF communication circuits fading is endemic. Fading must be reduced to some acceptable minimum for good communications. Fading on HF is due to the variations in the ionosphere, with skywave received signal intensities varying from minute to minute, day to day, month to month, and year to year.

Consider an HF system designed for a 95% path (propagation) reliability, a high design goal on long circuits. The median received signal level intensity must be increased on the order of 8 dB to overcome slow variations of skywave field intensity, 11 dB to overcome rapid variations of skywave field intensity, and 13 dB to overcome variations in atmospheric noise. Therefore a good operational figure is that we need at least 32 dB above median signal level—the level that will provide good communication 50% of the time—to give us margin over fades for good communication 95% of the time. We shall discuss other approaches to meet this figure further on in our discussion. Our concern now is the primary approach—choice of the right frequency.

When dealing with HF propagation and the choice of operating frequency, the terms MUF, FOT, and LUF come into play. These terms are used on the CRPL predictions.

MUF and LUF are the upper and lower limiting frequencies for skywave communications between points X and Y. MUF is the median maximum usable frequency and LUF the lowest usable frequency. Suppose that we chose the MUF or LUF as an operating frequency. We would be operating near boundary limits, regions of heavy fade. Depending on conditions, one or the other or both may drop out (i.e., communications would be lost entirely). Therefore there is some other frequency above the LUF, usually just somewhat below the MUF, that is an optimum frequency. Use of this frequency, the FOT (fréquence optimum de travail), OWF in English (optimum working frequency), will provide a received signal with less fading and minimize frequency changes. CRPL predictions give MUF, LUF, and FOT. The OWF is equal to 85% of the maximum usable frequency for the F_2 layer.

For the second approach in the determination of operating frequency, the operator consults the CRPL prediction and chooses an assigned frequency as close as is available to the OWF.

Another concept that comes into play at this point will be used later in system design. This is the angle of incidence, or the angle at which the transmitted ray or beam strikes the ionosphere. Here we want to determine the maximum usable frequency, and from it we can determine the incidence angle. This can be derived from the following formula:

$$f_0 = f_n \sec i \tag{4.1}$$

where f_0 = maximum usable frequency at oblique angle i

      f_n = maximum frequency that will be reflected back at vertical incidence

      i = angle of incidence (i.e., the angle between the direction of propagation and a line perpendicular to the earth)

As sec i increases, the angle of incidence i is increased. For most effective HF transmission it is desirable to strive for the maximum concentration of energy (i.e., the transmitted ray) at low angles with respect to the horizon because higher frequencies can be used which are less affected by absorption than lower frequencies. Thus we design for low take-off angles (i.e., high angles of incidence), particularly on longer paths.

Important natural phenomena that affect the ionosphere and, likewise, skywave propagation are as follows:

- Sunspots
- Magnetic storms
- Sudden ionospheric disturbances (SID)
- Sporadic E (layer)

Sunspot activity greatly influences skywave HF propagation. It can cause the loss of long-haul HF communication for extended periods or permit phenomenal world-wide communication with relatively low power and homemade installations for weeks at a time. Sunspot activity vastly increases ultraviolet radiation. The number and intensity of sunspots are indicative of ion density in the ionosphere. Thus, in turn, sunspot activity is a measure of the probability of skywave communication. Sunspot activity is cyclic, with a full cycle being on the order of 11.1 years. The activity is measured by the Wolf *sunspot number*. High numbers indicate high activity, and low numbers, low activity.

Magnetic storms are associated with solar activity and are most likely to occur during periods of high sunspot numbers. They reoccur in 27-day cycles, the rotation period of the sun. Magnetic storms are so named because the solar radiation seems to be deflected by the earth's magnetic field. Therefore their effects are experienced most severely in the regions of the magnetic poles or in the northern and southern auroral zones. Magnetic storms tend to come on suddenly, last for several days, and slowly subside. During this period skywave communication may be spotty or cut entirely. Magnetic storms often occur from 18 to 36 h after a sudden ionospheric disturbance.

A SID is an occasional daytime phenomenon rendering HF skywave communication impossible or nearly so. It is believed that a SID is caused by a chromospheric eruption on the sun resulting in a marked increase in ion density in the D region with a moderate increase in the E region, increasing their absorptive properties.

Sporadic E propagation occurs in the E region but cannot be accounted for by normal E propagation theory. One theory is that it is caused by particle radiation caused by meteor bursts (i.e., meteors entering the atmosphere). Sporadic E has been found to permit skywave propagation from 25 to 50% of the time on frequencies up to 15 MHz.

4.4 BASIC HF RADIO SYSTEMS

The most elementary HF installation operating on a point-to-point basis requires as a minimum a transmitter at one end and a receiver at the other. Each must have an antenna and each must employ some sort of signal transducer, say, a microphone and headsets.

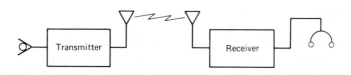

This configuration is valid for one-way transmission such as that of broadcasting.

The radio amateur requires a half-duplex arrangement where he can talk to a distant acquaintance and listen as well. His equipment arrangement will appear as follows:

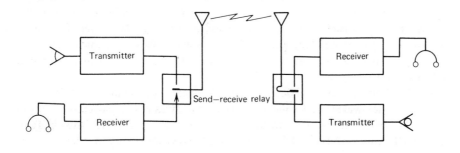

In this case, while one transmits the other receives. When invited to do so, the transmitting station ceases and switches to receive (i.e., he switches his antenna to receive and stops transmitting), and the other station goes into a transmit condition.

For commercial or military point-to-point service a full-duplex operation is necessary. This means that both ends transmit and receive simultaneously. With sufficient frequency separation between transmit and receive frequencies and some good filters, this can be done with only one antenna on each end. However, most land installations use at least two antennas.

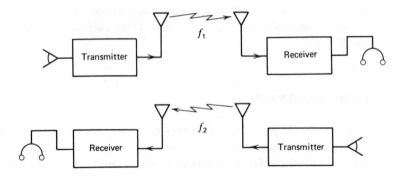

4.5 PRACTICAL HF POINT-TO-POINT COMMUNICATION

4.5.1 General

Reliable HF point-to-point facilities are complex, often occupying more than one site. In such cases the transmitters are separated physically from their receiving counterparts by 3–45 km so that noise from high-power transmitters will not affect reception.

4.5.2 Emission Types

Table 4.1 provides a list of emission types used on HF. For many years the backbone of HF transmission was what we familiarly call *continuous wave* (CW). Information is transmitted by turning on and off (keying) the carrier. The carrier is keyed using short and long pulses called dots and dashes. An alphanumeric code is used, the international Morse code (see Table 4.2). CW transmission is listed as A1 in Table 4.1.

Standard teletypewriter service is also used widely. For the most part for single-channel service wide-shift frequency shift keying (FSK) is employed. FSK is discussed in Chapter 8. The standard frequency shift is ±425 Hz. This is listed in Table 4.1 as F1 emission. Table 4.3 shows comparative emissions and their assigned bandwidths.

HF is a highly congested band. Conservation of bandwidth is extremely important. The goal of HF transmission is to consume minimum bandwidth and achieve maximum energy transfer in a heavily fading environment.

4.5.3 Recommended Modulation (Emission Type)

SSBSC systems, including independent sideband (ISB) systems, for multichannel application are recommended. This recommendation is made according to the

Table 4.1 Designation of Specific Emission Types

Desig-nation	Modulation Type	Type of Transmission	Supplementary Characteristics
A0	Amplitude modulation	With no modulation	None
A1	"	Telegraphy without the use of a modulating audio frequency (by on-off keying)	None
A2	"	Telegraphy by the on-off keying of an amplitude-modulating audio frequency or audio frequencies, or by the on-off keying of the modulated emission (special case: an unkeyed emission amplitude modulated)	None
A3	"	Telephony	Double sideband
A3A	"	Telephony	Single sideband, reduced carrier
A3J	"	Telephony	Single sideband suppressed carrier
A3B	"	Telephony	Two independent sidebands
A4	"	Facsimile (with modulation of main carrier either directly or by a frequency modulated subcarrier)	None
A4A	"	Facsimile (with modulation of main carrier either directly or by a frequency modulated subcarrier)	Single sideband, reduced carrier
A5C	"	Television	Vestigial sideband
A7A	"	Multichannel voice frequency telegraphy	Single sideband, reduced carrier
A9B	"	Cases not covered by the above, e.g., a combination of telephony and telegraphy	Two independent sidebands
F1	Frequency (or phase) modulation	Telegraphy by frequency shift keying without the use of a modulating audio frequency: one of two frequencies being emitted at any instant	None
F2	"	Telegraphy by the on-off keying of a frequency modulating audio frequency or by the on-off keying of a frequency modulated emission (special case: an unkeyed emission, frequency modulated)	None

(Continued)

Table 4.1 (*Continued*)

Desig-nation	Modulation Type	Type of Transmission	Supplementary Characteristics
F3	"	Telephony	None
F4	"	Facsimile by direct frequency modulation of the carrier	None
F5	"	Television	None
F6	"	Four-frequency duplex telegraphy	None
F9	Frequency (or phase) modulation	Cases not covered by the above, in which the main carrier is frequency modulated	None
P0	Pulse modulation	A pulsed carrier without any modulation intended to carry information (e.g., radar)	None
P1D	"	Telegraphy by the on-off keying of a pulsed carrier without the use of a modulating audio frequency	None
P2D	"	Telegraphy by the on-off keying of a modulating audio frequency or audio frequencies, or by the on-off keying of a modulated pulsed carrier (special case: an unkeyed modulated pulsed carrier)	Audio frequency or audio frequencies modulating the amplitude of the pulses
P2E	"	"	Audio frequency or audio frequencies modulating the width (or duration) of the pulses
P2F	"	"	Audio frequency or audio frequencies modulating the phase (or position) of the pulses
P3D	"	Telephony	Amplitude modulated pulses
P3E	"	Telephony	Width (or duration) modulated pulses
P3F	"	Telephony	Phase (or position) modulated pulses
P3G	"	Telephony	Code modulated pulses (after sampling and quantization)

Table 4.1 (*Continued*)

Desig-nation	Modulation Type	Type of Transmission	Supplementary Characteristics
P9	"	Cases not covered by the above in which the main carrier is pulse modulated	None
		Bandwidths Whenever the full designation of an emission is necessary, the symbol for that emission, as given above, shall be preceded by a number indicating in kilohertz the necessary bandwidth of the emission. Bandwidths shall generally be expressed to a maximum of three significant figures, the third figure being almost always a zero or a five.	

Source: *Radio Regulations*, Geneva, 1968, Article 2 (RR2-1).

following criteria when considering the transmission of voice or other complex analog signals:

- Spectrum conservative
- More efficient from the point of view of power
- Superior to other emission types during periods of poor propagation conditions

After a short description of SSB modulation, the traditional comparison with DSBEC will be made.

4.5.4 Single-Sideband Suppressed Carrier

An SSB signal is simply an audio frequency signal translated to the radio frequency (RF) spectrum. If we mix (modulate) two signals A and B in a linear device, we have the following appearing at the output:

Signal A
Signal B
Sum of signals, $A + B$
Difference of signals, $A - B$

Table 4.2 The International Morse Code

Character	International Morse	Character	International Morse
A	.—	.	.—.—.—
B	—...	;	—.—.—.
C	—.—.	,	.—.—
D	—..	:	———...
E		?	..——..
F	..—.	!	——..——
G	——.	'	.————.
H	....	-	—....—
I	..	/	—..—.
J	.———	Ā	.—.—
K	—.—	A or Å	.——.—
L	.—..	E	..—..
M	——	CH	————
N	—.	Ñ	——.——
O	———	Ö	———.
P	.——.	Ü	..——
Q	——.—	(OR)	—.———.
R	.—.	"	.—..—.
S	...	—	.——.—
T	—	SOS	...———...
U	..—	Attention	—.—.—
V	...—	CQ	—.—.——.—
W	.——	DE	—..
X	—..—	Go ahead	—.—
Y	—.——	Wait	.—...
Z	——..	Break	—...—.—
1	.————	Understand	..—.
2	..———	Error	....
3	...——	OK	.—.
4	....—	Separator, heading—	
5	.....	text and end of	
6	—....	message	—...—
7	——...	End of work	...—.—
8	———..		
9	————.		
0	—————		

Table 4.3   Emission Bandwidths (kHz)

Type of Service	Remarks	Bandwidth (kHz)
Double sideband radio-telephony	Speech grade quality at 100% modulation	6
Standard broadcast (AM)	High quality service	8-2*
Single sideband radio-telephony	Speech grade quality, carrier suppressed, 10, 20, or 40 dB, single channel	3
Single sideband radio-telephony	Speech grade quality, carrier suppressed, 10, 20, or 40 dB, two channel	6
Independent sideband radio-telephony	Speech grade quality, carrier suppressed, 10, 20, or 40 dB, four 3-kHz voice channels	12
Manual continuous wave radio telegraphy	30 words per minute	<2
Modulated manual continuous wave telegraphy	30 words per minute	2
Radio teleprinter frequency shift	Up to 150 words per minute	1.7
Radio teleprinter audio frequency shift, SSB	100 words per minute, single channel	<1
Radio teleprinter audio frequency shift, SSB	16 channels, 100 words per minute each channel, carrier suppressed, 20 or 40 dB	3
Frequency modulation broadcast service	Broadcast quality	150
Facsimile	AM subcarrier modulated	6
Facsimile-SSB	NBFM [a] modulation, carrier suppressed, 20 or 40 dB	3

[a] NBFM, narrow band frequency modulation.
* Depending on quality desired.

As an example, modulate (mix) a 4000-kHz RF carrier with a 1.000-kHz audio signal. Following the above rules, we would have:

4000 kHz (= 4,000,000 Hz)	Carrier
1.000 kHz (= 1000 Hz)	Modulating signal
4,000,000 Hz + 1000 Hz	Upper sideband
4,000,000 Hz - 1000 Hz	Lower sideband

By the use of filters one or the other sideband is suppressed. The modulating signal as a single entity disappears because the bandpass of the modulator filter is far above the modulating frequency. The carrier itself is suppressed by using a special balanced modulator.

Let us go through the exercise again, this time considering a voice channel occupying the audio spectrum of 300–3000 Hz. Here we must define boundaries, the highest modulating frequency (3000 Hz) and the lowest (300 Hz), as though they were simple sinusoidal tones. We use the same RF carrier, 4000 kHz. Modulation again is mixing or heterodyning. Thus we have the following:

$$4000 \text{ kHz} = 4,000,000 \text{ Hz} \qquad \text{RF carrier}$$

$$\left.\begin{array}{l} 4,000,000 \text{ Hz} - 300 \text{ Hz} \\ 4,000,000 \text{ Hz} - 3000 \text{ Hz} \end{array}\right\} \text{Difference or lower sideband}$$

$$\left.\begin{array}{l} 4,000,000 \text{ Hz} + 300 \text{ Hz} \\ 4,000,000 \text{ Hz} + 3000 \text{ Hz} \end{array}\right\} \text{Sum or upper sideband}$$

The resulting lower sideband will then be a frequency of 3,999,700 Hz which represents the 300-Hz modulating tone and a frequency of 3,997,000 Hz representing the 3000-Hz modulating tone. Note that for the lower sideband an inversion has taken place. This means that the higher modulating frequency became the lower RF frequency, and the lower became the higher. Such a sideband is called an inverted sideband.

The upper sideband will be the frequencies 4,003,000 Hz and 4,000,300 Hz. This is called an erect or upright sideband. The lowest modulating frequency is the lowest RF frequency, and the highest is the highest RF frequency.

As mentioned earlier, in most modern SSB systems the carrier is not transmitted. It is suppressed by 50 or 60 dB at the transmitter. At the SSB receiver the carrier is reinserted locally as closely as possible to the original suppressed carrier frequency. Another approach is to suppress the carrier only about 20 dB, and use the remaining reduced carrier (pilot carrier) as a frequency reference at the receiver. This pilot carrier technique is discussed below.

4.5.5 SSB Operation and Its Comparison with Conventional AM

Before proceeding further, the reader should consult Figures 4.4 and 4.5, which are simplified functional block diagrams of a typical SSB transmitter and receiver, respectively.

TRANSMITTER OPERATION

A nominal 3-kHz voice channel amplitude modulates a 100-kHz* stable IF carrier; one sideband is filtered and the carrier is suppressed. The resultant sideband occupies the RF spectrum of 100 +3 kHz or 100 −3 kHz, depending on whether the upper or lower sideband is to be transmitted. This signal is then mixed with a local oscillator to provide output at frequency F_0. Thus the local oscillator output to the mixer if $F_0 + 100$ kHz or $F_0 - 100$ kHz. Note that frequency inversion takes

_____

*Other common IF frequencies are 455 and 1750 kHz.

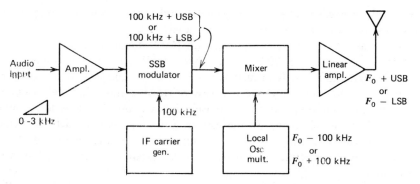

Figure 4.4 Simplified functional block diagram of a typical SSB transmitter (100 kHz IF). USB = upper sideband; LSB = lower sideband; F_0 = operating frequency; IF = intermediate frequency.

place when lower sidebands are selected (see Section 3.2, which discusses inverted sidebands). The output of the mixer is fed to the linear power amplifier (LPA). The output of the LPA is radiated by the antenna.

RECEIVER OPERATION

The incoming signal, consisting of a suppressed carrier plus a sideband, is amplified by one or several RF amplifiers, mixed with a stable local oscillator to produce an IF (assume again that the IF is 100 kHz*). Several IF amplifiers will then increase the signal level. Demodulation takes place by reinserting the carrier at IF. In this case it is from a stable 100-kHz oscillator, and detection is usually via a product detector. The output of the receiver is the nominal 3-kHz voice channel.

As we see, SSB is a special form of conventional AM, which has been around for a long time. Conventional AM transmits a carrier plus two sidebands. The transmitted intelligence is in the sidebands. The intelligence is identical in both sidebands. If the intelligence is identical, why not just transmit only one sideband? The carrier, which consumes two-thirds of our power on a system that is 100%

*Other common IF frequencies are 455 and 1750 kHz.

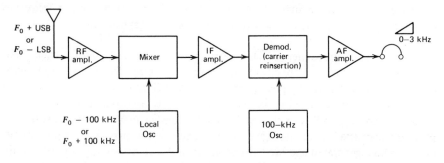

Figure 4.5 Simplified functional block diagram of a typical SSB receiver (100 kHz IF). USB = upper sideband; LSB = lower sideband; F_0 = operating frequency; IF = intermediate frequency.

modulated, is needed only in the demodulation process. Why not suppress the carrier and reinsert it at the receiver at a low level?

Consider an AM transmitter operating with 100% modulation, modulated with a 1000-Hz sine wave with 1500-W output. This 1500 W is allocated as follows: 1000 W in the carrier and 250 W in each sideband. It is the sidebands that contain the useful information, and only one is necessary (leaving aside phase relationships). It is apparent, therefore, that we are transmitting at a cost of 1500 W to communicate 250 W of useful energy. In the case of SSB only the one sideband is transmitted; the carrier is suppressed and the unwanted sideband eliminated. At the receiver the carrier is reinserted and its phase in relation to the received SSB signal is irrelevant.

Because only one sideband is being received, the effective receiver bandwidth is reduced to half in the case of SSB when compared to an equivalent AM system. This conserves spectrum and gives SSB a comparable 3-dB noise advantage over AM.

It can be shown using voice tests that under fading conditions SSB may give up to a 9-dB advantage over AM for equal power outputs.

4.5.6 Key to SSB Transmission

One of the most important considerations in the development of the SSB signal at the near end (transmitter) and its demodulation at the far end (receiver), other than linearity, carrier suppression, and unwanted sideband suppression, is that of accurate and stable frequency generation and reinsertion. It is equally important (frequency generation) at the transmitter as at the receiver (generation and reinsertion).

Most of us have heard the "monkey talk," or high or low pitched speech, just barely intelligible, on some poor telephone circuits or while "shortwave" listening on an HF receiver. This is the characteristic of SSB systems that are not operating on the same frequency and/or are not properly "frequency synchronized." The quotes are purposeful, as we shall see later.

Generally SSB circuits can maintain tolerable intelligibility when the transmitter or receiver are no more than 50 Hz out of "synchronization." Voice frequency carrier telegraph (VFTG) and low-speed data require "synchronization" of 2 Hz or better.

There are two approaches to achieve "synchronization": (1) pilot carrier operation, and (2) synthesized operation. In the first the carrier frequency is suppressed only 20 dB (10 dB during poor propagation conditions or through interference). The remaining carrier is called the pilot carrier and is used to actuate an automatic frequency control (AFC) circuit such that the tuning local oscillator centers the receiver's converted incoming signal on the center of the first IF. The pilot carrier may also be regenerated and used after conversion for reinsertion; this is seldom the case, however. It is more usual to use a fully independent locally generated reinsertion carrier. Its frequency is usually so low that even without any frequency control (oven) of the carrier oscillator, there will not be more than a fraction of a hertz of error in the received demodulated signal due to carrier reinsertion.

Synthesized transmission and reception are used widely today in HF. A synthesizer is no more than a highly stable frequency generator. It gives one or several simultaneous sinusoidal RF outputs on discrete frequencies in the HF range. In most cases it provides the frequency supply for all RF carrier needs in SSB applications. For example, it will supply:

- Transmitter IF carrier
- Transmitter local oscillator supply
- Receiver local oscillator supply (supplies)
- IF carrier reinsertion supply

The following are some of the demands we must place upon an HF synthesizer:

- Frequency stability
- Frequency accuracy
- A number of frequency increments
- Supplementary outputs
- Capability of being slaved to a frequency standard
- Spectral purity of RF output(s)

For VFTG* operation, as noted before the end-to-end frequency error may not exceed 2 Hz (± 1 Hz) to maintain a satisfactory error rate. If the HF transmitter and receiver were the only devices in a system to inject or cause frequency error, we could then allow ± 0.5-Hz error for each. Thus the frequency stability and accuracy must be maintained to ± 0.5 Hz under all conditions for synthesizers used at both ends of the circuit. First consider a system operating at 10 MHz (1×10^7 Hz). It would require a transmitter and receiver stability of $\pm 0.5 \times 10^{-7}$ or $\pm 5 \times 10^{-8}$. If the maximum operating frequency of the system were 30 MHz, then the stability would be three times more stringent than for the 10-MHz system, that is, $\pm 5/3 \times 10^{-8}$, or 1.66 parts in 10^8. Synthesizers are usually based on oven-controlled crystal oscillators. These crystals drift in frequency as they age. Thus the specification must state stability requirements over a time limit. Usually the time frame is the month. A requirement should also be stated for the periodic readjustment of the synthesizer to assure an accuracy with a known standard.

A synthesizer provides outputs on discrete RF frequencies. One requirement may be that outputs shall be from 3 to 30 MHz in 10-kHz increments. Another system may need a tighter requirement with frequency steps of 1 kHz, or even 1 Hz (or less) from 3 to 30 MHz.

Tight requirements increase the price of the equipment.

*VFTG = Voice frequency telegraph.

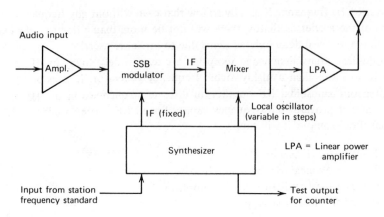

Figure 4.6 Transmitter.

4.5.7 Synthesizer Application Block Diagrams

Synthesizers, when used for HF SSB operation, provide stable frequency sources. Figure 4.6 shows a typical application for an SSB transmitter, Figure 4.7 that for an SSB receiver.

To simplify operation of the synthesizer, the frequency readout on the front panel is "offset" by the IF and is not the true local oscillator injection frequency (applicable for both transmitters and receivers), but the RF operating frequency. However the "output for counter" will read the local oscillator frequency, and the operator must subtract or add the IF (first IF) (depending on whether the local oscillator is offset above or below the operating frequency).

If the synthesizer is phase locked to a station frequency standard, it will take on

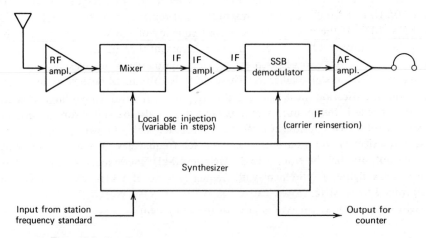

Figure 4.7 Receiver.

the stability of the standard. Some station frequency standards, such as a cesium atomic standard, have frequency stabilities better than 1×10^{-11} per month.

The use of synthesizers allows more efficient HF SSB operation by permitting equivalent full carrier suppression (-50 to -60 dB) for some additional cost. If an installation is to operate on a very limited number of RF frequencies, simple oven-controlled crystals may give a more economic solution. Such crystals now display stabilities on the order of 1×10^{-7} per month or better. This certainly is sufficient stability for voice operation of SSB.

4.6 LINEAR POWER AMPLIFIERS IN POINT-TO-POINT SSB SERVICE

The power amplifier in an SSB installation raises the power of a low-level signal input without distorting the signal. That is, the envelope of the signal output must be as nearly as possible an exact replica of the signal input. By definition the power amplifier that will perform this function is an LPA. For HF application peak envelope power outputs of LPAs are on the order of 1 to 10 kW, and in some installations 40- to 50-kW amplifiers are used. Air cooling is almost universal.

Desirable characteristics of an LPA are the following:

- High efficiency
- High gain
- Low grid-to-plate capacitance
- Good linearity at all operating frequencies

The most common tube used in the LPA is the tetrode,* which usually requires neutralization to prevent it from self-oscillating.

Automatic load control (ALC) is an important feature of an LPA. It provides the means of keeping the signal level adjusted so that the power amplifier works near its maximum capability without being overdriven on signal peaks. The ALC circuit receives its input from the envelope peaks in the power amplifier and uses its output to control the gain of the exciting signal. ALC circuits in transmitters are very similar in their function to AGC circuits in receivers. As in receiver AGC circuits, important parameters of ALCs are proper attack and release times. For voice operation a fast attack on the order of 10 ms is used to remove gain rapidly and avoid overload. A release of about 100 ms returns the gain to normal.

4.7 INTERMODULATION (IM) DISTORTION

IM distortion must be rigidly monitored and controlled, particularly on SSB/ISB transmitting installations. Measuring IM distortion is one quick method of determining linearity. IM distortion may be measured in two different ways:

*Solid-state LPAs are now available with up to 2-kW RF output.

- Two-tone test
- Tests using white noise loading

The two-tone test is carried out by applying two tones at the audio input of the SSB or ISB* transmitter. A $3:5$ frequency ratio between the two test tones is desirable to identify the IM products easily. In a 6-kHz input we could apply tones of 3 and 5 kHz, and to a 3-kHz channel, tones of 1500 and 2500 Hz.

The test tones are applied at equal amplitudes, and gains are increased to drive the transmitter to full power output. Exciter or transmitter output is sampled and observed in a spectrum analyzer. The amplitudes of all undesired products (see Section 1.9.6) and the carrier are measured in terms of decibels below either of the two equal-amplitude test tones. The decibel difference is the signal-to-distortion ratio. This should be at least 40 dB or better on HF SSB (ISB) systems. Normally the highest level IM product is the third-order product. This product is two times the frequency of one tone minus the frequency of the second tone. For example, two test tones are 1500 and 2500 Hz. Thus

$$2 \times 1500 - 2500 = 500 \text{ Hz} \quad \text{or} \quad 2 \times 2500 - 1500 = 3500 \text{ Hz}$$

and consequently, the third-order products will be 500 and 3500 Hz. The presence of IM products numerically lower than 40 dB indicates maladjustment or deterioration of one or several transmitter stages, or overdrive.

The white noise test for IM distortion more nearly simulates operating conditions of a complex signal such as voice. One simple approach is to load three of the four voice channels* in an ISB* system with white noise and measure the level of the noise in the idle channel. Next measure the noise level on the same channel with no noise loading on the other channels. This is the idle noise. The IM noise level is the difference between the level in the idle channel when noise loading is applied and that of idle noise.

4.8 HF ANTENNAS

4.8.1 General

Perhaps the most important element in an HF telecommunication system is the antenna. To achieve the 90% or better path reliability, great care must be taken in the selection and design of antennas.

For HF point-to-point operation two basic antennas types are available:

- Rhombic
- Log periodic (LP)

---

*Section 4.11 introduces the reader to ISB and three- and four-channel operation.

Both the rhombic and the LP antennas should be considered for selection on any path over 2500 km in length.

Precision-designed rhombics outperform any other antennas on point-to-point paths more than 4000 km long.

4.8.2 Basic Antenna Considerations

To design an HF antenna system for point-to-point service, we would want to consider the following:

1. Broadband nature of the antenna (VSWR characteristics)
2. Its efficiency and dissipation losses on transmission
3. Gain
4. Fixed or variable take-off angle
5. Polarization (in the case of LPs)
6. Height above ground
7. Land area required
8. Need of counterpoise (ground screen)
9. Transmission line requirements
10. Side lobe suppression
11. Transmit voltage standing-ware radio (VSWR)
12. Azimuth beamwidth
13. Cost

The rhombic and LP antennas are compared in Table 4.4 regarding the 13 points listed above.

4.8.3 Performance Characteristics

BANDWIDTH

The bandwidth of an antenna is the frequency range over which its performance does not deviate unacceptably from the optimum. Antenna performance is usually specified by the input impedance, gain, radiation pattern, and power-handling capacity.

IMPEDANCE

HF transmitters are designed to operate into a specific load resistance (usually 50 or 60 Ω unbalanced or 300 to 600 Ω balanced). A deviation of the actual load imped-

Table 4.4 Comparison Chart—Rhombic and LP Antennas

Consideration	Rhombic	LP
1. Broadband	1–2 octaves	3 octaves
2. Efficiency	50–70%	95%
3. Gain over an isotropic	8–20 dB	10–30 dB (vertical LP) 10–17 dB (horizontal LP).
4. Minimum take-off angle (TOA)	$5°$	$5°$
5. Polarization	Horizontal	Horizontal and vertical
6. Height	Fixed	Sloping
7. Land area	5–15 acres	2–5 acres
8. Need of counterpoise	No	Yes on many installations
9. Transmission line	Open wire; Usually uses 600-Ω balanced line	Coaxial 50-Ω coaxial/300-Ω balanced
10. Side lobe suppression	6 dB	14 dB
11. Transmit VSWR	2:1	2:1
12. Azimuth beam-width	$30 \pm 15°$	$7\text{-}23°$ (horizontal LP) $110\text{-}5°$ (vertical LP)
13. Cost	$12,000–24,000 exclusive of labor and land	$24,000–50,000 exclusive of labor and land

ance from the design value can be tolerated within limits. These are usually defined in terms of the VSWR, which is unity (1:1) for a perfect match and infinity (∞:1) for a complete mismatch. Most HF transmitters will tolerate a mismatch with a VSWR as high as 3:1, and the bandwidth of an antenna should not, therefore, include any frequency at which this value is exceeded.

In the case of reception, the variation of antenna impedance is of less importance, and higher VSWRs are more acceptable than they are for transmitting antennas.

GAIN

The gain of a transmitting antenna results from the concentration of radiation in wanted directions and the suppression of radiation in unwanted directions. Gain is often expressed in decibels relative to an isotropic source in free space, dBi. If the field strength provided by a given antenna is E_1 and the field strength provided by an isotropic source is E_2, the distance and total radiated power being the same in both cases, the gain is given by:

$$G_{\mathrm{dBi}} = 20 \log_{10} \left(\frac{E_1}{E_2} \right) \qquad (4.2)$$

The term *power gain* refers to the value of G if losses (including antenna radiation efficiency and feeder attenuation) are taken into account. If the losses are ignored, the value of G depends only on the nature of the radiation pattern and is then known as *directive gain*. It is important to note that power gain is of interest in the transmitting case where it is essential to radiate the maximum possible power. Directive gain is of interest in the receiving case where suppression of unwanted interference is essential. For instance, an antenna with a directive gain of 12 dBi, a radiation efficiency of 90% (−0.5 dB), and a feeder (transmission line) loss of 3 dB will have a power gain of 8.5 dBi. Such an antenna gives an increase in field strength of 8.5 dBi over an isotropic source when transmitting. The same antenna yields an improvement of approximately 12 dB in received signal-to-noise ratio when compared with an isotropic receiving antenna.

RADIATION PATTERN

A high-gain HF antenna will usually have subsidiary radiation lobes (known as side lobes) in addition to the main lobe used for communication. In the transmitting case large side lobes can result in a loss of antenna gain with a consequent decrease of received signal at the distant end and possible interference with cochannel or other near frequency emitters. In the receiving case large side lobes can result in a loss of directive gain and a consequent reduction in received signal-to-noise ratio. The level of the side lobes relative to the level of the main lobes should be reasonably low, at least 10 dB.

TAKE-OFF ANGLE (TOA)

Since an HF signal reaches the far-end point of reception by single or multiple reflection from the ionosphere, the TOA, defined as the vertical angle at which the transmitting antenna has maximum radiation, varies with path length. Table 4.5

Table 4.5 Some Basic HF Path Parameters

Path Length (km)	Mode (number of hops)	TOA (degrees)	Frequency Range (MHz)
100	1	80	2–7
200	1	70	2–7
400	1	55	2.5–8
800	1	34	3.5–11
1600	1	16	5–18
3200	2	16	5–18
6000	2	5	7–24

Source: TCI, Mountain View, CA.

shows typical values of path length, required TOA, mode (i.e., number of hops), and frequency range.

Table 4.5 also shows that a short path requires a high TOA and a comparatively low frequency, and that TOA decreases and frequency increases as the path length becomes greater.

For horizontally polarized antennas the TOA decreases as the height above ground increases and is given approximately by:

$$TOA = \sin^{-1}\left(\frac{\lambda}{4h}\right) \qquad (4.3)$$

where λ is the wavelength and h is the height above ground of the radiating elements (both in meters).

Using formula (4.3) in the following example it will be seen from Table 4.5 that an antenna height of 125 ft is required at 2 MHz (TOA 80°) and 180 ft is required at 5 MHz (TOA 16°). For vertically polarized antennas the TOA is virtually independent of antenna height and depends primarily on ground conductivity.

POWER-HANDLING CAPACITY

The current-carrying capacity and maximum voltage level of an antenna are specified in terms of the average and peak power, respectively. The average power rating is usually limited by the antenna's capability to dissipate heat. This is particularly true for inefficient antennas, such as rhombics, which have resistive elements (dissipation lines) that must dissipate a substantial fraction of the antenna input power. The peak power rating of an antenna is limited by the antenna's ability to withstand high voltages, which can generate arcs or corona discharge which in turn can break insulators or damage radiating elements.

POLARIZATION

Since HF signals become randomly polarized after reflection from the ionosphere, an HF antenna can be vertically polarized or horizontally polarized, irrespective of the polarization at the other end of the link. The choice of antenna polarization can therefore be made on grounds of practical convenience. One should note, however, that horizontally polarized antennas are "quieter" in receiving than vertically polarized antennas, which are less susceptible to atmospheric and man-made noise. In transmitting, horizontally polarized antennas also suffer from much less reduction in gain than vertically polarized antennas owing to poor ground conductivity.

LAND AREA AND OVERALL COST

In many HF installations a number of antennas need to be co-sited and must be properly spaced to avoid mutual interference. Large areas of comparatively flat land are often scarce and expensive, and it follows that the size of each antenna must be as small as possible for a given performance. The size of the site required will also depend on the number of antennas used, and wide-bandwidth designs (preferably covering the whole HF band from 2 to 30 MHz) will reduce this number to a minimum. It will, in many cases, be realized that the unit cost of an antenna is not the most important criterion in determining overall cost. A single compact design of antenna covering the entire HF band will inevitably be more economic than two to three simpler but larger narrow-bandwidth antennas, each of which will require a considerable land area.

4.8.4 Directional HF Transmitting Antennas

For point-to-point telecommunication links high-gain directional antennas are essential. In the past rhombic horizontally polarized antennas were widely used for this purpose. In recent years, however, LP antennas are superceding the rhombics in many applications because of their smaller size and often improved performance (see Table 4.4).

RHOMBIC ANTENNAS

The rhombic antenna comprises several horizontal long-wire radiators arranged in the form of a rhombus. One end of the rhombus is connected to the transmitter via a transmission line and the other end is terminated in a resistance. The antenna behaves approximately as a matched transmission line, and the terminating impedance has a value slightly greater than the characteristic impedance to compensate for the loss of energy by radiation. As will be seen, the rhombic is essentially a traveling-wave antenna, and, as is common with this type of antenna, the radiation

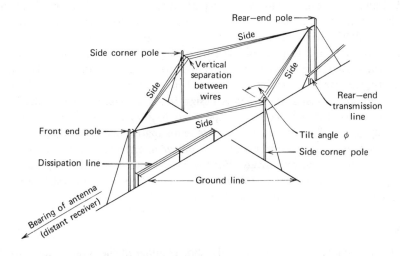

Figure 4.8 Rhombic antenna (transmitting).

Table 4.6   Dimensions of Standard Compromise Rhombics

Rhombic Type	Range (mi) (km)	Side Length (ft) (m)	Tilt Angle (°)	Ht (ft) (m)	Length End-to-End (ft) (m)	Width Side-to-Side (ft) (m)
A	>3000 (4800)	375 (115)	70	65 (20)	723 (222)	258 (82.4)
B	2000–3000 (3200–4800)	350 (107.7)	70	60 (18.5)	676 (208)	251 (77.2)
C	1500–2000 (2400–3200)	315 (96.8)	70	57 (17.5)	611 (188)	228 (70.2)
D	1000–1500 (1600–2400)	290 (89.2)	67.5	55 (16.9)	553 (170)	234 (72.0)
E	600–1000 (960–1600)	270 (83)	65	53 (16.3)	506 (155)	240 (73.8)
F	400–600 (640–960)	245 (75.3)	62.5	51 (15.7)	453 (139.3)	238 (73.2)
G	200–400 (320–640)	225 (69.2)	60	50 (15.3)	407 (125.2)	237 (73.0)

Source: Ref. 2.

efficiency suffers as some power is lost in the termination. The fact that the antenna behaves as a matched transmission line results in an input impedance that remains fairly constant (i.e., a VSWR of less than 2:1) over a very wide frequency band. However, the gain and radiation pattern vary considerably with frequency, and this results in a relatively low useful bandwidth of somewhat under 2 octaves. With the full HF band covering nearly 4 octaves, two or possibly three rhombics are therefore required to give complete frequency coverage. A further disadvantage is that the rhombic antenna requires from 5 to 15 acres of flat land. Despite its simplicity, the overall costs may be high if land costs are high. Figure 4.8 shows a typical rhombic transmitting antenna design in which multiple wire radiators are used to reduce the characteristic impedance to 600 Ω. To enable transmitter powers up to 10 kW to be used, an iron-wire dissipative line or high-power lumped resistors are used as the termination. For power in excess of 10 kW, an iron-wire dissipation line is used exclusively. For low power and for reception purposes the termination is usually a standard low wattage carbon resistor. Table 4.6 gives dimensions of standard compromise rhombics.

LP ANTENNAS

The LP antenna overcomes some of the shortcomings of the rhombic antenna. It is more compact and has good performance over a wider bandwidth than its rhombic

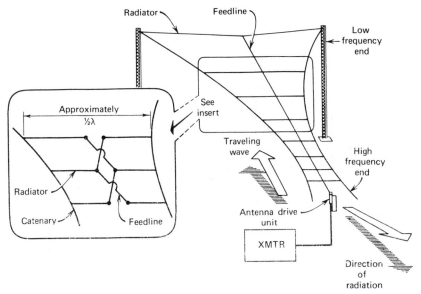

Figure 4.9 Simplified diagram of an LP antenna. (Courtesy Granger Associates, Menlo Park, CA.)

counterpart. The principles of the LP technique are illustrated in Figures 4.9 and 4.10. Figure 4.10 shows that the basic arrangement comprises a number of radiating elements with 180° transpositions between adjacent elements. The purpose of the transpositions is to form a beam radiating from the end of the antenna with the smallest elements. The physical lengths of the dipoles vary from a half-wavelength at the lowest frequency down to a half-wavelength at the highest frequency. The feed point is at the end with the smaller elements. The relationship between dipole lengths and spacings results in a system in which the radiation pattern and impedance repeat periodically with the logarithm of the frequency. As a result the impedance and the radiation pattern are, to a first approximation, independent of frequency, and the bandwidth of an LP antenna is limited only by the physical size and the number of dipoles used. At any given frequency, typically three to four dipoles resonate and form a "radiation center" which moves steadily along the antenna as frequency changes.

To achieve increased gain, two of the LP antennas illustrated in Figure 4.10 can be arrayed side by side. If the two LP antennas are arrayed so that they lie in the same plane, a modified arrangement (i.e., the patented *clamped-mode* principle) enables the increased gain to be achieved together with a simplified mechanical arrangement. The clamped-mode principle is illustrated in Figure 4.11, and it will be seen that half the wires are discarded. It can be shown that this simplification results in a negligible change in performance.

Figure 4.12 shows a practical realization of a clamped-mode LP antenna where a

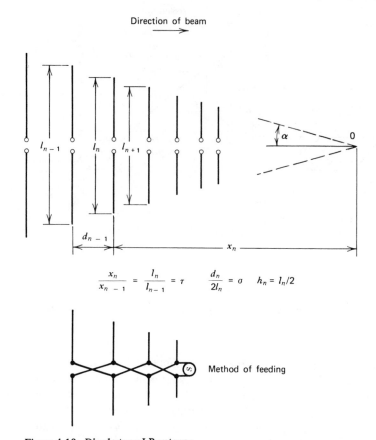

Direction of beam

$$\frac{x_n}{x_{n-1}} = \frac{l_n}{l_{n-1}} = \tau \qquad \frac{d_n}{2l_n} = \sigma \qquad h_n = l_n/2$$

Method of feeding

Figure 4.10 Dipole type LP antenna.

single guyed tower is used to support the antenna, rather than the two towers conventionally used.

The gain of an LP antenna can also be increased without mechanical complication by using the *extended-aperture* principle. This technique requires the insertion of series capacitors in the dipoles of a conventional half-wavelength LP antenna. As a result, the length of each dipole can be increased to a full wavelength at its resonant frequency without a reversal of phase of the radiating current. The extended-aperture technique decreases the radiation Q in the active region of the antenna. A greater number of elements are then excited, giving a further increase of gain without an increase in antenna dimensions.

Two or more antennas using the clamped-mode and extended-aperture techniques can be arrayed to provide even higher gain and configured to meet particular radiation pattern requirements. Several variations of the application of these techniques are shown in Figures 4.13 and 4.14.

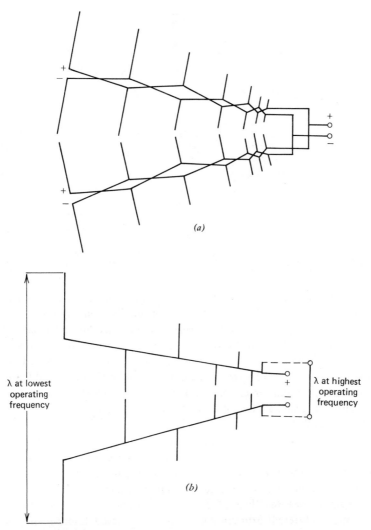

Figure 4.11 Clamped-mode LP antenna. (*a*) Side-by-side array of two half-wavelength LP antennas. (*b*) Full-wavelength-wide clamped-mode antenna equivalent to *a*. Plus and minus signs denote instantaneous phase. (Courtesy TCI, Mountain View, CA.)

VERTICALLY POLARIZED LP ANTENNAS

An LP antenna can be mounted either vertically or horizontally since, as has already been pointed out, the transmitter polarization can be selected on the grounds of practical convenience. In transmitting, finite earth conductivity can reduce radiated power by 4–8 dB. To overcome this significant loss in gain, a large ground screen must be used, or the antenna must be mounted close to and fire over a large body of salt water. Over a perfectly conducting earth a vertically polarized LP antenna

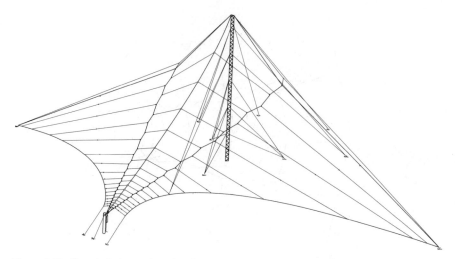

Figure 4.12 Practical clamped-mode LP antenna. (Courtesy TCI, Mountain View, CA.)

has a TOA of virtually zero. In practice, however, finite earth conductivity results in a TOA of around 15°. The use of a large ground screen will lower the TOA to about 5°, which may be necessary on very-long-distance links. The advantage of a vertically polarized LP antenna is that a low TOA can be obtained without the need for a very great height above ground. For example, a vertically polarized LP antenna covering the band of 3-30 MHz will have an overall height of around 55 m (179 ft) and will, with a ground screen, have a TOA of 5° at 10 MHz. A horizontally polarized LP antenna giving a TOA of 5° at 10 MHz would require an overall height of around 86 m (280 ft).

Even with a large ground screen a vertically polarized LP antenna suffers some loss of gain due to a finite ground conductivity. However, in addition to its advantage regarding a low TOA, without undue height, it will provide a wide azimuth beamwidth (around 90°) since it achieves gain by extension of the vertical aperture. A wide azimuth beamwidth is of importance in cases where a number of distant receiver sites at different azimuth angles are to be served. Shore-to-ship communication is a typical example of this type of application. Figure 4.13 shows three typical vertically polarized LP antenna arrays.

HORIZONTALLY POLARIZED LP ANTENNAS

The horizontally polarized LP antenna usually achieves its gain by extension of the horizontal aperture and may have a relatively narrow azimuthal beamwidth. This, of course, is no disadvantage in the case of point-to-point communication, and the horizontally polarized LP antenna has a number of very important advantages. Among these advantages are:

1. Finite ground conductivity does not result in any significant loss of radiation

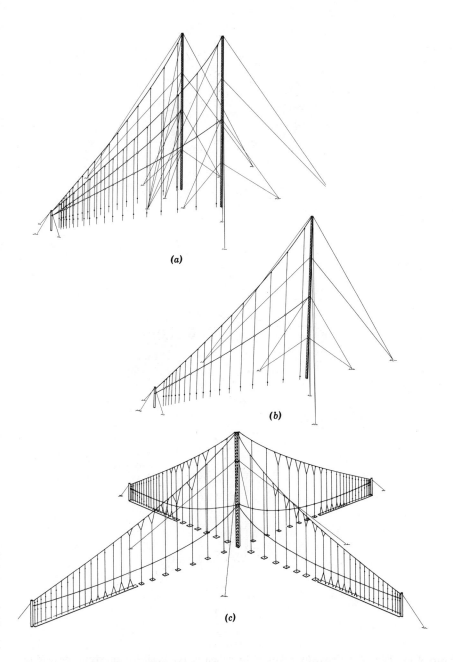

(a)

(b)

(c)

Figure 4.13 Typical vertically polarized LP antennas. (*a*) Array of two extended-aperture LP antennas. (*b*) Extended-aperture LP antenna. (*c*) Array of four LP antennas to provide steerable beams. (Courtesy TCI, Mountain View, CA.)

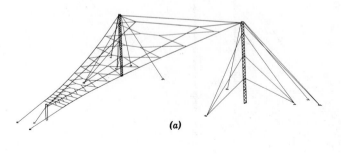

(a)

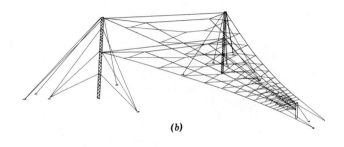

(b)

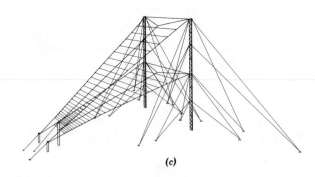

(c)

Figure 4.14 Typical horizontally polarized LP antennas. (*a*) Single-curtain clamped-mode LP antenna; gain 16 dBi, azimuth beamwidth 38°. (*b*) Two vertically stacked clamped-mode LP antennas; gain 18 dBi, azimuth beamwidth 38°. (*c*) Two vertically stacked transposed dipole LP antennas; gain 15 dBi, azimuth beamwidth 68°. (Courtesy TCI, Mountain View, CA.)

160

efficiency at low TOAs. Further, horizontally polarized LP antennas can be stacked vertically, resulting in higher antenna gains.

2. The height of the active region can be arranged to provide optimum TOA at each operating frequency.

3. Unlike vertically polarized LP antennas, there is no vertical null, and effective communications at relatively short distances is possible.

Summarizing, the horizontally polarized LP antenna can be designed for high gain and for short-, medium-, and long-distance application. Except in the rare cases where a very low TOA is required at a low frequency in the HF band, it is an ideal antenna for general applications. Figure 4.14 shows typical horizontally polarized LP antennas.

4.8.5 HF Receiving Antennas

For high-performance receiving systems the various LP antennas discussed previously can be used to achieve high directive gain. However, for many applications it is necessary to use smaller antennas due to the need to optimize equipment cost-performance, to reduce the required land area for installation, or both. The disadvantage of electrically small receiving antennas which do not have costly tunable matching circuits is that their efficiency is comparatively low and that they tend to have higher VSWRs. These characteristics result in higher noise figures, sometimes on the order of 50 dB. In the 2–32-MHz HF band this generally poses no serious problem because the receiving system noise level is insignificant compared to the external noise such as man-made, atmospheric, and galactic noise. Moreover, because the radiation resistance of an electrically small antenna is low, the mutual effects, due to coupling between a number of antennas, can be ignored, and elements can be arrayed to provide sufficient directivity using simple array theory. A vertically polarized, balanced and screened loop constructed from a large low-inductance aluminum tube forms an ideal basic element for a receiving array. A broadband matching network is used to provide an optimum impedance match for maximum power transfer and efficiency over the whole HF band.

When a number of loops are arranged to provide a specified directive radiation pattern, the feed lines from each element can be combined in the appropriate phases. Usually a number of loops are arranged linearly to provide either a unidirectional or a bidirectional end-fire array. For a unidirectional array, the signal from each loop is fed directly to a delay line, and a hybrid network is then used to combine the signals from the delay lines. For a bidirectional array, the feed lines from the loops are connected to a "beam-forming" unit in which the signal from each loop is first split and then fed through two separate delay circuits. The two sets of delay circuits are then combined in separate hybrids to generate independent beams in opposite directions along the axis of the array.

For both the unidirectional and the bidirectional arrangements, a broadband preamplifier with a 30-dB gain immediately following each hybrid combiner network is necessary. The preamplifier is situated at the point where beam-forming is accomplished, and this ensures the maximum possible signal-to-noise ratio. The purpose of the preamplifier is to compensate for attenuation in the main feeder if the loss is high enough to make receiver front-end noise a limiting factor in determining signal-to-noise ratio. In compact installations where the feeder loss is low, the preamplifier may be omitted, which eliminates the possibility of damage and intermodulation caused by nearby radiation from transmitters.

The loops can be mounted permanently on wooden posts or, for transportable systems, on metal tripods. In designing a linear array of loops, the following rules should be followed:

- To obtain a good front-to-back ratio, the array should be a half-wavelength long at the lowest frequency.
- To suppress unwanted lobes (i.e., high angle lobes which tend to reduce the main beam sensitivity and increase interference), the spacing between loops should not exceed a half-wavelength at the highest frequency.
- The gain of an end-fire array increases as the square root of the electrical length.

4.8.6 Transmission Lines

The characteristic impedance of a transmission line is given by the following equations. For balanced open-wire two-wire lines,

$$Z_0 = 267 \log_{10}\left(\frac{b}{a}\right) \ (\Omega) \tag{4.4}$$

For unbalanced coaxial lines,

$$Z_0 = \frac{138}{\epsilon} \log_{10}\left(\frac{D}{d}\right) \ (\Omega) \tag{4.5}$$

where b = center-to-center spacing between lines
a = radius of each conductor
D = inner diameter of the outer conductor
d = outer diameter of the inner conductor
ϵ = relative dielectric constant (= 1 for air)

Because of the logarithmic term, the characteristic impedance varies slowly with the physical dimensions of the transmission line. Practical impedances are 300–600 Ω for open-wire lines and 50–75 Ω for coaxial lines.

Many modern HF installations use coaxial cables for powers up to 10 kW or more. At higher powers, large-diameter coaxial lines are available, but open-wire lines are often used because of their lower cost. An advantage of coaxial cables is that they can be buried, providing a measure of physical protection. Air dielectric coaxial lines are pressurized to prevent condensation of moisture. Coaxial lines readily adapt to modern antenna switching matrices. Directional couplers are used on coaxial lines to facilititate VSWR and power measurements. Wideband transformers (balun transformers) adapt unbalanced to balanced lines.

A common balanced line is the 600-Ω open-wire line which has a characteristic impedance of 600 Ω and consists of two 6-guage copper wires spaced 12 in. apart. Such lines are less expensive than their coaxial line counterparts and have less RF attenuation at the higher frequencies.

4.9 HF FACILITY LAYOUT

HF point-to-point transmission facilities may consist of one-site, two-site, or three-site configurations. At small installations, where one or perhaps two transmitters operate simultaneously on a full-duplex basis, the companion receivers may be co-sited with the transmitter(s) if care is taken to reduce transmitter spurious outputs, and at least a 10% frequency separation is maintained between all transmitter and receiver frequencies that could simultaneously be in use.

Larger HF point-to-point facilities require a minimum separation of the transmitter location from the receiver location. Transmitter spurious outputs must not interfere with receiver operations, and separations greater than 3–5 km are required.

Often a three-site configuration is used, and both transmitter and receiver antenna sites require many acres of land. Transmitting antennas should be separated one from another by at least 300 m in the direction of transmission and by at least 80 m at the sides to reduce IM interference. Two or three antennas are often used for each transmit link, and twice that number may be required at the receiver site when space diversity is used. For good space diversity reception, diversity antennas should be separated by at least six wavelengths at the lowest operating frequency. The third site of a three-site configuration, called the communication relay center (CRC), carries out the control, switching, and message processing activities. Such a site is usually conveniently located in or near a population center or major military base with easy access to landline communication facilities (telephone, telegraph, and data). The sites are interconnected by a multichannel telephone link, either multipair or coaxial cable, or radiolink.

Transmitting and receiving sites require level ground with good ground conductivity, especially at the receiver site. The receiver site must be located far from sources of radiofrequency interference (RFI). Power lines have been found to be particularly troublesome and it is also advisable that the receiver site be inspected periodically to ensure sufficient ignition noise suppression and that all metal, such as doors, benches, and tables, be well grounded. Grounding is also important at transmitter sites to eliminate stray RF coupling.

4.10 GREAT-CIRCLE BEARING AND DISTANCE

For siting antennas on HF installations and for system calculations, great-circle
bearings and distances are used. This involves relatively simple concepts of solid
geometry and trigonometry. The method of calculation is described in several publications. One of the most readily available is *Reference Data for Radio Engineers*
(Ref. 1, sec. 28-9).

Another means of determining great-circle bearings and distances is by use of
certain navigation tables and charts available from the U.S. Navy Hydrographic
Office. Some of these are listed below:

Navigation Tables for Navigators and Aviators: HO 206

Dead Reckoning Altitude and Azimuth Table: HO 211

Great Circle Charts:
    HO Chart 1280, North Atlantic Ocean
    HO Chart 1281, South Atlantic Ocean
    HO Chart 1282, North Pacific Ocean
    HO Chart 1283, South Pacific Ocean
    HO Chart 1284, Indian Ocean

4.11 INDEPENDENT SIDEBAND (ISB) TRANSMISSION

Modern point-to-point HF with voice or composite voice–data–facsimile transmission requirements normally uses some form of ISB transmission. ISB utilizes SSB
techniques in that the carrier is suppressed but affords transmission of separate
information in each sideband. Up to four nominal 3-kHz voice channels can be
transmitted simultaneously on an ISB system. The limit of four voice channels in
a 12-kHz total bandpass is set by international agreement.

For ISB transmission the upper sideband by convention is referred to as the *A*
side and the lower sideband as the *B* side. If a particular circuit transmits all four
voice channels, the on-air transmitted spectrum is as shown in Figure 4.15. Figure
4.16 shows a simplified block diagram of an ISB exciter. The receiving side is similar for the conventional SSB receiver. It provides two independent audio outputs,
the information transmitted on the upper and lower sidebands.

The reader will take note that the *A* input and the *B* input to the transmitter each
contain two voice channels in the 6-kHz bandpass. One method of providing the
two-channel multiplexed inputs is to use a voice multiplexer as a separate device
ahead of the balanced modulator. Such a multiplexer passes the inboard channel
(*A*1 or *B*1) directly and mixes the outboard channel (*A*2 or *B*2) with a 6425-Hz
carrier; the lower sideband is selected and then combined with the direct channel.
At the distant end a demultiplex operation is carried out to separate the translated
channel and provide two normal 3-kHz voice channel outputs.

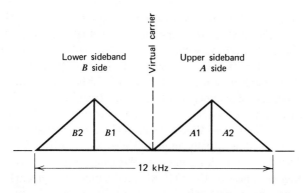

Figure 4.15   Voice channel assignment, standard four-channel ISB emission.

Another approach to HF voice channel multiplexing is to carry out the multiplexing at the transmitter IF using the proper subcarrier derived from a common synthesizer. The subcarrier for the outboard channel $B2$ is the virtual center frequency or carrier minus 6425 Hz, and for the $A2$ channel, the center frequency plus 6425 Hz. Care must be taken that $B2$ be an erect channel and $A2$ an inverted channel. At the receiver demultiplexing is carried out in the IF as well. Carrier insertion is the IF center frequency plus 6425 Hz for the $A2$ channel, and the IF center frequency minus 6425 Hz for the $B2$ channel. All reinsertion frequencies derive from a common receiver synthesizer.

Any compatible type of information may be placed on the 3-kHz voice channel for transmission. Commonly transmitted are voice, telegraph/data, and facsimile. Special problems for the transmission of data/telegraph in HF are discussed in Chapter 8.

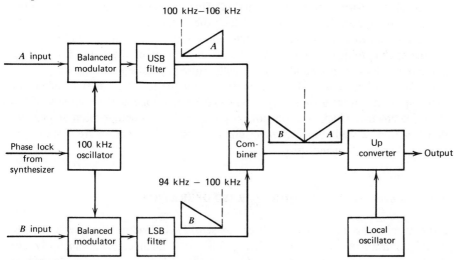

Figure 4.16 Simplified block diagram of ISB exiter.

The severe fading in HF coupled with the continuous variation of the voice level itself have limited the quality of telephone service that can be provided. Signal levels may vary 60 or 70 dB in very short periods. With full carrier suppression AGC on receivers does not function well. If one of the four available channels has either FM facsimile or telegraph/data tones, which have a fairly constant amplitude, AGC can be derived from such a channel, and level control can be fairly effective.

Older systems and some systems still in existence use a device called a *voice-operated automatic gain device* (VOGAD). A VOGAD is supplied for each demodulated voice channel and is usually mounted near the receiver. By detecting the average level of the voice syllables, some effective gain control is provided.

The outstanding weakness of the VOGAD (and similar equipment) is that it has no continuous-level signal with which to sense continuously the variations of the audio level and afford the necessary gain control. One approach is to split the voice channel providing a small bandpass at channel center which carries a sinusoidal tone with which AGC can act. This equipment is usually combined with the two-wire-four-wire HF telephone terminal. Such a terminal is normally configured as follows:

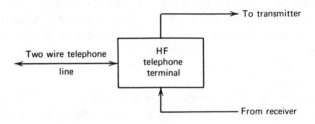

The gain control tone described above also provides a control on the voice terminal to control the transmit-to-receive function. This is done by frequency shifting the tone. A shift upward in frequency will lock the distant end in a receive condition and a shift downward in frequency will allow the distant end to transmit. This eliminates the problem of voice terminal lockup common on older systems. Lockup in the transmit or receive condition is usually caused by noise in the system.

The latest refinement of the tone control technique for gain has the trade name Lincompex. It utilizes a frequency modulated control tone at the high end of the HF voice channel which controls not only the level but a compression and expansion function as well. Voice channel compression and expansion are discussed in Chapter 3 under compandors.

4.12 TRANSMITTER LOADING FOR ISB OPERATION

ISB transmitters as multichannel devices are sensitive to loading. Consider a 10,000-W ISB transmitter. While transmitting four voice channels simultaneously, we would ideally allow sufficient drive (excitation) such that each voice channel on the air provides 10,000/4 or 2500 W of power. We speak here of peak power or, more properly, peak envelope power (PEP). This would be a simple matter if we transmit only a simple sinusoidal signal in each voice channel. Outside of FSK, the transmitted signal applied to each voice channel is more complex. The drive on each voice channel input must be backed off, so on the average, each channel contributes less than 2500 W of on-air power. This is due to a peaking factor. If voice is being transmitted on all four ISB voice channels with an average PEP of 2500 W output from each, at some time several or all channels may reach a voice peak simultaneously and the 10,000 W would be well exceeded, probably causing the transmitter to blow a circuit breaker, arc over tubes, and so on. Thus backoff of input signals must be carried out to ensure that the equipment stays within the peaking limits reasonably. It also must be remembered that the ratio of peak power to average power of a pure sine wave is $2:1$, that is, the average power of a pure sine wave is one-half the peak power. A more detailed discussion of loading and peak factors is given in Chapter 3.

4.13 DIVERSITY TECHNIQUES ON HF

Two types of diversity reception are in fairly common use in HF systems today:

- Space diversity
- In-band frequency diversity

In practice diversity improvement is realized to the greatest extent only on digital systems. For composite voice–telegraph systems, diversity combining is used on the digital channel only.

For space diversity two antennas are used on each link. For full space diversity improvement the two antennas must be separated by at least six wavelengths (6λ). Combining is usually of the selective type and is normally carried out in the voice frequency carrier telegraph equipment located at the CRC. A typical space diversity system is shown in Figure 4.17. Such a system may provide an improvement of bit error rate on digital transmission (usually carried in the $A1$ channel) from one to two orders of magnitude.

The reason for the improved error rate is that seldom is a fade correlated on antennas separated by at least six wavelengths. In other words, a signal received on one antenna may be in a deep fade, whereas on the second antenna the signal level will be high provided the fading follows a Rayleigh distribution.

In-band frequency diversity is somewhat less attractive as a diversity option and is effective on frequency selective fading only. It has been found that if two telegraph (data) channels derived from a single digital source (i.e., transmitting identical information) are separated by at least 600 Hz, an equivalent of a full order of diversity improvement may be derived on HF systems regarding frequency selective fading. The nominal maximum number of voice frequency telegraph (VTFG) channels on

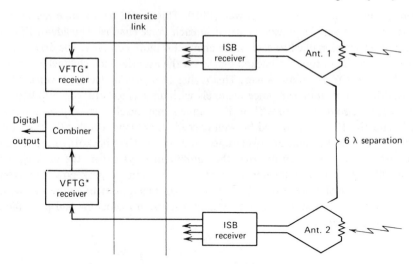

Figure 4.17 Typical space diversity system.* VF telegraph (a digital demodulator); see Chapter 8.

HF is 16.* The subcarrier arrangement is such that the system may be reconfigured to eight channels of information, permitting the pairing of channels and a 600-Hz separation between each pair member. A switch at the receiving terminal pairs already existing groups of pairs such that the system goes into a form of quadruple diversity when in-band frequency diversity is added to space diversity. (See Chapter 8 for more details on data/telegraph transmission.)

4.14 IONOSPHERIC SOUNDERS

Selection of the OWF (FOT) of a long-haul point-to-point HF link is normally a "by guess and by God" procedure even with the use of CRPL forecasts, particularly during periods of high sunspot activity. Synchronized oblique ionospheric sounding removes much of the guesswork. Such a sounding system is shown schematically in Figure 4.18.

Oblique sounding as an adjunct to communications is carried out by a separate, pulsed radio system that operates in parallel with an HF radiolink or links. The system consists of a sounder transmitter at one end of a link and a receiver and display at the other end. The transmitter and receiver are clock controlled with an accuracy of a few milliseconds per week. The sounder transmitter uses a separate antenna, whereas the receiver may be multicoupled or share the same antenna with the communications service.

Sounding is carried out by the transmission of RF pulses on discrete frequencies. The transmission of pulses is a form of frequency scanning. One such equipment

---

*The reader should note that some present-day systems using 120-Hz tone frequency spacing handle up to 22 telegraph channels. 120-Hz spacing is discussed in Chapter 8. Such a system is not suitable for tone diversity.

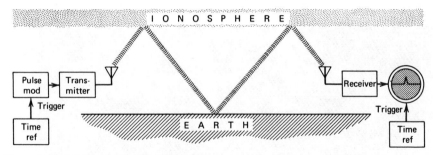

Figure 4.18 Synchronized oblique sounding.

now in operation initiates a scan sequence with synchronized clocks at both ends of the link. The transmitter and receiver step in synchronization across the HF band at a rate of 10 frequency channels per second. The pulse train from the sounder transmitter consists of two RF pulses on each channel from the lowest frequency (4 MHz) to the highest (32 MHz), which are radiated by a broadband antenna to the distant end illuminating the ionosphere along the path. The distant end receiver displays the received pulses on an A-scan storage tube on which the horizontal axis is the frequency scale, marked off linearly from 4 to 32 MHz. The vertical scale is marked off in relative time delay in milliseconds.

Under usual circumstances not all the transmitted pulses are detected by the sounder receiver and displayed. Those not received include frequencies below the LUF and above the MUF. The display at the receiver is called an ionogram. The relative time delay is indicative of the number of hops a signal takes to arrive at the distant end. Multipath is evident in the ionogram by signal smearing or stretching. A typical ionogram is shown in Figure 4.19.

It would appear that the equivalent frequency scanning used by ionospheric sounders would cause significant interference to other HF services. Synchronized oblique sounders employ pulses on the order of 1 ms in length with a *pulse repetition frequency* (PRF) of 20 pps. Channel spacing is on the order of 20 times (at midrange) the bandwidth of a typical communication system. It is impossible to hear the sounder on conventional HF receivers unless co-located with the sounder transmitter. The pulsewidth is considerably shorter than half of the shortest pulsewidth normally used on HF data circuits (see Chapter 8). Thus chances of interference are minimal.

It has been stated previously in this chapter that one of the most important considerations in HF pertains to frequency selection. Long-term propagation predictions are indispensable for circuit engineering. This involves decisions regarding antenna elevation and gain, diversity, TOA, and so on. The day-to-day and hour-by-hour selection of the MUF is the immediate need of the operator. There is considerable variation between actual up-to-the-minute ionospheric conditions and those predicted three months previously. The oblique ionospheric sounder is a tool that gives the operator real-time information on ionospheric conditions and may significantly reduce propagation outage on HF.

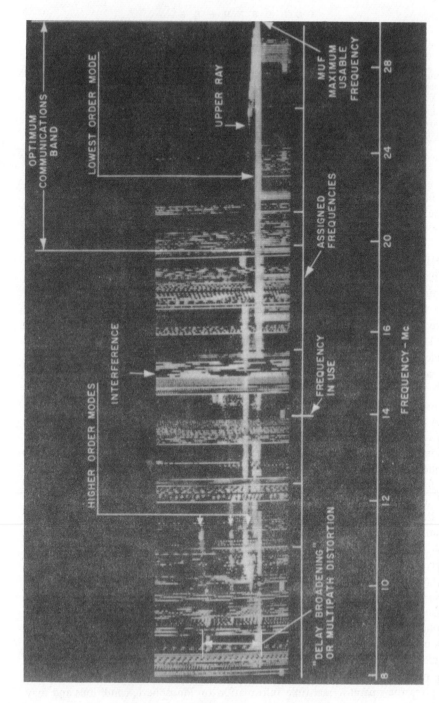

Figure 4.19 Typical HF ionogram from an oblique sounding system. (Courtesy Granger Associates, Menlo Park, CA.)

REFERENCES AND BIBLIOGRAPHY

1. *Reference Data for Radio Engineers*, 6th ed., Howard W. Sams, Indianapolis, IN, 1977.
2. *Electrical Communication Systems–Radio*, *U.S. Army Tech. Manual TM11 486–6*, June 27, 1958.
3. *Fundamentals of Single Sideband*, 3rd ed., Collins Radio Co., Cedar Rapids, IO.
4. F. Barghausen *et al.*, *"Predicting Long-Term Parameters of HF Sky-Wave Telecommunication Systems,"* ERL-110-ITS Environmental Science Services Administration, Boulder, CO, May 1969.
5. *Modern HF Antennas–Selection and Application*, 3rd ed., Granger Asso., Menlo Park, CA.
6. *The Use of Ionosphere Sounders to Improve H-F Communications*, Tech. Bull. 3, Granger Asso., Menlo Park, CA, Mar. 1966.
7. *The Application Engineering Manual*, Defense Communications Agency, Washington DC, 1965.
8. *Uses of Time and Frequency in H-F Communications*, ITT Communication Systems, 1962.
9. *MF/HF Communication Antennas*, Addendum No. 1 to DCAC 330-175-1 U.S. Department of Defense, Washington, DC, 1966.
10. *Handbook of Engineering, Siting and Installation of HF Rhombic Antennas*, T.O. 31R-1-7, U.S. Department of Defense (USAF), Washington, DC.
11. *Siting of H-F Facilities*, T.O. 31-1-13, U.S. Department of Defense (USAF), Washington, DC.
12. *Preliminary Analysis of Adaptive Communications*, U.S. Department of Defense, Washington, DC, 1963.
13. W. Henneberry, *Single Sideband Handbook*, Technical Materiel Corp., Mamaroneck, NY, 1963.
14. *SSB Communications*, NavShips 93225, U.S. Government Printing Office, Washington, DC.
15. *Ionospheric Radio Propagation*, National Bureau of Standards, Monograph 80, U.S. Government Printing Office, Washington, DC.
16. *Handbook for CRPL Predictions Based on Numerical Methods of Mapping*, NBS Standard Handbook No. 90, Boulder, CO, 1962.
17. W. Sichak and R. Adams, "The Radio Path in Communication Systems Planning," North Jersey Section of the IEEE (session 8), Seminar on Overall Communication System Planning, 1964.
18. "Recommendations and Reports of the CCIR 1978," XIV Plenary Assembly, Kyoto, 1978, Vols. III and VI.
19. "Radio Regulations," ITU General Secretariat, Geneva, 1976.
20. "Methods of Improving the Performance of HF Digital Radio Systems," NTIS PB80-128 606, U.S. NTIA, Boulder, CO, Oct. 1979.

5 RADIOLINK SYSTEMS (LINE OF SIGHT MICROWAVE)

5.1 INTRODUCTION

Let us define radiolink systems as those that fulfill the following requirements:

1. Signals follow a straight line or line-of-sight (LOS) path.
2. Signal propagation is affected by free-space attenuation and precipitation.
3. Use of frequencies greater than 150 MHz, thereby permitting transmission of more information per RF carrier by use of a wider information baseband.
4. Use of angle modulation (i.e., FM or PM), or spread-spectrum and time-sharing techniques.

Tables 5.1*a* gives conventional letter designations for microwave bands and 5.1*b* shows typical assignments for the frequency region through 13 GHz. Radio transmission above 13 GHz is covered in Chapter 10.

A valuable characteristic of LOS transmission is that we can predict the level of a signal arriving at a distant receiver with known accuracy.

5.2 LINK ENGINEERING

Engineering a radiolink system involves the following steps:

1. Selection of sites (radio equipment plus tower locations) that are in line-of-sight of each other.
2. Selection of an operational frequency band from those set forth in Table 5.1*b*, considering RF interference environment and legal restraints.
3. Development of path profiles to determine radio tower heights. If tower heights exceed a certain economic limit, then step 1 must be repeated, bringing the sites closer together or reconfiguring the path, usually along another route. In making a profile, it must be taken into consideration that microwave energy is

172

Table 5.1a Letter Designations for Microwave Bands

Subband	Frequency (GHz)	Wavelength (cm)	Subband	Frequency (GHz)	Wavelength (cm)
	P Band			*X* Band (*continued*)	
	0.225	133.3	*d*	6.25	4.80
	0.390	76.9	*b*	6.90	4.35
			r	7.00	4.29
	L Band		*c*	8.50	3.53
	0.390	76.9	*l*	9.00	3.33
p	0.465	64.5	*s*	9.60	3.13
c	0.510	58.8	*x*	10.00	3.00
l	0.725	41.4	*f*	10.25	2.93
y	0.780	38.4	*k*	10.90	2.75
t	0.900	33.3			
s	0.950	31.6		*K* Band	
x	1.150	26.1		10.90	2.75
k	1.350	22.2	*p*	12.25	2.45
f	1.450	20.7	*s*	13.25	2.26
z	1.550	19.3	*e*	14.25	2.10
			c	15.35	1.95
	S Band		*u* [b]	17.25	1.74
	1.55	19.3	*t*	20.50	1.46
e	1.65	18.3	*q* [b]	24.50	1.22
f	1.85	16.2	*r*	26.50	1.13
t	2.00	15.0	*m*	28.50	1.05
c	2.40	12.5	*n*	30.70	0.977
q	2.60	11.5	*l*	33.00	0.909
y	2.70	11.1	*a*	36.00	0.834
g	2.90	10.3			
s	3.10	9.67		*Q* Band	
a	3.40	8.32		36.0	0.834
w	3.70	8.10	*a*	38.0	0.790
h	3.90	7.69	*b*	40.0	0.750
z [a]	4.20	7.14	*c*	42.0	0.715
d	5.20	5.77	*d*	44.0	0.682
			e	46.0	0.652
	X Band				
	5.20	5.77		*V* Band	
a	5.50	5.45		46.0	0.652
q	5.75	5.22	*a*	48.0	0.625
y [a]	6.20	4.84	*b*	50.0	0.600

Table 5.1a *(Continued)*

Subband	Frequency (GHz)	Wavelength (cm)	Subband	Frequency (GHz)	Wavelength (cm)
V Band *(continued)*			*W* Band		
c	52.0	0.577		56.0	0.536
d	54.0	0.556		100.0	0.300
e	56.0	0.536			

Source: Ref. 1.
[a] *C* band includes S_z through X_y (3.90–6.20 GHz).
[b] K_1 band includes K_u through K_q (15.35–24.50 GHz).

- Attenuated or absorbed by solid objects
- Reflected from flat conductive surfaces such as water and sides of metal buildings
- Diffracted around solid objects
- Refracted or bent by the atmosphere; often the bending is such that the beam may be extended beyond the optical horizon

Table 5.1b. Microwave Radiolink Frequency Assignments for Fixed Service[a]

General Frequency Assignments

450–470 MHz	
890–960 MHz	5925–6425 MHz
1710–2290 MHz	(7250)7300–8400 MHz
2550–2690 MHz	10550–12700 MHz
3700–4200 MHz	14400–15250 MHz

Specific Frequency Assignments, United States

Service	GHz	Service	GHz
Military	1.710–1.850	Common carrier (space)	5.925–6.425
Operational fixed	1.850–1.990	Operational fixed	6.575–6.875
Studio transmitter link	1.990–2.110	Studio transmitter link	6.875–7.125
Common carrier	2.110–2.130	Military	7.125–7.750
Operational fixed	2.130–2.150	Military	7.750–8.400
Common carrier	2.160–2.180	Common carrier	10.7–11.7
Operational fixed	2.180–2.200	Operational fixed	12.2–12.7
Operational fixed (TV only)	2.500–2.690	CATV studio transmitter link (CARS)[b]	12.7–12.95
Common carrier (space)	3.700–4.200	STL	12.95–13.2
Military	4.400–5.000	Military	14.4–15.25

[a]Point-to-point communications and some other nonmobile applications.
[b]CATV = community antenna television; CARS = community antenna radio service.

4. Path calculations. After setting a propagation reliability expressed as a percentage of time, the received signal will be above a certain threshold level. Often this level is the FM improvement threshold of the FM receiver. To this level a margin is set for signal fading under all anticipated climatic conditions.

5. Making a path survey to ensure correctness of steps 1–4. It also provides certain additional planning information vital to the installation project or bid.

6. Equipment configuration to achieve the fade margins set in step 4 most economically.

7. Establishment of a frequency plan and necessary operational parameters.

8. Installation.

9. Beam alignment, equipment lineup, checkout, and acceptance by a customer.

Reference will be made, where applicable, to these steps so that the reader will be exposed to practical radiolink problems.

LOS microwave is often used synonymously with radiolink. Radiolink is preferred in this text because

- Microwave is harder to define, even LOS microwave.
- The term *radiolink* is more universal, particularly outside the United States, for the material that this chapter covers.

5.3 PROPAGATION

5.3.1 Free-Space Loss

Consider a signal traveling between a transmitter at A and a receiver at B:

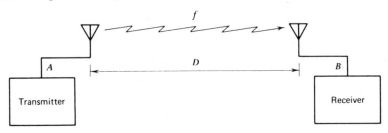

The distance between antennas is D and the frequency of transmission is f. Let D be in kilometers and f in megahertz; then the free-space loss in decibels may be calculated with the following formula:

$$L_{dB} = 32.44 + 20 \log D + 20 \log f \tag{5.1a}$$

If D is in statute miles, then

$$L_{dB} = 36.6 + 20 \log D + 20 \log f \tag{5.1b}$$

(all logarithms are to be base 10).

Suppose that the distance separating A and B were 40 km ($D = 40$). What would the free-space path loss be at 6 GHz ($f = 6000$)?

$$L = 32.44 + 20 \log 40 + 20 \log 6 \times 10^3$$

$$= 32.44 + 20 \times 1.6021 + 20 \times 3.7782$$

$$= 32.44 + 32.042 + 75.564$$

$$= 140.046 \text{ dB}$$

Let us look at it another way. Consider a signal leaving an isotropic antenna.* At one wavelength (1λ) away from the antenna, the free-space attenuation is 22 dB. At two wavelengths (2λ) it is 28 dB; at four wavelengths it is 34 dB. Every time we double the distance, the free-space attenuation (or loss) is 6 dB greater. Likewise, if we halve the distance, the attenuation (or loss) decreases by 6 dB.

Figure 5.1 relates free-space path loss to distance at eight discrete frequencies between 450 and 14,800 MHz.

---

*An isotropic antenna is an ideal antenna with a reference gain of 1 (0 dB). It radiates equally in all directions (i.e., is perfectly omnidirectional).

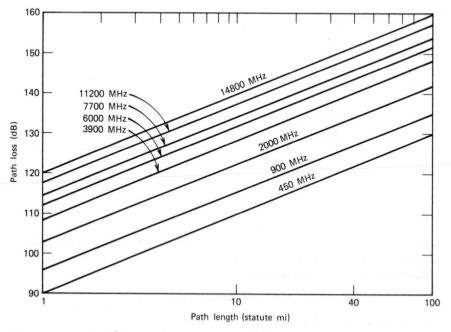

Figure 5.1 Path loss in decibels related to path length in statute miles for eight discrete radiolink frequencies.

5.3.2 Bending of Radio Waves above 100 MHz from Straight-Line Propagation

Radio waves traveling through the atmosphere do not follow true straight lines. They are refracted or bent. They may also be diffracted.

The velocity of an electromagnetic wave is a function of the density of the media through which it travels. This is treated by Snell's law, which provides a valuable relationship for an electromagnetic wave passing from one medium to another (i.e., from an air mass with one density to an air mass with another density). It states that the ratio of the sine of the angle of incidence to the sine of the angle of refraction is equal to the ratio of the respective velocities in the media. This is equal to a constant which is the refractive index of the second medium relative to the first medium.

The absolute refractive index of a substance is its index with respect to a vacuum and is practically the same value as its index with respect to air. It is the change in the refractive index that determines the path of an electromagnetic wave through the atmosphere, or how much the wave is bent from a straight line.

If radiowaves above 100 MHz traveled a straight line, the engineering of LOS microwave (radiolink) systems would be much easier. We could then accurately predict the height of the towers required at repeater and terminal stations and exactly where the radiating device on the tower should be located (steps 2 and 3, Section 5.2). Essentially what we are dealing with here, then, is a method to determine the height of a microwave radiator (i.e., an antenna or other radiating device) to permit reliable radiolink communication from one location to another.

To determine tower height, we must establish the position and height of obstacles in the path between stations with which we want to communicate by radiolink systems. To each obstacle height, we will add earth bulge. This is the number of feet or meters an obstacle is raised higher in elevation (into the path) owing to earth curvature or earth bulge. The amount of earth bulge in feet at any point in a path may be determined by the formula

$$h = 0.667\, d_1 d_2 \qquad\qquad (5.2)$$

where d_1 = distance from the near end of the link to the point (obstacle location), and d_2 = distance from the far end of the link to the obstacle location.

The equation will become more useful if it is made directly applicable to the problem of ray bending. As the equation is presented above, the ray is unbent or a straight line.

Atmospheric refraction may cause the ray beam to be bent toward the earth or away from the earth. If it is bent toward the earth, it is as if we shrank the earth bulge or lowered it from its true location. If the beam is bent away from the earth, it is as if we expanded the earth bulge or raised it up toward the beam above its true value. This lowering or raising is handled mathematically by adding a factor K to the earth bulge equation. It now becomes

$$h_{\mathrm{ft}} = \frac{0.667\, d_1 d_2}{K} \qquad (d_1 \text{ and } d_2 \text{ in mi}) \qquad (5.3\mathrm{a})$$

$$h_{\mathrm{m}} = \frac{0.078\, d_1 d_2}{K} \qquad (d_1 \text{ and } d_2 \text{ in km}) \qquad (5.3\mathrm{b})$$

and

$$K = \frac{\text{effective earth radius}}{\text{true earth radius}} \qquad (5.4)$$

If the K factor is greater than 1, the ray beam is bent toward the earth, which essentially allows us to shorten radiolink towers. If K is less than 1, the earth bulge effectively is increased, and the path is shortened or the tower height must be increased. Figure 5.2 gives earth curvature, in feet, for various values of K.

Many texts on radiolinks refer to normal refraction, which is equivalent to a K factor of $\frac{4}{3}$ or 1.33. It follows a rule of thumb that applies to refraction in that a propagated wave front (or beam) bends toward the region of higher density, that is, toward the region having the higher index of refraction. Older texts insisted that the K factor should nearly always be $\frac{4}{3}$. Care should be taken when engineering radiolinks that the $\frac{4}{3}$ theory (standard refraction or normal refraction) not be accepted "carte blanche" on many of the paths likely to be encountered. However, $K = \frac{4}{3}$ may be used for gross planning of radiolink systems. Figure 5.3 may be used to estimate tower heights with $K = \frac{4}{3}$ for smooth earth paths (no obstacles besides midpath earth bulge).

Another factor must be added to the obstacle height to obtain an effective obstacle height. This is the Fresnel clearance. It derives from electromagnetic wave

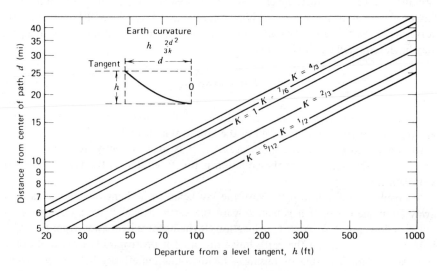

Figure 5.2 Earth curvature or earth bulge for various K factors.

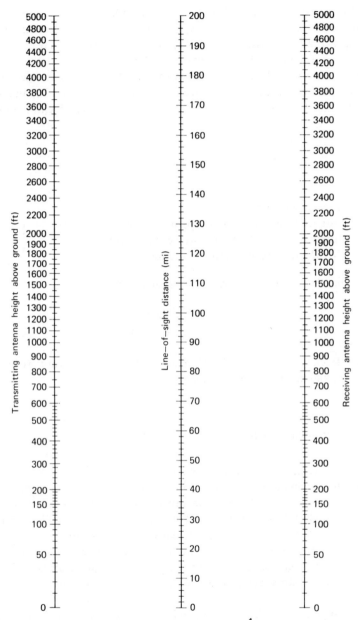

Figure 5.3 LOS distance for smooth spherical earth with $K = \frac{4}{3}$; distance in statute miles, tower heights in feet.

theory that a wave front (which our ray beam is) has expanding properties as it travels through space. These expanding properties result in reflections and phase transitions as the wave passes over an obstacle. The result is an increase or decrease in the received signal level. The amount of additional clearance that must be allowed to avoid problems with the Fresnel phenomenon (diffraction) is expressed in Fresnel zones. Optimum clearance of an obstacle is accepted as 0.6 of the first Fresnel zone radius. The first Fresnel zone radius, in feet, may be calculated with the following formula:

$$R_t = 72.1 \sqrt{\frac{d_1 d_2}{FD}} \qquad (5.5a)$$

where F = frequency of signal (GHz)
   d_1 = distance from transmitter to path obstacle (statute mi)
   d_2 = distance from path obstacle to receiver (statute mi)
   $D = d_1 + d_2$, total path length (statute mi)

To determine the first Fresnel zone radius when using metric units:

$$R_m = 17.3 \sqrt{\frac{d_1 d_2}{FD}} \qquad (5.5b)$$

where F = frequency of signal (GHz), d_1, d_2, and D are the same as in formula (5.5a) but in kilometers, and R_m is in meters.

5.3.3 Path Profiling–Practical Application (Step 3)

After tentative terminal or repeater sites have been selected, path profiles are plotted on rectangular graph paper. Obstacle information is taken from topographical maps. For the continental United States the best topographical maps available are from the U.S. Geological Survey. They are $7\frac{1}{2}$-min maps of latitude and longitude with a scale of 1:24,000, where 1 in. = 2000 ft, and 15-min maps with a scale of 1:62,500, where 1 in. = approx 1 mi. 30-min and 1-deg maps are also available, but their scales are not fine enough for path profile application. Many areas of Canada are covered by maps with scales of 1:50,000 (1 in. = 0.79 mi) and 1:63,360 (1 in. = 1 mi).

Profiles are made on available linear graph paper; 10 divisions per inch is suggested. Any convenient combination of vertical and horizontal scales may be used. For paths 30 mi or less in length, 2 mi to the inch plotted on the horizontal scale is suggested. For longer paths graph paper may be extended by trimming and pasting. For the vertical scale, 100 ft to the inch is satisfactory for fairly flat country, where there is no more than 800 ft change in altitude along the path; for hilly country,

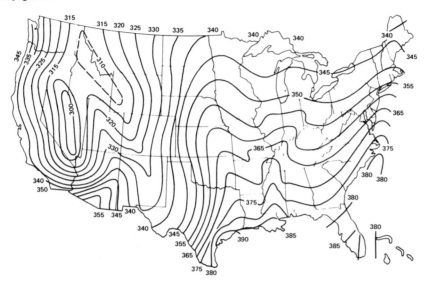

Figure 5.4 Sea level refractivity index N_0 for the continental United States: maximum for worst month (August).

200 ft to the inch and for mountainous country, 500 or 1000 ft to the inch are appropriate. One suggestion to preserve a proper relationship between height and distance is that if the distance scale is doubled, the height scale should be quadrupled. All heights should be shown with reference to mean sea level (MSL), but the base (0 elevation on the graph paper) need not be MSL, but the lowest elevation of interest on the profile.

On the topographical map draw a straight line connecting the two adjacent radio-link sites. Carefully trace with your eye or thumb down the line from one site to the other, marking all obstacles or obstructions and possible points of reflection, such as bodies of water, marshes, or desert areas, assigning consecutive letters to each obstacle.

Plot the horizontal location of each point on the graph paper. Mark the path midpoint, which is the point of maximum earth bulge and should be marked as an obstacle. Determine the K factor by one of the following methods:

1. Refer to a sea level refractivity profile chart (see Figure 5.4). Select the appropriate N_0 for the area of interest. Apply the N_0 selected to Figure 5.5, which gives the midpath elevation and refractivity N_0. Read off the corresponding K factor.

2. Lacking a refractivity contour chart for the area, such as shown in Figure 5.4, plot using the three K factors 1.33, 1.0, and 0.5. A later field survey will help to decide which factor is valid. For instance, in coastal regions, over-water paths, and damp regions assume the lowest value. In most dry regions (nondesert) the so-called normal value (1.33) may be assumed. Table 5.2 will help as a guide to determining the K factor.

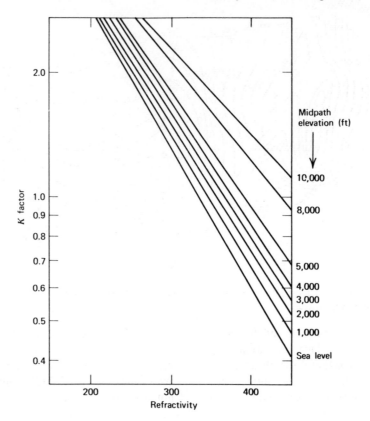

Figure 5.5 K factor scaled for midpath elevation above mean sea level.

For each obstacle point compute d_1, the distance to one repeater site, and d_2, the distance to the other. Compute the equivalent earth curvature EC for each point with equation (5.3):

$$EC = \frac{0.667 \, d_1 d_2}{K}$$

Compute the Fresnel zone clearance using Figure 5.6 and 5.7. Figure 5.6 gives the Fresnel clearance for midpath for various bands of frequencies. For other obstacle points compute the percentage of total path length. For instance, on a 30-mi path, midpoint is 15 mi (50%). However, a point 5 mi from one site is 25 mi from the other, or represents $\frac{5}{30}$ and $\frac{25}{30}$, or $\frac{1}{6}$ and $\frac{5}{6}$. Converted to percentage, $\frac{1}{6} = 16.6\%$ and $\frac{5}{6} = 83\%$. Apply this percentage on the X axis of Figure 5.7. In this case 16.6% or 83% is equivalent to 76% of midpath clearance. If midpath Fresnel clearance were 40 ft, only 30.4 ft of Fresnel clearance would be required at the 5-mi point.

Table 5.2 K Factor Guide[a]

	Propagation Conditions				
	Perfect	Ideal	Average	Difficult	Bad
Weather	Standard atmosphere	No surface layers or fog	Substandard, light fog	Surface layers, ground fog	Fog moisture over water
Typical	Temperate zone, no fog, no ducting, good atmospheric mix day and night	Dry, mountainous, no fog	Flat, temperate, some fog	Coastal	Coastal, water, tropical
K factor	1.33	1–1.33	0.66–1.0	0.66–0.5	0.5–0.4

[a] For 99.9–99.99 path reliability.

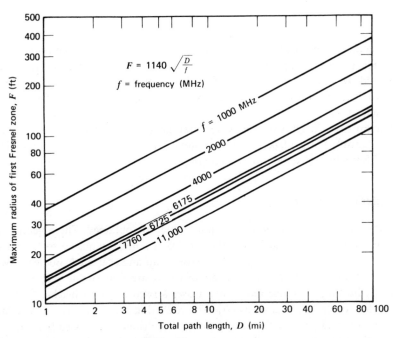

$$F = 1140 \sqrt{\frac{D}{f}}$$

f = frequency (MHz)

Figure 5.6 Midpath Fresnel clearance for first Fresnel zone. 0.6 of this value is used in the calculations or the value is calculated in conjunction with Figure 5.7. (Courtesy GTE Lenkurt Inc., San Carlos, CA.)

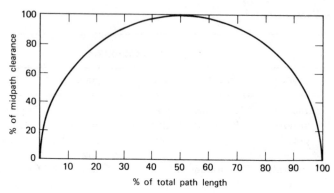

Figure 5.7 Conversion from midpath Fresnel zone clearance for other than midpath obstacle points.

Set up a table on the profile chart as follows:

Obstacle Point	d_1	d_2	EC	F (Fresnel)
A				
B				
C				
D				
E				

On the graph paper plot the height above sea level of each obstacle point. To this height add EC (earth curvature), the F (Fresnel zone clearance), and if tree growth exists, add 40 ft for trees and 10 ft for additional growth. If undergrowth exists, assign 10 ft for vegetation. A path survey will confirm or deny these figures or will permit adjustment.

Minimum tower heights may now be determined by drawing a straight line from site to site through the highest obstacle points. These often cluster around midpath. Figure 5.8 is a hypothetical profile exercise.

5.3.4 Reflection Point

From the profile, possible reflection points may be obtained. The objective is to adjust the tower heights such that the reflection point is adjusted to fall on land area where the reflected energy will be broken up and scattered. Bodies of water and other smooth surfaces cause reflections that are undesirable. Figure 5.9 will assist in adjusting the reflection point. It uses a ratio of tower heights h_1/h_2, and the shorter tower height is always h_1. The reflection area lies between a K factor of grazing ($K = 1$) and a K factor of infinity. The distance expressed is always from h_1, the shorter tower. By adjusting the ratio h_1/h_2 the reflection point can be moved. The objective is to ensure that the reflection point does not fall on an area

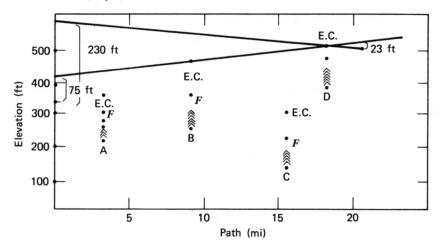

Figure 5.8 Practice profile.

of smooth terrain or on water, but rather on a land area where the reflected energy will be broken up or scattered (such as wooded areas).

For a highly reflective path, space diversity operation may be desirable to minimize the effects of multipath reception (see Section 5.5.6).

5.4 PATH CALCULATIONS (STEP 4)

5.4.1 General

The next step in path engineering is to carry out path calculations. Essentially this entails the determination of equipment parameters and configurations to meet a

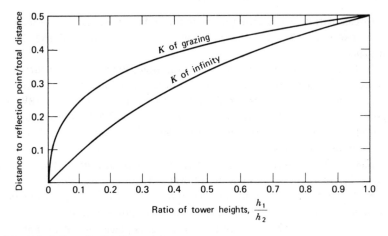

Figure 5.9 Calculation of reflection points.

Table 5.3   Reliability Versus Outage Time

Reliability (%)	Outage Time (%)	Outage Time		
		Per Year	Per Month (avg.)	Per Day (avg.)
0	100	8760 h	720 h	24 h
50	50	4380 h	360 h	12 h
80	20	1752 h	144 h	4.8 h
90	10	876 h	72 h	2.4 h
95	5	438 h	36 h	1.2 h
98	2	175 h	14 h	29 min
99	1	88 h	7 h	14 min
99.9	0.1	8.8 h	43 min	1.4 min
99.99	0.01	53 min	4.3 min	8.6 s
99.999	0.001	5.3 min	26 s	0.86 s

minimum performance requirement. Such performance requirements are usually related to noise in an equivalent voice channel or they are related to the signal-to-noise ratio, or both. Either way, the requirement is given for a percentage of time. For a single path this may be stated as 99, 99.9, or 99.99% of the time said performance exceeds a certain minimum. This is often called propagation reliability. Table 5.3 states reliability percentages versus outage time, where the outage time is the time that the requirement will not be met.

5.4.2   Basic Path Calculations

FM RECEIVER THRESHOLD—THE STARTING POINT

For path calculations the starting point is the FM receiver. The following question must be answered: what signal level coming into the receiver will give the desired performance?

For the basic approach the case must be simplified. Figure 5.10 is a curve comparing input carrier level to output signal-to-noise ratio of a typical radiolink FM receiver. Assume that the noise threshold for this receiver is at −122.5 dBW. Thus at this point the input carrier-to-noise ratio is zero. At −112.5 dBW input carrier level, the carrier-to-noise ratio is 10 dB. The carrier-to-noise ratio in decibels is shown below the input carrier level in Figure 5.10.

The relationship between the input carrier-to-noise ratio of an FM receiver and the output signal-to-noise ratio is as follows. At noise threshold, for every 1-dB increase of carrier-to-noise ratio, the signal-to-noise ratio of the output increases approximately 1 dB. This occurs up to a carrier-to-noise ratio of 10 dB or a little greater (11 or 12 dB), when a "capture effect" takes place and the output signal-to-noise ratio suddenly jumps to 30 dB. This point of "capture" is called the FM improvement threshold and is shown in Figure 5.10. Beyond this point the signal-to-noise ratio at the receiver output improves again by 1 dB for every increase of 1 dB of input carrier-to-noise ratio, up to a point where compression starts to take effect (saturation). For instance, if the input carrier-to-noise ratio were 15 dB, we might expect the output signal-to-noise ratio to be approximately 35 dB. This assumes that the FM system has been "adjusted" properly (e.g., there is sufficient deviation to effect FM improvement; see Section 5.5.3).

For many radiolink systems an input carrier-to-noise ratio of 10 dB is used as a starting point for path calculations. The reason is obvious, for it is at this point that we start to get the improved signal-to-noise ratio. For future discussion in this chapter we will use the same point of departure. We can calculate this point. It is the noise threshold level, calculated in Chapter 1, plus 10 dB. The FM improvement threshold is

$$\text{Input level (dBW)} = 10 \log kTB_{IF} + NF_{dB} + 10 \text{ dB} \qquad (5.6)$$

where T = noise temperature (K)
B_{IF} = noise bandwidth (bandwidth of the IF) (Hz)
k = Boltzman's constant (1.38×10^{-23} W-s/deg)
NF = receiver noise figure (dB)

Simplified for an uncooled receiver (i.e., T = 290 K or 62°F),

$$\text{Input level (dBW)} = -204 \text{ dBW} + NF + 10 \text{ dB} + 10 \log B_{IF} \qquad (5.7)$$

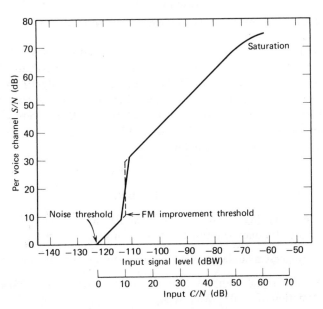

Figure 5.10 Input carrier-to-noise ratio C/N in decibels versus output signal-to-noise ratio S/N (per voice channel) for a typical FM radiolink receiver.

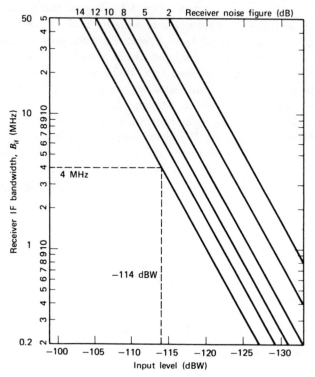

Figure 5.11 FM improvement threshold for several receiver NFs.

Figure 5.11 gives threshold values for several receiver NFs. However, the threshold may be computed rapidly by directly subtracting the receiver NF, the 10 dB, and then working with the bandwidth. If the bandwidth were only 10 Hz, we would subtract (algebraically add) 10 dB; if it were 20 Hz, we would subtract 13 dB; 100 Hz, 20 dB; 1000 Hz, 30 dB; 10 MHz, 70 dB; 20 MHz, 73 dB; and so on.

Consider an uncooled receiver with a 14-dB NF and 4-MHz IF bandwidth. What would the FM improvement threshold be?

$$\text{Threshold}_{dBW} = -204 + 14 + 10 + 66$$

$$= -114 \text{ dBW}$$

This is the necessary input level to reach an FM improvement threshold. The required IF bandwidth may be estimated by Carson's rule, which states:

$$B_{IF} = 2 \text{ (highest modulating baseband frequency + peak FM deviation)}$$

Example. A video signal with a 4.2-MHz baseband modulates an FM transmitter with a peak deviation of 5 MHz. What is B_{IF}?

$$B_{IF} = 2(4.2 \text{ MHz} + 5 \text{ MHz})$$

$$= 18.4 \text{ MHz}$$

The FM improvement threshold may be calculated when given the peak deviation and the bandwidth of the information baseband as well as the receiver NF (or noise temperature; see Section 1.12).

For the most basic path calculation, consider a receiver with an FM improvement threshold of -114 dBW, a free-space path loss of 140 dB, an isotropic antenna* at both ends, and lossless transmission lines. What would the transmitter output have to be to provide a -114-dBW input level to the receiver?

$$\text{Output dBW} = -114 + 140 = +26 \text{ dBW}$$

$$+26 \text{ dBW} \simeq 400 \text{ W}$$

For this case both antennas have unity gain and the transmission lines are lossless. Now extend the example for 2.0-dB line losses and 20-dB antenna gain at each end:

$$\text{Output dBW} = -114 \text{ dBW} + 2.0 - 20 \text{ dB} + 140 - 20 \text{ dB} + 2.0$$

$$= -10 \text{ dBW} = 0.1 \text{ W}$$

Path calculations, therefore, deal with adding gains and losses to arrive at a specified system performance.

As the discussion progresses, we shall see many means to attain gain and the various ways losses occur. Figure 5.12 shows graphically the gains and losses as a radiolink signal progresses from transmitter to receiver. Section 5.8 deals with detailed path calculations.

*Unrealistic, but useful for discussion.

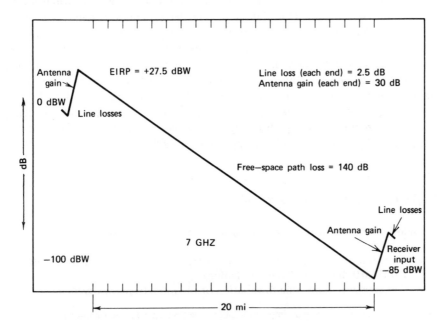

Figure 5.12 Radiolink gains and losses (simplified). Transmitter output = 0 dBW.

5.4.3 The Mechanism of Fading—An Introductory Discussion

GENERAL

Up to this point we have considered that the received signal level remains constant. Truly on most paths, particularly on shorter ones, at lower frequencies, much of the time this holds true. When a receive level varies from the free-space calculated level for a given far-end transmitter output, the result is called fading. Fading due to propagation mechanisms involves refraction, reflection, diffraction, scattering, focusing attenuation, and other miscellaneous causes.* Such factors, when associated with fading, relate to certain conditions classified as meteorological phenomena and terrain geometry. The radiolink engineer must be alert to these factors when planning specific links and during site survey phases.

The following paragraphs cover the two general types of fading, multipath and power.

MULTIPATH FADING

This type of fading is due to interference between a direct wave and another wave, usually a reflected wave. The reflection may be from the ground or from atmospheric sheets or layers. Direct path interference may also occur. It may be caused by surface layers of strong refractive index gradients or horizontally distributed changes in the refractive index.

Multipath fading may display fades in excess of 30 dB for periods of seconds or minutes. Typically this form of fading will be observed during quiet, windless, and foggy nights, when temperature inversion near the ground occurs and there is not enough turbulence to mix the air. Thus stratified layers, either elevated or ground based, are formed. Two-path propagation may also be due to specular reflections from a body of water, salt beds, or flat desert between transmitting and receiving antennas. Deep fading of this latter type usually occurs in daytime on over-water paths or other such paths with a high ground reflection. Vegetation or other "roughness" found on most radiolink paths breaks up the reflected components, rendering them rather harmless. Multipath fading at its worst is independent of obstruction clearance, and its extreme condition approaches a Rayleigh distribution.

POWER FADING

Dougherty (Ref. 26) defines power fading as a:

 . . . partial isolation of the transmitting and receiving antennas because of

- Intrusion of the earth's surface or atmospheric layers into the propagation path (earth bulge or diffraction fading)
- Antenna decoupling due to variation of the refractive index gradient

*Above 10 GHz rainfall is also an important factor; see Section 5.11 and Chapter 10.

- Partial reflection from elevated layers interpositioned between terminal antenna elevations
- "Ducting" formations containing only one of the terminal antennas.
- Precipitation along the propagation path . . .

Power fading is characterized by marked decreases in the free-space signal level for extended time periods. Diffraction may persist for several hours with fade depths of 20–30 dB.

5.5 FM RADIOLINK SYSTEMS

5.5.1 General

Figures 5.13*a* and 5.13*b* are simplified block diagrams of a radiolink transmitter and a radiolink receiver. The type of modulation used is FM. The input information to the transmitter (baseband) and the output from the receiver are considered to be groupings of nominal 4-kHz telephone channels in an SSBSC FDM configuration. This type of multiplexing is discussed in Chapter 3. Orderwire plus alarm information is inserted from 300 Hz to 12 kHz for many installations accepting an FDM multitelephone channel input. Thus the composite baseband will be made up of the following:

300 Hz–12 kHz	Orderwire/alarms
12 kHz–60 kHz	A 12-channel group of FDM multichannel information

or

300 Hz–12 kHz	Orderwire/alarms
60 kHz–*	Supergroups/mastergroups of multichannel information

or

300 Hz–12 kHz	Orderwire/alarms
312 kHz–*	Supergroups/mastergroups of multichannel information

Other systems use a 12-kHz spectrum with an SSB signal at 3.25 MHz for alarms and orderwire. Video plus program channel subcarriers is another information input which will be discussed later. The orderwire/alarms input for video is necessarily inserted above the video in the baseband.

*See Chapter 3 for specific baseband makeup.

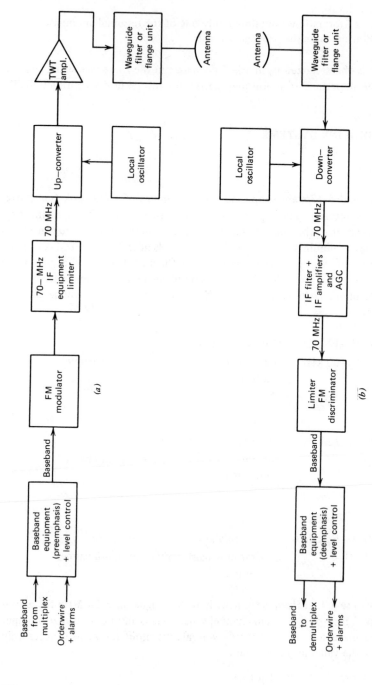

Figure 5.13 (a) Typical radiolink terminal transmitter. (b) Typical radiolink terminal receiver.

192

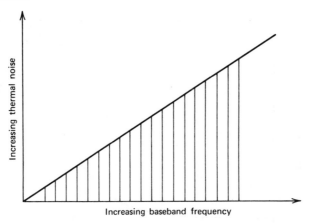

Figure 5.14 Sketch of increasing thermal noise from the output of an FM receiver in a system without preemphasis–deemphasis.

5.5.2 Preemphasis–Deemphasis

The output characteristics of an FM receiver without system preemphasis–deemphasis, with an input FM signal of a modulation of uniform amplitude, are such that it has linearly increasing amplitude with increasing baseband frequency. This is shown in Figure 5.14. Note the ramplike or triangular noise of the higher frequency baseband components. The result is decreasing signal-to-noise ratio with increasing baseband frequency. The desired receiver output is a constant signal-to-noise ratio across the baseband. Preemphasis at the transmitter and deemphasis at the receiver achieve this end.

Preemphasis is accomplished by increasing the peak deviation during the FM modulation process for higher baseband frequencies. This increase of peak frequency deviation is done in accordance with a curve designed to effect compensation for the ramplike noise at the FM receiver output. CCIR has recommended standardization on the curve shown in Figure 5.15 for multichannel telephony (CCIR Rec. 275-2), and for video transmission a preemphasis curve may be found in CCIR Rec. 405-1 (Figure 5.16).

The preemphasis characteristic is achieved by applying the modulating baseband to a passive network that "forms" the input signal. At the receive end after modulation, the baseband signal is applied to a deemphasis network which restores the baseband to its original amplitude configuration. This is shown diagrammatically in Figure 5.17.

5.5.3 The FM Transmitter

The combined baseband (orderwire and FDM multiplex line frequency) frequency modulates a wave. The modulated wave is at IF or converted to IF, and thence

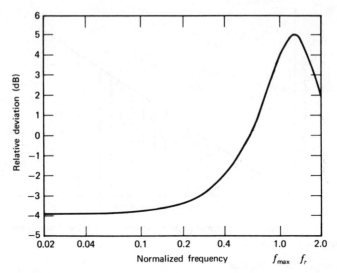

Figure 5.15 Preemphasis characteristic for multichannel telephony (CCIR Rec. 275; Courtesy ITU–CCIR.)

up-converted to the output frequency. The up-converted signal may be applied directly to the antenna system for radiation or amplified further. The additional amplification is usually carried out by a traveling wave tube (TWT) for radiolink systems. Outside of the TWT, modern radiolink transmitters are all solid state. However, some equipment is still available in which the input signal directly modulates a klystron whose output is radiated.

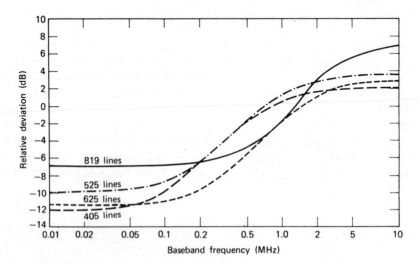

Figure 5.16 Television preemphasis characteristics for 405-, 525-, 625-, and 819-line systems. 0 dB corresponds to a deviation of 8 MHz for a 1-V peak-to-peak signal (CCIR Rec. 405; Courtesy ITU–CCIR.)

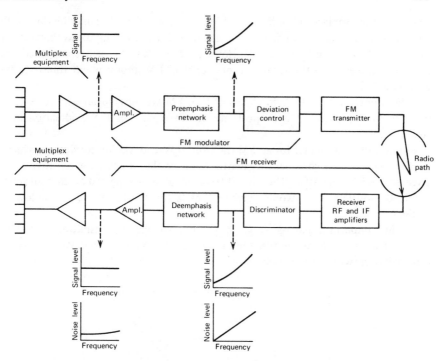

Figure 5.17 Simplified block diagram of an FM radiolink system showing the effects of preemphasis on the signal and thermal noise.

Radiolink transmitter outputs have fairly well standardized as follows:

0.1 W   −10 dBW   (usually for transmitters operating at higher
                                frequencies)

1.0 W   0 dBW

10 W   +10 dBW   (for transmitters operating at lower frequencies
                                or those with a TWT final amplifier)

The percentage of modulation or the modulation index is a most important parameter for radio transmitters. The percentage of modulation is used to describe the modulation in AM transmitters. The modulation index is the equivalent parameter for FM transmitters. The modulation index for frequency modulation is

$$M = \frac{F_d}{F_m} \tag{5.8}$$

where F_d = frequency deviation, and
    F_m = maximum significant modulation frequency.

Keep in mind that the frequency deviation is a function of the input level to the transmitter—the modulation input.

One of the primary advantages of FM over other forms of modulation is the improvement of the output signal-to-noise ratio for a given input, as illustrated in Figure 5.18 (also see Figure 5.10). However, this advantage is accomplished only if the signal level into the receiver is greater than FM improvement threshold (see Section 5.4.2).

If the modulation index is in excess of 0.6, FM systems are superior to AM, and the rate of improvement is proportional to the square of the modulation index (Ref. 13).

An important parameter for an FM transmitter is its deviation sensitivity, which is usually expressed in volts per megahertz. Suppose that a deviation sensitivity were given as 0.05 V rms/MHz. Then if a 0.05-V signal level appeared at the transmitter input, the output wave would deviate 1 MHz (or 1 MHz above and below the carrier).

Another important parameter given for FM wideband transmitters is the rms deviation per channel at the test tone level. Table 5.4 relates channel capacity and rms deviation (without preemphasis).

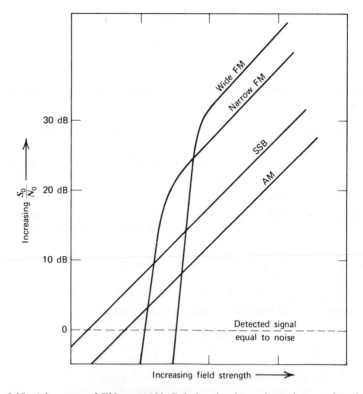

Figure 5.18 Advantage of FM over AM. Relative signal-to-noise ratios as a function of field strength. (Courtesy Institute of Telecommunication Sciences, Office of Telecommunications, U.S. Department of Commerce.)

Table 5.4 Frequency Deviation Without Preemphasis

Maximum Number of Channels	RMS Deviation per Channel[a] (kHz)
12	35
24	35
60	50, 100, 200
120	50, 100, 200
300	200
600	200
960	200
1260	140, 200
1800	140
2700	140

Source: CCIR Rec. 404-2.
[a] For 1 mW, 800-Hz tone at a point of 0 reference level.

For systems using SSBSC FDM basebands, peak deviation D may be calculated as follows, for $N = 240$ or more,

$$D = 4.47\,d\left(\log^{-1}\frac{-15 + 10\log N}{20}\right) \tag{5.9a}$$

or for N = between 12 and 240,

$$D = 4.47\,d\left(\log^{-1}\frac{-1 + 4\log N}{20}\right) \tag{5.9b}$$

where D = peak deviation (kHz)
d = per channel rms test tone deviation (kHz)
N = number of SSBSC voice channels in the system

(See Section 3.4.4 for more explanation of the loading of FDM systems.)

Example. A 300-channel radiolink system would have a 200-kHz/channel rms deviation according to Table 5.4. Then

$$D = (4.47)\,(200)\left(\log^{-1}\frac{9.77}{20}\right) = 2753 \text{ kHz}$$

The question arises, how much deviation will optimize FM transmitter operation. We know that as the input signal to the transmitter is increased, the deviation is increased. Again, as the deviation increases, the FM improvement threshold becomes more apparent. In effect we are trading off bandwidth for thermal noise improvement. However, with increasing input levels, the IM noise of the system begins to increase. There is some point of optimum input to a wideband FM transmitter where the thermal noise improvement in the system has been optimized and where the IM noise is not excessive. This concept is shown in Figure 5.19.

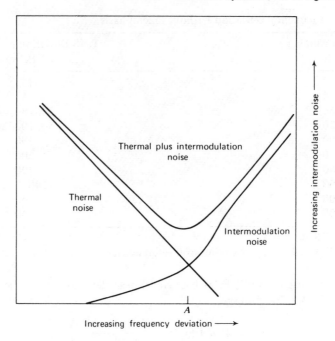

Figure 5.19 Optimum setting for deviation is shown by letter A.

Frequency stability requirements for FM systems are far less severe than for equivalent SSB systems (see Chapter 4); ±0.001%/month is usually sufficient.

5.5.4 The Antenna System

GENERAL

For conventional radiolinks the antenna sybsystem offers more room for trade-off to meet minimum system requirements than any other subsystem. Basically the antenna system looking outward from a transmitter must have

- Transmission line (waveguide or coaxial line)
- Antenna: a reflecting surface or device
- Antenna: a feed horn or other feeding device

In addition the antenna system may have

- Circulators
- Directional couplers
- Phasers
- Passive reflectors
- Radomes

ANTENNAS

Below 700 MHz antennas used for point-to-point radiolink systems are often a form of yagi and are fed with coaxial transmission lines. Above 700 MHz some form of parabolic reflector-feed arrangement is used. 700 MHz is no hard and fast dividing line. Above 2000 MHz the transmission line is usually waveguide. The same antenna is used for transmission and reception (see Section 5.5.6).

An important antenna parameter is its radiation efficiency. Assuming no losses, the power radiated from an antenna would be equal to the power delivered to the antenna. Such power is equal to the square of the rms current flowing on the antenna times a resistance, called the radiation resistance:

$$P = I_{rms}^2 R \qquad (5.10)$$

where P = radiated power (W)
    R = radiation resistance (Ω)
    I = current (A)

In practice, all the power delivered to the antenna is *not* radiated in space. The radiation efficiency is defined as the ratio of the power radiated to the total power delivered to an antenna.

To derive a more realistic equation to express power, we divide the resistance, which we shall call the terminal resistance, into two component parts, namely, R = radiation resistance and R_1 = equivalent terminal loss resistance, so that

$$P = I_{rms}^2 (R + R_1) \qquad (5.11)$$

and

$$\text{Radiation efficiency } (\%) = \frac{R}{R + R_1} \times 100 \qquad (5.12)$$

Antenna gain is a fundamental parameter in radiolink engineering. Gain is conventionally expressed in decibels and is an indication of the antenna's concentration of radiated power in a given direction. Antenna gain expressed anywhere in this work is gain over an isotropic.* An isotropic is a theoretical antenna with a gain of 1 (0 dB). In other words, it is an antenna that radiates equally in all directions.

For parabolic reflector type antennas, gain is a function of the diameter of the parabola D and the frequency f. Theoretical gain is expressed by the formula

$$G_{dB} = 20 \log F_{MHz} + 20 \log_{10} D_{ft} - 52.6 \qquad (5.13)$$

for a 54% surface efficiency.

Figure 5.20 is a graph from which the gain of a parabolic reflector type of antenna may be derived for several discrete reflector diameters.

Note. In practice we assume surface efficiencies of usually around 55% for radiolink systems. Chapter 7 discusses antennas with improved efficiencies.

*Often expressed in dBi.

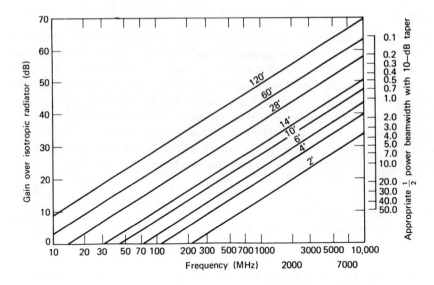

Figure 5.20 Parabolic antenna gains for discrete reflector diameters. (Courtesy Institute of Telecommunication Sciences, Office of Telecommunications, U.S. Department of Commerce.)

For uniformly illuminated parabolic reflectors, an approximate relation between half-power beamwidth and gain is expressed by the following:

$$Q \cong \frac{142}{\sqrt{G}} \quad (\text{degrees}) \tag{5.14}$$

where Q = half-power beamwidth and

$$\sqrt{G} = \text{antilog}\left(\frac{G_{dB}}{20}\right)$$

In practice absolutely uniform illumination is not used so that side lobe levels may be reduced using a tapered form of illumination. Expect beamwidths to be 0.1–0.2° wider in practical applications.

Beamwidths are narrow in radiolink systems. Table 5.5 illustrates their narrowness. From this it is evident that considerable accuracy is required in pointing the antenna at the distant end.

Table 5.5 Antenna Beamwidths

Gain (dB)	Half-Power Beamwidth
30	5°
35	3°
44	1°

POLARIZATION

In radiolink systems antennas use linear polarization and, depending on the feed, may be horizontally or vertically polarized, or both. That is, an antenna radiating and receiving several frequencies at once often will radiate adjacent frequencies with opposite polarizations, or the received polarization is opposite to the transmit. Isolation of 26 dB or better may be expected between polarizations, and on well designed installations 35 dB, allowing closer interworking at installations using multi-RF carrier operation.

5.5.5 The FM Receiver

In most applications the radiolink receiver shares the same antenna and waveguide as its companion transmitter to or from a common distant end. Figure 5.13*b* is a simplified block diagram of a typical radiolink receiver (ideal configuration). The receiver may or may not be connected to the common waveguide manifold by means of a circulator. It also may use a bandpass or preselector filter. Both the circulator and the preselector filter reduce the effects of adjacent transmitter energy to a negligible amount in the receiver front end. Radiolink receivers operating at lower frequencies use coaxial transmission lines instead of waveguide.

From the manifold or via a circulator and/or preselector the incoming FM signal next looks into a mixer or down-converter. This unit heterodynes the received signal with the local oscillator signal to produce an IF. Most installations have standardized on a 70-MHz IF. However, CCIR discusses a 140-MHz IF for systems designed to carry voice channels in excess of 1800 in a standard CCITT FDM configuration.

From the mixer or down-converter output, the IF is fed through several amplification stages, often through a phase equalizer to correct delay distortion introduced by IF (and RF) filters. IF gains commonly are on the order of 80–90 dB.

The output of the IF is fed to a limiter–discriminator which is the FM detector or demodulator. The output of the discriminator is the composite baseband made up of the information baseband plus orderwire and alarm signals.

After demodulation the composite baseband is passed through a deemphasis network (see Figure 5.17) and is amplified. Thence the composite signal is split, the information baseband being directed to the demultiplex equipment and the orderwire (service channel) and alarms to the orderwire equipment and alarm display.

5.5.6 Diversity Reception

GENERAL

Various types of diversity reception are used widely on point-to-point HF systems, on transhorizon microwave (tropo), and to an increasing extent on radiolinks (LOS microwave). Diversity is attractive for the following reasons:

- Tends to reduce depth of fades on combined output.
- Provides improved equipment reliability (if one diversity path is lost due to equipment failure, other path(s) remain in operation).
- Depending on the type of combiner in use, the combined output signal-to-noise ratio is improved over that of any single signal path.

Diversity reception is based on the fact that radio signals arriving at a point of reception over separate paths may have noncorrelated signal levels. More simply, at one instant of time a signal on one path may be in a condition of fade while the identical signal on another path may not.

First one must consider what are separate paths and how "separate" must they be (CCIR Rep. 376-3). The separation may be in

- Frequency
- Space (including angle of arrival and polarization)
- Time (a time delay of two identical signals on parallel paths)
- Path (signals arrive on geographically separate paths)

The most common forms of diversity in radiolink systems are those of frequency and space. A frequency diversity system utilizes the phenomenon that the period of fading differs for carrier frequencies separated by 2–5%. Such a system employs two transmitters and two receivers, with each pair tuned to a different frequency (usually 2–3% separation since the frequency band allocations are limited). If the fading period at one frequency extends for a period of time, the same signal on the other frequency will be received at a higher level, with the resultant improvement in propagation reliability.

As far as equipment reliability is concerned, frequency diversity provides a separate path, complete and independent, and consequently one whole order of reliability has been added. Besides the expense of the additional equipment, the use of additional frequencies without carrying additional traffic is a severe disadvantage to the employment of frequency diversity, especially when frequency assignments are even harder to get in highly developed areas where the demand for frequencies is greatest.

One of the main attractions of space diversity is that no additional frequency assignment is required. In a space diversity system, if two or more antennas are spaced many wavelengths apart (in the vertical plane), it has been observed that multipath fading will not occur simultaneously at both antennas. Sufficient output is almost always available from one of the antennas to provide a useful signal to the receiver diversity system. The use of two antennas at different heights provides a means of compensating, to a certain degree, for changes in electrical path differences between direct and reflected rays by favoring the stronger signal in the diversity combiner.

Diversity combiners are discussed below. The antenna separation required for optimum operation of a space diversity system may be calculated using the following formula:

$$S = \frac{3\lambda R}{L} \tag{5.15}$$

where S = separation (m)
   R = effective earth radius (m)
   λ = wavelength (m)
   L = path length (m)

However, any spacing between 100 and 200λ is usually found to be satisfactory. The goal in space diversity is to make the separation of diversity antennas such that the reflected wave travels a half-wavelength further than the normal path. In addition, CCIR states that for acceptable space diversity operation the spacing should be such that the value of the space correlation coefficient does not exceed 0.6. (CCIR Rep. 376-3).

Figure 5.21a is a simplified block diagram of a radiolink frequency diversity system and Figure 5.21b that of a space diversity system. Polarization diversity is discussed in Section 6.9.6. Figure 5.22 shows approximate interference fading for

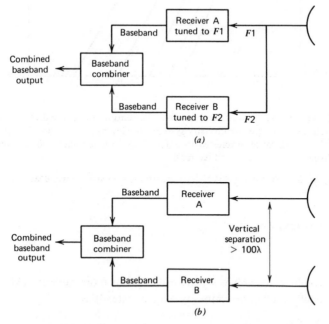

Figure 5.21 (a) Simplified block diagram of a frequency diversity configuration $F_1 < F_2 + 0.02\,F_2$. (b) Simplified block diagram of a space diversity configuration.

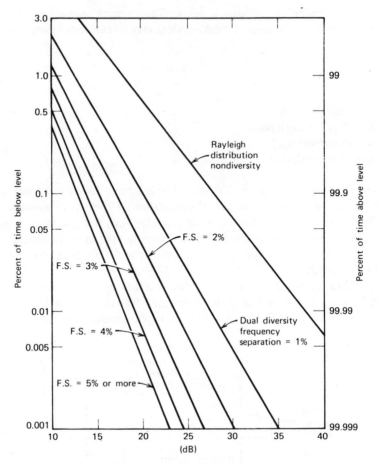

Figure 5.22 Approximate interference fading distribution for a nondiversity system with Rayleigh fading versus frequency diversity systems for various percentages of frequency separation (F.S.). (*Note.* Signal level referred to unfaded level for radiolinks; signal level referred to median for tropospheric scatter; see Chapter 6).

nondiversity versus frequency diversity systems for various percentages of frequency separation.

DIVERSITY COMBINERS

General

A diversity combiner combines signals from two or more diversity paths. Combining is traditionally broken down into two major categories:

- Predetection
- Postdetection

The classification is made according to where in the reception process the combining takes place. Predetection combining takes place in the IF. However, at least one system* performs combining at RF. With the second type, combining is carried out at baseband (i.e., after detection).

For predetection combining, phase control circuitry is required unless some form of path selection is used.

Figure 5.23*a* and 5.23*b* shows simplified functional block diagrams of radiolink receiving systems using predetection and postdetection combiners.

Types of Combiners

Three types of combiners find more common application in radiolink diversity systems. These are

- Selection combiner
- Equal-gain combiner
- Maximal-ratio combiner (ratio squared)

The selection combiner uses but one receiver at a time. The output signal-to-noise ratio is equal to the input signal-to-noise ratio from the receiver selected for use at the time.

The equal-gain combiner simply adds the diversity receiver outputs, and the output signal-to-noise ratio of the combiner is

$$\frac{S_o}{N_o} = \frac{S_1 + S_2}{2N} \tag{5.16}$$

where N = receiver noise.

The maximal-ratio combiner uses a relative gain change between the output signals in use. For example, let us assume that the stronger signal has unity output and the weaker signal has an output proportional to gain G. It then can be shown that $G = S_2/S_1$ such that the signal gain is adjusted to be proportional to the ratio of the input signals. We then have

$$\left(\frac{S_o}{N_o}\right)^2 = \left(\frac{S_1}{N}\right)^2 + \left(\frac{S_2}{N}\right)^2 \tag{5.17}$$

where N = receiver noise.

For the signal-to-noise ratio equation for the latter two combiners, we assume the following:

- All receivers have equal gain.
- Signals add linearly; noise adds on an rms basis.
- Noise is random.

---

*System manufactured by STC (U.K.).

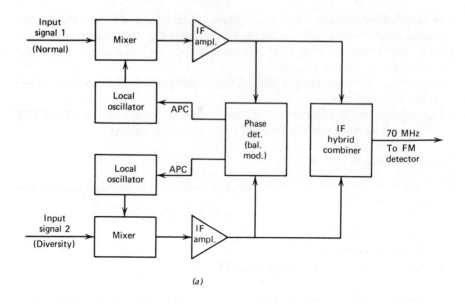

(a)

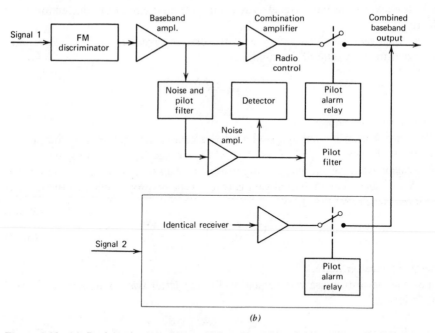

(b)

Figure 5.23 (*a*) Predetection combiner. APC = automatic phase control. (*b*) Postdetection combiner (maximum ratio squared).

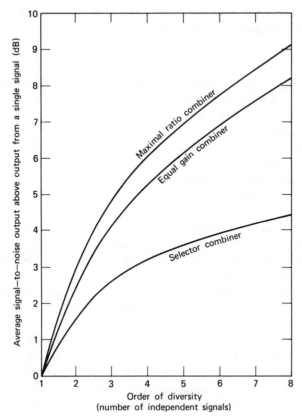

Figure 5.24 Signal-to-noise ratio improvement in a diversity system for various orders of diversity. (Courtesy Institute of Telecommunication Sciences, Office of Telecommunications, U.S. Department of Commerce.)

- All receivers have equal noise outputs N.
- The output (from the combiner) signal-to-noise ratio S_o/N_o is a constant.

Figure 5.24 shows graphically a comparison of the three types of combiners. We see the gain that we can expect in output signal-to-noise ratio for various orders of diversity for the three types of combiners discussed, assuming a Rayleigh distribution. The order of diversity refers to the number of independent diversity paths. If we were to use space *or* frequency diversity alone, we would have two orders of diversity. If we used space *and* frequency diversity, we would then have four orders of diversity.

The reader should bear in mind that the efficiency of diversity depends on the correlation of fading of the independent diversity paths. If the correlation coefficient is zero (i.e., there is no relationship in fading for one path to another), we can expect maximum diversity enhancement. The efficiency of a diversity system drops by half with a correlation coefficient of 0.8, and nearly full efficiency can be expected with a correlation coefficient of 0.3.

PILOT TONES

A radio continuity pilot is provided between radio terminals which are independent of the multiplex pilot tones. The pilot or pilots provided are used for:

- Gain regulating
- Monitoring
- Frequency comparison
- Measurement of level stability
- Control of diversity combiners

The latter application involves the simple sensing of continuity by a diversity combiner. The presence of the continuity pilot tells the combiner that the diversity path is operative. The problem is that the most commonly used postdetection combiners, the maximal ratio type, selection type, and others, use noise as the means to determine the path contribution to the combined output. The path with the least noise, as in the case of the maximal ratio combiner, provides the greatest path contribution.

If, for some reason, a path were to fail, it would be comparatively noiseless and would provide 100% contribution. Thus we would have a no-signal output. In this case the continuity pilot tells the combiner that the path is a valid one.

Pilots are inserted prior to modulation. They are stopped (eliminated) at baseband output and are reinserted anew if another radiolink is added. Table 5.6, taken from CCIR Rec. 401-2, provides a list of recommended radio continuity pilots.

5.5.7 Transmission Lines and Related Devices

WAVEGUIDE AND COAXIAL CABLE

Waveguides may be used on installations operating above 2 GHz to carry the signal from the radio equipment to the antenna (and vice versa). For those systems operating above 4 GHz it is mandatory from a transmission efficiency point of view. Coaxial lines are used on those systems operating below 2 GHz.

From a systems engineering aspect, the concern is loss with regard to a transmission line. Figures 5.25 and 5.26 identify several types of commonly used waveguides and coaxial cable with loss versus frequency.

Waveguide installations are always maintained under dry air or nitrogen pressure to prevent moisture condensation within the guide. Any constant positive pressure up to 10 lb/in.^{2} (0.7 kg/cm^2) is adequate to prevent "breathing" during temperature cycles.

Table 5.6 Radio Continuity Pilot Tones

System Capacity (Channels)	Limits of Band Occupied by Telephone Channels (kHz)	Frequency Limits of Baseband (kHz)[a]	Continuity Pilot Frequency (kHz)	Deviation (rms) Produced by the Pilot (kHz)[b, e]
24	12–108	12–108	116 or 119	20
60	12–252	12–252	304 or 331	24, 50, 100[c]
	60–300	60–300		
120	12–552	12–552	607[d]	25, 50, 100[c]
	60–552	60–552		
300	60–1300	60–1364	1499, 3200[f] or 8500[f]	100 or 140
600	60–2540	60–2792	3200 or 8500	140
	64–2660			
960 ⎫ 900 ⎭	60–4028 ⎫ 316–4188 ⎭	60–4287	4715 or 8500	140
1260 ⎫ ⎪ 1200 ⎭	60–5636 ⎫ 60–5564 ⎬ 316–5564 ⎭	60–5680	⎰ 6199 ⎱ 8500	100 or 140 140
1800	312–8204	300–8248	9203	100
	316–8204			
2700	312–12,388	308–12,435	13,627	100
	316–12,388			
Television			⎰ 8500 ⎱ 9023[g]	140 100

Source: CCIR Rec. 4012.

[a] Including pilot or other frequencies which might be transmitted to line.

[b] Other values may be used by agreement between the administrations concerned.

[c] Alternative values dependent on whether the deviation of the signal is 50, 100, or 200 kHz (Rec. 404–2).

[d] Alternatively, 304 kHz may be used by agreement between the administrations concerned.

[e] This deviation does not depend on whether or not a preemphasis network is used in the baseband.

[f] For compatibility in the case of alternate use with 600-channel telephony systems and television systems.

[g] The frequency 9023 kHz is used for compatibility purposes between 1800-channel telephony systems and television systems when the establishment of multiple sound channels so indicates.

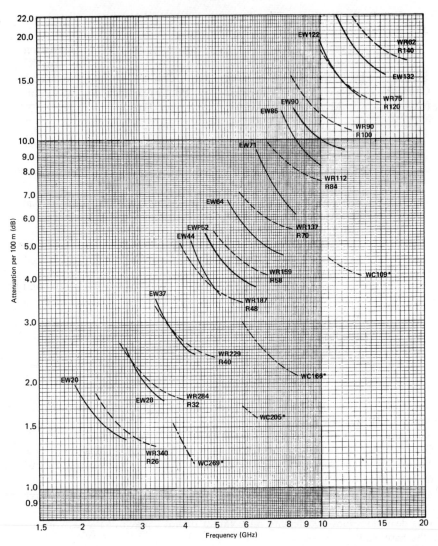

Figure 5.25 Waveguide attenuation.* Add 0.3 dB to allow for top and bottom transition. (Courtesy Andrew Corp.)

Waveguides may be rectangular, elliptical, or circular. Nearly all older installations used rectangular waveguide exclusively. Today its use is limited to routing in tight places where space is limited. However, bends are troublesome and joints add 0.06-dB loss each to the system. Optimum electrical performance is achieved by using the minimum number of components. Therefore it has become the practice that wherever a single length is required, elliptical waveguide is used from the antenna to the radio equipment without the addition of miscellaneous flex-twist or rigid sections which are used for rectangular waveguide.

Circular waveguide offers generally lower loss plus dual polarized capability such that only one waveguide run up the tower is necessary for dual polarized installations. Circular waveguide is used when the run is long so that excessive loss is not introduced.

TRANSMISSION LINE DEVICES

A ferrite load isolator is a waveguide component which provides isolation between a single source and its load, reducing ill effects of VSWR and often improving stability as a result. Most commonly load isolators are used with transmitting sources absorb-

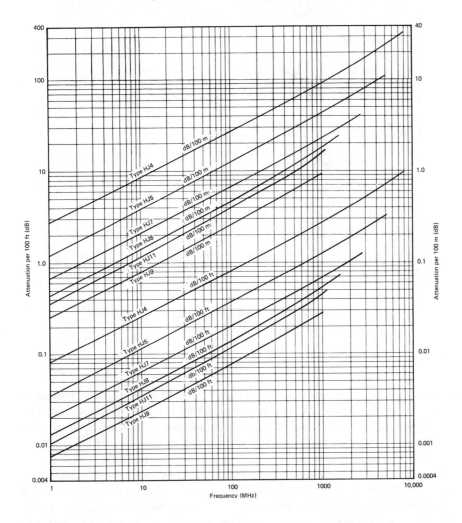

Figure 5.26 Attenuation per unit length for certain types of coaxial cable. Attenuation curves based on VSWR 1.0:1, 50-Ω, copper conductors. For 75-Ω cables multiply by 0.95. (*a*) Air dielectric. (*b*) Foam dielectric. (Courtesy Andrew Corporation).

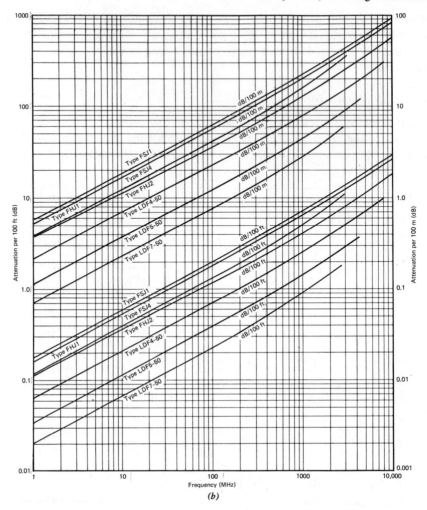

(b)

Figure 5.26 *(Continued)*

ing much of the reflected energy from high VSWR. Owing to the ferrite material with its associated permanent magnetic field, ferrite load isolators have a unidirectional property. Energy traveling toward the antenna is relatively unattenuated, whereas energy traveling back from the antenna undergoes fairly severe attenuation. The forward and reverse attenuations are on the order of 1 and 40 dB, respectively.

A waveguide circulator is used to couple two or three microwave radio equipments to a single antenna. The circulator shown in Figure 5.27 is a four-port device. It consists essentially of three basic waveguide sections combined into a single assembly. The center section is a ferrite nonreciprocal phase shifter. An external permanent magnet causes the ferrite material to exhibit phase shifting characteristics. Normally an antenna transmission line is connected to one arm and either three radio equipments are connected to the other three arms, or two equipments and a shorting

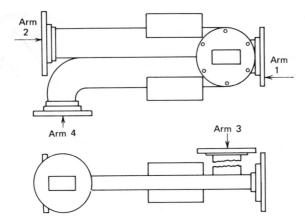

Figure 5.27   Waveguide circulator (four ports).

plate. Attenuation in a clockwise direction from arm to arm is low, on the order of 0.5 dB, whereas in the counterclockwise direction from arm to arm it is high, on the order of 20 dB.

A power splitter is a simple waveguide device that divides the power coming from or going to the antenna. A 3-dB power split divides power in half; such a device could be used, for instance, to radiate the power from a transmitter in two directions. A 20-dB power split, or 30 dB, has an output that serves to sample the power in a transmission line. Often such a device has directional properties (therefore called a directional coupler) and is used for VSWR measurements allowing measurement of forward and reverse power.

Magic tees or hybrid tees are waveguide devices used to connect several equipments to a common waveguide run.

PERISCOPIC ANTENNA SYSTEMS

Instead of a long waveguide run up a radiolink tower, a periscopic antenna system uses a short run to an antenna mounted near the tower, usually on the roof of the repeater building. A plane or semicurved reflector is mounted on the tower. The antenna is aimed upwards at the reflector, and the focused energy radiated by the antenna is reflected to the distant end.

A reflector will intercept all the main beam of radiated energy of a parabolic antenna if

- The reflector is situated in the near field.
- The cross-sectional area of the reflector is equal or slightly greater than the projected area of the antenna.

The further the reflector is moved from the near-field boundary, the larger the cross section of the beamwidth and the larger the size of the reflector required to properly reflect all the beam energy. Figure 5.28 gives some near-field boundaries.

Reflectors so used may give a gain or a loss in system (path) calculations, depend-

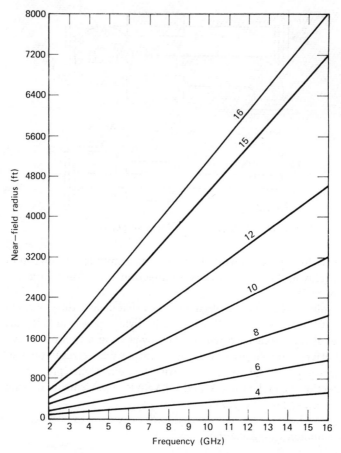

Figure 5.28 Near-field radius for parabolic antennas. Curves are for parabolic antennas with diameters in feet. $R = 2D^2/\lambda$, where D = dish diameter (ft) and λ = wavelength (ft); or $R = 0.002D^2F_{MHz}$. (Courtesy Collins Radio Group, Rockwell International Corporation).

ing on the operating frequency, the separation distance from the antenna to the reflector, and the size of the reflector. Figure 5.29 gives gain or loss in decibels for various reflector–antenna–frequency configurations. Often the curving is done after orientation on the distant end to improve energy focus.

5.6 LOADING OF A RADIOLINK SYSTEM

5.6.1 General

Noise on a radiolink system can essentially be broken down into IM noise and residual noise (basically thermal noise). The important fact is that the IM noise increases with load (i.e., increased traffic), and when a certain "break point" of load-handling capability is exceeded, IM noise becomes excessively high. On the other hand, residual noise is not affected by the amount of traffic or the traffic load of the system. FM radio overcomes much of the residual noise that appears in a radio system by distributing it over a wide radio bandwidth. Such an exchange of bandwidth for lower noise is also proportional to the signal level appearing at the system receiver. At periods of low signal (i.e., during a radio fade) thermal or residual noise becomes important.

5.6.2 Noise Power Ratio (NPR)

NPR has come into wide use in FM radiolink systems to quantify total noise. NPR is the decibel ratio of the noise level in a measuring channel with the baseband fully loaded with white noise to the level in that channel with all of the baseband noise loaded except the measuring channel.

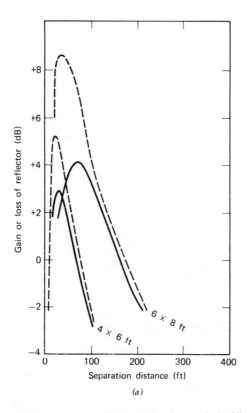

Figure 5.29   Periscopic reflector gain. Solid lines = flat reflector; dashed lines = curved reflector. (*a*) With 2-ft parabolic dish. (*b*) With 4-ft parabolic dish. (*c*) With 6-ft parabolic dish (*d*) With 8-ft parabolic dish.

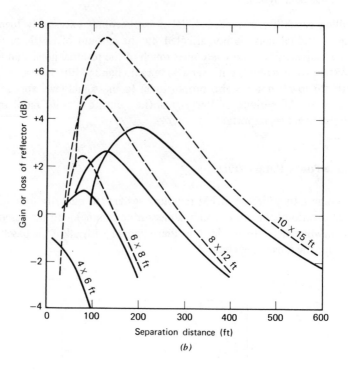

(b)

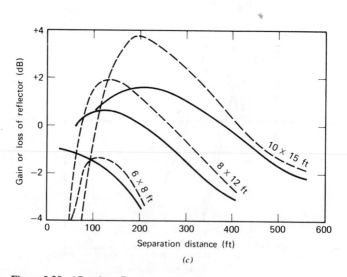

(c)

Figure 5.29 (Continued)

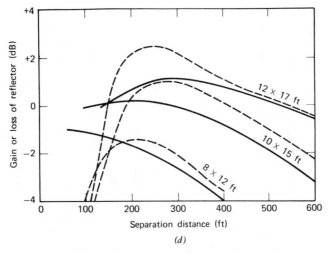

Figure 5.29 (*Continued*)

If the loading is removed from the system, all that remains in the measuring channel is thermal noise. Thermal noise, regarding such measurements as these, is often specified in terms of *baseband intrinsic noise ratio* (BINR). BINR is defined as the decibel ratio of noise in a test channel with the baseband fully loaded (i.e., bandstop filters switched out) to noise in the test channel with all noise loading removed. The difference between NPR and BINR therefore indicates the amount of noise present in the system due to IM distortion and crosstalk (Ref. 32).

5.6.3 Basic NPR Measurement

To measure NPR on a baseband-to-baseband basis, a radiolink transmitter is connected back to back with a receiver using proper waveguide attenuators to simulate real path conditions. A white noise generator is connected to the transmitter baseband input. The white noise generator produces a noise spectrum that approximates a spectrum produced in a multichannel (FDM) multiplex system. The output noise level from the generator is adjusted to a desired composite noise baseband power. A notched filter is then switched in to clear a narrow slot in the spectrum of the noise signal, and a noise analyzer is connected at the output of the system. The analyzer is used to measure the ratio of the noise in the illuminated (noise loaded) section of the baseband to the noise power in the cleared slot. The slot noise level is equivalent to the total noise (residual plus IM) that is present in the slot bandwidth. Slot bandwidths are the width of a standard voice channel and are taken at the upper, middle, and lower portions of the baseband. Table 5.7, taken from CCIR Rec. 399-4, gives standards for slot locations for various FDM channel configurations.

Table 5.7 Recommended NPR Measurement Frequencies

System Capacity (Channels)	Limits of Band Occupied by Telephone Channels (kHz)	Effective Cutoff Frequencies of Band-limiting Filters (kHz)		Frequencies of Available Measuring Channels (kHz)
		High Pass	Low Pass	
60	60–300	60 ± 1	300 ± 2	70 270
120	60–552	60 ± 1	552 ± 4	70 270 534
300	60–1300 / 64–1296	60 ± 1	1296 ± 8	70 270 534 1248
600	60–2540 / 64–2660	60 ± 1	2600 ± 20	70 270 534 1248 2438
960	60–4028 / 64–4024	60 ± 1	4100 ± 30	70 270 534 1248 2438 3886
900	316–4288	316 ± 5	4100 ± 30	534 1248 2438 3886
1260	60–5636 / 60–5564	60 ± 1	5600 ± 50	70 270 534 1248 2438 3886 5340
1200	316–5564	316 ± 5	5600 ± 50	534 1248 2438 3886 5340
1800	312–8120 / 312–8204 / 316–8204	316 ± 5	8160 ± 75	534 1248 2438 3886 5340 7600
2700	312–12 336 / 316–12 388 / 312–12 388	316 + 5	12 360 ± 100	534 1248 2438 3886 5340 7600 11 700

Source: CCIR Rec. 399-4, Geneva, 1978.

218

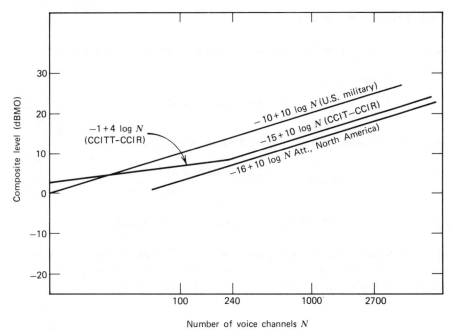

Figure 5.30 Curves for the loading of multichannel communication systems.

The composite noise power is taken from one of the following formulas for N telephone channels in an FDM (SSBSC) configuration (see Chapter 3). When $N <$ 240 channels,

$$P_{\mathrm{dBmo}} = -1 + 4 \log_{10} N \qquad (5.18a)$$

When $N > 240$ channels,

$$P_{\mathrm{dBmo}} = -15 + 10 \log_{10} N \qquad (5.18b)$$

For certain U.S. military systems with heavy data usage,

$$P_{\mathrm{dBmo}} = -10 + 10 \log_{10} N \qquad (5.18c)$$

Figure 5.30 shows curves for multichannel loading for all three applications above plus that for Holbrook and Dixon (Ref. 11). (See Chapter 3, particularly Section 3.4.4, for more information on loading FDM systems.)

$$\mathrm{NPR} = \text{composite noise power (dB)} - \text{noise power in the slot (dB)} \qquad (5.19)$$

A good guide for the NPR for high-capacity radiolink systems with a 300-channel capacity (1 hop) should be 55 dB; for 1200 voice channels, in excess of 50 dB.

5.6.4 Derived Signal-to-Noise Ratio

Given the NPR of a system, we can then compute the per-voice-channel test-tone-to-noise ratio:

$$\frac{S}{N} = \text{NPR} + \text{BWR} - \text{NLR} \tag{5.20}$$

where BWR = bandwidth ratio and NLR = noise load ratio.

$$\text{BWR} = 10 \log \frac{\text{occupied baseband bandwidth}}{\text{voice channel bandwidth}} \tag{5.21}$$

$$\text{NLR} = P \text{ [taken from load equation (5.18)]}$$

The signal-to-noise ratio as given is unweighted. For F1A or psophometric weighting add 3 dB (when reference is 1000 Hz). For an 800-Hz reference, add 2.5 dB. Figure 5.31 is an example of this application for a 240-channel system.

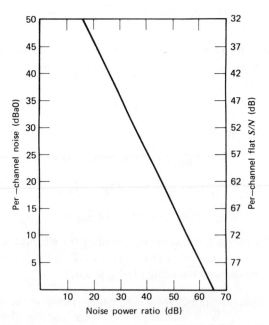

Figure 5.31 Conversion curve for noise power ratio to signal-to-noise ratio for a 240-channel system (Courtesy Marconi Instruments.)

5.6.5 Conversion of Signal-to-Noise Ratio to Channel Noise

Using the signal-to-noise ratio calculated above,

$$\text{Noise power (dBa0)} = 82 - \frac{S}{N} \tag{5.22a}$$

$$\text{Noise power (dBrnC)} = 88.5 - \frac{S}{N} \tag{5.22b}$$

$$\text{Noise power (pW)} = \frac{10^9}{\text{antilog}_{10}\, S/N} \tag{5.22c}$$

$$\text{Noise power (pWp)} = \frac{10^9 \times 0.56}{\text{antilog}_{10}\, S/N} \tag{5.22d}$$

The following approximations within about 0.2 dB may be made of the voice channel signal-to-noise ratio S/N when the NPR is given (Ref. 32).

For $N < 240$ channels,

$$\frac{S}{N}\text{(unweighted)} = \text{NPR} + 6\log_{10} N + 2.1 \text{ dB}$$

$$\frac{S}{N}\text{(psophometric)} = \text{NPR} + 6\log_{10} N + 4.6 \text{ dB}$$

$$\frac{S}{N}\text{(C-message)} = \text{NPR} + 6\log_{10} N + 3.6 \text{ dB}$$

For $N > 240$ channels,

$$\frac{S}{N}\text{(unweighted)} = \text{NPR} + 16.3 \text{ dB}$$

$$\frac{S}{N}\text{(psophometric)} = \text{NPR} + 18.8 \text{ dB}$$

$$\frac{S}{N}\text{(C-message)} = \text{NPR} + 17.8 \text{ dB}$$

5.7 OTHER TESTING TECHNIQUES

5.7.1 Out-of-Band Testing

NPR tests require taking a system out of traffic. Out-of-band testing permits a continuous monitor of IM and residual noise without interrupting normal traffic. In this arrangement normal traffic constitutes the loading of the system, and a noise receiver continuously monitors the noise in a channel that is approximately 10% outside the normal baseband. Bandstop filters at the out-of-band frequency slots are placed permanently in the circuit at the transmitter. The noise receiver indication then is due only to system IM and residual noise.

Often a three-channel pen recorder is used, where the first pen will show the variations of a system noise with time. The second pen is used to measure traffic level, and the third pen measures the AGC voltage of the radiolink receiver in question.

5.8 DETAILED PATH CALCULATIONS

Figure 5.32 is a path calculation sheet. It illustrates one way of performing detailed calculations. Section 5.4 discusses simple calculations. Before proceeding, a review of that section is advised.

Figure 5.32   Radiolink path calculations sheet.

Site . . . .   To site . . . .
Equipment type . . . .   Frequency (MHz) . . . . . .   Receiver noise figure . . dB
Type of traffic . . . . .   Highest baseband frequency . . . . .
Frequency deviation . . B_{IF} . . . .
Diversity or hot standby . . . .   Radio pilot . . . .
Reference path profile . . . .   True bearing A–B . . . .°
Path length (statute mi or km) . . .   True bearing B-A . . . .°

Step		Add or Subtract	Unit	Comments
A	Path loss (free space)	Minus	dB	
B	Connector losses (sum)	Minus	dB	Sum of transmit and receive
C	Circulator losses (sum)	Minus	dB	Sum of transmit and receive
D	Power split losses (if any)	Minus	dB	
E	Directional coupler losses (sum)	Minus	dB	Sum of transmit and receive
F	Transmission line losses (sum)	Minus	dB	Sum of transmit and receive
G	Other losses	Minus	dB	
H	Other losses	Minus	dB	
I	Sum of losses		dB	
J	Transmit power	Plus	dBW	Power output to flange
K	Transmit antenna gain	Plus	dB	
L	Transmit reflector (gain or loss)	Plus or minus	dB	Gains are +, losses are −
M	Receive antenna gain	Plus	dB	
N	Receive reflector (gain or loss)	Plus or minus	dB	
O	Sum of gains		dB	
P	Add step I + step O		dBW	Input level to receiver
Q	Receiver noise threshold		dBW	
R	FM improvement threshold (pWp)		dBW	R = 10 dB above Q
S	Fade margin to FM improvement		dB	Step P – step R
T	Unfaded signal-to-noise ratio		dB	

Figure 5.32 *(Continued)*

Step		Add or Subtract	Unit	Comments
U	Preemphasis improvement		dB	
V	Diversity improvement		dB	
W	Channel or video *S/N* (weighting)		dB	
X	Calculated path reliability		%	
Y	Channel noise contribution (radio)		pWp	(see CCIR Rec. 395-1)

Step A.   Enter path loss in decibels as calculated in Section 5.3.1.

Step B.   One rule of thumb is to add 0.06 dB of connector loss per joint. If we had 10 joints in a run, we would enter 0.6 dB. Sum for both transmit and receive.

Step C.   Insert circulator insertion loss. 0.5 dB is often a good figure. It is the sum of near-end transmit and far-end receive circulator insertion losses. See Section 5.5.7.

Step D.   Insert insertion loss in desired direction plus split attenuation. Sum transmit and receive.

Step E.   Insert insertion loss of directional couplers. Sum transmit and receive installations. See Section 5.5.7.

Step F.   Enter the sum of near-end transmit and far-end receive transmission line losses. See Section 5.5.7 and Figures 5.25 and 5.26. A good conversion figure to remember is 1 dB/100 ft = 3.28 dB/100 m.

Steps G, H.   Enter here other losses, if any. For instance, you may be called on to use a transition from rectangular to circular waveguide or from coaxial cable to waveguide. Enter those losses here for the transitions.

Step I.   Step *I* = sum of steps *A–H*.

Step J.   Refer to Section 5.5.3. Keep in mind that power outputs are usually to the flange of the output device. In other words, care should be taken in reading equipment manufacturers' specifications.

Step K.   Insert transmit antenna gain. Be sure the gain is given in dBi or decibels over an isotropic and not over a dipole. See Section 5.5.4 and Figure 5.20.

Step L.   If a reflector is used (i.e., a periscopic system), insert the gain as a + entry or the loss as a − entry. See Section 5.5.7.

Steps M, N.   The same as steps *K* and *L*, respectively, but for the receive side.

Step O.   Add the gains. If the reflector(s) give losses, add algebraically.

Step P.   The result of this addition (always add algebraically) will give the input to the receiver.

Step Q.   Calculate the receiver noise threshold. An explanation of this calculation is given in Section 5.4.2.

Step R.   Add −10 dB algebraically to the level in step *Q* and insert this figure. See Section 5.4.2.

Step S.   Add algebraically step *P* − step *R*. In other words, this is the number of decibels that the unfaded input level to the receiver, step *P*, is above step *R*, the FM improvement threshold, where the psophometrically weighted channel noise is in pWp. See Figure 5.34 and Section 5.6.4 and 5.6.5. The fade margin should be adjusted in accordance with Sections 5.4.3 and 5.9. A more direct approach is to adjust the figure as a first step. To economically achieve the required fade margin, we adjust steps *K*, *L*, *M*, and *N*. Assuming that the figure is too low, antenna sizes can be increased; losses can be reduced using a lower loss waveguide. Other steps which may be taken are to increase transmitter output by using a different transmitter, by using a receiver with an improved NF.

Figure 5.32 (*Continued*)

Step T.	Enter the unfaded signal-to-noise ratio. If we can assume a 1-for-1 increase (in decibels) of carrier-to-noise ratio for signal-to-noise ratio, once the signal input is above FM improvement threshold, then signal-to-noise ratio is approximately equal to carrier-to-noise ratio + 20 dB. A more accurate result can be found using information in Section 5.6.
Step U.	The preemphasis improvement, in decibels, can be taken from Figure 5.33 and is discussed in Section 5.5.2.
Step V.	The diversity improvement, in decibels, may be taken from information provided in Section 5.5.6 and Figure 5.24.
Step W.	The channel or video signal-to-noise ratio (unfaded) derives from steps *T-V* and is discussed in Section 5.6.
Step X.	Enter path reliability. This is usually taken to be the percentage of time that a path (hop) will be above FM improvement threshold. Refer to Section 5.10. Step *X* may have to be adjusted upwards when total line reliability is considered. If this is the case, step *S*, the fade margin, will have to be increased accordingly.
Step Y.	Enter the channel noise contribution (radio). See Section 5.6. This figure is added to the multiplex (if any is on the hop) noise for total hop noise contribution. Total link and system noise is covered in Section 5.18.

The following is an explanation of each entry or step in the path calculation process.

Enter the names of the two sites which the radiolink interconnects. Enter equipment type, operating frequency, and receiver NF.

Enter the type of traffic, such as FDM-SSBSC or video plus program channels. This will determine the type of preemphasis (see Section 5.5.2). Enter highest modulating baseband frequency. This will determine preemphasis network values and preemphasis improvement (step *U* and Figures 5.15 and 5.16). The noise "equalizing" of preemphasis adds 1–3 dB to the channel signal-to-noise ratio. This

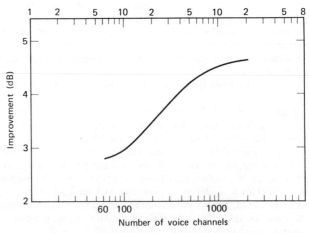

Figure 5.33 Preemphasis improvement in decibels for a given number of FDM voice channels.

improvement may be found using Figure 5.33. Enter the peak frequency deviation (see Section 5.5.3) and B_{IF} (see 5.4.2).

Enter whether or not diversity is used, or enter if hot standby is used (see Sections 5.5.6 and 5.12). Enter radio pilot frequency (Section 5.5.6).

For the record, and to properly associate this calculation sheet with its companion profile for the same link, enter identification of the related path profile. From the profile we can get the path length, which will be entered below. Enter true bearing of station A to station B and of station B to station A. This information will be obtained from the maps used while preparing the profile.

5.9   DETERMINATION OF FADE MARGIN (PART OF STEP 4, PARA 5.2)

Fading is a random increase in path loss during abnormal propagation conditions. During such conditions path loss may increase 10, 20, 30 dB or more for short periods. The objective of this subsection is to assist in setting a margin or overbuild in system design to minimize the effects of fading.

The factors involved in fading phenomena are many and complex. They have been discussed in some detail in Section 5.4.3. The principal types of fading below 10 GHz is multipath fading. CCIR Rep. 784 (Ref. 3) states that

multipath effects due to atmosphere increase slowly with frequency, but much more rapidly with path length (the deep fading probability follows an approximate law $f \cdot d^{3.5}$), where f is the carrier frequency and d is distance.

Determining the fade margin without resorting to live path testing is not an easy matter for the radio system engineer. One approach is to simply assume what is often considered the worst fading condition on a single radiolink hop. This is the familiar Rayleigh fading which can be summarized as follows:

Single-Hop Propagation Reliability (%)	Fade Margin (dB)
90	8
99	18
99.9	28
99.99	38
99.999	48

For other propagation reliability values simple interpolation can be used. For instance, for a hop where a 99.95% reliability is desired, the fade margin required for Rayleigh fading would be 33 dB.

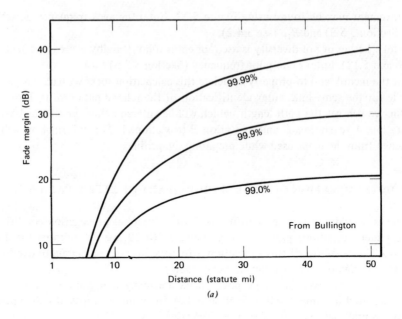

(a)

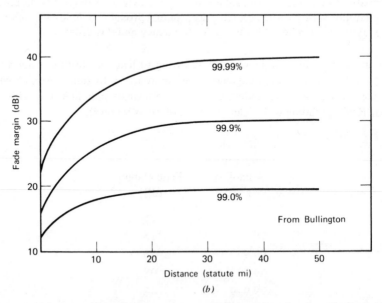

(b)

Figure 5.34 Fade margin related to distance or length of hop for three frequency bands. (*a*) 2 GHz. (*b*) 6 GHz. (*c*) 11.2 GHz (Ref. 20).

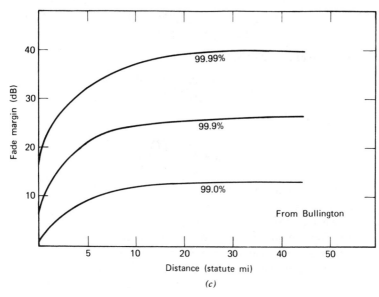

Figure 5.34 (*Continued*)

This expresses worst case fading and does not take into account the variables mentioned in the CCIR report, namely, hop length (distance) and frequency. Adding margin costs money, and it would be desirable to reduce the margin yet maintain hop reliability. Here we can resort to the derived Bullington curves of Figure 5.34 which give required fade margins for several reliabilities for the frequency bands more commonly used on radiolinks.

Further refinements have been made in the methodology of estimating fade margins. Experience in the design of many radiolink systems has shown that the incidence of multipath fading varies not only as a function of path length and frequency but also as a function of climate and terrain conditions. It has been found that hops over dry, windy, mountainous areas are the most favorable, displaying a low incidence of fading. Worst fading conditions usually occur in coastal areas that are hot and humid, and inland temperate regions are somewhere in between the extremes. Flat terrain along a radiolink path tends to increase the probability of fading, while irregular hilly terrain, especially with vegetation, tends toward lower incidence of fading (e.g., depth and frequency of events).

W. T. Barnett and A. Vigants of Bell Telephone Laboratories (Ref. 2) have developed an empirical method to further refine the estimation of fade margins. The percentage we have been using really expresses an availability A of a radiolink path or hop, sometimes called *time availability*. Subtracting the percentage from 1.000 gives the path unavailability u. As a percentage, $A = 100\ (1 - u)$. Following Barnett's argument (in general), let U_{ndp} be the nondiversity annual outage probability and r the fade occurrence factor,

$$r = \frac{\text{actual fade probability}}{\text{Rayleigh fade probability}}$$

If F is the fade margin in decibels, then

$$r = \frac{\text{actual fade probability}}{10^{-F/10}}$$

For the worst month,

$$r_m = a \times 10^{-5} \left(\frac{f}{4}\right) D^3$$

where D = path length (statute mi)
  f = frequency (GHz)
  F = fade margin (dB)
  a = 4 for very smooth terrain, over water, flat desert
    = 1 for average terrain with some roughness
    = 0.25 for mountains, very rough or very dry terrain

Over a year,

$$r_{yr} = br_m$$

where b = 0.5 for hot, humid coastal areas
    = 0.25 for normal, interior temperate or subarctic areas
    = 0.125 for mountainous or very dry but nonreflective areas

$$U_{ndp} = r_{yr}(10^{-F/10}) = br_m(10^{-F/10})$$
$$= 2.5 \, abfD^3(10^{-F/10})(10^{-6})$$

Example. Given a 25-mi path with average terrain but with some roughness in an inland temperate climate, operating at a frequency of 6.7 GHz with a desired propagation reliability of 99.95%. What fade margin should be assigned to the link?

$$U_{ndp} = 1 - \text{percentage}$$
$$= 1 - 0.9995$$
$$= 0.0005$$

Then

$$0.0005 = 2.5 \, abfD^3(10^{-F/10})(10^{-6})$$
$$= 2.5(1)(0.25)(6.7) \, 25^3 \, (10^{-F/10})(10^{-6})$$
$$F = 21 \text{ dB}$$

It is good design practice to add a miscellaneous loss margin to the derived fade margin by whatever method the derivation has been made. This additional margin is required to account for minor antenna misalignment and system gain degradation (e.g., waveguide corrosion, transmitter output and receiver NF degradations due to aging). The Defense Communications Agency of the U.S. Department of Defense recommends 6 dB (Ref. 35) for this additional margin.*

Diversity also reduces fade margin requirements as established above; it tends to reduce the depth of fades. Figure 5.22 shows how fade margins may be reduced with diversity. For instance, assume a Rayleigh distribution (random fading) with a 35-dB fade margin with no diversity. Then using frequency diversity with a frequency separation of 2%, only a 23.5-dB fade margin would be required to maintain the same path reliability (i.e., the percentage of time the level is exceeded).

5.10 SYSTEM AND LINK RELIABILITY

A planning engineer is interested in system reliability. He is more interested in setting a limit, expressed as a percentage, that a telephone conversation will be degraded by noise or will drop out entirely. He may set a reliability of 99.9% on a link between points X and Y. Suppose that this link is composed of 10 radiolink hops. What then will be the per-hop reliability? We consider here only propagation outages, not equipment outages (i.e., failures).

The method described here to assign a per-hop reliability percentage assumes that fades on separate hops will not be correlated. Thus the outage time of each individual hop will add directly to the total system outage time. This is the worst condition because there usually is some correlation of fades on different paths.

Example 1. With a system reliability of 99.9% we will have 0.1% outage per year. This corresponds to 0.1% for 10 radiolinks in tandem or 0.01% per hop. Thus the per-hop reliability required is 99.99% to maintain a system reliability of 99.9%.

Example 2. Assume a system reliability of 99.99%. The system consists of seven hops in tandem. 99.99% is equivalent to 0.01% outage, which corresponds to a per-hop outage of $0.01/7 = 0.00143\%$, which is equivalent to 99.9986% reliability per hop.

CCIR in Rec. 557 defines its availability objective for radio relay systems over a 2500-km hypothetical reference circuit (HRC) as 99.7% using a time frame of 1 year or more. The resulting unavailability (0.3%) is made up of three components: (1) outage due to power failures, (2) outage due to equipment failure, and (3) outage due to propagation. It is reasonable to allow about 50% of the outage time over a year for power failures and equipment failures and about 50% for propagation outage when a large number of radiolinks are analyzed and an average is taken (see CCIR Ref. 445-2). Then considering propagation alone, the system should have an availability (reliability) of 99.85% apportioned across the 2500-km system. This

*For more information on fading consult (Ref. 36).

provides an initial guideline to establish a per-hop propagation reliability for a particular system in question.

5.11 RAINFALL AND OTHER PRECIPITATION ATTENUATION

For radiolink systems rainfall and other precipitation attenuation are not significant below 10 GHz. The cause of attenuation of microwave energy passing through rain is attributed to both absorption and scattering. Absorption attenuation is predominant in clouds and fogs. Scattering attenuation is predominant in rain. Worst case cloud and fog attenuation is 1/10 that of rainfall attenuation below 15 GHz. Fog, cloud, and water vapor attenuation are covered in Chapter 10.

Rainfall attenuation is a function of rainfall density, usually given in millimeters per hour, and frequency. Figure 5.35 is a curve giving attenuation in decibels per mile for typical rainfall densities as a function of frequency.

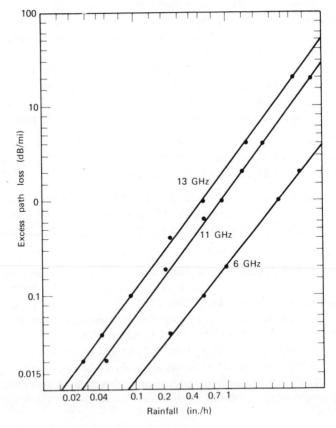

Figure 5.35 Excess path loss due to rainfall (Ref. 37).

We are dealing here not with the total quantity of rain, but with periods of heavy intensity. One short cloudburst can cause more outage than weeks of light drizzle.

For the continental United States extensive statistical data are available on rainfall.* The country has been divided into zones, and the incidence of rainfall and cloudburst activity is given statistically. This information is mandatory for good planning of radiolinks in the 11–13 and 15-GHz bands. In certain areas we can expect an increase of free-space attenuation of 3 dB/mi for that part of a path affected by the downfall or cloudburst. For example, in the Florida panhandle, by one set of standards, radiolink sections (hops) have been limited to 8 mi (12.8 km) due to rain attenuation. In this particular case the system was designed for CATV operation in the 12.9-GHz band. A more in-depth approach to the rainfall problem is presented in Chapter 10.

5.12 HOT-STANDBY OPERATION

Radiolinks often provide transport of multichannel telephone service and/or point-to-point broadcast television on high-priority backbone routes. A high order of route reliability is essential. Route reliability depends on path reliability (propagation) and equipment/system reliability. The first item, path reliability, has been discussed above; namely, a systematic overbuild is used to overcome fades or to minimize their effects.

The second item, equipment/system reliability, deals with making the equipment and system as a whole more reliable by minimizing system downtime to maximize equipment availability. One way to do this most effectively is to provide a parallel system. We did this in frequency diversity. There all active equipment operated in parallel with two distinct systems carrying the same traffic. This is expensive, but necessary if the reliability is desired. Here we speak about route reliability.

*Available from U.S. Weather Bureau and U.S. Government Printing Office.

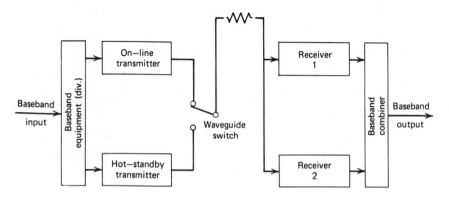

Figure 5.36 One-for-one hot-standby configuration (using a combiner).

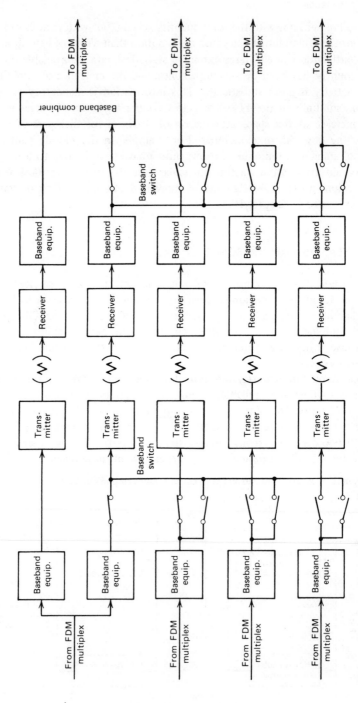

Figure 5.37 One-for-n hot-standby configuration. (A 1-for-4 configuration is shown.)

232

Often the additional frequency assignments to permit operation in frequency diversity are not available. The equivalent equipment reliability may be achieved by the use of a hot-standby configuration. As the expression indicates, hot standby is the provision of a parallel equipment configuration such that it can be switched in nearly instantaneously on the failure of operating equipment. The switchover takes place usually on the order of microseconds, often in less than 10 μs. The changeover of a transmitter and/or receiver line can be brought about by a change, over a preset amount, in one of the following values. For a transmitter,

- Frequency
- RF power
- Demodulated baseband (radio) pilot level

and for a receiver,

- AGC voltage
- Squelch
- Incoming pilot level

Hot-standby protection systems provide sensing and logic circuitry for the control of waveguide switches (or coaxial cable switches) on transmitters and output signals on receivers. The use of a combiner is common on the receiver side with both receivers on line at once.

There are two approaches to the use of standby equipment. These are called one-for-one and one-for-n. One-for-one provides one full line of standby equipment for each operational system (see Figure 5.36). One-for-n provides only one full line of equipment for n operational lines of equipment, $n \geqslant 2$. (see Figure 5.37). One-for-one is more expensive but provides a higher order of reliability. Its switching system is comparatively simple. One-for-n is cheaper, with only one line of spare hot-standby equipment for several operational lines. It is less reliable (i.e., suppose there were two equipment failures in n), and switching is considerably more complex.

5.13 RADIOLINK REPEATERS

Up to this point the only radiolink repeaters discussed have been baseband repeaters. Such a repeater fully demodulates the incoming RF signal to baseband. In the most simple configuration, this demodulated baseband is used to modulate the transmitter used in the next link section. This type of repeater also lends itself to dropping and inserting voice channels, groups, and supergroups. It may also be desirable to demultiplex the entire baseband down to voice channel for switching, and insert and drop a new arrangement of voice channels for multiplexing. The new baseband would then modulate the transmitter of the next link section. A simplified block diagram of a baseband repeater is shown in Figure 5.38a.

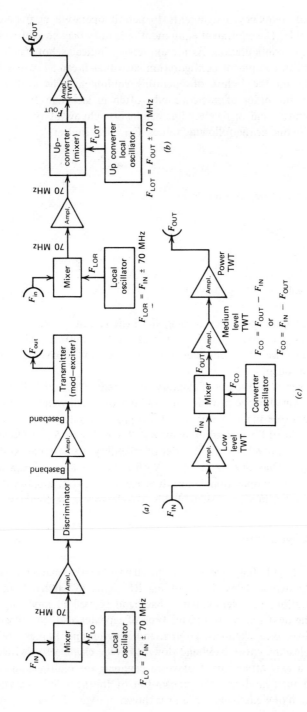

Figure 5.38 Radiolink repeaters. (*a*) Baseband repeater. (*b*) IF heterodyne repeater. (*c*) RF heterodyne repeaters. F_{IN} = input frequency to receiver; F_{OUT} = output frequency of the transmitter; F_{CO} = output frequency of the converter local oscillator; F_{LOR} = frequency of the receiver local oscillator; F_{LOT} = frequency of the transmitter local oscillator; F_{LO} = frequency of the local oscillator; TWT = traveling wave tube.

Two other types of repeaters are also available: the IF heterodyne repeater (Figure 5.38b) and the RF heterodyne repeater (Figure 5.38c). The IF repeater is attractive for use on long backbone systems where noise and/or differential phase and gain should be minimized.

Generally a system with fewer modulation–demodulation stages or steps is less noisy. The IF repeater eliminates two modulation steps. It simply translates the incoming signal to IF with the appropriate local oscillator and a mixer, amplifies the derived IF, and then up-converts it to a new RF frequency. The up-converted frequency may then be amplified by a TWT amplifier.

An RF heterodyne repeater is shown in Figure 5.38c. With this type of repeater amplification is carried out directly at RF frequencies. The incoming signal is amplified, up or down converted and amplified again, usually by a TWT, and then reradiated. RF repeaters are troublesome in their design in such things as sufficient selectivity, limiting and automatic gain control, and methods to correct envelope delay. However, some RF repeaters are now available, particularly for operation below 6 GHz.

5.14 FREQUENCY PLANNING (STEP 7)

5.14.1 General

To derive maximum performance from a radiolink system, the systems engineer must set out a frequency usage plan that may or may not have to be approved by the local administration.

The problem has many aspects. First, the useful RF spectrum is limited, from above direct current to about 150 GHz. The frequency range of discussion for radiolinks is essentially from the VHF band at 150 MHz (overlapping) to the millimeter region of 15 GHz. Second, the spectrum from 150 MHz to 15 GHz must be shared with other services such as radar, navigational aids, research (i.e., space), meterological, and broadcast. For point-to-point communications, we are limited by international agreement to those bands shown in Table 5.1.

Although many of the allocated bands are wide, some up to 500 MHz in width, FM by its very nature is a wideband form of emission. It is not uncommon to have an RF bandwidth B_{RF} = 25 or 30 MHz for just one emission. Guard bands must also be provided. These are a function of the frequency drift of transmitters as well as "splatter" or out-of-band emission, which in some areas is not well specified.

Occupied bandwidth has been specified in Section 5.4.2, Carson's rule. This same rule is followed by the U.S. regulatory agency, the Federal Communications Commission (FCC).

5.14.2 Spurious Emission

The American National Standards Institute, in its memorandum to the U.S. Defense Communication Agency, quotes the following at the "RF interface" (Ref. 14):

The spurious emissions, referred to the RF carrier frequency, shall not exceed the following values, excluding oscillator leak:

$$F_0 \pm 70 \text{ MHz}, -24 \text{ dBm}$$

$$F_0 \pm 140 \text{ MHz}, -54 \text{ dBm}$$

Harmonics of carrier frequency, -24 dBm

Local Oscillator Leakage
The amount of local oscillator power, as measured at the branching network/ transmission line point, shall not exceed -60 dBm.

Bandwidth
Each radio channel bandpass filter shall have a bandwidth not less than 40 MHz (at the 3-dB points relative to band center), and the transmission loss shall not vary by more than 0.1 dB over the central 10 MHz.

5.14.3 Radio Frequency Interference (RFI)

On planning a new radiolink system or on adding RF carriers to an existing installation, careful consideration must be given to the RFI of the existing (or planned) emitters in the area. Usually the governmental authorizing agency has information on these and their stated radiation limits. Typical limits have been given above. Equally important as those limits is that of antenna directivity and side lobe radiation. Not only must the radiation of other emitters be examined from this point of view, but also the capability of the planned antenna to reject unwanted signals. The radiation pattern of all licensed emitters should be known. Convert the lobe level in the direction of the planned installation to EIRP in dBW. This should be done for all interference candidates within interference frequency range. For each emitter's EIRP, compute a path loss to the planned installation to determine interference. Such a study could well affect a frequency plan or antenna design.

Nonlicensed emitters should also be looked into. Many such emitters may be classified as industrial noise sources such as heating devices, electronic ovens, electric motors, or unwanted radiation from your own and other microwave installations (i.e., radar harmonics). In the 6-GHz band a coordination contour should be carried out to verify interference from earth stations (see CCIR Rep. 448 and Rec. 359-4). For a general discussion on the techniques for calculating interference noise in radiolink systems see CCIR Rep. 388-3.

Another RFI consideration is the interference that can be caused by radiolink transmitting systems with the fixed satellite service in those frequency bands that are shared with that service. Guidance on this potential problem is provided by CCIR Rec. 406-4. Some of the major points of the CCIR recommendation are summarized below:

1. In those shared bands between 1 and 10 GHz the maximum EIRP of a radio-link transmitting system should not exceed +55 dBW, and the input power to the transmitting antenna should not exceed +13 dBW. As far as practicable, sites for

new radiolink transmitting stations where the EIRP exceeds +35 dBW should be selected so that the direction of maximum radiation of any antenna will be at least 2° away from a geostationary satellite orbit. In special situations the EIRP should not exceed +47 dBW for any antenna directed within 0.5° of a geostationary orbit and +47 to +55 dBW on a linear dB scale (8 dB per angular degree) for any antenna beam directed between 0.5° and 1.5° of a geostationary satellite orbit.

2. In those shared bands between 10 and 15 GHz the maximum EIRP of a radiolink transmitting system should not exceed +55 dBW, and the input power to the antenna system should not exceed +10 dBW, and as far as practicable, transmitting stations where the EIRP exceeds +45 dBW should be selected so that the direction of maximum radiation of any antenna will be at least 1.5° away from the geostationary satellite orbit.

5.14.4 Overshoot

Overshoot interference may occur when radiolink hops in tandem are in a straight line. Consider stations A, B, C, and D in a straight line, or that a straight line on a map drawn between A and C also passes through B and D. Link $A-B$ has frequency F_1 from A to B. F_1 is reused in the direction C to D. Care must be taken that some of the emission F_1 on the $A-B$ hop does not spill into the receiver at D. Reuse may even occur on an A, B, and C combination, so F_1 at A to B may spill into a receiver at C tuned to F_1. This can be avoided, provided stations are removed from the straight line. In this case the station at B should be moved to the north of a line A to C, for example.

5.14.5 Transmit–Receive Separation

If a transmitter and receiver are operated on the same frequency at a radiolink station, the loss between them must be at least 120 dB. One way to assure the 120 figure is to place all "go" channels in one-half of an assigned band and all "return" channels in the other. The terms "go" and "return" are used to distinguish between the two directions of transmission.

5.14.6 Basis of Frequency Assignment

"Go" and "return" channels are assigned as in the preceding section. For adjacent RF channels in the same half of the band, horizontal and vertical polarizations are used alternately. To carry this out we may assign, as an example, horizontal polarization H to the odd-numbered channels in both directions on a given section and vertical polarization V to the even-numbered channels. The order of isolation between polarizations is on the order of 26 dB, but often specified as 35 dB or more.

Table 5.8   Frequency Frogging at Radiolink Repeaters

	Minimum Separation (MHz)	
Number of Voice Channels	2000–4000 MHz	6000–8000 MHz
120 or less	120	161
300 or more	213	252

Source: *USAF Pub. TO 31R5-1-9*.

In order to prevent interference between antennas at repeaters between receivers on one side and transmitters in the same chain on the other side of the station, each channel shall be shifted in frequency (called frequency "frogging") as it passes through the repeater station. Recommended shifts of frequency as shown in Table 5.8.

5.14.7   IF Interference

Care must be taken when assigning frequencies of transmitter and receiver local oscillators, as to whether these are placed above or below the desired operating frequency. Avoid frequencies that emit the received channel frequency F_R and check those combinations of $F_R \pm 70$ MHz for equipment with 70-MHz IFs or $F_R \pm 140$ MHz when 140-MHz IFs are used. Often plots of all station frequencies are made on graph paper to assure that forbidden combinations do not exist. When close frequency stacking is desired, and/or nonstandard IFs are to be used, the system designer must establish rules as to minimum adjacent channel spacing and receive–transmit channel spacing. A listing of CCIR recommendations regarding channel spacing is given in the next section.

5.14.8   CCIR Recommendations

Regarding frequency assignments, Table 5.9 is presented as a guide to the relevant CCIR recommendations for radiolink systems.

5.15   ALARM AND SUPERVISORY SYSTEMS

5.15.1   General

Many radiolink sites are unattended, and many others remain unattended for weeks or months. To assure improved system reliability, it is desirable to know the status of unattended sites at a central, manned location. This is accomplished by means of a fault-reporting system. Such fault alarms are called status reports. The radiolink sites originating status reports are defined as reporting locations. A site which

TABLE 5.9 CCIR Recommendations for Preferred RF Channel Arrangements for Radio-Relay Systems Used for International Connections[a, b]

Recommendation	Maximum Capacity in Analog Operation of Each Radio Carrier (telephone channels or the equivalent)	Capacity of Each Digital Channel[c]	Preferred center Frequency f_0[d] (MHz)	Width of RF Band Occupied (MHz)
283-3	60/120/300/960[e]	Low, medium	1,808	200
			2,000	200
			2,203	200
			2,586	200
			1,903	400
382-2	600/1800		2,101	400
			4,003.5[f]	400[f]
383-1	600/1800		6,175	500
384-2	1260/2700		6,770	680
385-1	60/120/300		7,575	300
386-1	300/960[g]		8,350[g]	300[g]
387-3	600/1800	Low, medium	11,200	1000
497-1	960	Medium	12,996[h]	500

Source: CCIR Opinion 14-3. (Courtesy ITU–CCIR.)

[a]The recommendations referred to apply to LOS and near LOS systems. For transhorizon systems it has not yet been possible to formulate preferred channel arrangements, but the attention of the Administrative Radio Conference is drawn to CCIR Rec. 388 and Rep. 286.

[b]Attention should also be drawn to CCIR Rec. 389-2, Study Programme 4A-1/9, and to CCIR Rep. 284-1.

[c]The definition of the terms low- and medium-capacity digital systems is given in CCIR Rep. 378-3.

[d]Other center frequencies may be used by agreement between the administrations concerned.

[e]The 960-channel capacity can only be used with the center frequency 2586 MHz.

[f]In some countries, mostly in a large part of region 2 and in certain other areas, a reference frequency f_r = 3700 MHz is used at the lower edge of a band 500 MHz wide (see annex to CCIR Rec. 382-2).

[g]In some countries a maximum capacity of 1800 telephone channels or the equivalent on each RF carrier may be used with a preferred center frequency of 8000 MHz. The width of the RF band occupied in 500 MHz (see CCIR Rec. 386-1, section 7, and annex).

[h]Reference frequency.

receives and displays such reports is defined as a supervisory location. This is the standard terminology of the industry. Normally supervisory locations are those terminals that terminate a radiolink section. Status reports may also be required to be extended over a wire circuit to a remote location.

5.15.2 Monitored Functions

The following functions at a radiolink site, which is a reporting location (unmanned), may be desirable for status reports:

Equipment Alarms

Loss of receive signal

Loss of pilot (receiver)

High noise (receiver)

Power supply failure

Loss of modulating signal

TWT overcurrent

Low transmitter output

Off-frequency operation

Hot-standby actuation

Site Alarms

Illegal entry

Commercial power failure

Low fuel supply

Standby power unit failure

Standby power unit on line

Tower light status

Often alarms are categorized into "major" (urgent) and "minor" (nonurgent) in accordance with their importance. A major alarm may be audible as well as visible on the status panel. A minor alarm may then show only as an indication on the status panel.

The design intent is to make all faults binary: a light is either on or off, the receive signal level has dropped below -100 dBW or the noise is above so many picowatts; the power is below -3 dB of its proper level, and so on. By keeping all functions binary, using relay closure (or open) or equivalent transistor circuitry, the job of coding alarms for transmission is much easier. Thus all alarms are of a "go, no-go" nature.

5.15.3 Transmission of Fault Information

Common practice today is to transmit fault information in a voice channel associated with the service channel groupings of voice channels (see Section 5.5.1). Binary information is transmitted by VF telegraph equipment using FSK or tone-on, tone-off, as described in Section 8.13. Depending on the system used, 16, 18, or 24 tone channels may occupy the voice channel assigned. A tone channel is assigned to each reporting location (i.e., each reporting location will have a tone transmitter operating on the specific tone frequency assigned to it). The supervisory location will have a tone receiver for each reporting (unmanned site under its supervision) location.

At each reporting location the points described in Section 5.15.2 are scanned every so many seconds (between 2 and 85) and the information from each monitor or scan point is time division multiplexed in a simple code. The pulse output from each tone receiver at the supervisory location represents a series of reporting information on each remote unmanned site. The coded sequence is demultiplexed and displayed on a status panel.

A simpler method is the tone-on, tone-off method. Here the presence of a tone indicates a fault in a particular time slot; in the other method, it is indicated by the absence of a tone. A device called a fault-interrupter panel is used to code the faults so that different faults may be reported on the same tone frequency.

5.15.4 Remote Control

Through a similar system operating in the opposite direction the supervisory station can control certain functions at reporting locations via a VFTG tone link (see Chapter 8), with a tone frequency assigned to each separate reporting location. In this case the VFTG is almost always FSK. If only one function is to be controlled, such as turning on tower lights, then a mark condition would represent lights on, and a space, lights off. If more than one condition is to be controlled, then coded sequences are used to energize or deenergize the proper function at the remote reporting location.

5.16 ANTENNA TOWERS AND MASTS

5.16.1 General

Two types of towers are used for radiolink systems: guyed and self-supporting. However, other natural or man-made structures should also be considered or at least taken advantage of. Radiolink engineers should consider mountains, hills, and ridges so that tower heights may be reduced. They should also consider office buildings, hotels, grain elevators, high-rise apartment houses, and other steel structures (e.g., the sharing of a TV broadcast tower) for direct antenna mounting. For tower heights of 30-60 ft, wooden masts are often used.

One of the most desirable construction materials for a tower is hot dipped galvanized steel. Guyed towers are usually preferred because of overall economy and versatility. Although guyed towers have the advantage that they can be placed closer to a shelter or building than self-supporting types, the fact that they need a larger site may be a disadvantage where land values are high. The larger site is needed because additional space is required for installing guy anchors. Table 5.10 shows approximate land areas needed for several tower heights.

Tower foundations should be reinforced concrete with anchor bolts firmly embedded. Economy or cost versus height trade-offs usually limit tower heights to no more than 300 ft (188 m). Soil bearing pressure is a major consideration in tower construction. Increasing the foundation area increases soil bearing capability or equivalent design pressure. Wind loading under no-ice (i.e., normal) conditions is usually taken as 30 lb/ft^2 for flat surfaces. A design guide (EIA RS-222-B) indicates that standard tower foundations and anchors for self-supporting and guyed towers should be designed for a soil pressure of 4000 lb/ft^2 acting normal to any bearing area under specified loading.

Table 5.10 Minimum Land Area Required for Guyed Towers

Tower Height (ft)	Area Required [a] (ft)				
	80% Guyed	75% Guyed	70% Guyed	65% Guyed	60% Guyed
60	87 × 100	83 × 96	78 × 90	74 × 86	69 × 80
80	111 × 128	105 × 122	99 × 114	93 × 108	87 × 102
100	135 × 156	128 × 148	120 × 140	113 × 130	105 × 122
120	159 × 184	150 × 174	141 × 164	132 × 154	123 × 142
140	183 × 212	178 × 200	162 × 188	152 × 176	141 × 164
160	207 × 240	195 × 226	183 × 212	171 × 198	159 × 184
180	231 × 268	218 × 252	204 × 236	191 × 220	177 × 204
200	255 × 296	240 × 278	225 × 260	210 × 244	195 × 226
210	267 × 304	252 × 291	236 × 272	220 × 264	204 × 236
220	279 × 322	263 × 304	246 × 284	230 × 266	213 × 246
240	303 × 350	285 × 330	267 × 308	249 × 288	231 × 268
250	315 × 364	296 × 342	278 × 320	254 × 282	240 × 277
260	327 × 378	308 × 356	288 × 334	269 × 310	249 × 288
280	351 × 406	330 × 382	309 × 358	288 × 332	267 × 308
300	375 × 434	353 × 408	330 × 382	308 × 356	285 × 330
320	399 × 462	375 × 434	351 × 406	327 × 376	303 × 350
340	423 × 488	398 × 460	372 × 430	347 × 400	321 × 372
350	435 × 502	409 × 472	383 × 442	356 × 411	330 × 381
360	447 × 516	420 × 486	393 × 454	366 × 424	339 × 392
380	471 × 544	443 × 512	414 × 478	386 × 446	357 × 412
400	495 × 572	465 × 536	425 × 502	405 × 468	375 × 434
420	519 × 599	488 × 563	456 × 527	425 × 490	393 × 454
440	543 × 627	510 × 589	477 × 551	444 × 513	411 × 475

Tower Height (ft)	Area Required (acre)				
	80% Guyed	75% Guyed	70% Guyed	65% Guyed	60% Guyed
60	0.23	0.21	0.19	0.17	0.15
80	0.38	0.34	0.30	0.26	0.23
100	0.56	0.50	0.44	0.39	0.34
120	0.77	0.69	0.61	0.53	0.46
140	1.03	0.91	0.80	0.70	0.61
160	1.31	1.16	1.03	0.90	0.77
180	1.63	1.45	1.27	1.11	0.96
200	1.99	1.76	1.55	1.35	1.16
210	2.18	1.93	1.70	1.48	1.27
220	2.38	2.11	1.85	1.61	1.39
240	2.81	2.49	2.18	1.90	1.63
250	3.04	2.69	2.36	2.05	1.76
260	3.27	2.89	2.54	2.21	1.90

Table 5.10   (*Continued*)

Tower Height (ft)	Area Required (acre)				
	80% Guyed	75% Guyed	70% Guyed	75% Guyed	60% Guyed
280	3.77	3.33	2.92	2.54	2.18
300	4.30	3.80	3.33	2.89	2.49
320	4.87	4.30	3.77	3.27	2.81
340	5.48	4.84	4.24	3.65	3.15
350	5.79	5.11	4.48	4.88	3.33
360	6.12	5.40	4.73	4.10	3.52
380	6.79	5.99	5.25	4.55	3.90
400	7.50	6.62	5.79	5.02	4.30
420	8.24	7.27	6.36	5.52	4.73
440	9.03	7.96	6.96	6.03	5.17

[a]Preferred area is a square using the larger dimension of minimum area. This will permit orienting tower in any desired position.

5.16.2   Tower Twist and Sway

As any other structure, a radiolink tower tends to twist and sway due to wind loads and other natural forces. Considering the narrow beamwidths referred to in Section 5.5.4 (Table 5.5), with only a little imagination we can see that only a very small deflection of a tower or antenna will cause a radio ray beam to fall out of the reflection face of an antenna on the receive side or move the beam out on the far-end transmit side of a link.

Twist and sway, therefore, must be limited. Table 5.11 sets certain limits. The table has been taken from EIA RS-222B. From the table we can see that angular deflection and tower movement are functions of wind velocity. It should also be noted that the larger the antenna, the smaller the beamwidth, besides the fact that the sail area is larger. Thus the larger the antenna (and the higher the frequency of operation), the more we must limit the deflection.

To reduce twist and sway, tower rigidity must be improved. One generality we can make is that towers that are designed to meet required wind load or ice load specifications are sufficiently rigid to meet twist and sway tolerances. One way to increase rigidity is to increase the number of guys, particularly at the top of the tower. This is often done by doubling the number of guys from three to six.

5.17   PLANE REFLECTORS AS PASSIVE REPEATERS

A plane reflector as a passive repeater offers some unique advantages. Suppose that we wish to provide multichannel telephone service to a town in a valley, and

Table 5.11 Nominal Twist and Sway Values for Microwave Tower–Antenna–Reflector Systems[a]

A Total Beamwidth of Antenna or Passive Reflector between Half-Power Points (°)	Tower Mounted Antenna		Tower Mounted Passive Reflector		
	B Limits of Movement of Antenna Beam with Respect to Tower (±°)	C Limits of Tower Twist or Sway at Antenna Mounting Point (±°)	D Limits of Movements of Passive Reflector with Respect to Tower (±°)	E Limits of Tower Twist at Passive Reflector Mounting Point (±°)	F Limits of Tower Sway at Passive Reflector Mounting Point (±°)
14	0.75	4.5	0.2	4.5	4.5
13	0.75	4.5	0.2	4.5	4.3
12	0.75	4.5	0.2	4.5	3.9
11	0.75	4.5	0.2	4.5	4.6
10	0.75	4.5	0.2	4.5	3.3
9	0.75	4.5	0.2	4.5	3.9
8	0.75	4.2	0.2	4.5	2.6
7	0.6	4.1	0.2	4.5	2.3
6	0.5	4.0	0.2	4.3	2.1
5	0.4	3.4	0.2	3.7	1.8
4	0.3	3.1	0.2	3.3	1.6
3.5	0.3	2.9	0.2	2.9	1.4
3.0	0.3	2.3	0.1	2.5	1.2
2.5	0.2	1.9	0.1	2.1	1.0
2.0	0.2	1.5	0.1	1.7	0.9
1.5	0.2	1.1	0.1	1.2	0.6

Table 5.11 (*Continued*)

1.0	0.1	0.9	0.1	0.9	0.5
0.75[b]	0.1	0.7	0.1	0.7	0.4
0.5[b]	0.1	0.4	0.1	0.4	0.2

Source: *EIA RS-222B*.

[a] The values are tabulated as a guide for systems design and are based on the values that have been found satisfactory in the operational experience of the industry. These data are listed for reference only.

[b] These deflections are extrapolated and are not based on experience of the industry.

Notes

1. Half-power beamwidth of the antenna to be provided by the purchaser of the tower.

2. a. The limits of beam movement resulting from an antenna mounting on the tower are the sum of the appropriate figures in columns *B* and *C*.

   b. The limits of beam movement resulting from twist when passive reflectors are employed are the sum of the appropriate figures in columns *D* and *E*.

   c. The limits of beam movement resulting from sway when passive reflectors are employed are twice the sum of the appropriate figures in columns *D* and *F*.

   d. The tabulated values in columns *D*, *E*, and *F* are based on a vertical orientation of the antenna beam.

3. The maximum tower movement shown above (4.5°) will generally be in excess of that actually experienced under conditions of 20 lb/ft² wind loading.

4. The problem of linear horizontal movement of a reflector-parabola combination had been considered. It is felt that in a large majority of cases, this will present no problem. According to tower manufacturers, no tower will be displaced horizontally at any point on its structure more than 0.5 ft/100 ft of height under its designed wind load.

5. The values shown correspond to 10 db gain degradation under the worst combination of wind forces at 20 lb/ft². This table is meant for use with standard antenna-reflector configurations.

6. Twist and sway limits apply to 20 lb/ft² wind load only, regardless of survival or operating specifications. If there is a requirement for these limits to be met under wind loads greater than 20 lb/ft², such requirements must be specified by the user.

(Courtesy Electronics Industry Assoc., Washington, D.C.)

a mountain is nearby with poor access to its top. The radiolink engineer should consider the use of a passive repeater as an economic alternative. A prime requirement is that the plane reflector be within line-of-sight of the terminal antenna in town as well as line-of sight of the distant radiolink station. Such a passive repeater installation may look like the following example where a = the net path loss in decibels:

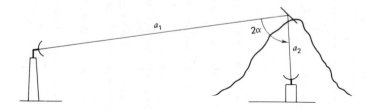

$$a = G_t + G_r + G_A - a_1 - a_2 \qquad (5.23)$$

where a_1 = path loss on path 1 (dB)
$\quad\; a_2$ = path loss on path 2 (dB)
$\quad\; G_t$ = transmitting antenna gain
$\quad\; G_r$ = receiving antenna gain
$\quad\; G_A$ = passive reflector gain

Let us concern ourselves with G_A for the moment. The gain of a passive reflector results from the capture of a ray beam of RF energy from a distant antenna emitter and the redirection of it toward a distant receiving antenna. The gain of the passive reflector is divided into two parts: (1) incoming energy and (2) redirected or reflected energy. The gain for (1) and (2) is

$$G_A = 20 \log \frac{4\pi A \cos \alpha}{\lambda^2} \qquad (5.24)$$

where α = one-half the included horizontal angle between incident and reflected
$\qquad$ wave
$\quad\; A$ = the surface area (ft^2 or m^2). If A is in square feet, then λ must also be in
$\qquad$ feet.

The reader will find the following relationship useful:

$$\text{wavelength (ft)} = \frac{985}{F}$$

where F is in megahertz. Passive reflector path calculations are not difficult. The first step is to determine if the shorter path (a_2 path in the figure above) places the passive reflector in the near field of the nearer parabolic antenna. Remember that when we dealt with periscopic systems, the reflector was always in the near field (see Section 5.5.7). To determine whether near field or far field, solve the following formula:

$$\frac{1}{K} = \frac{\pi\lambda d'}{4A}$$

If the ratio $1/K$ is less than 2.5, a near-field condition exists; if $1/K$ is greater than 2.5, a far-field condition exists. d' = length of the path in question (i.e., the shorter distance).

For the far-field condition, consider paths 1 and 2 as separate paths and sum their free-space path losses. Determine the gain of the passive plane reflector using formula (5.24). Sum this gain with the two free-space path losses algebraically to obtain the net path loss.

For the near-field condition, where $1/K$ is less than 2.5, the free-space path loss will be that of the longer hop (e.g., a_1 above) Algebraically add the repeater gain (or loss) which is determined as follows. Compute the parabola/reflector coupling factor l:

$$l = D' \sqrt{\frac{\pi}{4A}}$$

where D' = diameter of parabolic antenna (ft)
$\quad A$ = effective area of passive reflector (ft^2)

Figure 5.39 is now used to determine near-field gain or loss. The $1/K$ value is on the abscissa, and l is the family of curves shown.

Example 1. *Far-field:* A plane passive reflector 10 × 16 ft, or 160 ft^2, is erected 21 mi from one active site and only 1 mi from the other. $2\alpha = 100°$, $\alpha = 50°$. The operating frequency is 2000 MHz. By formula the free-space loss for the longer path is 129.5 dB, and for the shorter path it is 103 dB.

Calculate the gain G_A of the passive plane reflector, formula (5.24):

$$G_A = \frac{20 \log 4 \times 160 \cos 50°}{(985/2000)^2} = 20 \log 5.340$$

$$= 74.6 \text{ dB}$$

Net path loss = $-129.5 - 103 + 74.6 = -157.9$ dB

Example 2. *Near-field:* The passive reflector selected in this case is 24 × 30 ft. The operating frequency is 6000 MHz. The long leg is 30 mi and the short leg, 4000 ft. A 10-ft parabolic antenna is associated with the active site on the short leg. (6 GHz is approximately equivalent to 0.164 ft.) Determine $1/K$:

$$\frac{1}{K} = \frac{\pi\lambda d}{4A} = \frac{\pi(0.164)(4000)}{4 \times 720} = 0.717$$

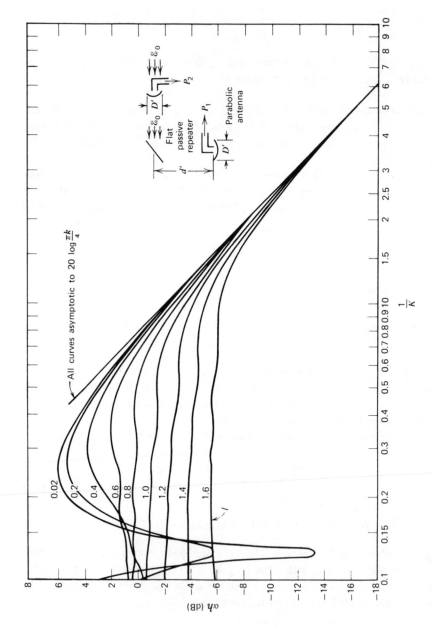

Figure 5.39 Antenna-reflector efficiency curves. (Courtesy Microflect, Inc., Salem, OR.)

Note that this figure is less than 2.5, indicating the near-field condition. Calculate l:

$$l = D' \sqrt{\pi/4A} = 10 \sqrt{\pi/4 \times 720} = 0.33$$

Using these two inputs, the values of l and $1/K$, we go to Figure 5.39 and find the net gain of the system to be +0.2 dB. The net free-space loss is then +0.2 - 142.3 = -142.1 dB. The free-space loss of the 30-mi leg is 142.3 dB.

5.18   NOISE PLANNING ON RADIOLINKS

CCIR Rec. 393-3 allots 10,000 pW psophometrically weighted noise on a 2500-km reference circuit carrying FDM telephony. Hence total noise accumulation on a per-kilometer basis is

$$\frac{10{,}000 \text{ pWp}}{2{,}500 \text{ km}} = 4 \text{ pWp/km}$$

Of the 10,000 pW, 2500 is allotted to terminal equipment (i.e., the multiplex equipment; see Chapter 3) and 7500 pW to line equipment. In the case of radiolinks, the line equipment is the radio equipment. Thus

$$\frac{7500 \text{ pWp}}{2500 \text{ km}} = 3 \text{ pWp/km}$$

This is the maximum noise accumulation permitted on a per-kilometer basis. (See CCIR Rec. 395-2 as well.)

Note.   To obtain pWp, when "flat" noise is given in a 3.1-kHz channel, reduce the figure in picowatts by 2.5 dB (see Section 1.9.6).

For this discussion we are dealing with thermal noise and IM noise which may be considered additive. For basic discussions on noise refer to Sections 1.9.6 and 5.6.5.

Here we assume that a system is loaded as indicated in Section 5.6.2 such that NPR is optimized, that is, loaded (i.e., the baseband input level) with sufficient deviation to minimize the effects of thermal noise, but not loaded to such an extent that noise is excessive. The curves in Figure 5.30 provide data on this sort of loading.

For noise engineering real circuits, CCIR Rec. 395-2 offers the following guidelines (quoted in part):

CCIR UNANIMOUSLY RECOMMENDS

1. That, in circuits established over real links which do not differ appreciably from the hypothetical reference circuit, the psophometrically weighted* noise power at a point of zero relative level in the telephone channels of frequency-division multiplex radio-relay systems of length L, where L is between 280 and 2500 km, should not exceed:

*The level of uniform-spectrum noise power in a 3.1-kHz band must be reduced by 2.5 dB to obtain the psophometrically weighted noise power.

1.1 $3L$ pW mean power in any hour

1.2 $3L$ pW one-minute mean power for more than 20% of any month

1.3 47,500 pW one-minute mean power for more than $(L/2500) \times 0.1\%$ of any month

It is recognized that the performance achieved for very short periods of time is very difficult to measure precisely, and that, in a circuit carried over a real link, it may, after installation, differ from the planning objective.

2. That circuits to be established over real links, the composition of which, for planning reasons, differs substantially from the hypothetical reference circuit, should be planned in such a way that the psophometrically weighted noise power at a point of zero relative level in a telephone channel of length L, where L is between 50 and 2500 km, carried in one or more baseband sections of frequency-division multiplex radio links, should not exceed:

2.1 For 50 km $\leq L \leq$ 840 km:

2.1.1 $3L$ pW + 200 pW one-minute mean power for more than 20% of any month

Table 5.12 Example of Noise Calculation on Radiolinks

Radiolink, FM

One modulation section consisting of ____ hops

Number of modulation sections	(1)	1			
Number of hops	(2)	6			
Measuring channel (kHz)	(3)	Lower	Center	Upper	
Basic noise					
Thermal receiver noise	(4)	–	-72.2^a	-71.2^a	dBm0p
	(5)	–	60^a	76^a	pW0p
Basic noise of the RF equipment	(6)	–	130^a	120^a	pW0p
Basic noise of the modem equipment	(7)	–	65^a	60^a	pW0p
Total thermal noise under no-fading	(8)	–	255	256	pW0p
condition (sum of 5 + 6 + 7)	(9)	–	-65.9	-65.9	dBm0p
IM noise					
RF and IF equipment	(10)	–	389^a	360^a	pW0p
Modem	(11)	–	65^a	60^a	pW0p
Total IM noise	(12)	–	454	420	pW0p
(sum of 10 + 11)	(13)	–	-63.4	-63.8	dBm0p
Total noise under no-fading conditions	(14)	–	709	676	pW0p
(sum of 8 + 12)	(15)	–	-61.6	-61.7	dBm0p
Permissible total noise	(16)	–	840	840	pW0p
	(17)	–	-60.8	-60.8	dBm0p
Permissible rise in receiver thermal	(18)	–	131	164	pW0p
noise (difference of 16 - 14)					
Permissible thermal receiver noise	(19)	–	191	240	pW0p
(sum of 18 + 5)	(20)	–	-67.2	-66.2	dBm0p
Average fading margin (difference of	(21)	–	5	5	dB
20 - 4)					

Source: Ref. 37.

[a]Values taken to demonstrate the procedure of calculation.

    2.1.2 47,500 pW one-minute mean power for more than $(280/2500) \times$ 0.1% of any month when L is less than 280 km, or more than $(L/2500) \times 0.1\%$ of any month when L is greater than 280 km

2.2 For 840 km $< L \leq$ 1670 km:

    2.2.1 $3L$ pW + 400 pW one-minute mean power for more than 20% of any month

    2.2.2 47,500 pW one-minute mean power for more than $(L/2500) \times$ 0.1% of any month

2.3 For 1670 km $< L \leq$ 2500 km:

    2.3.1 $3L$ pW + 600 pW one-minute mean power for more than 20% of any month

    2.3.2 47,000 pW one-minute mean power for more than $(L/2500) \times$ 0.1% of any month

3.   That the following notes should be regarded as part of the recommendation:

Note 1.   Noise in the frequency-division multiplex equipment is excluded. On a 2500 km hypothetical reference circuit the CCITT allows 2500 pW mean value for this noise in any hour.

For Notes 2–4, see CCIR Rec. 395-2.

Considering that the hypothetical reference circuit described in CCIR Rec. 393-3 is divided into nine homogeneous sections alloting 840 pW per section (approximately), then each section may be treated as in Table 5.12 (taken from Table 4, sec. B.IV.4). It should be noted that row 21, the 5-dB fade margin, is academic. As we established previously in Sections 5.8 and 5.9, each system must be examined under real conditions. Besides, the margin of fade in Table 5.12 is for six hops in tandem and assumes noncoincident fading in the individual hops. Nonetheless, the table offers a good approach to noise engineering of radiolink systems.

REFERENCES AND BIBLIOGRAPHY

1. *Reference Data for Radio Engineers,* 6th ed., Howard W. Sams, Indianapolis, IN, 1977.
2. "Engineering Considerations for Microwave Communication Systems," Lenkurt Electric Corp., San Carlos, CA, 1975.
3. "Recommendations and Reports of the CCIR 1978," XIV Plenary Assembly, Kyoto, 1978, Vol. IX.
4. CCITT Orange Books, Geneva, 1976, Vol. III, G recommendations.
5. D. M. Hamsher, Ed., *Communication System Engineering Handbook,* McGraw-Hill, New York, 1967.
6. *Collins CEL,* 19 and 22, Collins Radio Co., Dallas, TX.
7. *Microwave Radio Relay Systems,* T.O. 31R5-1-9, U. S. Department of Defense (USAF), Washington, DC, April 1, 1965.
8. W. Oliver, *White Noise Loading of Multichannel Communication Systems,* Marconi Instruments, Sept. 1964.
9. *Electrical Communications Systems Engineering–Radio,* U.S. Army TM 11-486-6, U.S. Department of Defense, Washington, DC.
10. *Transmission Systems for Communications,* 4th ed., Bell Telephone Laboratories, American Telephone and Telegraph Company, New York, 1971.

11. B. D. Holbrook and J. T. Dixon, "Load Rating Theory for Multichannel Amplifiers," *Bell Sys. Tech. J.,* Vol. 18, 624–644, Oct. 1939.

12. E. F. Plarman and J. J. Tary, *Required Signal-to-Noise Ratios, RF Signal Power and Bandwidth for Multichannel Radio Communication Systems,* Tech. Note 100, National Bureau of Standards, Boulder, CO, Jan. 1962.

13. A. P. Barkhausen *et al., Equipment Characteristics and Their Relationship to Performance for Tropospheric Scatter Communication Circuits,* Tech. Note 103, National Bureau of Standards, Boulder, CO, Jan. 1963.

14. Microwave Line-of-Sight Systems to Participating Companies," U.S. Standards Institute Memo. 40, Sept. 1966.

15. F. E. Terman, *Radio Engineering,* 4th ed., McGraw-Hill, New York.

16. D. C. Livingston, *The Physics of Microwave Propagation,* GTE Monograph, General Telephone and Electronics, 1967.

17. *Microflect Passive Repeater Engineering Manual,* No. 161, Microflect Inc., Salem, OR, 1962.

18. *Jerrold Path Calculations,* Jerrold Electronics Corp., Philadelphia, PA, 1967.

19. EIA RS-195B, RS-203, RS-222B, and RS-250B. Electronic Industries Association, Washington, DC.

20. K. Bullington, "Radio Propagation Fundamentals," *Bell Sys. Tech. J.,* May 1957.

21. J. Jasik, *Antenna Engineering Handbook,* McGraw-Hill, New York.

22. R. L. Marks *et al., Some Aspects of the Design of FM Line-of-Sight Microwave and Troposcatter Systems,* USAF Rome Air Development Center, NY, U.S. Technical Information Service, AD 617-686, Springfield, VA, Apr. 1965.

23. J. Fagot and P. Magne, *Frequency Modulation Theory,* Pergamon Ps, London, 1961.

24. *Andrew Catalog 30,* Andrew Corp., Orland Park, IL, 1979.

25. *Collins Telecommunication Equipment Catalog 2,* Collins Radio Co., Dallas, TX.

26. H. T. Dougherty, "A Survey of Microwave Fading Mechanisms, Remedies and Applications," ESSA Tech. Rep. ERL 69-WPL 4, Boulder, CO, Mar. 1968.

27. P. F. Panter, *Communication Systems Design—Line-of-Sight and Troposcatter Systems,* McGraw-Hill, New York, 1972.

28. R. F. White, *Reliability in Microwave Communication Systems—Prediction and Practice,* Lenkurt Electric Corp., San Carlos, CA, 1970.

29. A. P. Barsis *et al.,* "Analysis of Propagation Measurements over Irregular Terrain in 76 to 9200 MHz Range," ESSA Tech. Rep. ERL 114-ITS 82, Boulder, CO, Mar. 1969.

30. K. W. Pearson, "Method for the Prediction of the Fading Performance of the Multisection Microwave Link," *Proc. IEE,* Vol. 112, July 1965.

31. R. G. Medhurst, "Rainfall Attenuation of Centimeter Waves: Comparison of Theory and Measurement," *IEEE Trans. Ant. Propag.,* July 1965.

32. M. J. Tant, *The White Noise Book,* Marconi Instruments, Ltd., St. Albans, England.

33. H. Brodhage and W. Hormuth, *Planning and Engineering Radio Relay Links,* Siemens/ Heyden & Sons, London, 1977.

34. Topographic Maps, U.S. Geological Survey, Arlington, VA.

35. "Design Objectives for DCS LOS Digital Radio Links," Eng. Pub. DCEC EP 27-77, Defense Communications Engineering Center, Washington, DC, 1977.

36. A. G. Longly and R. K. Reasoner, *Comparison of Propagation Measurements with Predicted Values in the 20–10,000-MHz Range,* ESSA Tech. Rep. ERL 148-ITS 97, Institute of Telecommunication Sciences, Boulder, CO, Jan. 1970.

37. J. W. Ryde and D. Ryde, *Attenuation of Centimeter Waves by Rain, Hail, Fog and Clouds,* General Electric Company, Wembly, U.K., 1945.

38. *Economics and Technical Aspects of the Choice of Transmission Systems,* ITU.

6 | TROPOSPHERIC SCATTER

6.1 INTRODUCTION

Tropospheric scatter is one method of propagating microwave energy beyond LOS or "over the horizon." Communication systems utilizing the tropospheric scatter phenomena handle from 12 to 240 FDM telephone channels. Well-planned tropospheric scatter links may have propagation reliabilities on the order of 99.9% or better. These reliabilities are comparable to those of radiolink systems (LOS microwave) discussed in the preceding chapter. In fact, the discussion of tropospheric scatter is a natural extension of Chapter 5.

Tropospheric scatter takes advantage of the refraction and reflection phenomena in a section of the earth's atmosphere called the troposphere. This is the lower portion of the atmosphere from sea level to a height of about 11 km (35,000 ft). UHF signals are scattered in such a way as to follow reliable communications on hops up to 640 km (400 mi). Long distances of many thousands of kilometers may be covered by operating a number of hops in tandem. The North Atlantic Radio System (NARS) of the U.S. Air Force is an example of a lengthy tandem system. It extends from Canada to Great Britain via Greenland, Iceland, the Faeroes, and Scotland. A mix of radiolinks (LOS microwave) and tropospheric scatter is becoming fairly common. The Canadian National Telephone Company (CNT) operates such a system in the Northwest Territories. The Bahama Islands are interconnected for communications by a mix of radiolinks, tropospheric scatter, and HF.

Tropospheric scatter systems generally use transmitter power outputs of 1 or 10 kW, parabolic type antennas with diameters of 4.5m (15 ft), 9m (30 ft), or 18 m (60 ft), and sensitive (uncooled) broadband FM receivers with front-end NFs on the order of 2.0-4.0 dB. A tropospheric scatter installation is obviously a bigger financial investment than a radiolink (LOS microwave) installation. Tropospheric scatter, however, has many advantages for commercial application that could well outweigh the issue of high cost. These advantages are summarized as follows:

1. Reduces the number of stations required to cover a given large distance when compared to radiolinks. Tropospheric scatter may require from one third to one tenth the number of stations as a radiolink system over the same path.

2. Provides reliable multichannel communication across large stretches of water (e.g., over inland lakes, to offshore islands, between islands) or between areas separated by inaccessible terrain.

3. May be ideally suited to meet toll-connecting requirements of areas of low population density.

4. Useful when radio waves must cross territories of another political administration.

5. Requires less maintenance staff per route-kilometer than conventional radio-link systems over the same route.

6. Allows multichannel communication with isolated areas, especially when intervening territory limits or prevents the use of repeaters.

7. Desirable for multichannel communications in the tactical military field environment for links from 30 to 200 mi long (50–340 km).

6.2 THE PHENOMENON OF TROPOSPHERIC SCATTER

There are a number of theories explaining over-the-horizon communications by tropospheric scatter. One theory postulates atmospheric air turbulence, irregularities in the refractive index, or similar homogeneous discontinuities capable of diverting a small fraction of the transmitted radio energy toward a receiving station. This theory accounts for the scattering of radio energy in a way much as fog or moisture seems to scatter a searchlight on a dark night. Another theory is that the air is stratified into discrete layers of varying thickness in the troposphere. The boundaries between these layers become partially reflecting surfaces for radio waves and thereby scatter the waves downward over the horizon.

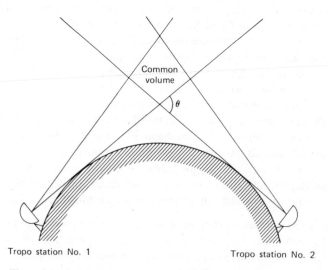

Figure 6.1 Tropospheric scatter model. θ = scatter angle.

Figure 6.1 is a simple diagram of a tropospheric scatter link showing two important propagation concepts. These are as follows:

- Scatter angle, which may be defined as either of two acute angles formed by the intersection of the two portions of the tropospheric scatter beam (lower boundaries) tangent to the earth's surface. Keeping the angle small effectively reduces the overall path attenuation.
- Scatter volume or "common volume" is the common enclosed area where the two beams intercept.

6.3 TROPOSPHERIC SCATTER FADING

Fading is characteristic of tropospheric scatter. It is handy to break fading in tropospheric scatter systems into two types, slow and fast fading. Expressed another way, these are long-term (slow) and short-term variations in the received signal level.

When referring to tropospheric scatter received signal levels, we usually use the median received level as reference. In general the hourly median and minute median are the same. In Chapter 5, the reference level was the unfaded signal level, which turned out to approximate sufficiently the calculated level under no-fade conditions. Such a straightforward reference signal level is impossible in tropospheric scatter because a tropospheric scatter signal is in a constant condition of fade. Thus for path calculations and path loss, we refer to the long-term median, usually extended over the whole year.

At any one moment a received tropospheric scatter signal will be affected by both slow and fast fading. It is believed that fast fading is due to the effects of multipath (i.e., due to a phase incoherence at various scatter angles). Fast fading is treated statistically "within the hour" and has a Rayleigh distribution with a sampling time of 1–7 min, although in some circumstances it has been noted up to 1 h. The fading rate depends on both frequency and distance or length of hop.

The U.S. National Bureau of Standards (NBS) describes long-term variations in signal level as variations of *hourly median* values of transmission loss. This is the level of transmission loss that is exceeded for a total of one half of a given hour. A distribution of hourly medians gives a measure of long-term fading. Where these hourly medians are considered over a period of 1 month or more, the distribution is log normal.

In studying variations in tropospheric scatter transmission path loss (fading), we have had to depend on empirical information. The signal level varies with the time of day, the season of year, and the latitude, among other variables. To assist in the analysis and prediction of long-term signal variation, the hours of the year have been broken down into eight time blocks, given in Table 6.1.

Most commonly we refer to time block 2 for a specific median path loss. Time block 2 may be thought of as an average winter afternoon in the temperate zone of the northern hemisphere.

It should be noted that signal levels average 10 dB lower in winter than in sum-

Table 6.1 Time Block Assignments

Time Block	Month	Hours
1	Nov.–April	0600–1300
2	Nov.–April	1300–1800
3	Nov.–April	1800–2400
4	May–Oct.	0600–1300
5	May–Oct.	1300–1800
6	May–Oct.	1800–2400
7	May–Oct.	0000–0600
8	Nov.–April	0000–0600

mer, and that morning or evening signals are at least 5 dB higher than midafternoon signals. Slow fading is believed due to changes in path conditions such as atmospheric changes (e.g., a change in the index of refraction of the atmosphere).

6.4 PATH LOSS CALCULATIONS

Tropospheric scatter paths typically display considerably larger losses when compared to radiolink (LOS microwave) paths. Losses up to 260 dB are not uncommon. There are a number of acceptable methods of estimating path losses for tropospheric scatter systems. One such method is outlined in CCIR Rep. 238-3, and another is described in NBS Tech Note 101 (Ref. 3). These are more commonly known as the CCIR method and the NBS method. Their approach to the problem is somewhat similar. In the following paragraphs we will summarize some of the more important aspects of the NBS method as described in Ref. 5. The procedure has been considerably simplified and abbreviated for this discussion.

The objective is to predict a long-term path loss that will not be exceeded for specified time availabilities, such as 50, 90, 99, 99.9, or 99.99% of the time. In the previous chapter on radiolinks we referred to this as propagation reliability.

The NBS method describes how to calculate these losses with a 50% probability (confidence level) of being the correct prediction for a path in question. Then it shows how to systematically add margin to assure improved probability that the prediction will be correct or more than the minimum necessary on a particular path. This probability of prediction is called *service probability*. It is common with tropospheric scatter path calculations to show a 50 and a 95% service probability. Service probability indicates the confidence level of the prediction.

6.4.1 Mode of Propagation

An over-the-horizon microwave path can and often does display two modes of propagation, diffraction and tropospheric scatter. In most cases either one or the other will predominate. Particularly on shorter paths the possibility of the diffraction mode should be investigated. Experienced engineers in over-the-horizon systems can often identify which propagation mode can be expected during the path profile and path survey phases of the link engineering effort. In the absence of this expertise, the following criteria may be used as an aid to identify the principal propagation mode:

1. The distance at which diffraction and tropospheric scatter losses are approximately equal is $65(100/f)^{1/3}$ km, where f = radio frequency in MHz. For path lengths less than this value, diffraction will be the predominant mode, and vice versa.

2. For paths having angular distances of 20 mrad or more, the diffraction mode may be neglected and the path can be considered to be operating in the troposcatter mode. (Angular distance is explained below.)

6.4.2 Basic Long-Term Tropospheric Scatter Transmission Loss

Following the NBS method, we determine the basic long-term tropospheric scatter transmission loss L_{bsr}:

$$L_{bsr} = 30 \log f - 20 \log d + F(\theta d) - F_0 + H_0 + A_a \qquad (6.1)$$

where f = operating frequency (MHz)
    d = great-circle path length (km)
    $F(\theta d)$ = attenuation function (dB)
    F_0 = scattering efficiency correction factor
    H_0 = frequency gain function
    A_a = atmospheric absorption factor from Figure 6.2

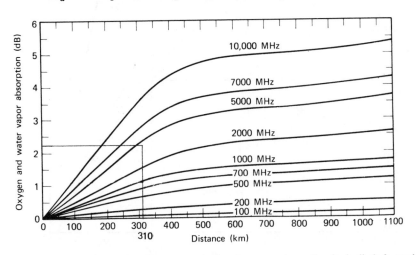

Figure 6.2 Determination of median oxygen and water vapor absorption in decibels for various operating frequencies when path length is given (for August, Washington, DC). (Courtesy Institute of Telecommunication Sciences, Office of Telecommunications, U.S. Department of Commerce.)

We have simplified our procedure by neglecting the frequency gain function and the scattering efficiency correction factor.

The numerical values of the first two terms of equation (6.1) are determined by substituting the assigned frequency in megahertz of the radio system to be installed (see Section 6.9.2) and the great-circle distance in kilometers. The third term requires some detailed discussion.

6.4.3   Attenuation Function

The attenuation function $F(\theta d)$, is derived from Figure 6.3. θd is the product of the angular distance (scatter angle) in radians and the great-circle path length in kilometers. The following is an abbreviated method of approximating the scatter

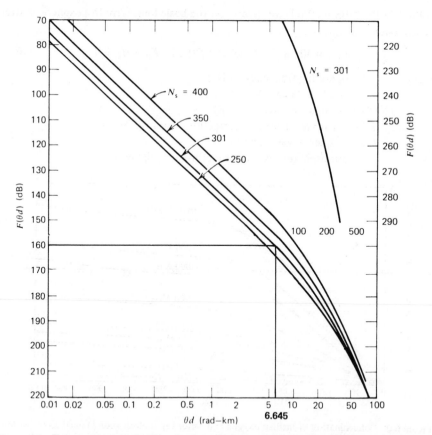

Figure 6.3   Attenuation function for the determination of scatter loss. (Courtesy Institute of Telecommunication Sciences, Office of Telecommunications, U.S. Department of Commerce.)

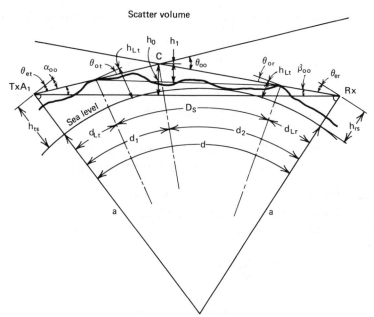

Distances are measured in kilometers along a great-circle arc:

$$\theta_{oo} = \frac{D_S}{a} + \theta_{ot} + \theta_{or} = \frac{d}{a} + \theta_{et} + \theta_{er}$$

Figure 6.4 Tropospheric scatter path geometry.

angle θ. We assume that a path profile has been carried out (see Section 5.3.3). Arbitrarily one site is denoted the transmitter site t and the other site the receiver site r. From the profile the horizon location in the direction of the distant site and its altitude above mean sea level (MSL) are determined as well as its distance from its corresponding site. For all further calculations distances and altitudes (elevations) are measured in kilometers and angles in radians. It is important to use only these units throughout. Figure 6.4 will assist in identifying the following distances, elevations, and angles. Let

d = great-circle distance between transmitter and receiver sites

d_{Lt} = distances from transmitter site to transmitter horizon

d_{Lr} = distance from receiver site to receiver horizon

h_{ts} = elevation above MSL to center of transmitting antenna (km)

h_{rs} = elevation above MSL to center of receiving antenna (km)

h_{Lt} = elevation above MSL of transmitter horizon point (km)

h_{Lr} = elevation above MSL of receiver horizon point (km)

N_0 = surface refractivity corrected for MSL (For the continental United States use Figure 5.4, and for other locations use Ref. 1, Sec. 28–24.)

Adjust the surface refractivity N_0 for the elevation of each site by the following formula:

$$N_{ts,rs} = N_0 \exp(-0.1057 h_{ts}, h_{rs}) \qquad (6.2)$$

Compute N_s:

$$N_s = \frac{N_{ts} + N_{rs}}{2} \qquad (6.3)$$

Calculate the effective earth radius by the following formula:

$$a = a_0 (1 - 0.04665 e^{0.005577 N_s})^{-1} \qquad (6.4)$$

where a_0 = true earth radius, 6370 km. If $N_s = 301$, then $a = 8500$ km, which is the familiar $\frac{4}{3}$ earth radius case alluded to in Chapter 5.

Calculate the antenna take-off angles at each site by the following formulas:

$$\theta_{et} = \frac{h_{Lt} - h_{ts}}{d_{Lt}} - \frac{d_{Lt}}{2a} \qquad (6.5a)$$

$$\theta_{er} = \frac{h_{Lr} - h_{rs}}{d_{Lr}} - \frac{d_{Lr}}{2a} \qquad (6.5b)$$

Calculate the scatter angle components α_0 and β_0 by the following formulas:

$$\alpha_0 = \frac{d}{2a} + \theta_{et} + \frac{h_{ts} - h_{rs}}{d} \qquad (6.6a)$$

$$\beta_0 = \frac{d}{2a} + \theta_{er} + \frac{h_{rs} - h_{ts}}{d} \qquad (6.6b)$$

Calculate the scatter angle (often called *angular distance*) θ_0 by the following formula:

$$\theta_0 = \alpha_0 + \beta_0 \text{ (rad)} \qquad (6.7)$$

Multiply θ_0 in radians by the path length in kilometers. This is θd.

Determine $F(\theta d)$ from Figure 6.3 using the product θd calculated above, and interpolate, if necessary, for the value of N_s taken from equation (6.3).

L_{bsr} is now calculated by equation 6.1, neglecting terms F_0 and H_0.

6.4.4 Basic Median Transmission Loss

The predicted median long-term transmission loss $L_n(0.5, 50)$, abbreviated $L_n(0.5)$, for the appropriate climatic region n is related to L_{bsr} by the following formula:

$$L_n(0.5) = L_{bsr} - V_n(0.5, d_e) \qquad (6.8)$$

where $L_n(0.5)$ = predicted transmission loss (in dB) exceeded by half of all hourly medians, and thus the yearly median value, and $V_n(0.5, d_e)$ = variability of the median value about the basic long-term transmission loss L_{bsr} for the appropriate climatic region n and the effective distance d_e. NBS has established eight climatic regions for the world as follows:

1. Continental temperature. Large land mass, 30-60°N latitude, 30-60°S latitude.

2. Maritime temperature overland. In this region prevailing winds, unobstructed by mountains, carry moist maritime air inland. 20-50°N, 20-50°S latitudes, typified by United Kingdom, West Coast of North America and Europe, and northwestern coastal areas of Africa.

3. Maritime temperate oversea. Fully over-water paths in temperate regions.

4. Maritime subtropical overland. 10-30°N, 10-30°S latitudes, near the sea with defined rainy and dry seasons.

5. Maritime subtropical. Latitudes same as region 4. Over-water paths. However, valid curves are not available due to lack of empirical data for this region. Use region 3 or region 4, whichever is more applicable.

6. Desert Sahara. Regions with year-round semi-arid conditions.

7. Equatorial. ±20° latitude from the equator, characterized by monotonous heavy rains and high average summer temperatures.

8. Continental subtropical. Usually 20-40°N latitude, an area of monsoons with seasonal extremes of summer rainfall and winter drought.

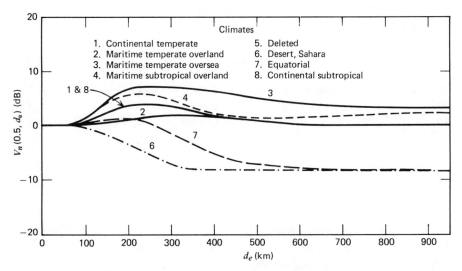

Figure 6.5 Function $V_n(0.5, d_e)$ for eight climatic regions (Ref. 5).

Select the most appropriate region (n) for the path in question, then compute the effective distance d_e. To calculate d_e, effective antenna heights are required, namely, h_{te} and h_{re}. These heights are functions of the average elevation of the terrain between each antenna and its respective radio horizon in the direction of the distant end of the path. For smooth earth conditions (i.e., typically an overwater path or a hypothetical path with no obstacles except central earth bulge) h_{te} and h_{re} are the respective elevations of each site above MSL. Under real overland conditions, the effective height is the average height above MSL of the central 80% between the antenna and its radio horizon.

Calculate d_L by the following formula:

$$d_L = 3\sqrt{2h_{te} \times 10^3} + 3\sqrt{2h_{re} \times 10^3} \qquad (6.9)$$

Calculate d_{s1} by the following formula:

$$d_{s1} = 65\left(\frac{100}{f}\right)^{1/3} \qquad (6.10)$$

where f = frequency (MHz).

There are two cases to calculate d_e. Use whichever is applicable.

1. If $d \leqslant d_L + d_{s1}$, then

$$d_e = \frac{130d}{(d_L + d_{s1})} \qquad (6.11a)$$

2. If $d > d_L + d_{s1}$, then

$$d_e = 130 + d - (d_L + d_{s1}) \qquad (6.11b)$$

With the climatic region n determined and the effective distance d_e calculated, derive V_n in decibels from Figure 6.5. Calculate $L_n(0.5)$ using formula (6.8). The value $L_n(0.5, 50)$ represents the long-term median path loss for a 50% time availability and 50% service probability.

6.4.5 The 50% Service Probability Case

The next step is to extend the time availability to the specified or desired value for the tropospheric scatter path in question. Often it is convenient to state time availability for a number of percentages as follows:

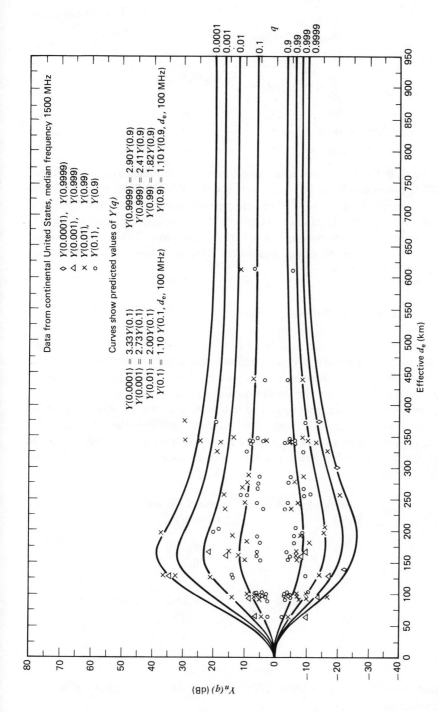

Figure 6.6 Long-term power fading, continental temperate climate, for frequency greater than 1000 MHz (Ref. 5).

263

Time Availability q (%)	Path Loss
50	$L_n(0.5, 50)$
90	A
99	B
99.9	C
99.99	D

Values for $A, B, C,$ and D in the path loss column are determined by adding a factor to $L_n(0.5, 50)$ called $Y_n(q, 50, d_e)$, where $q = 0.9, 0.99, 0.999,$ and 0.9999. Y_n values are derived from curves for the appropriate climatic region and frequency band. One example family of curves is presented in Figure 6.6, where n is region 1, the continental temperate region, and for frequencies above 1 GHz. Y_n is derived for several values of q using the appropriate effective distance d_e of the tropospheric scatter path under study.

6.4.6 Improving Service Probability

Under the values of path loss calculated in the previous section, only half of the paths installed for a specific set of conditions would have a measured long-term path loss equal to or less than those calculated. By definition, this is a service probability of 50%. To extend the service probability (i.e., improve the confidence level), the following procedures should be followed. Again we are dealing only with long-term power fading. The basic data required are the values obtained for $Y_n(q, d_e)$ and the standard normal deviate Z_{mo} for the service probability desired. Several standard normal deviates and their corresponding service probabilities are provided below.

Service Probability (%)	Standard Normal Deviate Z_{mo}
50	0
60	0.3
75	0.7
80	0.85
90	1.3
95	1.65
99	2.35

Calculate the path-to-path variance $\sigma_c^2(q)$ where q is the corresponding (or desired) time availability:

$$\sigma_c^2(q) = 12.73 + 0.12\, Y_n(q, d_e) \tag{6.12}$$

Calculate the prediction error $\sigma_{rc}(q)$ by the following formula:

$$\sigma_{rc}(q) = \sqrt{\sigma_c^2(q) + 4} \tag{6.13}$$

Calculate the product of Z_{mo} and $\sigma_{rc}(q)$. This value is now added to the path loss value for the corresponding time availability q given in the previous section.

6.4.7 Example Problem

Assume smooth earth conditions (i.e., no intervening obstacles besides earth bulge) and calculate the path loss from Newport, NY, to Bedford, MA (U.S.A). The great-circle distance between the sites is 310.5 km.

Site elevation, Newport, $h_{ts} = 2000$ ft (0.61 km)
Site elevation, Bedford, $h_{rs} = 100$ ft (0.031 km)
$N_0 = 310$
Operating frequency 4700 MHz
$d_{Lt} = 102$ km (smooth earth); $d_{Lr} = 22$ km (smooth earth) (see Fig. 5.3)
$h_{Lt}, h_{Lr} = 0.0$ km (smooth earth, by definition)
$N_{st} = 291$; $N_{sr} = 309$
$N_s = 300$
$a = 8500$ km
$\theta_{et} = 0.0119$ rad; $\theta_{er} = -0.0027$ rad
$\alpha_0 = 0.008$ rad; $\beta_0 = 0.0134$ rad
$\theta = 0.0214$ rad
$\theta d = 0.0214 \times 310.5 = 6.645$ km-rad

From Figure 6.3, $F(\theta d) = 160$ dB.
  Calculate L_{bsr}:

$$
\begin{array}{ll}
30 \log F = 110.16 \text{ dB} & \\
-20 \log d = -49.84 \text{ dB} & \\
F(\theta d) = 160 \text{ dB} & \\
A_a = 2.2 \text{ dB} & \text{(from Figure 6.2)} \\
\hline
L_{bsr} = 222.52 \text{ dB} &
\end{array}
$$

Calculate d_L and d_{sL}:

$$h_{te} = 0.609 \text{ km} \qquad d_{sL} = 65(100/F)^{1/3}$$
$$h_{re} = 0.061 \text{ km} \qquad F = 4700 \text{ MHz}$$
$$d_L = 98 \text{ km} \qquad d_{sL} = 18 \text{ km}$$

Calculate d_e:

$$d_e = 130 + 310.5 - 98 - 18 = 324.5 \text{ km}$$

Determine V_n from Figure 6.5:

$$V_n = 3.7 \text{ dB}, \qquad n = \text{region 1}$$

Then

$$L_n(0.5) = 219.52 - 3.7 \text{ dB} = 215.82 \text{ dB}$$

This is the predicted path loss for a 50% time availability and 50% service probability. All further path loss calculations are based on this value.

Make up a path loss distribution table similar to that shown in Section 6.4.5 for the 50% service probability case as follows:

Time Availability q (%)	$Y_1(q)$ (dB)	Path Loss (dB)
50	0	218.82
90	6	224.82
99	12	230.82
99.9	15	233.82
99.99	17.5	236.32

The values of $Y_1(q)$ were taken from Figure 6.6.

Prepare a second table for the 95% service probability case:

Time Availability q (%)	$Z_{mo}\sigma_{rc}(q)$ (dB)	Path Loss (dB)
50	6.95	225.77
90	7.79	232.61
99	9.9	240.72
99.9	11.24	245.06
99.99	13.16	249.48

6.5 APERTURE-TO-MEDIUM COUPLING LOSS

Some tropospheric scatter link designers include aperture-to-medium coupling loss as another factor in the path loss equation, and others prefer to include it as another loss in the path calculation portion of link design as if it were a waveguide loss or similar. In any event this loss must be included somewhere.

Aperture-to-medium coupling loss has sometimes been called the *antenna gain degradation*. It occurs because of the very nature of tropospheric scatter in that the antennas used are not doing the job we would expect them to do. This is evident if we use the same antenna on a LOS (or radiolink) path. The problem stems from the concept of the common volume. High-gain parabolic antennas used on tropospheric scatter paths have very narrow beamwidths (see Section 6.9.3). The tropospheric scatter loss calculations consider a larger common volume than would be formed by these beamwidths. As the beam becomes more narrow due to the higher gain antennas, the received signal level does not increase in the same proportion as it would under free-space (LOS) propagation conditions. The difference between the free-space expected gain and its measured gain on a tropospheric scatter hop is called the aperture-to-medium coupling loss. This loss is proportional to the scatter angle θ and the beamwidth Ω. The beamwidth may be calculated from the formula

$$\Omega = \frac{7.3 \times 10^4}{F \times D_r} \qquad (6.14)$$

where F = carrier frequency (MHz)

   D_r = antenna reflector diameter (ft)

The ratio θ/Ω is computed, and from this ratio the aperture-to-medium coupling loss may be derived from Table 6.2.

Example. Calculate the aperture-to-medium coupling loss for two 30-ft antennas,

Table 6.2 Aperature-to-Medium Coupling Loss

Coupling Loss (dB)	Antenna Beam-width Ratio, θ/Ω	Coupling Loss (dB)	Antenna Beam-width Ratio, θ/Ω
0.18	0.3	2.95	1.4
0.40	0.4	3.22	1.5
0.60	0.5	3.55	1.6
0.90	0.6	3.80	1.7
1.10	0.7	4.10	1.8
1.20	0.75	4.25	1.9
1.40	0.8	4.63	2.0
1.70	0.9	4.90	2.1
1.95	1.0	5.20	2.2
2.2	1.1	5.48	2.3
2.42	1.2	5.70	2.4
2.75	1.3	6.00	2.5

Source: Ref. 11.

one at each end of the path, with a 2° scatter angle and a 900-MHz operating frequency. The beamwidth Ω is

$$\Omega = \frac{7.3 \times 10^4}{30 \times 900}$$

$$= 2.7°, \text{ equivalent to 46 mrad}$$

The scatter angle is

$$\theta = 2°, \text{ equivalent to 35 mrad}$$

Thus the ratio $\theta/\Omega = 35/46$, or approximately 0.75. From Table 6.2 this is equivalent to a loss of 1.2 dB.

CCIR Rep. 238-3 suggests another approach to the calculation of aperture-to-medium coupling loss using solely antenna gains G_t and G_r:

$$\text{Aperture-to-medium coupling loss (dB)} = 0.07 \exp\left[0.055(G_t + G_r)\right] \quad (6.15)$$

where G_t = gain of the transmitting antenna (dB)
G_r = gain of receiving antenna (dB)

Using the previous example of two 30-ft antennas and 900-MHz operating frequency, $G_t = G_r = 36$ dB and $G_t + G_r = 72$ dB, and

$$\text{Aperture-to-medium coupling loss (dB)} = 0.07 \exp\left[0.055 \times 72\right]$$
$$= 3.67 \text{ dB}$$

6.6 TAKE-OFF ANGLE (TOA)

The TOA is probably the most important factor under control of the engineer who selects a tropospheric scatter site in actual path design. The TOA is the angle between a horizontal ray extending from the radiation center of an antenna and a ray extending from the radiation center of the antenna to the radio horizon. Figure 6.7 illustrates the definition.

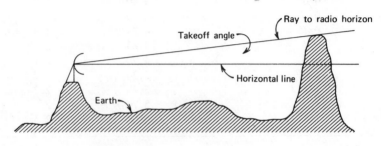

Figure 6.7   Definition of TOA.

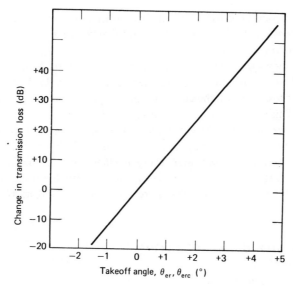

Figure 6.8 Effect of TOA on transmission loss.

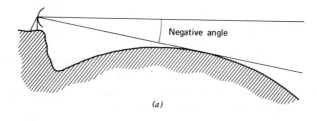

(a)

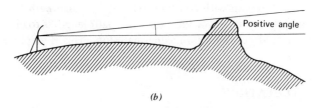

(b)

Figure 6.9 (a) A more desirable site regarding TOA. (b) A less desirable site regarding TOA. (Note that the beam should actually just clear the obstacle.)

269

The TOA is computed by means of path profiling several miles out from the candidate site location. It then can be verified by means of a transit siting. Path profiling is described in Section 5.3.3.

Figure 6.8 shows the effect of TOA on transmission loss. As the TOA is increased, about 12 dB of loss is added for each degree increase in TOA. This loss shows up in the scatter loss term of the equation for computing the median path loss. [equation (6.1)]. This approximation is valid at $0°$ in the range of +10 to $-10°$.

The advantage of siting a tropospheric scatter station on as high a site as possible is obvious. The idea is to minimize obstructions to the horizon in the direction of the "shot." As we shall see later, every decibel saved in median path loss may represent a savings of many thousands of dollars. Thus the more we can minimize the TOA, the better. Negative TOAs are very desirable. Figure 6.9 illustrates this criterion.

6.7 OTHER SITING CONSIDERATIONS

6.7.1 Antenna Height

Increasing the antenna height decreases the TOA, in addition to the small advantage of getting the antenna up and over surrounding obstacles. Raising an antenna from 20 ft above the ground to 100 ft above the ground provides something on the order of less than 3-dB improvement in median path loss at 400 MHz and about 1 dB at 900 MHz (Ref. 4).

6.7.2 Distance to Radio Horizon

The radio horizon may be considered one more obstacle which the tropospheric scatter ray beam must get over. Varying the distance to the horizon varies the TOA. If we maintain a constant TOA, the distance to the horizon can vary widely with insignificant effect on the overall transmission loss.

6.7.3 Other Considerations

If we vary the path length with constant TOA, the median path loss varies about 0.1 dB/mi. The primary effect of increasing the path length is to change the TOA, which will notably affect the total median path loss. This is graphically shown in Figure 6.8.

6.8 PATH CALCULATIONS

This section provides information to assist in determining the basic transmission parameters of a specific tropospheric scatter hop to meet a set of particular transmission objectives, usually related to an overall system plan. Such objectives may be found in the CCIR, namely, CCIR Recs. 395-1, 396-1, and 397-2. Another objective used by the U.S. Department of Defense is that found in (Ref. 6, sec. 4.2.2.1). These performance objectives deal with noise in the voice channel, stating that it should not exceed a particular level during a particular percentage of time. It is recommended that the reader review Section 5.6.5. Here we showed how the carrier-to-noise ratio can be related to the signal-to-noise ratio. If the carrier-to-noise ratio can be related to the desired time frame, then, as a consequence, the signal-to-noise ratio may be related to the same time percentage.

Path calculations here are based on the same criterion and starting point as in Chapter 5, namely, the thermal noise in the far-end receiver. Thus we start with -204 dBW as the thermal noise absolute floor value for the perfect receiver at room temperature with 1 Hz of bandwidth. The receiver thermal noise threshold is then calculated by algebraically adding the receiver NF in decibels and $10 \log B_{IF}$, the IF bandwidth in hertz (see Section 1.9.6).

Example. Let us now consider the example given in Section 6.4.7 with the following additional information:

Operating frequency 1000 MHz, but path losses as in example

60 VF channels to be transmitted (i.e., highest modulating frequency of 300 kHz); thus B_{IF} = 2.5 MHz

Receiver NF 3 dB

Path (propagation) reliability 99% (time availability)

Service probability 50%

Disregard antenna-to-medium coupling loss

From this information we can calculate the receiver noise threshold (refer to Section 5.4.2):

$$\text{Noise threshold} = -204 + 10 \log B_{IF} + 3$$

$$= -204 + 10 \log 2.5 + 10^6 + 3$$

$$= -204 + 10 \times 6.398 + 3$$

$$= -137 \text{ dBW}$$

(i.e., where the carrier-to-noise ratio exactly equals the thermal noise of the receiver; by definition, the noise threshold).

CCIR Rec. 395-1 recommends a thermal plus IM noise accumulation of noise at 3 pW/km. Our path is 310.5 km long. Therefore we may accumulate 310.5×3 pW, or 931.5 pWp of noise. It should be noted that CCIR refers to noise psophometrically weighted; thus all figures are in pWp.

931.5 pWp equates to 29.7 dBrnC (Table 1.4) for a signal-to-noise ratio of 59 dB. This conversion is made using formula (5.22a).

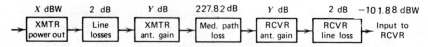

Figure 6.10 Network analogy of tropospheric scatter path analysis.

From a given signal-to-noise ratio S/N we can derive a carrier-to-noise ratio C/N as follows:

$$\frac{S}{N} = \frac{C}{N} + FM_{dB} + D_{im} - L_f + P_{im} \qquad (6.16)$$

where FM_{dB} = FM improvement factor (dB)

D_{im} = diversity improvement factor (dB)

L_f = noise load factor NLR (dB), derived in Section 5.6

P_{im} = preemphasis improvement factor (dB) taken from Figure 5.33 and discussed in Section 5.5.3

For the present assume FM_{dB} to be the traditional 20 dB implied in Section 5.4.2. Let

$$D_{im} = 7.2 \text{ dB (refer to Section 5.5.6) (Ref. 5)}$$

$$L_f = -1 + 4 \log N \text{ (we use CCIR loading)}$$

$$= -1 + 4 \log 60$$

$$= 6.12 \text{ dB}$$

$$P_{im} = 2.8 \text{ dB (from Figure 5.33), using 300 kHz as the highest modulating frequency}$$

Thus

$$\frac{S}{N} = 59 = \frac{C}{N} + 20 + 7.2 - 6.12 + 2.8 \text{ (all in dB)}$$

Table 6.3 Equipment Selection Table

Trans-mitter Power (kW)	$-X$ (dBW)	Antenna Reflector Diameters, Transmit and Receive (ft)	Y-dB Antenna Gain for 1 Antenna (dB)	$2Y$-dB Gains for Both Antennas (dB)	Sum of Gains to Attain Level (dBW)
2	+33	120	48.5	97	+130
10	+40	90	45.5	91	+131
20	+43	60	44	88	+131
50	+47	52	41.5	83	+130

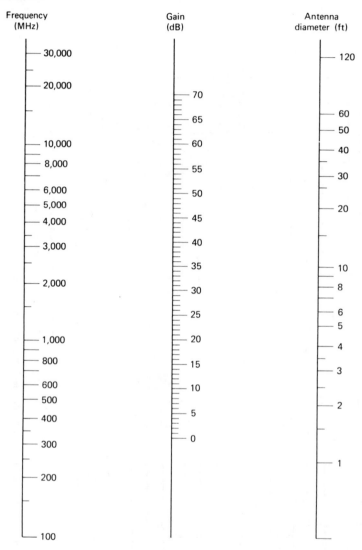

Figure 6.11   Antenna gain nomogram.

$$59 = \frac{C}{N} + 23.88$$

$$\frac{C}{N} = 35.12 \text{ dB}$$

This carrier-to-noise ratio of 35.12 dB must be maintained 99% of the time.

Select from Section 6.4.7, using the 50% service probability case, the tropo-

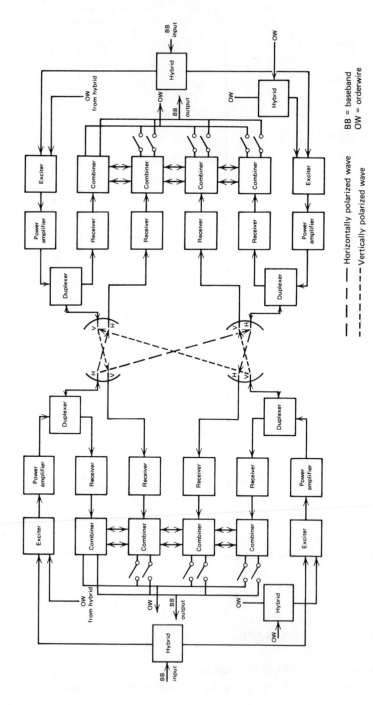

Figure 6.12 Simplified functional block diagram of a quadruple-diversity tropospheric scatter configuration.

—— —— —— Horizontally polarized wave
--------- Vertically polarized wave

BB = baseband
OW = orderwire

274

spheric scatter path loss for a time availability of 99%. This value is 227.82 dB. We want a carrier-to-noise ratio of 35.12 dB into the tropospheric scatter receiver. We also know that the receiver noise floor is −137 dBW. Thus the signal level should be 35.12 dB higher than this value, or −137 dBW + 35.12 dB = −101.88 dBW.

To achieve this level, allowing a total of 4 dB for line losses (2 dB at each end), we must select appropriate antenna sizes for the transmit and receive antennas and a transmitter output. We can reduce the problem to a set of networks in series as shown in Figure 6.10.

From the network in series in Figure 6.10 a simple formula may be derived:

$$-X \text{ dBW} + 2 \text{ dB} - Y \text{ dB} + 227.82 \text{ dB} - Y \text{ dB} + 2 \text{ dB} = -101.88 \text{ dBW}$$

or

$$-X \text{ dBW} - 2Y \text{ dB} + 231.82 \text{ dB} = -101.88 \text{ dBW}$$

$$X \text{ dBW} + 2Y \text{ dB} = 130 \text{ dBW}$$

From this we can build a table of values for X and Y to meet the required equivalent EIRP of + 130 dBW. Table 6.3 shows various combinations of standard tropospheric scatter transmitter power outputs as well as the antenna gains at each end to achieve the desired receiver input level. The antenna gains and reflector diameters are taken from Figure 6.11, the antenna gain nomogram. As we proceed in the discussion on tropospheric scatter, we can see how to make the most economical selection. For example, one possibility, which would ease the path gain requirement, would be to adopt a method of threshold extension to reduce overall costs.

The sample path shown used very tight requirements regarding noise (i.e., CCIR Rec. 395-1). Tropospheric scatter paths often are engineered to considerably reduced requirements such as described in CCIR Rec. 397-2, which allows deeper fades (Ref. 18).

6.9 EQUIPMENT CONFIGURATIONS

6.9.1 General

As indicated in Section 6.8, tropospheric scatter equipment must be configured in such a way as to (1) meet path requirements and (2) be an economically viable installation. All tropospheric scatter installations use some form of diversity (see Section 5.5.6). Except for some military transportable tropospheric scatter systems, quadruple diversity is the rule in nearly every case. A typical quadruple diversity tropospheric scatter system layout is shown in Figure 6.12. It is made up of identifiable sections as follows:

- Antennas, duplexer, transmission lines
- Modulator-exciters and power amplifiers

- Receivers, preselectors, and threshold extension devices
- Diversity and diversity combiners

Through proper site selection and system layout involving these four categories, realistic tropospheric scatter systems can be set up on paths up to 250 statute mi (400 km) in length.

6.9.2 Tropospheric Scatter Operational Frequency Bands

Tropospheric scatter installations commonly operate in the following frequency bands:

350–450 MHz	755–985 MHz
1700–2400 MHz	4400–5000 MHz

The reader should also consult CCIR Rec. 388 and CCIR Reps. 285-2 and 286.

6.9.3 Antennas, Transmission Lines, Duplexer, and Other Related Transmission Line Devices

ANTENNAS

The antennas used in tropospheric scatter installations are broadband high-gain parabolic reflector devices. The antennas covered here are similar in many respects to those discussed in Section 5.5.4, but have higher gain and therefore are larger and considerably more expensive. As we discussed in Section 5.5.4, the gain of this type of antenna is a function of the reflector diameter. Table 6.4 gives some typical gains for several frequency bands and several standardized reflector diameters. Figure 6.11 is a nomogram from which gain in decibels can be derived given the operating frequency and the diameter in.feet of the parabolic reflector. A 55% efficiency is assumed for the antenna. It should be noted that the tendency today is to improve feed methods, particularly where "decibels are so expensive," such as in the case of tropospheric scatter and earth station installations. Improved feeds illuminate the reflector more uniformly and reduce spillover, with the consequent improvement of antenna efficiency. For example, for a 30-ft reflector operating at 2 GHz, improving the efficiency from about 55% to 61% will increase the gain of the antenna about 0.5 dB.

It is desirable, but not always practical, to have the two antennas (as shown in Figure 6.12) spaced not less than 100 wavelengths apart to assure proper space diversity operation. Antenna spillover (i.e., radiated energy in side lobes and back lobes) must be reduced to improve radiation efficiency and to minimize interference with simultaneous receiver operation and with other services.

The first side lobe should be down (attenuated) at least 23 dB and the rest of the

Table 6.4 Some Typical Antenna Gains

Reflector Diameter (ft)	Frequency (GHz)	Gain (dB)
15	0.4	23
	1.0	31
	2.0	37
	4.0	43
30	0.4	29
	1.0	37
	2.0	43
	4.0	49
60	0.4	35
	1.0	43
	2.0	49
	4.0	55

unwanted lobes down at least 40 dB from the main lobe. Antenna alignment is extremely important because of the narrow beamwidths. These beamwidths are usually less than $2°$ and often less than $1°$ at the half-power points (see Section 5.5.4).

A good VSWR is also important, not only from the standpoint of improving system efficiency, but also because the resulting reflected power with a poor VSWR may damage components further back in the transmission system. Often load isolators are required to minimize the damaging effects of reflected waves. In high-power tropospheric scatter systems these devices may even require a cooling system.

A load isolator is a ferrite device with approximately 0.5 dB insertion loss. The forward wave (the energy radiated toward the antenna) is attenuated 0.5 dB; the reflected wave (the energy reflected back from the antenna) is attenuated more than 20 dB.

Another important consideration in planning a tropospheric scatter antenna system is polarization (see Figure 6.12). For a common antenna the transmit wave should be orthogonal to the receive wave. This means that if the transmitted signal is horizontally polarized, the receive signal should be vertically polarized. The polarization is established by the feeding device, usually a feed horn. The primary reason for using opposite polarizations is to improve isolation, although the correlation of fading on diversity paths may be reduced. A figure commonly encountered for isolation between polarizations on a common antenna is 35 dB. However, improved figures may be expected in the future.

TRANSMISSION LINES

In selecting and laying out transmission lines for tropospheric scatter installations, it should be kept in mind that losses must be kept to a minimum. That additional

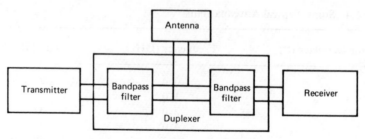

Figure 6.13 Simplified block diagram of a duplexer.

fraction of a decibel is much more costly in tropospheric scatter than in radiolink installations. The tendency, therefore, is to use waveguide on most tropospheric scatter installations because of its lower losses than coaxial cable. Waveguide is universally used above 1.7 GHz.

Transmission line runs should be less than 200 ft (60 m). The attenuation of the line should be kept under 1 dB from the transmitter to the antenna feed and from the antenna feed to the receiver, respectively. To minimize reflective losses, the VSWR of the line should be 1.05:1 or better when terminated in its characteristic impedance. Figures 5.25 and 5.26 show several types of transmission lines commercially available.

THE DUPLEXER

The duplexer is a transmission line device which permits the use of a single antenna for simultaneous transmission and reception. For tropospheric scatter application a duplexer is a three-port device (see Figure 6.13) so tuned that the receiver leg appears to have an admittance approaching (ideally) zero at the transmitting frequency. At the same time the transmitter leg has an admittance approaching zero at the receiving frequency. To establish this, sufficient separation in frequency is required between the transmitted and received frequencies. Figure 6.13 is a simplified block diagram of a duplexer. The insertion loss of a duplexer in each direction should be less than 0.5 dB. Isolation between the transmitter port and the receiver port should be better than 30 dB. High-power duplexers are usually factory tuned. It should be noted that some textbooks call the duplexer a diplexer.

6.9.4 Modulator–Exciter and Power Amplifier

The type of modulation used on tropospheric scatter transmission systems is commonly FM.* As our discussion develops, keep in mind that tropospheric scatter systems are high-gain low-noise extensions of the radiolink systems discussed in Chapter 5.

*Note: Digital tropo is becoming an important contender today in military systems. See chapter 15.

The tropospheric scatter transmitter is made up of a modulator–exciter and a power amplifier (see Figure 6.12). The power outputs are fairly well standardized at 1, 2, 10, 20, and 50 kW. For most commercial applications the 50-kW installation is not feasible from an economic point of view. Installations that are 2 kW or below are usually air-cooled. Those above 2 kW are liquid-cooled, usually with a glycol–water solution using a heat exchanger. If klystron power amplifiers are used, such tubes are about 33% efficient. Thus a 10-kW klystron will require at least 20 kW of heat exchange capacity.

The transmitter frequency stability (long-term) should be ±0.001%. Spurious emission should be down better than 80 dB below the carrier output level. Pre-emphasis is used as described in Section 5.5.2 and depends on the highest modulating frequency of the applied baseband.

The baseband configuration of the modulating signal, depending on the number of channels to be transmitted, is selected in the spectrum of 60–552 kHz (CCITT supergroups 1 and 2—see Section 3.3.6). However, CCITT subgroup A, 12–60 kHz, is often used as well. For longer route tropospheric scatter systems, the link design engineer may tend to limit the number of voice channels, selecting a baseband configuration that lowers the highest modulating frequency to be transmitted as much as possible. This tends to increase equivalent overall system gain by reducing the bandwidth of the IF [B_{IF} equation (6.17)], which is equivalent to reducing the RF bandwidth.

The modulator injects an RF pilot tone which is used for alarms at both ends as well as to control far-end combiners. 60 kHz is common in U.S. military systems. CCIR recommends 116 or 119 kHz for 24-channel systems, 304 or 331 kHz for 60-channel systems, and 607 (or 304) kHz for 120-channel systems (CCIR Rec. 401-2).

The modulator also has a service channel input. This is covered in CCIR Rec. 400-2. It recommends the use of the band, 300–3400 Hz. U.S. military systems often multiplex more than one service orderwire in the band 300–12,000 Hz. Often one of these channels may be used to transmit fault and alarm information.

The power amplifier should come equipped with a low-pass filter to attenuate second harmonic output by at least 40 dB and third harmonic output by at least 50 dB.

6.9.5 The FM Receiver Group

The receiver group in tropospheric scatter installations usually consists of two or four identical receivers in dual or quadruple diversity configurations, respectively. Receiver baseband outputs are combined in maximal ratio square combiners. See Section 5.5.6 for a discussion of combiners and the function of the ratio-square type combiner. A simplified functional block diagram of a typical quadruple diversity receiving system is shown in Figure 6.14.

The receiver noise threshold may be computed as follows:

$$\text{Noise threshold (dBW)} = 10 \log kTB_{IF} + \text{NF} \qquad (6.17)$$

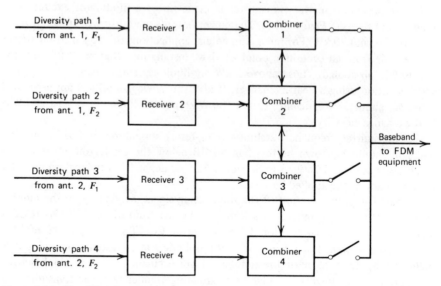

Figure 6.14 Simplified functional block diagram of a quadruple-diversity tropospheric scatter receiving system. F_1 = frequency 1; F_2 = frequency 2.

where k = Boltzmann's constant, 1.38×10^{-23} J K
$\quad T = 290$ K
$\quad B_{IF}$ = IF bandwidth (Hz)
$\quad$ NF = receiver noise figure (dB)

Typical receiver front-end NFs are given in Table 6.5. Table 6.6, which provides maximum IF bandwidths B_{IF} for several voice channel configurations, will be helpful in calculating some receiver noise thresholds when receiver front-end NFs are given. For our discussion, the receiver front-end NF will be the NF for the entire receiver. The figures in Table 6.6 were taken from (Ref. 19).

Having B_{IF} and NF, we can now calculate the noise threshold for each of the three voice channel configurations given using Tables 6.5 and 6.6. Table 6.7 tabulates the three bandwidths for each value of NF given in Table 6.5, giving the equivalent noise

Table 6.5   Typical Receiver Front-End NFs

Frequency Band (MHz)	Type	NF (dB)
350–450	Bipolar	1.0
775–985	Bipolar	1.0
1700–2400	GaAs FET	1.1
4400–5000	GaAs FET	1.5

Table 6.6 Maximum IF Bandwidths for Several Voice Channel
Configurations

Number of Voice Channels in FDM Baseband	Maximum IF Bandwidth B_{IF} (Hz)
36	3×10^6
60	6×10^6
120	10×10^6

threshold. The derived formula (i.e., -204 dBW + NF + 10 log B_{IF}) assumes un-cooled (room temperature) receivers.

From Table 6.7 it is obvious that to achieve FM improvement threshold, the receiver must have an input carrier-to-noise ratio equivalent to adding +10 dB algebraically to the noise threshold. For instance, if a receiver has a 3-dB noise figure, a B_{IF} of 3 MHz, its noise threshold is -136.2 dBW, and its FM improvement is 126.2 dBW.

Table 6.7 Noise and FM Thresholds for Several Receiver Figures
at Three IF Bandwidths

NF (dB)	Bandwidth (MHz)	Noise Threshold (dBW)	FM Improvement Threshold (dBW)
1.0	3	-138.2	-128.2
	6	-135.2	-125.2
	10	-133.0	-123.0
1.5	3	-137.7	-127.7
	6	-134.7	-124.7
	10	-132.5	-122.5
2.0	3	-137.2	-127.2
	6	-134.2	-124.2
	10	-132.0	-122.0
2.5	3	-136.7	-126.7
	6	-133.7	-123.7
	10	-131.5	-121.5
3.0	3	-136.2	-126.2
	6	-133.3	-123.2
	10	-131.0	-121.0
3.5	3	-135.7	-125.7
	6	-132.7	-122.7
	10	-130.5	-120.5
4.0	3	-135.2	-125.2
	6	-132.2	-122.0
	10	-130.0	-120.0

Another method of improving the equivalent equipment gain on a path is to use threshold extension techniques. The FM improvement threshold of a receiver can be "extended" by using a more complex and costly demodulator, called a *threshold extension demodulator.* The amount of improvement that can be expected using threshold extension over conventional receivers is on the order of 7 dB.* Thus for the above example, where the FM improvement threshold was −126.2 dBW without extension, with extension it would be −133.2 dBW.

Threshold extension works on an FM feedback principle which reduces the equivalent instantaneous deviation, thereby reducing the required bandwidth B_{IF}, which in turn effectively lowers the receiver noise threshold. A typical receiver with a threshold extension module may employ a tracking filter which instantaneously tracks the deviation with a steerable bandpass filter having a 3-dB bandwidth of approximately four times the top baseband frequency. The control voltage for the filter is derived by making a phase comparison between the feedback signal and the IF input signal.

6.9.6 Diversity and Diversity Combiners

Some form of diversity is mandatory in tropospheric scatter. Most present-day operational systems employ quadruple diversity. There are several ways of obtaining some form of quadruple diversity. One of the most desirable is shown in Figures 6.12, 6.14, where both frequency and space diversity are utilized. For the frequency diversity section, the system designer must consider the aspects of frequency separation illustrated in Figure 5.22. Space diversity is almost universally used, but the physical separation of antennas is normally in the horizontal plane with a separation distance greater than 100, and preferably 150 wavelengths.

Frequency diversity, although very desirable, often may not be permitted owing to RFI considerations. Another form of quadruple diversity, perhaps better defined as quasi-quadruple diversity, involves polarization, or what some engineers call *polarization diversity.* This is actually another form of space diversity and has been found not to provide a complete additional order of diversity. However, it often will make do when the additional frequencies are not available to implement frequency diversity.

Polarization diversity is usually used in conjunction with conventional space diversity. The four space paths are achieved by transmitting signals in the horizontal plane from one antenna and in the vertical plane from a second antenna. On the receiving end two antennas are used, each antenna having dual polarized feed horns for receiving signals in both planes of polarization. The net effect is to produce four signal paths that are relatively independent.

A discussion of diversity combiners is given in Section 5.5.6. There the feasibility of the maximal-ratio-square combiner is demonstrated, and consequently, it is the most commonly used combiner on tropospheric scatter communication systems.

---

*Assuming a modulation index of 3.

6.10 ISOLATION

An important factor in tropospheric scatter installation design is the isolation between the emitted transmit signal and the receiver input. Normally we refer to the receiver sharing a common antenna feed with the transmitter.

A nominal receiver input level for military tropospheric scatter systems is -80 dBm (Ref. 6) for design purposes. If a transmitter has an output power of 10 kW or +70 dBm and transmission line losses are negligible, then isolation must be greater than 150 dB.

To achieve overall isolation such that the transmitted signal interferes in no way with receiver operation when operating simultaneously, the following items aid the required isolation when there is sufficient frequency separation between transmitter and receiver.

- Polarization
- Duplexer
- Receiver preselector
- Transmit filters
- Normal isolation from receiver conversion to IF

6.11 INTERMODULATION (IM)

NPR measurements (see Section 5.6.2) are a good indication of operational IM distortion capabilities of tropospheric scatter equipment. When the NPR is measured on a back-to-back basis with 120-channel loading, we could expect a value of 55 dB. Once the same equipment is placed in operation on an active path, the NPR from the near-end transmitter to the far-end receiver may drop as low as 47 dB. The deterioration of the NPR is due to IM noise that can be traced to the intervening medium. It is just this IM distortion brought about in the medium that limits useful transmitted bandwidths in tropospheric scatter systems.

The bandwidth that a tropospheric scatter system can transmit without excessive distortion is related to the multipath delays experienced. These delays depend on the size of the scatter volume. The common volume is determined by antenna size and scattering characteristics.

6.12 MAXIMUM FEASIBLE MEDIAN PATH LOSS

A figure in decibels can be derived for a maximum median path loss that is feasible by considering the following problem:

Given a path with quadruple diversity reception
120-ft dishes

50-kW power amplifiers

12 FDM/SSB voice channels transmitted with the highest modulating frequency of 60 kHz

Receiver front-end NF 2.5 dB

Index of modulation 3

Aperture-to-medium coupling loss 2 dB

Line losses 2 dB at each end

Thus the peak deviation is 60 X 3 or 180 kHz. From this we calculate

$$B_{IF} = 2(FM_p + 2F_{bb}) = 2(180 + 2 \times 60) = 600 \text{ kHz}$$

that is, "twice the peak deviation plus four times the highest modulating frequency." We now calculate the receiver noise threshold from equation (6.17):

$$-204 + 10 \log 600 \times 10^3 + 2.5 = -143.7 \text{ dBW}$$

Assume that the FM improvement threshold is 10 dB above this figure, or -133.7 dBW.

Calculate the EIRP from the transmit antenna:

Transmitter power output, $= 10 \log 50 \times 10^3$ dBW	47 dBW
Gain of 120-ft dish at 2 GHz	+ 55 dB
	+102 dBW
Line and other insertion losses	- 2 dB
	+100 dBW

Calculate the receiver gain equivalent or the antenna gain at the receiver minus line losses and aperture-to-medium coupling loss, or 55 - 2 - 2 = 51 dB.

Consider the following networks in series to assist us in solving the problem.

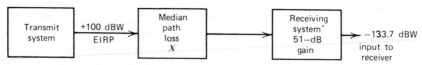

Antenna and transmission line only.

Thus the problem comes down to the following, denoting by X the maximum feasible tropospheric scatter path loss:

$$+ 100 \text{ dBW} - X + 51 \text{ dB} = -133.7 \text{ dBW}$$

that is, the minimum input level to the receiver. Then

$$X = 284.7 \text{ dB}$$

50-kW transmitters and 120-ft parabolic dishes are not feasible on a commercial basis. By reducing the installation size to 60-ft dishes and 10-kW amplifiers, an equivalent figure can be calculated which is indicative of a commercially feasible maximum.

Thus the transmit system has an EIRP of +40 dBW + 49 dB – 2 dB (transmitter output plus antenna gain minus lines losses) = +87 dBW. The receiving system equivalent gain will be reduced by 6 dB from the previous example, or 45 dB (i.e., 51 dB – 6 dB). Then we have

$$87 \text{ dBW} - X + 45 \text{ dB} = -133.7 \text{ dBW}$$

and

$$X = 265.7 \text{ dB (see Table 6.9)}$$

6.13 TYPICAL TROPOSPHERIC SCATTER PARAMETERS

Table 6.8 presents some of the more important path parameters for several operational tropospheric scatter paths. Note that TOAs are not given, nor is the tropospheric scatter path loss. Now consult Table 6.9, which supports the preceding section. This table compares measured tropospheric scatter path loss with cal-

Table 6.8 Some Typical Troposcatter System Parameters

Path Distance (km)	Frequency Band (MHz)	Transmitter Power (kW)	Antenna Diameter (m)[a]	Diversity	Channel Capacity	Comments
150–250	1000	1	9	4	72	
	1000	2	9	2	84	
	1000	2	14	4	108	
	1000	10	18	4	132	Parametric amplifier
	4000	5	9	4	24	
250–320	1000	10	18	2	36	
	1000	10	18	4	72–240	
	2000	10	9	4	72	
	2000	10	18	4	36	
320–420	1000	0.5	18	2	6	
	1000	10	18	2	36	
	1000	10	18	4	24–48	
	2000	10	18	4	36	
320–500	1000	10	18	4	24	Parametric amplifier
500–600	1000	10	37	4	24	Parametric
700–900	1000	50	37	4	24	amplifier
	1000	57	37	4	24[b]	

[a] Nominal diameter of parabolic reflector.

[b] Can support only 12 operational channels with a bit error rate on data dystems in excess of 1×10^{-3} indicative of noise hits due to fades.

Table 6.9   Tropospheric Scatter Path Loss—Measured Values Versus Calculated Values Using Three Calculation Methods

World Area	Path Length (km)	Frequency (MHz)	Predicted Tropospheric Scatter Path Loss (dB)			Measured Loss
			NBS 101	CCIR	Yeh	
Northeastern United States	275	460	191.8	192.4	200.9	196.0
United Kingdom	275	3480	218.7	220.3	221.6	220.9
Japan	300	1310	204.1	206.0	207.2	211.0
Southeast Asia	595	1900	261.0	254.8	243.9	260.9
South Caribbean	314	900	184.7	186.0	186.8	189.0
Canada, New York State	465	468	218.2	218.7	223.9	220.6
South Asia	738	1840	255.0	254.5	236.8	245.7

Source: Ref. 16.

culated values using three of the accepted, tropospheric scatter path loss calculation methods.

6.14   FREQUENCY ASSIGNMENT

The problem of frequency assignment in tropospheric scatter systems is similar to that of radiolink systems (see Section 5.14). The problem with tropospheric scatter becomes more complex because of the following:

- Radiated power is much greater (on the order of 30–60 dB greater).
- Nearly all installations are quadruple diversity.
- Receivers are more sensitive, with front-end NFs of about 3 dB versus about 10 dB for radiolink receivers.

Furthermore, splatter must be controlled so as not to affect other nearby services. The splatter may be a result of side lobe radiation or from radiation on unwanted frequencies.

For CCIR references, the reader may wish to consult CCIR Rec. 283-2. CCIR Reps. 285-2 and 286 offer some guidance on frequency arrangement.

To reduce splatter from harmonics, (Ref. 19)* recommends a transmitter low-pass filter attenuating second-harmonic output at least 40 dB and third harmonics at least 50 dB. (Ref. 19) also specifies the following:

- The minimum separation between transmit and receive carrier frequency of the same polarization on the same antenna shall be 120 MHz.

*This reference is given for guidance only. It is superceded by the Mil-STD-188-100 series, in which these recommendations have been omitted.

- The minimum separation between transmit and receive carrier frequency at a single station shall be 50 MHz, but in any case an integral multiple of 0.8 MHz.
- To avoid interference within a single station, separation of the transmit and receive frequencies shall not be near the first IF of the receiver.
- The minimum separation of transmit (receive) carrier frequencies is 5.6 MHz on systems with 36 FDM voice channels or less, 11.2 MHz for 60 voice channels, and 16.8 MHz for 120 voice channels (B_{IF} assumed as in Table 6.6).

This document further states that frequency channels shall be assigned on a hop-by-hop basis such that the median value of an unwanted signal in the receiver shall be at least 10 dB below the inherent noise of the receiver (i.e., noise threshold) when using the same or adjacent frequency channels in two relay sections.

REFERENCES AND BIBLIOGRAPHY

1. *Reference Data for Radio Engineers*, 6th ed., Howard W. Sams, Indianapolis, IN, 1977.
2. R. L. Freeman, "Multichannel Transmission by Tropospheric Scatter," *Telecommun. J.* (ITU), Geneva, June 1969.
3. P. L. Rice *et al.*, *Transmission Loss Predictions for Tropospheric Scatter Communication Circuits*, Tech. Note 101, as revised, National Bureau of Standards, Boulder, CO, Jan. 1967.
4. K. O. Hornberg, *Siting Criteria for Tropospheric Scatter Propagation Communication Circuits*, Memo. Rep. PM-85-15, National Bureau of Standards, Boulder, CO, April 1959.
5. "General Engineering—Beyond-Horizon Radio Communications," USAF T.O. 31Z-10-13, U.S. Department of Defense, Washington, DC, Oct. 1971.
6. Military Standard Mil-Std-188-313, U.S. Department of Defense, Washington, DC, Dec. 1973.
7. A. P. Barghausen *et al.*, *Equipment Characteristics and Their Relationship to Performance for Tropospheric Scatter Communication Circuits*, Tech. Note 103, National Bureau of Standards, Boulder, CO, Jan. 1963.
8. R. L. Marks *et al.*, *Some Aspects of the Design for FM Line-of-Sight Microwave and Troposcatter Systems*, USAF Rome Air Development Center, NY, U.S. Technical Information Service, AD 617-686, Springfield, VA, Apr. 1965.
9. E. F. Florman and J. J. Tory, *Required Signal-to-Noise Ratios, RF Signal Power and Bandwidth for Multichannel Radio Communication Systems*, Tech. Note 100, National Bureau of Standards, Boulder, CO, Jan. 1962.
10. A. P. Barsis *et al.*, *Predicting the Performance of Long Distance Tropospheric Communication Circuits*, Rep. 6032, National Bureau of Standards, Boulder, CO, Dec. 1958.
11. *Forward Propagation Tropospheric Scatter Communications Systems*, Handbook for Planning and Siting, USAF T.O. 31R-5-1-11, U.S. Department of Defense, Washington, DC, as revised Nov. 30, 1959.
12. E. D. Sunde, "Digital Troposcatter Transmission and Modulation Theory," *Bell Sys. Tech. J.*, Vol. 43, Jan, 1964.
13. E. D. Sunde, "Intermodulation Distortion in Analog FM Tropospheric Scatter Systems," *Bell Sys. Tech. J.*, Jan. 1964.
14. P. F. Panter, *Communication Systems Design—Line-of-Sight and Troposcatter Systems*, McGraw-Hill, New York, 1972.

15. *Naval Shore Electronics Criteria–Line-of-Sight and Tropospheric Scatter Communication Systems*, Navelex 0101.112, Department of the Navy, Washington, D.C., May 1972.

16. R. Larsen, "A Comparison of Some Troposcatter Prediction Methods," Conference Paper, IEE Conference on Tropospheric Radio Wave Propagation, Sept.-Oct. 1968.

17. CCIR Rep. 233-1, Vol. II, CCIR, Oslo, 1966.

18. CCIR Rep. 397-2, Vol. IV, CCIR, Oslo, 1966.

19. *DCS Engineering Installation Manual* DCAC 330-175-1, through Change 9, U.S. Department of Defense, Washington, DC.

20. "Facility Design for Tropospheric Scatter," Mil-Hdbk 417, U.S. Department of Defense, Washington, DC, Nov. 1977.

21. L. P. Yeh, "Tropospheric Scatter Communication Systems," presented at the ITU World Planning Committee Meeting, Mexico City, 1967.

7 | EARTH STATION TECHNOLOGY

7.1 INTRODUCTION

The U.S. Federal Communications Commission and a number of world bodies have come to accept the term *earth station* as a radio facility located on the earth's surface which communicates with satellites (or other space vehicles, for that matter). A *terrestrial station* is a radio facility on the earth's surface operating directly with other similar facilities on the earth's surface. Today the term earth station has come more to mean a radio station operating with other stations on the earth via an orbiting satellite relay.

This chapter deals with the design of those earth stations that operate with synchronous earth satellites. Such satellites orbit the earth with a period of 24 h. Thus they appear to be stationary over a particular geographic location on earth. The altitude of a synchronous satellite is 22,300 mi (35,900 km) above the surface of the earth.

7.2 THE SATELLITE

A communication satellite is an RF repeater.* It may be represented in its most simple configuration as shown in Figure 7.1. Theoretically three such satellites properly placed in synchronous orbit could provide 100% earth coverage from an earth station located anywhere on the earth's surface. This concept is shown in Figure 7.2.

7.3 EARTH–SPACE WINDOW

In selecting a band of frequencies that is optimum for earth–space communication, two important phenomena must be taken into account: atmospheric absorption and noise, both galactic and man-made. Thus the problem is to select a band of frequencies that will permit broadband communication where the earth–space signal will suffer minimum attenuation due to absorption and where the inherent

---

*This simplification will no longer be true with the advent of processing satellites. Chapter 15 describes one form of processing.

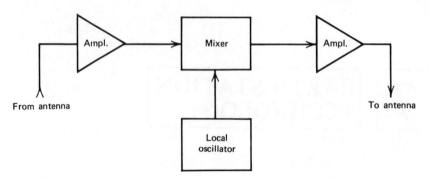

Figure 7.1 Simplified functional block diagram of the radio relay portion of a typical communication satellite.

noise level is least. Regarding absorption, an opening exists between 10 and 10,000 MHz. Such an opening is more commonly called a *window*. This same band is limited on the lower end by noise to frequencies above 1000 MHz. Therefore the optimum earth–space window is between 1000 and 10,000 MHz. This by no means indicates that earth–space communication cannot be and is not carried out on frequencies outside this region. The noise effect is shown in Figure 7.3 and the absorption effect in Figure 7.4*a*. Figure 7.4*b* shows the variation of attenuation due to precipitation and adds emphasis to the desirability of operating below 10 GHz. This point is delved into in Section 10.7.

In the 1000–10,000-MHz window, four bands, each 500 MHz wide, have been assigned for use with communication satellites. These bands are as follows:

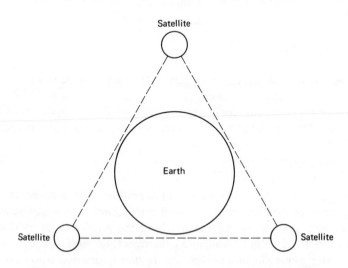

Figure 7.2 Three synchronous satellites properly placed can provide 100% earth coverage.

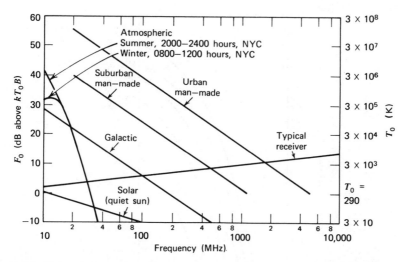

Figure 7.3 Noise window effect. Median values of average noise power expected from various sources, using omnidirectional antenna near earth's surface (Ref. 2).

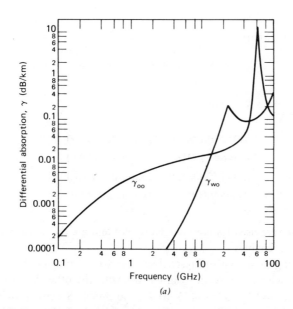

(a)

Figure 7.4 (*a*) Atmospheric absorption versus frequency. (*b*) Attenuation due to precipitation [from CCIR Rep. 234 (Oslo, 1966) and (Ref. 2)].

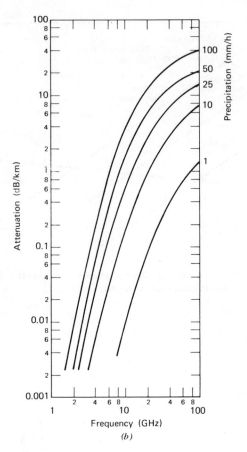

Figure 7.4 (*Continued*)

3700–4200 MHz	(Satellite–earth)
5925–6425 MHz	(Earth–satellite)
7250–7750 MHz	(Satellite–earth)
7900–8400 MHz	(Earth–satellite)

The first two bands are used for commercial service and are those that will be essentially covered in this chapter.

Frequency sharing exists with terrestrial services for all four of the bands. CCIR discusses ways in which one service can coexist with the other in its Recs. 382-2 (3700–4200 MHz), 383-1 (5925–6425 MHz), and 386-1. Specific problems arising from frequency sharing and the development of coordination contours are discussed in Section 7.16. Space communication has shown phenomenal growth since 1970 and will continue to grow for the coming years. This growth has been such that there is now a tendency of orbital crowding of geostationary satellites operating in the 4–6-GHz bands. The problem has been further multiplied by the proliferation of small-antenna earth stations with inherent wider beamwidths. This implies less satellites in the desirable "window" band due to the fact that greater angular separation between satellites is required to permit small antennas (e.g., less than 12 m in diameter) to discriminate satellites. Here we mean that they will suffer minimal interference from neighboring satellites operating in the same band. This is forcing the use of frequencies above 10 GHz. Several of these bands are listed below. They are valid for ITU region 2.

$$11.7-12.2 \text{ GHz} \quad \text{(Downlink)}$$
$$14.0-14.5 \text{ GHz} \quad \text{(Uplink)}$$
$$17.7-20.2 \text{ GHz} \quad \text{(Downlink)}$$
$$27.5-30.0 \text{ GHz} \quad \text{(Uplink)}$$

The propagation problems peculiar to these bands are discussed in Chapter 10.

7.4 PATH LOSS

Section 5.3.1 covered LOS* path loss. The earth–satellite–earth communication link may be considered a radiolink hop to an RF repeater (the satellite) and another hop from the satellite to earth. The path loss must be computed in both cases. Here we use equation (5.1):

$$L_{\text{dB}} = 36.6 + 20 \log D_{\text{mi}} + 20 \log F_{\text{MHz}}$$

where D = distance (statute mi) and F = operating frequency (MHz). For midband, 3950 MHz,

$$L_{\text{dB}} = 36.6 + 20 \log (22{,}300) + 20 \log (3950)$$
$$= 195.498 \text{ dB}$$

For the highest frequency in the 4-GHz band, 4200 MHz,

$$L_{\text{dB}} = 36.6 + 20 \log (22{,}300) + 20 \log (4200)$$
$$= 36.6 + 86.96 + 72.464 = 196.02 \text{ dB}$$

For 6175 MHz (midband),

$$L_{\text{dB}} = 36.6 + 20 \log (22{,}300) + 20 \log (6175)$$
$$= 199.37 \text{ dB}$$

---

*More properly defined as *free-space* path loss.

For 6425 MHz, the highest frequency in the 6-GHz band,

$$L_{dB} = 36.6 + 20 \log (22,300) + 20 \log (6425)$$

$$= 199.72 \text{ dB}$$

7.5 SATELLITE-EARTH LINK

By international agreement the maximum EIRP of a communication satellite is +32.1 dBW (Ref. 1) per carrier. This limit has been placed on satellite radiators to minimize interference with shared terrestrial services. The interference effects per se are dealt with in Section 7.16. The problem to be discussed here is how to derive a useful signal at the earth station on the downlink.

It will be seen that the downlink is the limiting factor on earth station design. The uplink (i.e., the link from earth to satellite) may take advantage of powerful earth station transmitters, high-gain narrow-beam antennas with little likelihood of interference with other services. On the other hand, the downlink is severely limited in radiated power owing to the possibility of interfering with terrestrial services. Thus basically we can say that it is the downlink around which earth stations are designed.

Let us consider the downlink and compute the signal level impinging on an earth station antenna. Use the worst path loss from Section 7.4. Therefore,

$$+32 \text{ dBW} - 196.02 = -164.02 \text{ dBW}$$

Such a broadband signal level is very low indeed, and the earth station must be designed using the most refined radiolink techniques to achieve a proper signal-to-noise ratio at the voice channel output of the demultiplexer.

The reader must remember that earth stations using frequency division multiple access (FDMA) are extensions of the technology developed in Chapters 3, 5, and 6. Briefly stated,

- The radio signal is frequency modulated by a multichannel telephone signal made up of FDM/SSB baseband.
- Because the signal traverses so little atmosphere, the received signal level is not affected by fading except under conditions of very heavy rain.

Now let us take the technology of Chapter 5, and, with a little imagination and reasoning, the so-called difficult concepts of earth stations come to light. Use Table 7.1 as the basis for this introductory discussion; it provides parameters of the INTELSAT IVA/V type of satellites. If -164 dBW impinges on an earth station antenna and, assuming for the discussion's sake that the receiving antenna had a zero gain, allowing a 10-dB NF for the receiver, and a 4-MHz IF bandwidth B_{IF} for a 60-channel system (item 3, Table 7.1), we can compute the receiver's FM improvement threshold [equation (5.7)]:

Table 7.1 INTELSAT IVA/V Characteristics FDM/FM Operation—Selected Capacities

1. Carrier capacity (VF channels)	12	24	60	132	612	972
2. Allocated satellite bandwidth (MHz)	1.25	2.5	5.0	10	20	36
3. Occupied satellite bandwidth (MHz)	1.125	2.0	4.0	7.5	17.8	36
4. Baseband occupancy (kHz)	12–60	12–108	12–252	12–552	12–2540	12–4028
5. Deviation, rms, 0 dBm0 test tone (kHz)	109	164	270	430	454	802
6. Multichannel rms deviation (kHz)	159	275	546	1020	1996	4417
7. Carrier-to-thermal noise power ratio at operating point (8000 + 200 pWp0) (dBW/K)	−154.7	−153	−149.9	−147.1	−134.2	−135.2
8. Carrier-to-noise ratio in occupied bandwidth (dB)	13.4	12.7	21.1	12.7	21.9	17.8

Source: INTELSAT BG-28-72E M/6/77 (Aug. 24, 1977).

$$\text{Threshold dBW} = -204 + 10 \log B_{IF} + 10 + NF_{dB}$$

$$= -204 + 66 + 10 + 10$$

$$= -118 \text{ dBW}$$

The next step is to compute the minimum gain of a parabolic antenna to bring the nominal signal level of -164 dBW to the FM improvement threshold of the receiver, or

$$-164 \text{ dBW} - (-118 \text{ dBW}) = 46 \text{ dB}$$

A 46-dB gain requires a reflector diameter of 22 ft (7 m) at 55% efficiency at 4 GHz.

Now open the bandwidth up to 36 MHz (Table 7.1, item 3). It is seen that the antenna system would require nearly 10 dB more gain, or about a gain of 56 dB. (Use the nomogram in Figure 6.11). Thus an antenna of almost 18.9 m (62 ft) would be necessary.

Let us stretch our imagination a little further by adding a second carrier to our receiver system. A power split would be needed at the antenna feed to allow the addition of a second receiver. This places at least a 3-dB loss in what was a lossless system requiring an increase of the antenna size to something nearer 27.4 m (90 ft).

The analogy ends here. Let us now consider a modern earth station. Obviously a major change will have to take place in our station design in order to have a station capable of receiving numerous RF carriers that may be assigned anywhere in the 500-MHz downlink bandwidth (i.e., from 3700 to 4200 MHz).

To improve system sensitivity, the size of the antenna could be increased and the receiver made more sensitive. Economics have dictated that antennas larger than the nominal 100-ft (30-m) dish have a cost trade-off that is untenable. Large antennas tend to increase in price exponentially as their size increases, as the following equation shows (Ref. 5):

$$C \sim D^{2.8} \qquad\qquad (7.1)$$

where C = cost and D = diameter of the parabolic antenna. Thus large commercial earth stations (i.e., INTELSAT A) have standardized on the nominal 100-ft (30-m) antenna.

7.5.1 Figure of Merit G/T

The figure of merit of an earth station G/T has been introduced into the technology to describe the capability of an earth station to receive a signal from a satellite. One of the basic publications covering earth stations operating in the INTELSAT network, BG-28-72E (Ref. 4) is quoted in part:

Antenna System, Gain-to-Noise Temperature Ratio. Approval of an earth station in the category of standard A earth stations will only be obtained if the following minimum condition is met in the direction of the satellite and for the polarizations chosen for the satellite series under clear sky conditions, in light wind, and for any frequency in the band 3.7–4.2 GHz:

$$\frac{G}{T} \geqslant 40.7 + 20 \log_{10} \left(\frac{f}{4}\right)^* \text{(dB/K)} \tag{7.2}$$

G is the antenna gain measured at the input of a low-noise amplifier expressed in dB relative to an isotropic radiator; and T is the receiving noise temperature referenced to the input of a low-noise amplifier, expressed in dB relative to 1 kelvin.

*f is the receiver frequency, expressed in GHz.

At exactly 4 GHz formula (7.2) could be stated in the following manner:

$$40.7 \text{ dB} = G_{dB} - 10 \log T \tag{7.3}$$

The earth station system designer must provide a G/T of 40.7 dB or better. The problem is how to achieve it. In order to meet the G/T requirement, a cryogenically cooled parametric amplifier (20 K) and an improved antenna with a 60-dB or greater gain are required. The parametric amplifier system noise temperature coupled with the remainder of the noise contributions provides a system noise temperature not in excess of 85.1 K. (*Note.* Noise temperature and NF are discussed in Section 1.13.)

One may reach this figure by following some simple manipulation of algebra. As stated,

$$G - 10 \log T = 40.7 \text{ dB}$$

Allow the antenna gain at 4 GHz to be 60 dB, which is a conservative figure for a nominal 100-ft (30-m) antenna, given the inherent efficiencies, when modern feed devices are used (Cassegrain). Now substitute 60 dB in the formula (7.3). Thus

$$60 - 10 \log T = 40.7 \text{ dB}$$

$$10 \log T = 60 - 40.7$$

$$= 19.3 \text{ dB}$$

Then

$$T = 81.5 \text{ K}$$

Many system designers want a certain field margin to the minimum G/T and design to 41.5 dB. Therefore

$$10 \log T = 60 - 41.5$$

$$= 18.5 \text{ dB}$$

Then

$$T = 70 \text{ K}$$

Consider point-to-point (radiolink) microwave at 4 GHz. Allow a 10-dB NF for a typical FM receiver operating in that band. The receiver noise temperature turns out to be approximately 2600 K. Noise temperature can be related to NF by formula (1.28):

$$\text{NF}_{dB} = 10 \log \left(1 + \frac{T}{290} \right)$$

The T we have been using is more commonly referred to as equivalent noise temperature and is the summation of all noise components in the receiving system. This concept has been with us all the time in Chapters 5 and 6. We have simply shunted it aside because receiver noise temperatures were so large (e.g., 2600 K) that the other noise components were insignificant. In earth station technology, the receiver noise temperature has been lowered so much that other contributors now become important.

The noise components comprising T may be broken down into four categories:

- Antenna noise, usually taken at the $5°$ elevation angle of the antenna, which is the worst allowable case. This noise includes antenna "spillover," galactic noise, and atmospheric noise.
- Passive component noise. This is the summation of the equivalent noise from the passive components before the incoming signal reaches the first active component.

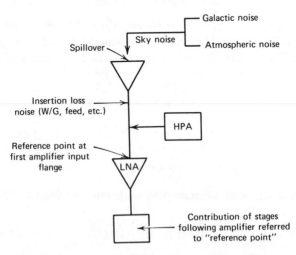

Figure 7.5 Graphical representation of noise contributors. HPA = high-power amplifier; LNA = low-noise amplifier.

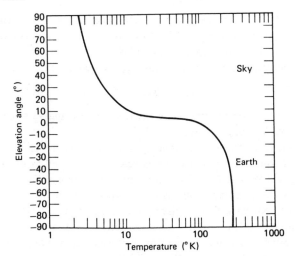

Figure 7.6 Approximate sky noise variation with antenna elevation angle (4 GHz).

- High-power amplifier (HPA) leakage noise.
- Sum of the excess noise contributions of the various active amplifying stages of the receiving system.

Figure 7.5 represents these noise contributors graphically. Figure 7.6 shows approximate sky noise variation with antenna elevation angle. At an angle of 5° we see that the sky noise temperature reaches the order of 25 K. It will also be seen that minimum antenna noise occurs when the antenna is at zenith (i.e., an elevation angle of 90°). *Note.* Elevation angles are referred to the horizon; thus the elevation angle would be 0° when the antenna is pointed directly at the horizon.

Antenna spillover refers to radiated energy from the antenna to the ground and scatter off antenna spars. In both cases noise generators are formed and must be considered in a "noise budget." Spars refer here to those metal elements used to support the antenna feeding device. The sum total of antenna noise may reach 39 or 40 K, 25 K of which is sky noise.

We now must calculate the system noise temperature T_{sys} of the several elements in tandem making up the receiver chain. This chain may be represented as follows:

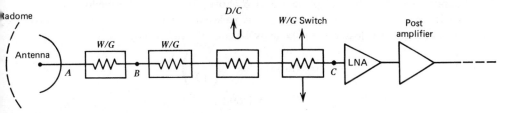

where A, B, and C are reference points or reference planes. A is at the base of the radiating element, perhaps a horn. B is at the base of the antenna pedestal, and C is the input point to the low-noise amplifier (LNA).

To calculate the system noise temperature T_{sys} we can say:

$$T_{sys} = T_{ant} + T_r \qquad (7.4)$$

where T_{ant} = antenna noise temperature discussed above and T_r = receiving system noise temperature. All temperatures are in kelvins.

The first step is to establish a reference point or reference plane. This is an arbitrary point from where the antenna gain is calculated as well as its noise temperature T_{ant}. T_{sys} will vary as G will vary, depending on that point of reference. We will find that as the reference point is moved further from the antenna feed, the gain will decrease, but so will the noise temperature. However, G/T for a given system will remain constant, no matter what the reference. In the above drawing each ohmic loss component is a noise generator, as is each active component such as the LNA and the postamplifier, mixer, IF amplifiers, and so on. Noise contributors to the left of the reference plane are included in T_{ant} in equation (7.4) and always include the sky noise. To the right of the reference plane, that is, down the system, away from the antenna, all noise contributors are included in T_r.

To differentiate between ohmic and nonohmic losses, consider that all devices with an insertion loss are in the ohmic category and all losses not associated with an insertion loss are nonohmic. Examples of nonohmic losses are free space loss and pointing loss.

The analysis to determine T_{sys} is a two-step operation, namely T_{ant} and T_r are calculated separately, and then the sum of the two is taken. This is simplified considerably by summing the ohmic losses involved separately for each case.

When calculating the noise contribution of an ohmic loss, which is given in the traditional unit of measurement, the decibel, we must convert the decibel value to its equivalent numerical ratio. From Chapter 1,

$$\text{Loss (or gain) (dB)} = 10 \log_{10} \left(\frac{P_1}{P_2} \right)$$

Let $P_1/P_2 = L$. Then

$$\text{Loss (ratio)} = \log_{10}^{-1} \left(\frac{L}{10} \right) \qquad (7.5)$$

Suppose that the reference plane was at B above, there was a radome loss of 1 dB, and the waveguide loss to the base of the pedestal was 1.3 dB. Calculate the loss ratio. First sum the losses, then divide by 10 and take the antilog:

$$\text{Loss (ratio)} = \log^{-1} \left(\frac{2.3}{10} \right)$$

$$= 1.698$$

Assume L_T to be the total loss of the antenna network, including the radome, expressed as a loss ratio. Then

$$T_{ant} = \frac{(L_T - 1) T_{amb} + T_s}{L_T} \tag{7.6}$$

where T_s = sky noise in kelvins and T_{amb} = ambient temperature in kelvins, traditionally given as 290 K (17°C). In practical systems it is wise to use the highest ambient expected (worst case). For military systems 55°C is often used, equating to 328 K.

The receiver noise temperature T_r is the total receiver noise obtained by referring the effects of the LNA contribution (and subsequent amplifiers or mixers) and the input circuit loss to the same reference plane as that of the antenna noise temperature. In other words, we are dealing with all the noise contributors to the right of the arbitrary reference plane (i.e., down the system).

When computing T_r we must utilize the traditional cascade formula for noise temperature:

$$T_T = T_1 + \frac{T_2}{G_1} + \frac{T_3}{G_1 G_2} + \frac{T_4}{G_1 G_2 G_3} + \frac{T_5}{G_1 G_2 G_3 G_4} + \cdots$$

where T_1 = noise temperature of noise contributor n and G_n = gain of contributor n, $n = 1, 2, \ldots$:

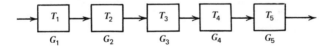

Remember that the loss of a device that attenuates a signal may be expressed as an equivalent gain which is less than 1.

The receiver noise temperature T_r is expressed as:

$$T_r = (L_i - 1) T_{amb} + T_{LNA} L_i + \frac{T_{pa} L_i}{G_{LNA}} + \cdots \tag{7.7}$$

where L_i = sum of the losses from the reference plane to the LNA input, where these losses are again expressed as a ratio, T_{LNA} = noise temperature in kelvins of the LNA, G_{LNA} = gain of the LNA, and T_{pa} = noise temperature in kelvins of the postamplifier, where applicable, or the mixer.

Example. Given a sky noise of 50 K, a waveguide loss of 0.2 dB to the base of the antenna pedestal, and another waveguide loss of 0.15 dB making up the waveguide from the pedestal base to the input of the LNA, a directional coupler insertion loss of 0.09 dB, a waveguide switch insertion loss of 0.07 dB, T_{LNA} = 105 K, G_{LNA} = 30 dB, and T_{pa} = 600 K.

1. What is T_{sys} when the reference plane is placed at the antenna pedestal base?
2. If the antenna gain is 40 dB, what is G/T, using the same reference plane as in question 1?

Calculate T_{ant}. Sum the ohmic losses to the reference plane. The waveguide loss to the base of the antenna pedestal is 0.2 dB. Then, from equation (7.5),

$$L_T = \log_{10}^{-1}\left(\frac{0.2}{10}\right)$$

$$= 1.047$$

Assume T_{amb} = 290 K. Then, from equation (7.6),

$$T_{ant} = \frac{(L_T - 1)\,T_{amb} + T_s}{L_T}$$

$$= \frac{(1.047 - 1)\,290 + 50}{1.047} = \frac{13.63 + 50}{1.047} = 60.77 \text{ K}$$

Calculate T_r. Sum the losses from the reference plane to the input of the LNA:

Waveguide loss	0.015 dB
Directional coupler loss	0.09
Waveguide switch	0.07
Total	0.175 dB

Then

$$L_i = \log_{10}^{-1}\left(\frac{0.175}{10}\right)$$

$$= 1.041$$

Using equation (7.7),

$$T_r = (L_i - 1)\,T_{amb} + T_{LNA}L_i + \frac{T_{pa}L_i}{G_{LNA}}$$

$$= (0.041)\,290 + 105\,(1.041) + \frac{600\,(1.041)}{1000}$$

$$= 121.8 \text{ K}$$

If T_{sys} = 182.57 K and G = 40 dB - 0.2 dB = 39.8 dB,

$$\frac{G}{T} = G - 10\log_{10}T_{sys}$$

$$= +17.18 \text{ dB}$$

It should be noted that in formula (7.7) G_{LNA} must be converted from its decibel value to its equivalent numerical value; in this case 30 dB is equivalent to 1000. Also, to calculate G/T, the raw antenna gain must be reduced by the losses incurred from the antenna radiating element to the reference plane selected; in this case a 0.2-dB loss was incurred.

7.5.2 The Ratio of Carrier-to-Thermal Noise Power

The carrier-to-thermal noise power ratio C/T provides an absolute measurement of carrier power regardless of bandwidth. It is related to the familiar carrier-to-noise ratio C/N in the following way:

$$\frac{C}{T}(dB) = \frac{C}{N}(dB) + 10 \log B + 10 \log k$$

where B = the bandwidth (Hz) and k = Boltzmann's constant.

This can be derived by letting

$$\frac{C}{N} = \frac{C}{kTB} \tag{7.8}$$

[This is a simple identity where we let $N = kTB$; see equation (1.14).]

$$\frac{C}{N}(dB) = \frac{C}{T}(dB) - 10 \log k - 10 \log B$$

Rearranging terms,

$$-\frac{C}{N} = -\frac{C}{T} + 10 \log k + 10 \log B$$

$$\frac{C}{T} = \frac{C}{N} + 10 \log k + 10 \log B \tag{7.9}$$

Table 7.1 gives some typical carrier-to-thermal noise power values for stations operating with INTELSAT IVA/V satellites.

Example. Let $C/N = 30$ dB, use item 3 of Table 7.1 for 60-channel operation, and let $10 \log k$ be the familiar -228.6 dBW/K (Section 1.13). Then

$$\frac{C}{T} = +30 - 228.6 + 66$$

$$= -132.6 \text{ dBW/K.}$$

The carrier-to-thermal noise power ratio for the tangential noise threshold, of course, is, where $C/N = 0$ dB,

$$\frac{C}{T} = -228.6 + 66$$

$$= -162.6 \text{ dBW/K}$$

7.5.3 Relating C/T to G/T

For the down link,

$$\frac{C}{T} = \text{EIRP (satellite)} - \text{path loss (4 GHz)} + \frac{G}{T} \text{(of earth station)} \qquad (7.10)$$

If +22.5 dBW is used for the EIRP of the satellite, -196 dB for the path loss, and +40.7 dB for the G/T of the earth station, then, for the example of Section 7.5.2,

$$\frac{C}{T} = +22.5 - 196 + 40.7$$

$$= -132.8 \text{ dBW/K}$$

Not included here is the EIRP adjustment for satellite backoff of power when more than one carrier is carried on a satellite transponder.

7.5.4 Deriving Signal Input from Illumination Levels

Many standards and specifications for earth stations give illumination levels or flux densities. From this information the earth station system engineer will want to derive a receive level at the feed of the earth station. The flux density is given in a level value per square meter or square foot.

A typical specification may state, "The specification in this exhibit shall apply for all illumination levels over the range of -150 dBW/m² to -116 dBW/m²." If the antenna described uses a 32-ft (10-m) parabolic reflector, what would be the input at the feed?

The first step is to compute the projected area. For the case of a parabolic antenna, this is a circle. For our 32-ft diameter antenna it would be a circle with a radius of 16 ft (5 m). For the area of the circle with radius 16 ft we have $\pi(16)^2$ or $256\pi =$ 804.25 ft² (78.5 m²). If our antenna were 100% efficient and illuminated with a signal with a flux density of -150 dBW/m², we would then multiply 78.5 by -150 dBW/m². Thus we have

$$10 \log 78.5 - 150 \text{ dBW/m}^2 = 18.95 - 150 = -131.05 \text{ dBW}$$

for the receive level at the feed for this hypothetical case.

These figures are for an antenna with 100% efficiency. Let us use something more practical, say, an antenna with 60% efficiency. Then we must multiply the projected area by 0.60 to get what is called the effective area:

$$78.5 \text{ m}^2 \times 0.60 = 47.10 \text{ m}^2$$

The input level at the feed will then be

$$+10 \log 47.1 - 150 \text{ dBW/m}^2 = 16.73 - 150 \text{ dBW}$$

$$= -133.27 \text{ dBW}$$

Another example is to use the higher flux density value from the above with the same antenna values. Thus

$$+10 \log 47.1 - 116 \text{ dBW/m}^2 = +16.73 - 116 \text{ dBW}$$

$$= -99.27 \text{ dBW}$$

Given the input level at the feed and subtracting the various line losses, the earth station engineer can easily calculate the input level to the receiver front end.

7.5.5 Station Margin

One of the major considerations when designing radiolink and tropospheric scatter systems is the fade margin. This is the signal level that was added in the system calculations to allow for fading. Often this value was on the order of 20–50 dB.

In other words the system in question was overbuilt in receive signal level by between 20 and 50 dB to overcome most fading conditions or to ensure that noise would not exceed a certain norm for a fixed time frame.

As we saw in Chapter 5, fading is caused by anomalies in the intervening medium between stations or by the reflected signal causing interference to the direct ray signal. There would be no fading phenomenon on a radio signal being transmitted through a vacuum. Thus satellite earth station signals are subject to fade only during the time they traverse the atmosphere. For this case most fading, if any, may be attributed to rainfall.

Margin or station margin is the additional design advantage that has been added to the station to compensate for deteriorated propagation conditions or fading. The margin designed into an LOS system is large and is achieved by increasing the antenna size, improving the receiver noise figure, or increasing the transmitter output power.

The station margin of a satellite earth station in comparison is small, on the order of 4–6 dB. Typical rainfall attenuation exceeding 0.01% of a year may be from 1 to 2 dB without a radome on the antenna and when the antenna is at a $5°$ elevation angle. As the antenna elevation increases to zenith, the attenuation notably decreases because the signal passes through less atmosphere. The addition of a radome could increase the attenuation to 6 dB or greater during precipitation. The receive station margin for an earth station, those extra decibels on the downlink, is often achieved by use of threshold extension demodulation techniques. These were discussed in Section 6.9.5. Uplink margin is provided by using larger transmitters and by increasing power output when necessary. A G/T in excess of 40.7 dB at the $5°$ elevation angle will also provide margin, but may prove quite expensive to provide, and the additional performance may be difficult to justify on an economic basis.

7.6 UPLINK CONSIDERATIONS

The uplink segment of an earth station in most cases is less critical from the point of view of earth station design. Usually the EIRP is specified for a particular satellite system, and these specifications must be carefully adhered to, particularly when the operation is in the FM/FDM mode, to avoid overload of a transponder since in most cases these transponders are shared by several users. Table 7.2 states EIRPs for INTELSAT IV and V global beam FDM/FM and TV carriers. The EIRP must also be adjusted for the elevation angle of the satellite from the particular earth station in question. For INTELSAT IV, IVA, and V satellites document BG-28-72E M/6/77 (Ref. 4) states, referring to the values of EIRP in Table 7.2:

> ... These values apply to regular carriers for earth stations with a $10°$ elevation angle and include a 1.0-dB mandatory margin. Correction factors necessary to compute the EIRP for earth stations with other elevation angles are computed by using the following formulas:
>
> $$-0.02(\alpha - 10) \text{ dB} \quad \text{for hemispheric (IVA and V) and zone (V)} \quad (7.11a)$$
>
> $$-0.06(\alpha - 10) \text{ dB} \quad \text{for global (IV, IVA, and V) and spot (IV)} \quad (7.11b)$$
>
> where α is the earth station elevation angle in degrees.

For some satellite systems the EIRP for an earth station is expressed as a level in dBW per voice channel. To determine the uplink carrier level for the composite of voice channels, one takes the required output per voice channel in dBW and adds logarithmically $10 \log N$, where N is the number of voice channels to be transmitted.

Suppose a specification required that an uplink EIRP be +61 dBW per voice channel and we wished to transmit 60 FDM channels on the link. Thus we would have

$$+61 \text{ dBW} + 10 \log 60 = 61 + 17.78 = +78.78 \text{ dBW}$$

Consider the nominal 100-ft (30-m) antenna to have a gain of 63 dB (at 6 GHz) and losses typically at 3 dB. What transmitter output power P_t is required?

$$\text{EIRP (dBW)} = P_t + G_{ant} - \text{line losses (dB)}$$

where P_t = output power of the transmitter (dBW) and G_{ant} = antenna gain (dB) (uplink).

Then in the example,

$$+78.78 \text{ dBW} = P_t + 63 - 3$$

$$P_t = +18.78 \text{ dBW}$$

$$= 75.6 \text{ W}$$

Table 7.2 INTELSAT IVA and V Required Earth Station EIRPs[a]
(Global Beam FDM/FM and TV Carriers)

| Bandwidth Unit (MHz) | Regular Carriers | |
	Carrier Size (channels)	EIRP (dBW)
1.25[b]	12	73.0
2.5	24	74.7
	36	77.7
	48	81.0
	60	83.7
5.0	60	77.8
	72	78.6
	96	82.2
	132	86.3
7.5	96	79.5
	132	81.8
	192	87.1
10.0	132	80.6
	192	83.3
	252	87.8
15.0	252	82.8
	312	85.2
20.0	432	86.6
25.0	432	85.1
36.0	972	90.1
	1092 [c]	93.6
17.5	432 [c, d]	88.0
17.5	TV[e]	88.0
30.0	TV[f]	88.0

Source: INTELSAT (Ref. 4).
[a] For 10° elevation angle, including 1 dB of mandatory margin.
[b] Approved for INTELSAT V only.
[c] Boxed channel numbers are not available for INTELSAT V.
[d] Contingency carrier.
[e] Half-transponder TV.
[f] Full transponder TV.

Typically for television transmission for INTELSAT V,

$$+88 \text{ dBW} = P_t + 63 - 3$$

$$P_t = +28 \text{ dBW}$$

$$= 631 \text{ W}$$

The earth station design engineer will probably be more concerned with trade-offs on the location of the transmitter power amplifier to minimize line losses, trade-offs on the use of TWTs, or klystrons in the power amplifier(s) considering problems of IM products under multicarrier operation.

7.7 MULTIPLE ACCESS

Our discussion in Sections 7.1 and 7.2 was extremely simplified. We should remain in that sphere of simplicity to describe the term "multiple access," which refers to the manner in which a number of stations may use a repeater (or any other input-output device) simultaneously. There are three methods now in use which permit this simultaneous multiusage. These are more commonly called FDMA, TDMA, and CDMA. The first, frequency division multiple access (FDMA), allows multiple access in the frequency domain. The second, time division multiple access (TDMA), permits multiple access in the time domain. The third method, code division multiple access (CDMA), (spread spectrum), operates in the time and frequency domains. It basically finds application with military satellite communications in a jamming environment and will not be discussed any further in this text.

FDMA

Allow the satellite repeater a 500-MHz bandwidth. Each earth station is assigned a segment of that usable bandwidth. Sufficient guard band is allocated between segments to assure that one user will not interfere with another by drifting into or splattering the other user's segment. Assign 2225 MHz as the frequency of the local oscillator in Figure 7.1. The band of frequencies transmitted up to the satellite, representing the input in the figure, is 5925–6425 MHz. Thus a 6000-MHz carrier received by the satellite repeater is retransmitted in the 4000-MHz band as 6000 – 2225 = 3775 MHz. Note that in the mixing operation we are taking the difference frequency. To prove the viability of the mixer, check the band-edge frequencies:

$$5925 - 2225 = 3700 \text{ MHz}$$

$$6425 - 2225 = 4200 \text{ MHz}$$

As we already know, 3700–4200 MHz is the band of frequencies transmitted from the satellite to the ground.

Consider the satellite antennas in all examples to be omnidirectional.* Now let us

*These are referenced in the literature as global beams or earth coverage (ec) antennas.

do a sample exercise with three earth stations within common view of a synchronous satellite which wish to intercommunicate. Assign each earth station a 132-channel configuration, and from Table 7.1 we see that the allocated bandwidth for 132-voice-channel configurations is 10 MHz. This includes the guard band. Situate one station in Buenos Aires, another in New York, and the third in Madrid, Spain. (For the sake of simplicity, we will work with only two supergroups of the 132-channel configuration, leaving CCITT subgroup A spare.) Assign the segments as follows:

New York	5930–5940 MHz
Madrid	5990–6000 MHz
Buenos Aires	6220–6230 MHz

for the uplink carriers for this hypothetical case. When converted in the satellite, the downlink carriers will then be

New York	3705–3715 MHz
Madrid	3765–3775 MHz
Buenos Aires	3995–4005 MHz

As we know from Table 7.1, the 10-MHz bandwidth, which includes guard bands, requires at the earth station receivers a $B_{IF} = B_{RF}$ of 7.5 MHz. Inside the 7.5-MHz bandwidth are contained two supergroups plus orderwire information. The baseband is transmitted by standard FM techniques, which have been covered in Chapter 5.

Look at the following salient facts for our hypothetical example:

- Each earth station transmitting to the satellite transmits one carrier up to the satellite, which carrier is reradiated omnidirectionally.
- That reradiated carrier may be received by any earth station within view (over 5° elevation angle) of the satellite.
- For this hypothetical case information is contained (i.e., in the baseband) on the uplink carrier from the New York facility for both Buenos Aires and Madrid, likewise at Buenos Aires for New York and Madrid, and so forth.

This concept is depicted in Figure 7.7. The baseband assignments are referenced to Table 7.1. The same concept can be expanded showing how groups may be broken out for distinct locations. It should be kept in mind that, although an earth station may transmit only one carrier up to the satellite, it must be prepared to receive at least one carrier for each distant location with which it wishes to communicate. In our hypothetical case, each station would provide one chain of equipment for the uplink and two separate chains of receiving equipments to receive and demodulate the two separate receive carriers from the two distant ends. Some earth stations intercommunicate with 10 or more distant stations, and thus require 10 chains of receiving equipment. They transmit to all 10 stations on one uplink carrier. This is just an expansion of the concept depicted in Figure 7.7. For instance, Madrid may

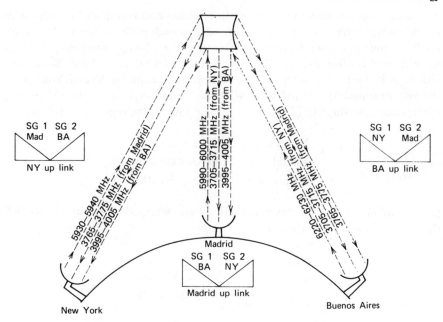

Figure 7.7 Simplified illustration of multiple access in the frequency domain (FDMA) (hypothetical).

wish to intercommunicate assigning groups in a 120-channel baseband, modulating just one uplink carrier. The uplink baseband assignment is shown in Figure 7.8. Keep in mind that this is just one more hypothetical case.

TDMA

TDMA operates in the time domain. Use of the satellite transponder is on a time-sharing basis. Individual time slots are assigned to earth stations operating with that transponder in a sequential order. Each earth station has full and exclusive use of the transponder bandwidth during its time-assigned segment. Depending on the bandwidth of the transponder and the type of modulation used, bit rates of 10 to over 100 megabits per second (Mbps) are used.

With TDMA operation earth stations use digital modulation (see Chapter 15) and transmit with bursts of information. The duration of the burst lasts for something less than the time period of the slot assigned. Timing synchronization is a major consideration.

A frame for TDMA operation may be defined as a repeating cycle of events. It contains one burst of data from each ground terminal accessing a TDMA transponder on the satellite. The frame period t is made up of the sum of the burst times from the accessing earth stations (one burst per station) plus the guard times between bursts. The frame concept in TDMA may be better understood by examining Figure

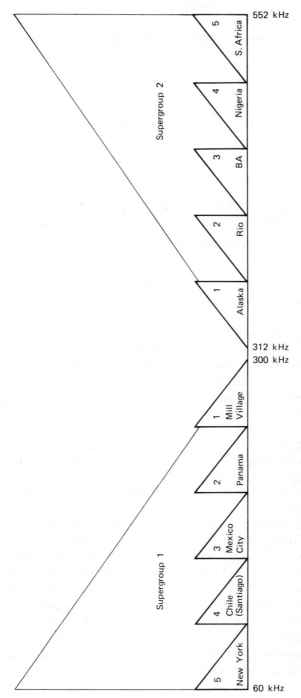

Figure 7.8 As a multiplex (FDM) baseband configuration might appear on an uplink carrier from Madrid, interconnecting with 10 distant points, assigning a 12-channel group to each point (CCITT subgroup A not assigned) (hypothetical).

311

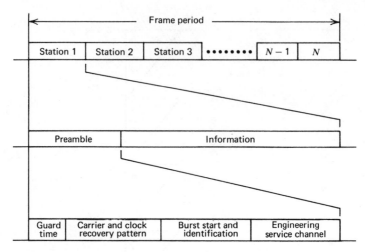

Figure 7.9 Example of TDMA burst format.

7.9. The sample frame in the figure is for earth stations 1, 2, and 3 to N. For present technology N can vary from 2 to about 15. Frame periods t may range from 125 μs to about 15 ms and up to several seconds for some low-speed data networks. Typical frame periods are 750 μs for INTELSAT and 250 μs for the Canadian TeleSat. However, we would expect individual burst lengths inside a frame to be different since different gross traffic loads are transmitted by each station. Reconfiguring the traffic load is easily accomplished by reallocating the burst length assigned to each station. This could be done dynamically as the busy hour shifted in an east–west direction through the passage of the day.

The reader will appreciate that timing is crucial to effective TDMA operation. The greater N becomes (i.e., the more stations operate in the frame period), the more clock timing affects the system. The secret lies in the *carrier and clock (timing) recovery pattern* (Figure 7.9). The literature often refers to this as carrier recovery/ bit timing recovery (CR/BTR). This sequence is used by the digital demodulator in each receiver to recover the local carrier and digital clock for coherent demodulation. For INTELSAT systems the CR/BTR contains 60 bits. This is followed by a 20-bit unique word (UW), which establishes an accurate time reference. In some systems the UW also identifies the station. In other systems, such as INTELSAT, the UW is followed by station identification code (SIC) of about 6 bits in length.

For successful TDMA service the system design and operation must assure that data bursts from one station do not overlap the preceding or succeeding bursts of the other earth terminals in the frame sequence. First is the system timing consideration. In this case the system is all the stations, 1 through N, which are accessing a particular transponder in the TDMA mode. To control system timing a reference burst is used. This timing burst may be the normal burst of an earth terminal assigned the status of master station. For instance, it could be station 1 in Figure 7.9, or it could be a special short reference burst from the master station inserted at the beginning of each frame. Each subsidiary station in the system receives this burst and adjusts its local clock according to the master timing information.

Another important consideration in TDMA timing involves two factors:

1. Each TDMA terminal has a different slant range from the satellite and thus a different propagation time.

2. Even though the satellites considered here are geostationary, they have some relative motion with respect to the earth's surface.

It is a fairly simple matter to compensate for the gross slant range differences between accessing terminals in the system. This can be a static or fixed compensation for the propagation time to the satellite. But there also must be small dynamic compensations for the relative satellite motion. This latter is a phase compensation. One method to carry out these dynamic adjustments it to phase lock to the master terminal burst, which is called closed-loop compensation. Another method is open-loop compensation where each terminal computes the satellite position using stored and periodically updated ephemeris data on the satellite.

The information block of a TDMA burst contains one or several of the following in a time-multiplexed format: digital data, facsimile, digitized voice such as pulse-code modulation (PCM) or delta modulation (DM see Chapter 9), or digitized video.

Commonly the digital modulation of the TDMA carrier is binary phase-shift keying (BPSK) or quaternary phase-shift keying (QPSK), although more exotic forms of modulation will probably be used in the future such as maximum-shift keying (MSK), coherent frequency-shift keying (CFSK), and offset QPSK (see Chapter 15). The reason for this is to better conserve spectrum (i.e., more bits per hertz of bandwidth) as well as, in the case of coding, to allow a lower received E_b/N_0 for a given error rate and thus save valuable satellite prime power. For those channels carrying speech, the capacity (i.e., the number of voice channels) will be increased by using digital speech interpolation (DSI); see (Ref. 20).

One basic reason for using TDMA is that it allows only one carrier to utilize the satellite transponder at a time. This has the advantage of eliminating the creation of IM products on the satellite due to simultaneous multicarrier operation. With FDMA several carriers (sometimes up to 10 or more) use a single transponder simultaneously. The use of TWTs as RF power amplifiers in satellite transponders is universal. TWTs are highly susceptible to the creation of IM products when operated at full excitation. To reduce IM products, the excitation of the TWT (the drive) must be cut back substantially as each additional carrier is added. The cutback is referred to as "backoff." Backoff reduces system efficiency notably, for the full output of the TWT cannot be utilized. With TDMA, on the other hand, only one carrier at a time occupies a transponder. Then by definition no IM products can be generated, allowing the TWT to be excited to its full power output.

To summarize the two basic methods of access, FDMA and TDMA, consider their following advantages and disadvantages:

Advantages	Disadvantages

FDMA

Advantages	Disadvantages
Easy to implement	IM problems in satellite TWTs with multicarrier operation
Well-known technology	Requires uplink power control for "backoff"
No network timing requirements	Satellite TWTs operate inefficiently with multicarrier operation

TDMA

Advantages	Disadvantages
Maximum use of satellite power	Stringent network timing requirements
Uplink power control not required	Analog signals require conversion to digital format
Transmission plan is straightforward and easy to implement	Interface with FDM terrestrial plant is expensive
Digital format is compatible with:	

- Source coding
- Error control coding (FEC)
- Digital terrestrial interface
- Demand access algorithms

Can work with a reasonable mix of earth stations, and adjustments can be made dynamically to varying traffic conditions for each access

7.8 DEMAND-ASSIGNMENT MULTIPLE ACCESS (DAMA)

7.8.1 Introduction

DAMA is a method of satellite access where voice channels are allocated to an earth station on demand. A pool of voice channels is available, and assignments from the pool are made on request. When a call has been completed on a particular channel, the channel is returned to the pool for reassignment.

DAMA is analogous to a telephone switch. When a subscriber goes off-hook, a line is seized; on dialing, a connection is made; and when the call is completed and there is an on-hook condition, the voice path through the switch is returned to "idle" and is ready for use by another subscriber.

DAMA is actually a subset of FDMA in that on the satellite a segment of transponder or an entire transponder is divided into single-channel frequency segments from 25 to 50 kHz wide. These segments may be accessed by all DAMA terminals on the ground which are serviced by the satellite DAMA transponder. Some method, of course, must be incorporated into the system to keep track in real time of the idle segments for their dynamic assignment on demand. This will be described below.

7.8.2 The Rationale of Demand Assignment

We might call demand assignment the antithesis of preassignment. FDMA and, for this argument, TDMA are preassignment techniques. Actually, from a network designer's point of view, preassigned satellite channels represent long-distance trunks which have been dimensioned for specific traffic loadings in erlangs [or cent-call-seconds (CCS)]. There is no flexibility of usage as a function of such traffic loading. The full efficiency of such circuit groups is only realized during the busy hour (BH). For the remainder of a working day the efficiency is reduced, and for other periods of the day preassigned channels are highly inefficient (i.e., very lightly loaded), carrying traffic loads far below capacity. (Note. Reference 16 chap. 1 presents an introductory discussion of traffic loading and trunk efficiency.)

DAMA provides a dramatic improvement of trunk efficiency as a function of traffic loading at the expense of the less efficient usage of satellite bandwidth when compared to conventional FDMA, for example. Suppose, for discussion's sake, that an earth station had to provide voice communications with 40 distant destinations with a traffic loading for each destination of 0.5 erlang and a grade of service of $p = 0.01$. Using the Erlang B formula (Ref. 16) with preassigned channels, four trunks would be required per destination and 4×40, or 160, trunks or satellite channels for the total configuration. In the DAMA mode of operation we consider total traffic. In this case $0.5 \times 40 = 20$ erlangs, and the Erlang B formula (or appropriate traffic tables) gives a requirement of 30 trunks or satellite channels. The improvement in efficiency for the example given is $160/30$, or 5.3. Consider another example. Suppose that there were 10 destinations that a particular earth station was required to serve with a traffic loading of 1.0 erlang per destination. For the preassigned case based on the Erlang B formula ($p = 0.01$), 50 conventional FDMA channels would be required; for the DAMA case only 18 would be needed. The major motivation for DAMA is the potential improvement in communication system capacity.

One rule of thumb suggests that DAMA should be considered in lieu of preassignment when a particular destination requires 12 circuits or less.

Another application of DAMA is to serve as overflow from the preassigned service. Suppose that service to a destination required 8.87 erlangs ($p = 0.01$). Rather than configuring 14 channels to that destination, we could use a 12-channel group with conventional FDMA, and the two remaining channels would be taken from the DAMA pool. If we could do this with a number of destinations to be served by a particular earth station, a DAMA facility (such as SPADE described below) would prove economical. There are savings not only in FDMA channel equipment, group, and possibly supergroup equipment, but also in charges for space segments that would otherwise be wasted.

7.8.3 DAMA Operation

There are three methods for handling a DAMA system:

- Polling method
- Random-access central-control method
- Random-access distributed-control method

The polling method is fairly self-explanatory. A master station "polls" all other stations in the system sequentially. When a positive reply is received, a channel is assigned accordingly. As the number of stations increases, the polling interval becomes longer, and the system tends to become unwieldy. With random-access central control the status of channels is coordinated by a central computer, which is usually located at the master earth station. Call requests (called call attempts in switching) are passed to the central processor via a digital orderwire (i.e., digitally over the radio service channel), and a channel is assigned, if available. Once the call is completed and the subscriber goes on-hook, the speech path is taken down and the channel used is returned to the demand-access pool. According to the system design, there are various methods to handle blocked calls [all trunks busy (ATB) in telephone switching], such as queuing and second attempts.

The distributed-control random-access method utilizes a processor controller at each earth station in the system. All earth stations in the network monitor the status of all channels via the continuous updating of channel status information by means of a digital orderwire circuit. When an idle circuit is seized, all users are informed of the fact, and the circuit is removed from the pool. Similar information is transmitted to all users when a circuit returns to the idle condition. The same problems arise regarding blockage (ATB) as in the central-control system. Distributed control is more costly, particularly in large systems with many users. It is attractive in the international environment because it eliminates the "politics" of a master station.

DAMA systems employ the technique of a single channel per carrier (SCPC), whereas conventional FDMA and TDMA normally are multichannel systems. Presently SCPC systems use FM or PSK modulation. FM systems operate on an analog basis, usually with preemphasis, threshold extension, and syllabic companding. PSK systems are digital in the PCM or DM format. Modulation rates for PCM are commonly 64 kbps per channel and with DM 32 or 40 kbps (see Chapters 11 and 15). A block diagram of a typical FM SCPC terminal is shown in Figure 7.10

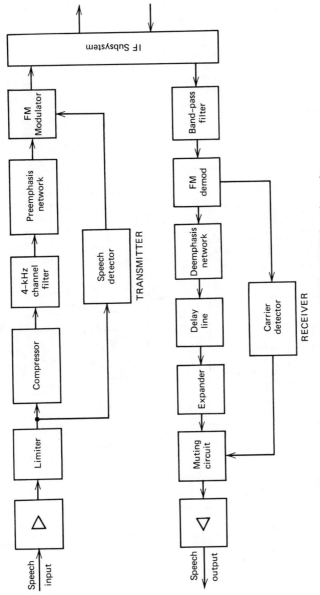

Figure 7.10 Block diagram of a typical FM SCPC station unit.

317

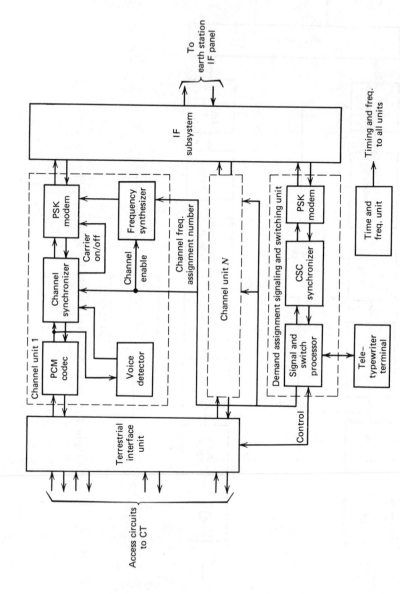

Figure 7.11 Block diagram of a SPADE terminal CT = international switching center; CSC = common signaling channel.

318

and that of a SPADE PSK terminal in Figure 7.11. SPADE is discussed further below.

7.9 FUNCTIONAL OPERATION OF A "STANDARD" EARTH STATION

7.9.1 The Communication Subsystem

Figure 7.12 is a simplified functional block diagram of an earth station showing the basic communication subsystem only. We shall use this figure to trace a signal through the station. Figure 7.13 is a more detailed functional block diagram of a typical earth station.

The operation of an earth station communication subsystem in the FDMA/FM

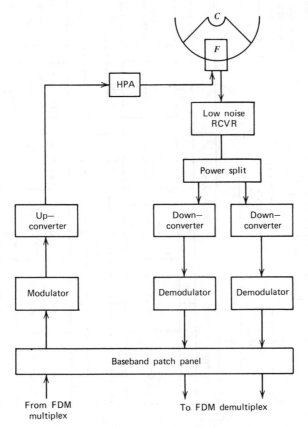

Figure 7.12 Simplified functional block diagram of earth station communication subsystem. F = feed; HPA = high-power amplifier; C = Cassegrain subreflector.

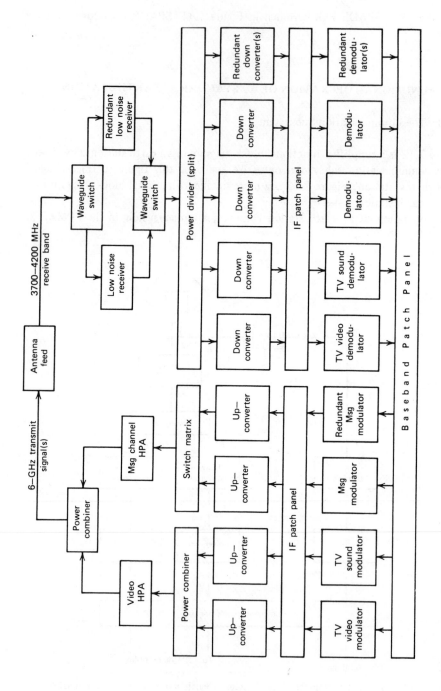

Figure 7.13 Detailed functional block diagram of communication subsystem.

320

mode varies little from that of a LOS radiolink system. The variances are essentially these:

- Use of cryogenically cooled low-noise preamplifier(s)
- An HPA with a capability of 0.5–1 kW output of power
- Larger high-efficiency antennas, feeds
- Careful design to achieve as low a noise as possible
- Use of a signal processing technique that allows nearly constant transmitter loading (i.e., spreading waveform)
- Use of threshold extension demodulators in some cases.

Now let us trace a signal through the communication subsystem typical of Figure 7.12. The FDM baseband is fed from the multiplex equipment through the baseband patch facility to the modulator. A spreading waveform signal is added to the very low end of the baseband, and the baseband signal is then shaped with a preemphasis network (described in Section 5.5.2). The baseband so shaped frequency modulates a carrier and the resultant is up-converted or multiplied to a 70-MHz IF. Patching facilities usually are available at the IF to loop back through the receiver subsystem or through a test receiver for local testing. The 70-MHz IF is then passed to an up-converter, which translates the IF to the 6-GHz operating frequency. The signal is then amplified by the HPA, directed to the feed, and radiated by the antenna.

On reception, the low-noise receiver looks at the entire 500-MHz band (i.e., from 3700 to 4200 MHz), amplifying the broadband signal from 20 to 40 dB. Often this signal is amplified again by a low-level TWT called a driver. This is done if comparatively long waveguide runs are required to connect the low-noise receiver to the down-converter and demodulators. The low-noise receiver is placed as close as possible to the feed. The remaining equipment in the receive chain may be located in another building. The low-noise receiver is cryogenically cooled,* usually with liquid helium, achieving a physical temperature of about 12-15K. The equivalent noise temperature of the receiver is usually on the order of 17–20K.

The comparatively high-level broadband receive signal is fed to a power split. There is one output from the split for every down-converter–demodulator chain. In addition there is often a test receiver available as well as one or several redundant receivers in case of the failure of an operational receiver chain. It should be kept in mind that each time the broadband incoming signal is split in two, there is a 3-dB loss due to the split, plus an insertion loss of the splitter. If the splitter has four outputs, the level going into a down-converter is something over 6 dB below the level at the input to the splitter. A splitter with eight outputs will cause a loss of something on the order of 10 dB.

Each down-converter is tuned to its 4-GHz operational frequency and converts it to a 70-MHz IF. Dual conversion is becoming more popular. One equipment on the

---

*On standard A earth stations.

market converts from 4 GHz to a first IF of 727 MHz and thence to 70 MHz as a second IF. Patching facilities are available for local loop back with the modulator.

The 70-MHz IF is then fed to the demodulator. The resulting demodulated signal, the baseband, is reshaped in the deemphasis network and the spreading waveform signal removed. The baseband output is fed to the baseband patch facilities and thence to the demultiplex equipment. At many earth station facilities, threshold extension demodulators are used in lieu of conventional demodulators to achieve "station margin." As of this writing, threshold extension techniques (see Section 6.9.5) are effective on the narrower B_{IF} used in the so-called *message service*. The terms "message demodulator," "message up-converter," and so on are used to indicate the equipment used to carry telephone traffic. This is in opposition to the equipment used to carry video picture and TV sound traffic.

7.9.2 Antenna Tracking Subsystem

Geostationary earth satellites tend to drift in small suborbits (figure eights). However, even with improved satellite station keeping, the narrow beamwidths encountered with "standard" 30-m (100-ft) antennas require precise pointing and subsequent tracking of the earth station antenna to maximize the signal on the satellite and from the satellite. The basic modes of operation to provide these capabilities are:

- Manual pointing
- Programmed tracking (open-loop tracking)
- Automatic tracking (closed-loop tracking)

Pointing deals with "aiming" the antenna initially on the satellite. Tracking keeps it that way. Programmed tracking may assume both duties. It points the antenna passively to a computed satellite position. With programmed tracking, the antenna is continuously pointed by interpolation between values of a precomputed time-indexed ephemeris. With adequate information as to the actual satellite position and true earth station position, pointing resolutions are on the order of 0.03-0.05°.

Manual pointing may be effective for initial satellite acquisition or "capture" for later automatic tracking. It is also effective for wider beam "nonstandard" earth stations when used with satellites that are keeping station well. Manual pointing is used on certain types of domestic satellite systems, particularly those with smaller antennas and therefore wider beamwidths.

Automatic tracking is used almost universally on "standard" earth stations. It actively seeks to maximize the received signal power by continuously comparing the relative signal amplitudes or phases of the satellite beacon signal of comparatively narrow RF beamwidth. One automatic tracking system discussed here requires an antenna feed modified for tracking, a sufficiently sensitive tracking receiver, a tracking signal comparator, and a servo unit.

One automatic high-accuracy tracking technique which is comonly used today is called *monopulse*. Monopulse may be of either the amplitude or phase type. A phase sensing antenna system may be constructed using two or more antenna apertures separated by several wavelengths. When the antenna is off the satellite-antenna axis, there will be a time delay or phase delay between the signals arriving at the different aperture feeds.

An amplitude sensing antenna system is constructed using a single antenna aperture having two or more closely spaced feeds. In an amplitude sensing system each of the closely spaced feeds produces a radiation pattern that is displaced from the antenna–satellite axis. This displacement angle is a function of the separation of the feed-horn phase centers and the focal point of the antenna aperture. The displaced patterns intersect on the antenna–satellite axis. Thus all received signals off-axis will induce unequal signal amplitude in the feeds. When the two signals are subtracted, a null will be produced only for those signals arriving along the axis line.

To achieve the two signal patterns, sum and difference signal envelopes must be produced. This is done in a comparator, which in its simplest form could consist of a folded T hybrid. The sum is obtained by vectorial addition in the hybrid sum arm and the difference by vectorial subtraction in the difference or orthogonal arm of the hybrid.

The practical monopulse system uses a more elaborate method for improving the resolution of azimuth and elevation information, but the basic sum and difference concept remains the same.

Figure 7.14 is a simplified functional block diagram of a monopulse tracking subsystem. The key to system operation is the monopulse tracking receiver. It deals with three signals, ΔAZ (azimuth), ΔEL (elevation), and Σ (sum). The Σ signal derives from coupling some output signal from the operational low-noise amplifier. The ΔAZ and ΔEL derive from the monopulse comparator associated with the

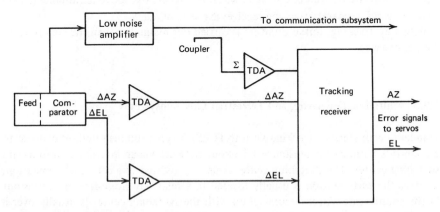

Figure 7.14 Simplified functional block diagram of a monopulse tracking subsystem. TDA = tunnel diode amplifier.

feed. The main function of the monopulse receiver is to detect the amplitude of the
Δ signals relative to the Σ signal. The nature of the desired control signal(s) after
detection is a dc null voltage which goes positive for an antenna pointing on one
side of the satellite–antenna axis and negative for a pointing error on the other side
of the axis. The output of the tracking receiver is the input error information for
the servo control subsystem for azimuth and elevation control of the antenna. The
tracking accuracy has been fairly well standardized at 0.1 antenna beamwidth for
winds at steady state of 30 mi/h with gusts to 45, and 0.15 beamwidth for 45 mi/h
steady-state winds gusting to 60 mi/h. At 4 GHz for a 60-dB gain antenna, half-
power beamwidth is about $0.18°$, dictating a tracking accuracy of 0.018 and $0.027°$,
respectively.

Step tracking is another method of autotracking which is much less expensive to
implement than the monopulse method described above. A step tracker could be
called a beam-seeking device and is used where the target dynamics are low. Step
tracking is particularly applicable for medium-size earth stations operating with geostationary satellites. It is used almost exclusively on MariSat terminals aboard ship.

A basic element of a step-track system is a microprocessor which is programmed
to periodically offset the antenna some seconds or tenths of seconds of arc in
azimuth and elevation. A sensing circuit on a receiver then indicates to the processor whether the signal level from the satellite has increased or decreased due to a
specific offset maneuver. If a decrease in level is noted, the microprocessor returns
the antenna to its azimuth or elevation position prior to offset. If the signal increases
in level, the antenna is offset again in the same direction until a decrease is noted,
then the antenna returns to its azimuth or elevation position just prior to the last
offset. Optimizing cycles may be programmed to occur every minute to every 100
minutes or more.

There is one important requirement that the receive signal being sensed be a full-
period (e.g., on all the time) constant-amplitude signal. This signal does not necessarily have to be the satellite beacon signal if it meets the above requirement. Thus
in many cases a separate tracking receiver is not required.

With step tracking, initial pointing is carried out manually, then the step-tracking
loop is engaged.

7.9.3 Multiplex, Orderwire, and Terrestrial Link Subsystems

Standard earth stations working with INTELSAT synchronous satellites connect to
a country's national long-distance telephone network via an international four-wire
switching center. The switching center is usually located in a center of population,
whereas the earth station is usually located in a relatively noise-free area, thus out
in the countryside. An interconnection with the switching center is usually over a
broadband transmission medium such as a radiolink.

Such a system is shown in Figure 7.15. Here a synchronous satellite interconnects

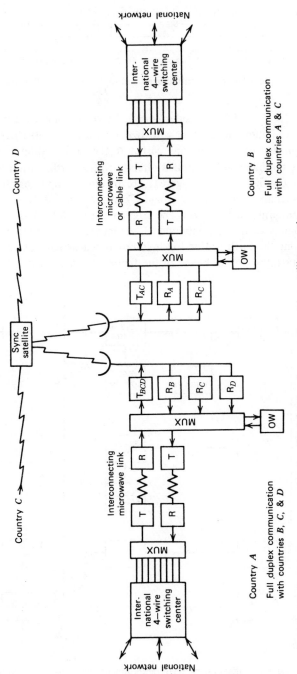

Figure 7.15 Interconnection of national toll networks via a synchronous satellite: earth station to switching center link-up problem. OW = orderwire; T = transmitter; R = receiver.

325

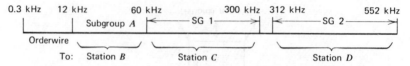

Figure 7.16 Configuration of transmitted baseband (uplink) from earth station A.

four countries: A, B, C, and D. Consider country A. The uplink is simple. The uplink baseband is configured at the four-wire switching center in its FDM multiplex equipment, transmitted on one radiolink carrier to the earth station, demodulated to baseband, and the baseband is applied directly to the earth station modulator. This baseband is configured as in Figure 7.16. In our example, earth station A receives three downlink carriers F_1, F_2, and F_3 in the band of 3700–4200 MHz. Each carrier is demodulated as shown. The problem is to pass the baseband of each of the three incoming carriers on only one RF carrier via a conventional microwave radiolink connecting the earth station to the international switching center.

Figure 7.17 illustrates the problem. In the figure the demodulator from B con-

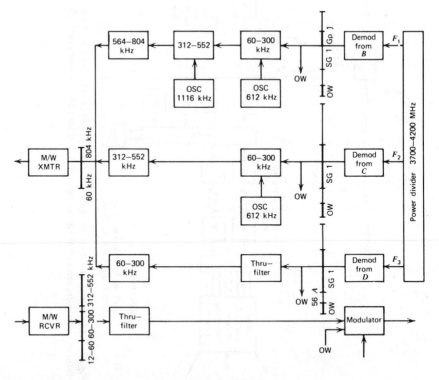

Figure 7.17 Simplified block diagram of reconfiguration of downlink FDM multiplex basebands for transmission over terrestrial link. OW = orderwire; OSC = oscillator; SG = supergroup; Gp = group; M/W = microwave (LOS).

tains in its baseband group 1 of supergroup 1 the 12 voice channels destined for country A. The demodulator from C (F_2 demodulator) contains in its demodulated baseband supergroup 1 (60 voice channels) destined for country A, and the demodulator from D contains supergroup 1 (60 voice channels).

Each baseband is processed by the multiplex equipment as shown in Figure 7.17. The output of the multiplex equipment is a single baseband containing supergroups 1, 2, and 3. It should be noted in the example shown that from the country with which there is 12 full-duplex channel communication, only group 1 of supergroup 1 is used. It has been placed in the supergroup 3 slot of the terrestrial microwave baseband, even though only one group of the supergroup carries active traffic.

The example shown is a very simplified one. As mentioned previously, many earth stations are designed to transmit several carriers for direct communication with 10 or more distinct geographical locations on the earth's surface via one synchronous earth satellite. This would require 10 or more separate receivers to receive 10 distinct carriers. The multiplex equipment required to place the channels, groups, and supergroups so received on one terrestrial system radio carrier can indeed be complex.

The orderwire facilities are transmitted for message carrier operation in the baseband segment form 300 Hz to 12 kHz, as shown in Figure 7.16. The facilities provided are a teleprinter using tone telegraph techniques in a "voice plus" arrangement. (Teleprinter, tone telegraph, and voice plus are discussed in Chapter 8.) Two voice channels are provided in each case, denoted by P_1 and P_2 (P_1 for 4–8 kHz and P_2 for 8–12 kHz) and are transmitted inverted. In each channel, the voice is transmitted from 300 to 2500 Hz. Five tone telegraph channels are available above 2500 Hz with center frequencies at 2.7, 2.82, 2.94, 3.06, and 3.18 kHz. The voice channels, speech portion, use in-band signaling, 2280 Hz. Video carriers do not transmit orderwire information directly on the carrier, but utilize a portion of the aural carrier. The transmission of TV is discussed below.

7.10 INTELSAT V

INTELSAT V is the latest operational satellite of the INTELSAT satellite series and will serve as the principal satellite type for international satellite communication links for the early and mid-1980s until it is replaced by INTELSAT VI in the late 1980s. The total possible voice channel capacity of INTELSAT V exceeds that of any predecessor satellite. The published capacity of each satellite is 14,500 VF channels, but its potential capacity is greater if all transponders were used on an FDMA basis, 972 channels each, with full frequency reuse. The design life of the satellite is 7 years.

The INTELSAT V communication subsystem operates in two frequency-band pairs, the familiar 6/4-GHz band and the 14/11-GHz band. The uplink for the latter pair occupies the band of 14,000–14,500 MHz, and the downlink is split occupying

Table 7.3　INTELSAT V Coverage Beams

Band	Coverage	Polarization	
		Uplink	Downlink
6/4 GHz	Earth	Left-hand circular	Right-hand circular
	West hemispheric	Left-hand circular	Right-hand circular
	East hemispheric	Left-hand circular	Right-hand circular
	Zone 1	Right-hand circular	Left-hand circular
	Zone 2	Right-hand circular	Left-hand circular
14/11 GHz	East spot[a]	Linear	Linear
	West spot[a]	Linear	Linear

[a]Polarization of the east spot coverage shall be orthogonal to that of the west spot coverage.

the bands of 10,950–11,200 and 11,450–11,700 MHz. The satellite utilizes these bands in seven distinct earth surface coverage areas, five in the 6/4-GHz band and two in the 14/11-GHz band. Reuse of the frequency bands between coverage areas is accomplished by means of spatial and/or polarization isolation. Table 7.3 summarizes available coverages and respective polarizations of each coverage beam.

The INTELSAT V transponder channelization plan is shown in Figure 7.18. Routing through the transponder can be switched by ground command, which provides a large number of combinations of paths through the satellite, giving various receive and transmit antennas connectivity with many of the transponders. This is shown in Figure 7.19, which is a signal flow diagram of the satellite's communications subsystem. Note the transponder numbering; it reflects nominal 36-MHz bandwidths. We find numbers 1–12 for the 6/4- and the 14/11-GHz band pairs, and these numbers correspond in Figures 7.18 and 7.19.

Tables 7.4 and 7.5 give the primary communication performance parameters of the satellite.

7.11　TELEPHONE CHANNEL NOISE BUDGET FOR AN INTELSAT LINK

Owing to the modulation and transmission equipment of the transmitting earth station and to the transmission and demodulation equipment of the receiving earth station, 1000 pW0p is allocated for noise in any telephone channel. This value also includes noise due to group delay.

INTELSAT offers the following breakdown of the earth station noise budget for guidance only:

Earth station transmitter noise　　　　　　　250 pW0p
excluding multicarrier IM
and group delay noise

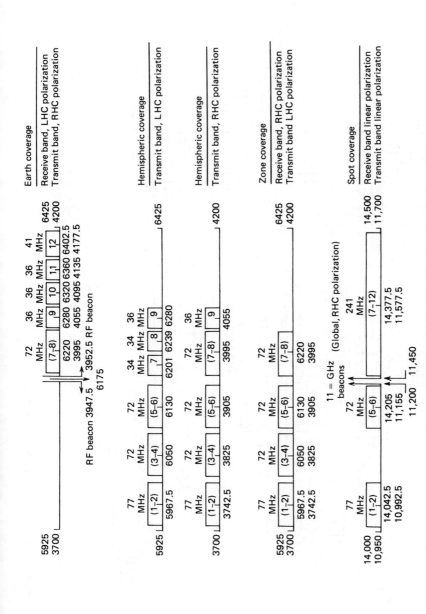

Figure 7.18 INTELSAT V transponder frequency plan. (Ref. 4; Courtesy of INTELSAT.)

329

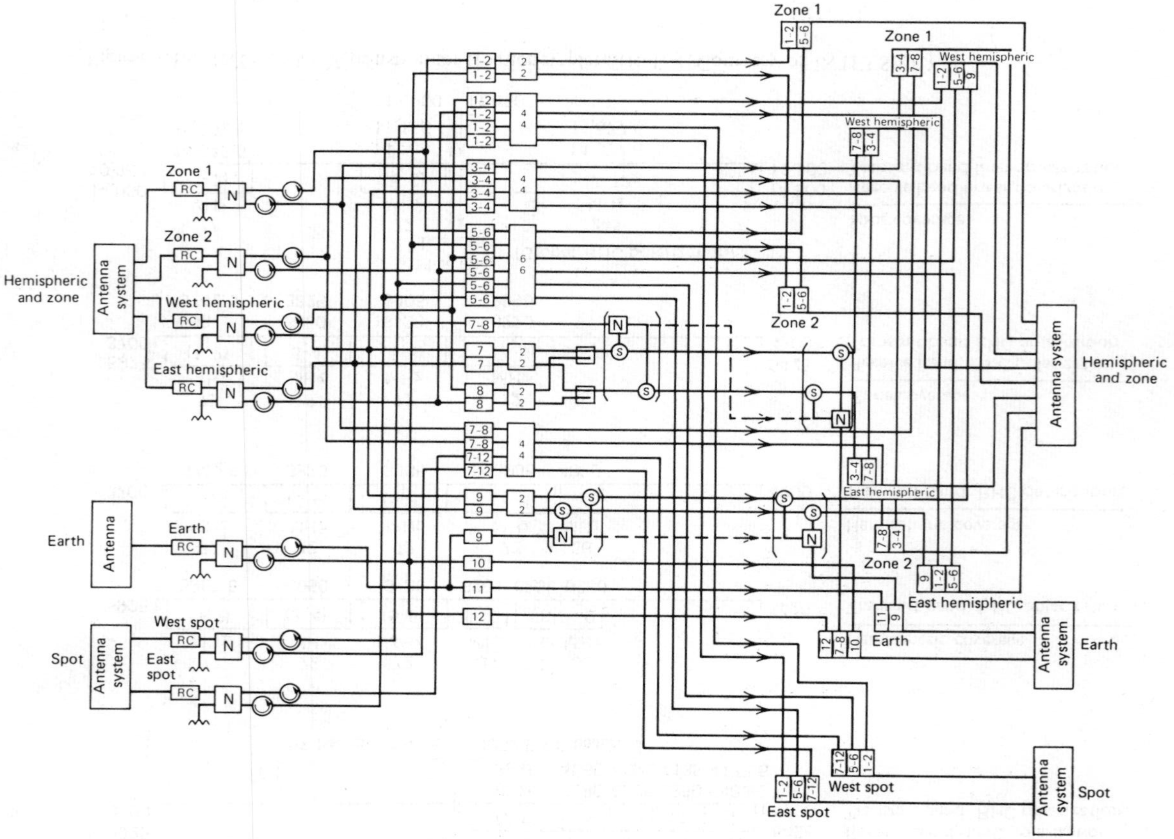

Figure 7.19 INTELSAT V signal flow diagram.

Table 7.4 Communication Performance Parameters for INTELSAT V

Polarization performance (voltage axial ratio for transmit and receive beams)	
Global	1.09
Hemispheric	1.09
Zone	1.09
Antenna isolation (dB) (minimum beam isolation for transmit and receive beams)	
Hemispheric to hemispheric	27
Hemispheric to zone	27
Zone to zone	27
Spot to spot	
$8°$ east–west separation	33
$6.5°$ east–west separation	27
Receive system G/T (dB/K)	
6-GHz global	-18.6
6-GHz hemispheric	-11.6
6-GHz zone	-8.6
14-GHz east spot	0.0
14-GHz west spot	3.3
EIRP (dBW)	
4-GHz global	
Channels 7–8	26.5
Channels 9, 10, 11, 12	23.5
4-GHz Hemispheric or zone	
Channels 1–2, 3–4, 5–6, 7–8	29.0
Channel 9	26.0
11-GHz east spot	
Channels 1–2, 5–6, 7–12	41.1
11-GHz west spot	
Channels 1–2, 5–6, 7–12	44.4

Table 7.5 Saturation Flux Density for INTELSAT V

	Saturation Flux Density (dBW/m^2)	
Uplink Beam	High Gain	Low Gain
6-GHz, all beams		
Channels 1–2, 3–4, 5–6, 7, 8, 7–8	-72.0 ± 2	7.5 dB higher
Channels 9, 10, 11, 12	-75.0 ± 2	7.5 dB higher
14-GHz east spot		
Channels 1–2, 5–6, 7–8, 7–12	-77.0 ± 2	5 dB higher
14-GHz west spot		
Channels 1–2, 5–6, 7–8, 7–12	-80.3 ± 2	5 dB higher

Noise due to total system group delay after any necessary equalization*	500 pW0p
Other earth station receiver noise	250 pW0p

An allowance of 1000 pW0p has been reserved for possible interference from terrestrial systems sharing the same frequency bands in accordance with CCIR recommendations.

The INTELSAT space segment allocates 8000 pW0p to up- and down-path thermal noise, transponder IM, earth station out-of-band emission, cochannel interference within the operating satellite, and interference from adjacent satellite networks. Within the 8000-pW0p allocation, an allowance of 500 pW0p is reserved for earth station RF out-of-band emission caused by multicarrier IM from other earth stations in the system.

The total INTELSAT system noise budget is therefore divided into three major portions:

Space segment	8,000 pW0p
Earth stations	1,000
Terrestrial interference	1,000
Total	10,000 pW0p

7.12 TV TRANSMISSION

Standard practice for the transmission of TV on the INTELSAT series of satellites is to transmit the video and audio portions of the TV signal separately. As shown in Table 7.2, the video EIRP required is considerably greater than that required for message carriers.

For INTELSAT IVA and V, transponder 12 (global beam) is used for TV transmission, both video and aural channels. Two 24-channel telephony carriers are provided for the aural channels, one for "go" and one for "return." The RF carrier assignment is:

Video carrier frequency	6403.0 MHz
"Go" (TV-associated) aural channel	6383.25 MHz
"Return" aural carrier frequency	6385.75 MHz

For this configuration the bandwidth of the video carrier is 30 MHz.

---

*Total group delay may be broken down after equalization at the earth stations as follows: 200 pW0p for earth station transmit path, 100 pW0p satellite intrinsic group delay/dual path, and 200 pW0p for earth station receive path.

An alternative configuration places two TV video signals on the transponder, each 17.5 MHz in bandwidth. In this case:

Channel 1   6390.75 MHz
Channel 2   6409.25 MHz

For this latter configuration two 24-channel carrier assignments have been made in transponders 10 (Atlantic region) and 11 (Indian Ocean region) to carry TV-associated audio for each of these video signals. The exact carrier assignments will be specified by INTELSAT. It should be noted that there is a possibility of modifying this latter configuration in the future for INTELSAT V operation to take advantage of the extra bandwidth in transponder 12 of that satellite series.

7.13 SPADE

The single channel per carrier PCM multiple-access demand assignment equipment (SPADE) is a good example of a DAMA system. SPADE channels can also be pre-assigned. SPADE was developed by INTELSAT and is in use on INTELSAT IVA and V satellites.

An INTELSAT IVA/V transponder has its nominal 36-MHz bandwidth divided into 45-kHz segments for SPADE operation, giving the transponder an 800-channel (one-way) capacity (e.g., 800×45 kHz = 36 MHz) or 400 full-duplex circuits. Thus SPADE is really of the FDMA technique discussed above. Channel encoding is PCM (64 kbps per channel), and four-phase CPSK is used for modulating each SPADE carrier.

Figure 7.20 illustrates the SPADE channelization plan. It will be noted that a SPADE network normally uses common channel signaling, which has a 160-kHz bandwidth located 18.045 MHz below the pilot frequency shown in the figure. The pilot is a constant carrier and is provided by the common signaling channel (CSC) reference station. It is used by all SPADE receivers in the network to control satellite and earth station downconverter local oscillator frequency drift. Channels 1 and 2 and channels 1' and 2' are left vacant to allow a wider bandwidth for the CSC. The first possible carrier is designated channel 3 and is located 112.5 kHz above the lower end of the band. The 399th channel is located (center frequency) 67.5 kHz below the pilot. The first possible channel above the pilot is designated channel 3' and is located 157.6 kHz above the pilot. Channel 399' is located 22.5 kHz inside the upper edge of the allocated band. Any two like-numbered channels (primed and unprimed) are 18.045 MHz apart and are used to form a duplex circuit. Thus there are really only 794 VF channels or 397 useful full-duplex telephone circuits available in the system (INTELSAT V).

Earth station EIRPs required for SPADE operation are as follows:

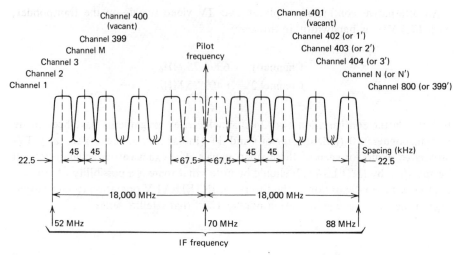

Figure 7.20 SPADE frequency plan at IF. (Ref. 4; Courtesy of INTELSAT.)

Standard A earth station   +63.0 dBW

Standard B earth station   +69.8 dBW

assuming a 10° antenna elevation angle. For conversion to other elevation angles, see Section 7.6. INTELSAT advises that a provision for adjustment of EIRPs over a range of 15 dB below the mandatory maximum values (given above) shall be provided in the earth station design.

Figure 7.11 is a functional block diagram of a SPADE terminal. Its operation is described below. Note that on the left-hand side of the figure is the interface equipment with the four-wire telephone circuits deriving from an international switching center (CT) and on the right-hand side are the 70-MHz inputs/outputs of the earth station up-converter/down-converter, respectively.

An incoming voice channel has its signaling information removed and is passed to the PCM codec (coder–decoder) where the voice channel is converted into a digital bit stream in PCM format. The format consists of a seven-level code and A-law companding (see Chapter 11 for a full discussion of PCM). The output of the transmit side of the codec is a bit stream of 64,000 bps which modulates a four-phase PSK modem (coherent) in a 38-kHz noise bandwidth (the 45 kHz mentioned earlier includes guard bands). The output may be on any one of the 800 frequencies in the "pool." Frequency control is by means of a synthesizer.

For the receive side of a full-duplex voice channel, the above process is reversed. The 70-MHz output of the down-converter is fed to the SPADE terminal input (receive) of the PSK modem. The receive section of the modem, the demodulator, provides an output which is a digital bit stream. It is fed to the decoder section of the codec where the digital-to-analog processing takes place. The resulting analog voice channel is fed to the terrestrial interface unit and thence to the CT after

Table 7.6   Technical Characteristics of SPADE

Channel encoding	PCM
Modulation	4-phase PSK (coherent)
Bit rate	64 kbps
Bandwidth per channel	38 kHz
Channel spacing	45 kHz
Stability requirement	±2 kHz (with AFC)
Bit error rate at threshold	10^{-4}
TDMA common signaling channel	
Bit rate	128 kbps
Modulation	2-phase PSK
Frame length	50 ms
Burst length	1 ms
Number of accesses	50 (49 stations + 1 reference)
Bit error rate	10^{-7}

proper signaling information has been restored. The basic technical characteristics of a SPADE voice channel are reviewed in Table 7.6.

How does the SPADE terminal know what frequency slot is vacant at a particular interval of time? This information is provided by the demand-assignment signaling and switching (DASS) unit. The terrestrial interface equipment routes signaling information to the DASS unit. The DASS unit uses a separate information channel for signaling and provides constant status updating information of the busy–idle condition of channels in the pool, so that busy frequencies become unavailable for assignment of new calls. DASS provides signaling interface on one common RF channel for all voice channels transmitted. This common signaling channel is actually a TDMA broadcast channel. It receives status updating information as well. Such a separate channel is required at all SPADE installations and operates on a TDMA basis, allowing up to 49 stations plus a reference station to access the channel in 50 successive 1-ms bursts per frame at a bit rate of 128 kbps. The channel modem is different from the voice channel modem. It uses biphase CPSK (see Chapter 8 for a discussion of modems). The bit error rate is on the order of 1×10^{-7}. The CSC channel operates at an RF level 7 dB higher than the ordinary SPADE voice channel to maintain the lower error rate at the higher bit rate.

7.14   INTELSAT STANDARD B EARTH STATIONS

The standard B earth station has somewhat relaxed technical performance requirements compared to the standard A earth station. Whereas a standard A earth station requires a G/T of 40.7 dB/K at 4 GHz, the standard B specification is stated as follows (Ref. 4.):

Approval of an earth station in the category of standard B earth stations will only be obtained if the following minimum condition is met in the direction of the satellite... under clear sky conditions, in light wind for any frequency in the band 3.7 to 4.2 GHz:

$$\frac{G}{T} \geq 31.7 \text{ dB} + 20 \log \frac{f}{4} \text{ (dB/K)} \qquad (7.12)$$

G = antenna gain measured at the input of a low-noise amplifier (dBi) and T = receiving system noise temperature referred to the input of the low-noise amplifier (dB/K).

At the time of preparation of this text the modulation methods shown below are approved for standard B earth stations which operate in the 6/4-GHz frequency bands. Carriers will be assigned to each satellite on an FDMA basis using one of the following:

- FDM/FM for TV-associated audio carriers only
- Preassigned SCPC/PCM/PSK four-phase for voice-activated telephony
- Preassigned SCPC/PSK four-phase for data service
- FM video for TV

The standard B earth station EIRP requirements are shown in Table 7.7. Adjustment of the EIRP values for angles of elevation other than $10°$ is made in accordance with the formulas given for the standard A earth station described previously.

If we were to allow 180 K for T_{sys}, then

$$\frac{G}{T} = 31.7 \text{ dB/K} = G - 10 \log 180 \text{ K}$$

Table 7.7   Required Standard B Earth Station EIRP per Carrier[a]

Type of Service	G/T of Receive Station (dB/K)	Maximum Standard B EIRP (dBW)
SCPC/PCM/PSK (four-phase)	40.7	63.0
SCPC/PSK (four-phase)	31.7	69.8
TV		
30 MHz	40.7	81.8
30 MHz	31.7	85.0
17.5 MHz	40.7	85.0
2.5-MHz FDM/FM		
(for TV only)	31.7 or 40.7	74.7

Source: INTELSAT (ref. 4).
[a]For $10°$ elevation angle, including 1 dB of mandatory margin.

and

$$G = 54.25 \text{ dB}$$

This corresponds to a parabolic dish diameter of about 53 ft (16.3 m), assuming an antenna efficiency of 60%. With T_{sys} of 180 K the use of a conventional uncooled (room temperature) low-noise amplifier is permitted. With INTELSAT V satellites no autotracking will be required due to the wider beamwidth when compared to the beamwidth of a 30-m dish of the standard A earth station. These three factors permit a drastic reduction in cost of the standard B earth station compared to the standard A station with a 30-m dish, cooled low-noise amplifier, and provisions of autotrack.

However, there is one drawback. Standard B users should expect to pay a larger space segment charge than standard A users for similar services.

7.15 REGIONAL SATELLITE COMMUNICATION SYSTEMS

7.15.1 Introduction

Regional or "domestic" satellite communication systems are attractive to "regions" of the world with a large community of interest. These regions may be just one country with a large geographical expanse such as the United States, India, or the Soviet Union, or a group of countries with some common interests such as Europe, Spanish-speaking South America, or Oceania. Most important is that the satellite communication system must compete favorably against its terrestrial counterpart, usually microwave radiolinks (Chapter 5) and coaxial cable systems (Chapter 9).

From a cost point of view, the large earth stations operating in the INTELSAT series can be justified. They are supported by significantly large toll revenues of transoceanic communications. Such earth stations compete with costly multichannel undersea cable systems, whereas the regional satellite system must compete with relatively inexpensive terrestrial systems.

The primary concern of the system design engineer is to reduce the cost of communication without reduction of performance. Up to this point in the text only telecommunication systems with multiple links in tandem have been treated. For analog systems with tandem links noise accumulates; for digital systems under the same circumstances, bit error rates (BER) tend to deteriorate. In very simplisitc terms we could state that for a voice circuit the following requirements must be met at the end-user:

- Analog VF channel: $S/N = 30$ dB or better
- Digital VF channel (64 kbps PCM): BER $= 1 \times 10^{-5}$ or better

With the above in mind, two approaches to domestic/regional satellite communication systems evolve:

- Trunking systems
- Direct-to-user (DTU) systems

INTELSAT and the U.S. COMSTAR are typical trunking systems, whereas the U.S. SBS (Satellite Business Systems) is primarily a DTU system. Trunking systems generally imply that the user at each end of a circuit is connected to an earth station via a segment of the telephone network. In this case two system elements affect end-user performance:

- The satellite link
- Those segments of the telephone network that connect users to the earth stations at each end

For DTU service, only the satellite links affect end-user service performance. From this we can see that a DTU system can permit somewhat lower satellite link performance than a trunking system because there is no added degradation to performance due to trunking elements such as switches, cable, and microwave links.

The other side of the argument is persuasive to trunking systems in that a large number of users can share the costs of a more sophisticated satellite terminal. For a DTU system, in theory, a single user pays the entire bill for a somewhat less sophisticated terminal. Such sophistication can vary from a terminal with a capability of tens of megabits per access down to individual household users with a TV-receive-only (TVRO) terminal, the former with costs in the millions down to the latter with costs now reaching several thousands of dollars.

Trunking systems too can vary. On one extreme we have COMSTAR which primarily provides ATT Longlines with high usage and direct long-distance (toll) trunks. At the other extreme is the thin route service provided by the Canadian Telesat series of satellites supplying telephone, Telex, and TV service to small communities in Canada's subarctic and arctic regions. In these remote locations SCPC techniques are used for telephony with terminals with 2–8 VF channel capability. Such small facilities have a G/T of +20 dB/K and an EIRP per SCPC carrier from +55 to +58 dBW. Telesat, like COMSTAR, also provides heavy route service between distant large population centers.

7.15.2 Rationale of Small Earth Stations

We argued the reason for large earth stations (i.e., 30-m parabolic reflector antennas, autotracking, and T_{sys} = 100 K) in Section 7.5. The requirements placed on such stations were essentially:

- 500-MHz receive bandwidth
- Full capability at relatively low elevation (look) angles (10°)

Just raising the minimum elevation angle to 25 or 30° can reduce the sky noise temperature by 15–20 K. This in itself will allow us to reduce the antenna reflector size from 30 to 25 m and still maintain a G/T of 40.7 dB/K.

Rather than requiring the full 500-MHz receiver bandwidth, let the receiver look at only one transponder, say 36 MHz, allowing us to reduce G in G/T by 10 log (500/36) or about 11 dB, thereby reducing the antenna reflector diameter by nearly a factor of 4. Now we can consider 18-ft or 5-m antenna systems.

For the case of DTU systems where the 10,000-pW0p noise contribution per voice channel can be relaxed, the antenna size may now be reduced to 3 m or less, depending, of course, on capacity and performance. In each of the latter examples T_{sys} can be permitted to increase, permitting the use of inexpensive transistor front-end receivers. Of course the wider antenna beamwidths can eliminate auto-track requirements entirely.

7.15.3 SBS–An Example of a DTU System

SBS is a good example of an advanced DTU domestic satellite communication system. It is a partnership arrangement of three U.S. corporations: ComSat General Corporation, International Business Machines (IBM) Corporation, and Aetna Casualty and Surety Company, and is designed to serve the various communication

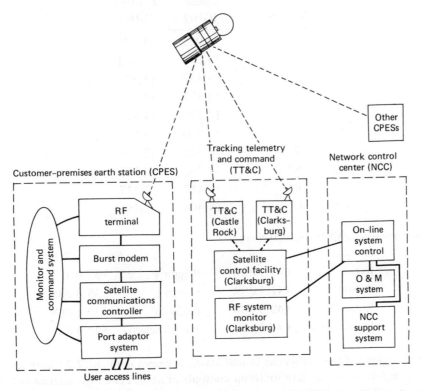

Figure 7.21 SBS functional block diagram. (Courtesy of Satellite Business Systems.)

needs of a large community of business and government users. The greatest benefit is derived from SBS by those of its customers who generate communication traffic at a number of locations geographically dispersed throughout the 48 contiguous states of the United States and whose requirements include the need for high-speed transmission of telecommunication traffic. SBS is the first commercial domestic satellite system to become operational that is fully TDMA.

Figure 7.21 gives a basic overview of SBS as a system. Three functional elements are shown, namely, customer-premises earth stations (CPES), tracking, telemetry, and command (TT&C), and the system management function. The latter two functions involve system administration and control such as monitoring and controlling satellites, customer earth station maintenance, and computerized billing for services.

The remainder of this section will deal with customer-premises earth stations and their operation with the satellite.

The communication subsystem in the SBS satellite has 10 active transponders, each assigned 49 MHz (channel spacing) with a usable bandwidth of 43 MHz. Each transponder traveling wave tube amplifier (TWTA) provides 23 W at its output. Transponder center frequencies are as follows:

Transponder	Uplink (GHz)	Downlink (GHz)
1	14.025	11.725
2	14.074	11.774
3	14.123	11.823
4	14.172	11.872
5	14.221	11.921
6	14.270	11.970
7	14.319	12.019
8	14.368	12.068
9	14.417	12.117
10	14.466	12.166

The transponder EIRP are as follows:

Region 1	+43.7 dBW	Region 4	+37.0 dBW
Region 2	+41.7 dBW	Region 5	+38.0 dBW
Region 3	+40.0 dBW	Region 6	+39.0 dBW
Los Angeles	+41.2 dBW	San Francisco	+42.0 dBW

These regions of the contiguous United States, as shown in Figure 7.22, correspond more or less to the satellite footprint contours of its area coverage antenna and, of course, will impact earth station design, as will be shown below.

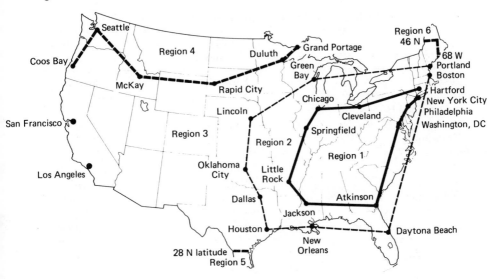

Figure 7.22   Illustrative example of SBS regions. (Courtesy of Satellite Business Systems.)

The satellite communication receiving system G/T are as follows:

Region 1	+2 dB/K	Region 4	−5.5 dB/K
Region 2	+0 dB/K	Region 5	−4.5 dB/K
Region 3	−2.5 dB/K	Region 6	−5.5 dB/K
Los Angeles	−0.3 dB/K	San Francisco	+0.5 dB/K

The satellite signal polarization is linear, transmit being horizontal and receive, vertical. Satellite station keeping is specified at ±0.05° maximum for both the north–south and the east–west directions. The satellite antenna has a shaped beam providing coverage of the 48 contiguous states of the United States. The antenna is a dual-surface offset parabolic reflector using separate multibeam field arrays. In the direction of maximum radiation the antenna gain is approximately 35 dB at 12 and 14 GHz. Refer to Figure 7.22 for illustrative contour coverage.

Earth stations in the SBS network are designed for unattended operation. Table 7.8 summarizes earth station antenna system parameters. 5-m antennas will be used in region 1 and a large portion of region 2. In the remaining regions 7-m antennas will be employed. Maximum HPA power output at an earth station is 500 W, but will usually be driven at a somewhat lower output so that each satellite access signal arrives at the satellite at the same signal level under clear sky conditions.

The heart of the SBS earth station is the TDMA burst modem. Modulation is QPSK, coding 2 bits per symbol (di-bit coding). The modem is capable of operating over an input range of 35–55 Mbps of raw data. During the initial operations phase of SBS, links will operate at a raw data rate of 43 Mbps. This rate may be increased

Table 7.8   SBS Earth Station Antenna Characteristics

Characteristics	Regions 1 and 2	Regions 3–6
Type	Parabolic	Parabolic
Diameter (nominal)	5 m (16 ft)	7 m (23 ft)
Gain (midband)		
Receive	53.8 dB	56.7 dB
Transmit	55.3 dB	58.2 dB
Polarization		
Receive	Horizontal	Horizontal
Transmit	Vertical	Vertical
Tracking mode	Fixed	Command track
G/T	+30.4 dB/K	+33.3 dB/K
EIRP (midband), main beam	+79.8 dBW	+82.7 dBW

to 48 Mbps as experience is gained with the system, particularly in the area of link margins. The functions of the burst modem include:

- Modulation and demodulation
- Generation of the preamble for each burst
- Overlay and removal, upon demodulation, of an energy dispersal sequence
- Acquisition and synchronization with each burst

RF digital channel performance objectives are shown in Table 7.9.

On speech telephony circuits 32 kbps per channel DM* modems are used. Such modulation provides good voice quality with a BER down to 1×10^{-2}.

When SBS customers require improved error performance on data circuits than that shown in Table 7.9, SBS offers an option of forward error correction (FEC) coding (Chapter 15), which will improve the BER to 1×10^{-7} or better 99.5% of the time.

Table 7.9   RF Channel Performance Objectives

Maximum Channel BER	Availability (%)
1×10^{-6}	95.0
1×10^{-4}	99.5
1×10^{-2}	99.8

*DM = delta modulation, see Chapter 11.

7.16 THE COORDINATION CONTOUR—INTRODUCTORY CONCEPTS

Many of the frequency bands used for satellite communications are bands shared with terrestrial services, particularly radiolinks (LOS microwave, Chapter 5). A typical pair of shared bands are the 4- and 6-GHz bands that we have been emphasizing previously in this chapter. The terms *coordination contour* and *coordination distance* deal with methods to ensure that one service does not interfere with the other. Coordination contour establishes around an earth station an area beyond which the possibility of mutual interference may be considered negligible. The coordination distance is the distance to that boundary at any given azimuth from the earth station in question.

For the coordination area two cases have to be considered:

- For the earth station receiving system, which can be interfered with by terrestrial stations
- For the earth station transmitting system, which is capable of interfering with terrestrial stations

A systemmatic method for the development of a coordination contour is described in detail with examples in CCIR Rep. 382-3. Other pertinent documents are CCIR Reps. 393-3, 614-1, and 387-3, CCIR Recs. 452-2, 406-4, 358-2, 558, 357-3, and 355-2, and CCIR Rep. 793.

7.17 SATELLITE LINK BUDGETS

The satellite link budget is simply a path calculation. It is very similar to the path calculation method described for radiolinks (LOS microwave) discussed in detail in Chapter 5. Two link budgets must be analyzed by the transmission system engineer for a new earth station installation, one for the uplink and the other for the downlink. By going through the link budget exercise, the transmission system engineer can adjust certain system parameters and station design to most economically meet specifications. These specifications are provided to the engineer at the outset of the station design process. They may include some of the following:

- Signal-to-noise ratio in a demodulated voice channel
- Signal-to-noise ratio (with weighting) in a demodulated video (TV) channel
- Signal-to-noise ratio in a demodulated (derived) program channel

For digital systems usually an error-rate parameter is given. Link budget analysis may be used to verify the feasibility of an earth station. If this is the case, we would expect satellite EIRP and G/T to be given. In other cases for the uplink, a power level in dBW/m^2 is given to saturate a satellite transponder. If this is the case, we must know the aperture area and antenna efficiency to develop an input level to the satellite LNA.

The receive carrier-to-noise density ratio C/N_0 in dB-Hz for the uplink and downlink can be calculated using the following formula:

$$\frac{C}{N_0} = \text{EIRP}_{dBW} + \frac{G}{T} - L_{path} - L_{atmos} - L_{point} - L_{polar} -$$

$$L_{rain} - (-228.6 \text{ dBW})^* \tag{7.13}$$

For the uplink case the EIRP in dBW would be the EIRP of the earth station in question; for the downlink case it would be the EIRP of the satellite transponder-antenna combination. For either case the G/T in dB/K is given.

L in equation (7.13) refers in each case to a specific loss. These are path loss, atmospheric absorption losses, pointing-tracking loss, polarization loss, and rainfall loss.

Path loss is the free space loss discussed in the beginning of this chapter. The distance values used in those calculations were for a geostationary satellite with the earth station antenna at zenith (i.e., elevation angle of 90°). Seldom, in real life conditions, are we so fortunate to have a condition where an earth station has the satellite directly above and thus would have an elevation angle of 90°. Almost

*Boltzmann's constant

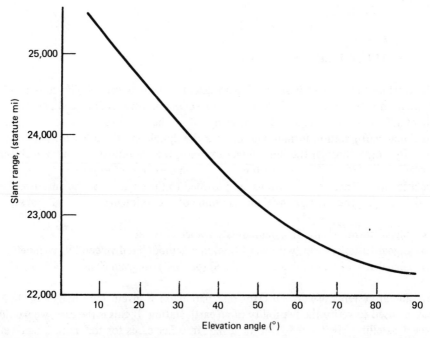

Figure 7.23 Slant range in statute miles to geostationary satellite when the elevation angle is given.

always the elevation angle is something other than $90°$ and, therefore, the distance, called here the slant range, is something greater than 22,300 statute mi. Using Figure 7.23 we can derive the slant range when the elevation angle is given. This value is then inserted in the free space loss equation for "distance" [see equation (5.1) and Section 7.4]. For instance, if the elevation angle is $22°$, the slant range is approximately 24,500 statute mi.

The next loss we encounter in equation (7.13) is the atmospheric losses, which are due to atmospheric gaseous absorption. Such losses are a function of frequency and elevation angle. The following are guidelines which may be used for gaseous absorption losses. At 4 GHz for elevation angles from zenith to $20°$, use 0.1 dB; for $10°$ use 0.2 dB, and for $5°$ use 0.4 dB. At 12 GHz for a $10°$ elevation angle, 0.6 dB is usual, at 18 GHz with a $45°$ elevation angle it is 0.6 dB, and with the same elevation angle at 30 GHz, it is 1.1 dB. All of these losses are for clear sky conditions.

There are two subsets of pointing error loss. The first is for earth station pointing in that seldom an earth station is pointed exactly toward the satellite. If no other information is available, use 0.25 dB for this value. The second is satellite pointing

Table 7.10 Space Budget Worksheet

Uplink		
EIRP earth station	+	dBW
Pointing loss, earth station	–	dB
Pointing loss, satellite (if applicable)	–	dB
Polarization loss	–	dB
Path loss (at elevation angle)	–	dB
Atmospheric gaseous absorption loss	–	dB
Signal level impinging on satellite antenna	(–)	dBW
Satellite G/T		dB/K
Boltzmann's constant – (–)	+228.6	dBW/Hz
C/N_0 (uplink)		dB/Hz
Downlink		
EIRP satellite	+	dBW
Pointing loss, satellite (if applicable)	–	dB
Pointing loss, earth station	–	dB
Polarization loss	–	dB
Path loss (at elevation angle)	–	dB
Atmospheric gaseous absorption loss	–	dB
Signal level impinging on earth station antenna		dBW
Earth station G/T		dB/K
Boltzmann's constant – (–)	+228.6	dBW
C/N_0 (downlink)		dB/Hz

error, and a loss must be assigned when a satellite uses an area coverage (spot beam) antenna. This loss is due to the fact that only a comparatively small area on the earth's surface is covered by the main beam full power. Usually a satellite footprint contour is available. Locate the earth station in question to see where it falls on the contour chart, often expressed at 1.0-, 2.0-, and 3.0-dB contour lines. This loss also compensates for the fact that satellite antenna pointing will not perfectly coincide with these contour lines.

Polarization loss is due to errors and twisting of the desired polarization by the atmosphere and Faraday rotation. If no other value is available, use 0.25 dB.

Table 7.10 provides a worksheet to calculate satellite link budgets for uplink and downlink. Equation (7.14) computes the total carrier-to-noise density ratio C/N_0. To derive C/N, subtract 10 log B, where B = earth station receiver bandwidth B_{1f} for analog systems. To derive E_b/N_0 subtract 10 log BR, where BR is the bit rate in bits per second. This latter is a measure of performance for digital systems. Either one or both links may determine the overall performance. It is usually assumed that the noise from all sources, including IM, is additive. Then

$$\left(\frac{C}{N_0}\right)_t^{-1} = \left(\frac{C}{N_0}\right)_u^{-1} + \left(\frac{C}{N_0}\right)_d^{-1} + \left(\frac{C}{N_0}\right)_i^{-1} \qquad (7.14)$$

where the subscripts t, u, d, and i apply to the calculated total, uplink, downlink, and IM carrier-to-noise density ratios.

7.18   THE IMPACT OF WARC-79 ON SATELLITE COMMUNICATIONS

The World Administrative Radio Conference (WARC) sets certain standards and requirements for radio transmitter operations such as operational bands and spurious emission. These conferences are held under the auspices of the International Telecommunication Union (ITU). The last conference was held in the fall of 1979 and had some notable impact on satellite communication. The Final Acts of WARC-79 will enter into force as of January 1, 1982, after ratification by members of the ITU. However, different effective dates will apply to the many specific provisions of the Final Acts, such as new frequency tolerances and spurious levels as well as what are termed "footnote provisions." Once ratified, the Final Acts of the WARCs are incorporated in the Radio Regulations. It should be noted that the previous discussion in this chapter treated satellite communications as it was prior to the incorporation of WARC-79 into the Radio Regulations, particularly in the area of frequency allocation. The discussion that follows treats regulatory issues subsequent to WARC-79.

The reader should note, for the proper interpretation of the following tables, that the ITU divides the world into three geographical regions. Region 1 contains Europe, Africa, and U.S.S.R., and the Peoples Republic of Mongolia; region 2 consists of North and South America and Greenland, and region 3 includes Asia, except for Asiatic Russia and Mongolia, as well as New Zealand, Australia, and Oceania.

Table 7.11 Post-WARC Fixed Satellite Service Allocations[a]

Below 35 GHz

Band	Earth to Space (GHz)	Bandwidth (MHz)	Band	Space to Earth (GHz)	Bandwidth (MHz)
S	2.655–2.690 R2[b],3[b]	35	S	2.5–2.535 R2[b],3[b]	35
				2.535–2.690 R2[b]	155
6	5.725–5.85 R1	125	4	3.4–4.2	800
	5.85–7.075	1225		4.5–4.8	300
8	7.9–8.4	500	7	7.25–7.75	500
	12.5–12.7 R1	500			
	12.7–12.75 R1,2	200			
14	12.75–13.25	50	11	10.7–11.7	1000
	14.0–14.5	500			
			12	11.7–12.3 R2[b,e]	600
				12.2–12.5 R3[b,d]	300
				12.5–12.75 R1,3	250
30	27.0–27.5 R2,3[e]	500	20	17.7–21.2	3500
	27.5–31.0	3500			

Above 35 GHz

Earth to Space (GHz)	Bandwidth (GHz)	Space to Earth (GHz)	Bandwidth (GHz)
42.5–43.5	1	37.5–40.5	3
47.2–49.2[e]	2		
49.2–50.2	1		
50.4–51.4	1		
71–74	3	81–84	3
74–75.5	1.5		

347

Table 7.11 (*Continued*)

Band	Earth to Space[a] (GHz)	Bandwidth MHz)	Band	Space to Earth (GHz)	Bandwidth (MHz)
	92–95	3		102–105	3
	202–217	15		149–164	15
	265–275	10		231–241	10
	Total bandwidth:	37.5		Total bandwidth:	34

Source: From (Ref. 21).
[a] Does not include bands that are limited to BSS feeder links.
[b] Limited to national and subregional systems.
[c] Upper band limit (12.3 GHz) may be replaced by a new value in the range 12.1–12.3 GHz at the 1983 RARC for region 2.
[d] Footnote allocation.
[e] Intended for use by, but not restricted to, BSS feeder links.
Letter R is indicative of ITU Region.

Table 7.12 Intersatellite Service Allocations,
Post-WARC

Band (GHz)	Bandwidth (GHz)
22.55–23.55	1
32.33	1
54.25–58.2	3.95
59–64	5
116–134	18
170–182	12
185–190	5
Total bandwidth	45.95

Table 7.11 presents post-WARC-79 frequency allocations for the fixed satellite
service (FSS) and Table 7.12 gives the post-WARC intersatellite allocations.

Several comments on the outcome of WARC-79 are presented below. Basically
the material has been abstracted from (Ref. 21). As we have pointed out, with the
congestion now present in the band pair 6/4 GHz, the band pair 14/11 GHz is starting to get further utilized. As a result, its bandwidth was exactly doubled by
WARC-79 by adding new downlink allocations at 10.7–10.95 and 11.2–11.45 GHz
and a new uplink allocation at 12.75–13.25 GHz. Thus a contiguous downlink
band was created (10.7–11.7 GHz) and a split uplink band (12.75–13.25 and 14.0–
14.5 GHz) whose two 500-MHz segments are separated by 750 MHz. In the United
States it is expected that domestic coordination restrictions on the downlink band
will limit applications to international systems.

Another change in the 12-GHz band is noted for the FSS downlink application,
with a footnote permitting transponders on FSS satellites to be used for broadcast
service satellite (BSS) TV transmission with an EIRP up to a maximum of +53 dBW
per channel, provided that these transmissions neither cause greater interference nor
require more interference protection than the coordinated FSS frequency assignment.

Region 3 was also allocated 12-GHz FSS downlink operation in a footnote, which
permitted the band of 12.2–12.5 GHz downlink, but its usage was restricted to
national and subregional systems, as it was for region 2.

Considering the regional 11.7–12.75-GHz allocations, there are no clearly defined
matching uplinks in the FSS. Perhaps the worldwide FSS uplink allocations of the
14–14.5 GHz band and possibly the 12.75–13.25-GHz band could be used for this
purpose. We foresee that intersystem coordination in certain segments of the geostationary orbit will be required since these bands are already matched to the
worldwide FSS downlink allocation of 10.7–11.7 GHz.

350

Earth Station Technology

7.19 INTELSAT VI

INTELSAT VI is a series of satellites designed to replace the INTELSAT V and VA series. They are scheduled for operation in the period of 1986/1987. INTELSAT VI will have 46 active transponders, as shown in Figure 7.24. The design of this new series of satellites takes maximum feasible advantage of frequency reuse providing a total available bandwidth of 3460 MHz. Transponders connected to hemispherical and zone antennas will have a 6× reuse factor, whereas those connected to the global beam antennas will have a 2× reuse factor. The result is a satellite which will provide 33,350 nominal 4-kHz voice channels for FDMA operation and 33,600 equivalent voice channels for TDMA operation. The planned capacity of each TDMA transponder is 2800 equivalent voice channels. However, depending on the earth stations involved, some of the TDMA transponders may achieve 3200-channel operation.

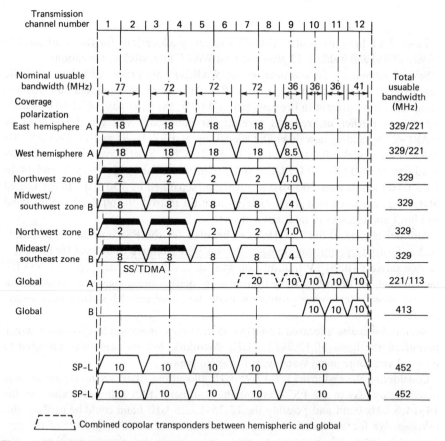

Figure 7.24 INTELSAT VI channelization plan. Figures shown within transponders are output power levels in watts. (Courtesy INTELSAT.)

As we can see in Figure 7.24, INTELSAT VI has 12 transponders dedicated to TDMA operation, more than half of its equivalent voice channel capacity. TDMA will incorporate DSI. DSI effectively increases the equivalent voice channel capacity of the TDMA transmission system by interleaving speech bursts from different terrestrial voice channels on the same satellite channel (interpolation). As brought up in Section 3.4, speech is bursty in nature, and when a composite of a number of "busy" voice channels operates conventionally, the effective utilization is only about 25%. DSI was developed to improve voice channel utilization, and in many respects it is similar to time assigned speech interpolation (TASI), which was widely used on multichannel (FDM) undersea cable systems. It should be noted that the inputs to the DSI processing modules are PCM (Chapter 11) in accordance with CCITT Rec. G.711, using A-law companding with alternate digit inversion.

The TDMA/DSI system provides high-quality service in accordance with CCIR Rec. 522. The principal features of the system are:

- The transmission bit rate is nominally 120 Mbps.
- The TDMA frame length is nominally 2 ms.
- Four-phase PSK modulation with coherent demodulation (CQPSK, Chapter 15) is used.
- Absolute encoding is employed (i.e., differential encoding is not employed).
- Forward error correction (FEC) is applied to selected traffic bursts. (FEC–BCH 128, 112, rate = 7/8, see Chapter 15.)

REFERENCES AND BIBLIOGRAPHY

1. Extraordinary Administrative Radio Conference (EARC), Geneva 1963, ITU, Final Acts.
2. *Reference Data for Radio Engineers*, 6th ed., Howard W. Sams, Indianapolis, IN, 1977.
3. CCITT Green Books, New Delhi, 1970, Vol. IV, Recs. 382-2, 383-1, and 386-1.
4. "Standard A Performance Characteristics of Earth Stations in the INTELSAT IV, IVA, and V Systems Having a G/T of 40.7 dB/K," BG-28-72E M/6/77, INTELSAT, Washington, DC, Aug. 1977.
5. F. D. Doidge, "Antenna Gain as It Applies to Satellite Communication Earth Stations," Edited Lectures, U.S. Seminar on Communication Satellite Earth Station Technology, May 1966, ComSat, Washington, DC.
6. M. Hoffman, "Antenna Noise Temperature," Edited Lectures, U.S. Seminar on Communication Satellite Earth Station Technology, May 1966, ComSat, Washington, DC.
7. R. L. Freeman, "An Approach to Earth Station Technology," *Telecommun. J.*, (ITU), June 1971.
8. B. Cooper, "Station Margin," Edited Lectures, U.S. Seminar on Communication Satellite Earth Station Technology, May 1966, ComSat, Washington, DC.
9. *A Satellite Time Division Multiple Access Experiment*, ICSC/T-17-6E W/1/67, Communication Satellite Corp., Washington, DC.
10. *Spade System Specifications*, ICSC 47-14, May 1970, Communication Satellite Corp., Washington, DC.

11. A. M. Werth, "Spade: A PCM FDMA Demand Assignment System for Satellite Communications," INTELSAT/IEE Symposium on Digital Satellite Communications, London, Nov. 1969.

12. Cohen and Steinmetz, "Amplitude and Phase Sensing Monopulse Systems Parameters," *Microwave J.,* Oct. 1959.

13. *INTELSAT IV System Specifications*, ICSC 45-13, Communication Satellite Corp., Washington, DC.

14. Stamininger and Jeffers, "Transmission System Planning for INTELSAT IV," Conference on Earth Station Technology, London, 1970.

15. R. R. Willett, "Some Basic Criteria for a Regional Communication Satellite System Operating above 10 GHz," Conference on Earth Station Technology, London, 1970.

16. R. L. Freeman, *Telecommunication System Engineering*, Wiley Interscience, New York, 1980.

17. R. G. Gould and Y. F. Lum, *Communications Satellite Systems: An Overview of the Technology*, IEEE Press, New York, 1975.

18. H. L. Van Trees, Ed., *Satellite Communications*, IEEE Press, New York, 1979.

19. *IEEE Transactions on Communications*, Special Issue on Satellite Communications, Vol. COM-27, Oct. 1979.

20. S. J. Campanella, "Digital Speech Interpolation," *ComSat Tech. Rev.,* Spring 1976.

21. C. Dorian *et al.,* "The 1979 World Administrative Radio Conference and Satellite Communications," *ComSat Tech. Rev.,* Vol. 10, Spring 1980.

22. *INTELSAT TDMA/DSA System Specification (TDMA/DSI Traffic Terminals*, BG-42-65E B/6/80, INTELSAT, Washington, DC, June 1980.

8 | THE TRANSMISSION OF DIGITAL DATA

8.1 INTRODUCTION

Up to this point in our discussion we have been dealing with analog transmission. An analog transmission system has an output at the far end which is a continuously variable quantity representative of the input. With analog transmission there is continuity (of waveform); with digital transmission there is discreteness.

The simplest form of digital transmission is binary, where an information element is assigned one of two possibilities. There are many binary situations in real life where only one of two possible values can exist; for example, a light may be either on or off, an engine is running or not, and a person is alive or dead.

An entire number system has been based on two values, which by convention have been assigned the symbols 1 and 0. This is the binary system, and its number base is 2. Our everyday number system has a base of 10 and is called the decimal system. Still another system has a base of 8 and is called the octal system.

The basic information element of the binary system is called the bit, which is an acronym for *b*inary dig*it*. The bit, as we know, may have the values 1 or 0.

A number of discrete bits can identify a larger piece of information, and we may call this larger piece a character. A code is defined by the IEEE as "a plan for representing each of a finite number of values or symbols as a particular arrangement or sequence of discrete conditions or events (Ref. 37)."

Binary coding of written information and its subsequent transmission have been with us for a long time. An example is teleprinter service (i.e., the transmission of a telegram).

The greater number of computers now in operation operate in binary languages; thus binary transmission fits in well for computer-to-computer communication and the transmission of data.

Although communication networks have been advocated which are built exclusively for digital transmission, and several such networks now exist, most data are transmitted over the existing telephone network facilities or other analog facilities that are based on the nominal 4-kHz telephone channel as the basic building block.

This chapter introduces the reader to the transmission of binary information or

353

data over telephone networks. It considers data on an end-to-end basis and the effects of the variability of transmission characteristics on the final data output. There is a review of the basics of the makeup of digital data signals and their application. Therefore the chapter endeavors to cover the entire field of digital data and its transmission over telephone network facilities. It includes a discussion of the nature of digital signals, coding, information theory, constraints of the telephone channel, modulation techniques, and the dc nature of data transmission.

No distinction is made between data and telegraph transmission. Both are binary, and often the codes used for one serve equally for the other. Likewise the transmission problems of data apply equally to telegraph. Telegraph communication is in message format (but not always), transmitted at rates less than 110 words per minute (wpm), and it asynchronous. Data transmission is most often synchronous (but not necessarily so), and usually is an alphanumeric mix of information, much of the time destined for computers [electronic data processing (EDP)]. Data often are transmitted at rates in excess of 110 wpm.

8.2 THE BIT AND BINARY CONVENTION

In a binary transmission system the smallest unit of information is the bit. As we know, either one or two conditions may exist, the 1 or the 0. We call one state a mark, the other a space. These conditions may be indicated electrically by a condition of current flow and no current flow. Unless some rules are established, an ambiguous situation would exist. Is the 1 condition a mark or a space? Does the no-current condition mean that a 0 is transmitted, or a 1? To avoid confusion and to establish a positive identity to binary conditions, CCITT Rec. V.1 recommends equivalent binary designations. These are shown in Table 8.1. If the table is adhered

Table 8.1 Equivalent Binary Designations

Active Condition	Passive Condition
Mark or marking	Space or spacing
Current on	Current off
+ voltage	− voltage
Hole (in paper tape)	No hole (in paper tape)
Binary "1"	Binary "0"
Condition Z	Condition A
Tone-on (amplitude modulation)	Tone-off
Low frequency (frequency shift keying)	High frequency
No phase inversion (differential phase shift keying)	Inversion of phase
Reference phase	Opposite to reference phase

Source: CCITT Rec. V.1.

to universally, no confusion will exist as to which is a mark, which is a space, which is the active condition, which is the passive condition, which is 1, and which is 0. It defines the *sense* of transmission so that the mark and space, the 1 and 0, will not be inverted. Data transmission engineers often refer to such a table as a table of *mark-space convention.*

8.3 CODING

8.3.1 Introduction to Binary Coding Techniques

Written information must be coded before it can be transmitted over a digital system. The discussion of coding below covers only binary codes. But before launching into coding itself, the term *entropy* is introduced.

Operational telecommunication systems transmit information. We can say that information has the property of reducing the uncertainty of a situation. The measurement of uncertainty is called entropy. If entropy is large, then a large amount of information is required to clarify a situation; if entropy is small, then only a small amount of information is required for clarification. Noise in a communication channel is a principal cause of uncertainty. From this we now can introduce Shannon's noisy channel coding theorem, stated approximately (Ref. 1, 41–42):

> If an information source has an entropy H and a noisy channel capacity C, then provided $H < C$, the output from the source can be transmitted over the channel and recovered with an arbitrarily small probability of error. If $H > C$, it is not possible to transmit and recover information with an arbitrarily small probability of error.

Entropy is a major consideration in the development of modern codes. Coding can be such as to reduce transmission errors (uncertainties) due to the transmission medium and even correct the errors at the far end. This is done by reducing the entropy per bit (adding redundancy). We shall discuss errors and their detection in greater detail in Section 8.4. Channel capacity is discussed in Section 8.10.

Now the question arises, how big a binary code? The answer involves yet another question, how much information is to be transmitted?

One binary digit (bit) carries little information; it has only two possibilities. If two binary digits are transmitted in sequence, there are four possibilities,

$$00 \quad 10$$
$$01 \quad 11$$

or four pieces of information. Suppose three bits are transmitted in sequence. Now there are eight possibilities:

000 100
001 101
010 110
011 111

We can now see that for a binary code the number of distinct information characters available is equal to 2 raised to a power equal to the number of elements or bits per character. For instance, the last example was based on a three-element code giving eight possibilities or information characters.

Characters				Code Elements[a]						
Letters Case	Communi-cations	Weather	CCITT #2[b]	START	1	2	3	4	5	STOP
A	–	↑			X	X				X
B	?	⊕			X			X	X	X
C	:	○				X	X	X		X
D	$	↗	WRU		X			X		X
E	3	3			X					X
F	1	→	Unassigned		X		X	X		X
G	&	↘	Unassigned			X		X	X	X
H	STOP[c]	↓	Unassigned				X		X	X
I	8	8				X	X			X
J	'	↙	Audible signal		X	X		X		X
K	(	←			X	X	X	X		X
L	)	↖				X			X	X
M	.	·					X	X	X	X
N	,	⦶					X	X		X
O	9	9						X	X	X
P	0	0				X	X		X	X
Q	1	1			X	X	X		X	X
R	4	4				X		X		X
S	BELL	BELL	,		X		X			X
T	5	5							X	X
U	7	7			X	X	X			X
V	;	⦶	=			X	X	X	X	X
W	2	2			X	X			X	X
X	/	/			X		X	X	X	X
Y	6	6			X		X		X	X
Z	"	+	+		X				X	X
BLANK		–								X
SPACE							X			X
CAR. RET.								X		X
LINE FEED						X				X
FIGURE					X	X		X	X	X
LETTERS					X	X	X	X	X	X

[a] Blank, spacing element; crosshatched, marking element.

[b] This column shows only those characters which differ from the American "communications" version.

[c] Figures case H(COMM) may be stop or +.

Figure 8.1 Communication and weather codes, CCITT international telegraph alphabet no. 2.

Another more practical example is the Baudot teleprinter code. It has five bits or information elements per character. Therefore the different or distinct graphics or characters available are $2^5 = 32$. The American Standard Code for Information Interchange (ASCII) has seven information elements per character, or $2^7 = 128$; so it has 128 distinct combinations of marks and spaces that are available for assignment as characters or graphics.

The number of distinct characters for a specific code may be extended by establishing a code sequence (a special character assignment) to shift the system or machine to uppercase (as is done with a conventional typewriter). Uppercase is a new character grouping. A second distinct code sequence is then assigned to revert to lowercase. As an example, the CCITT International Telegraph Alphabet (ITA) no. 2 code (Figure 8.1) is a five-unit code with 58 letters, numbers, graphics, and operator sequences. The additional characters (additional above $2^5 = 32$) come from the use of uppercase. Operator sequences appear on a keyboard as *space* (spacing bar), *figures* (uppercase), *letters* (lowercase), *carriage return*, *line feed* (spacing vertically), and so on.

When we refer to a 5-unit, 6-unit, 7-unit, or 12-unit code, we refer to the number of information units or elements that make up a single character or symbol, that is, we refer to those elements assigned to each character that carry information and that make it distinct from all other characters or symbols of the code.

8.3.2 Some Specific Binary Codes for Information Interchange

In addition to the ITA no. 2 code, some of the more commonly used codes are the field data code, the IBM data transceiver code (Figure 8.2), the American Standard Code for Information Interchange (ASCII) (Figure 8.3), the CCITT no. 5 code (Figure 8.4), the Extended Binary Coded Decimal Interchange Code (EBCDIC) (Figure 8.5), the Hollerith code (Figure 8.6), and the binary-coded decimal (BCD) code.

The field data code was adopted by the U.S. Army in 1969 and is now being replaced by ASCII. The field data code is an eight-bit code consisting of seven information bits and a control bit. With the 8 bits, 256 (2^8) bit patterns or permutations are available. Of these, 128 are available for assignment to characters. These 128 bit patterns are subdivided into two groups of 64 each. The groups are distinguished from one another by the value or state of a particular bit known as the control bit. Owing to this arrangement of bits and bit patterns, the field data code has parity check capability.

Parity checks are one way to determine if a character contains an error after transmission. We speak of even parity and odd parity. On a system using an odd parity check, the total count of 1's or marks has to be an odd number per character (or block) (e.g., it carries 1, 3, 5, or 7 marks or 1's). Some systems, such as the field data code, use even parity (i.e., the total number of marks must be an even number, such as, 2, 4, 6, or 8). The code is used by the U.S. Army for communication in

X O N R 7 4 2 1 ↓↓↓↓↓↓↓↓	0 0 0	0 0 1	0 1 0	0 1 1	1 0 0	1 0 1	1 1 0	1 1 1
0 0 0 0 0								
0 0 0 0 1								TPH/TGR
0 0 0 1 0								@
0 0 0 1 1				(NA)		(NA)	Space	
0 0 1 0 0								#
0 0 1 0 1				(NA)		(NA)	9	
0 0 1 1 0				(NA)		(NA)	8	
0 0 1 1 1		G	P		X			
0 1 0 0 0								(NA)
0 1 0 0 1				(NA)		(NA)	6	
0 1 0 1 0				(NA)		(NA)	5	
0 1 0 1 1		D	M		U			
0 1 1 0 0				(NA)		(NA)	3	
0 1 1 0 1		B	K		S			
0 1 1 1 0		A	J		/			
0 1 1 1 1	0							
1 0 0 0 0								Restart
1 0 0 0 1				SOC/EOC		$\overset{+}{0}$	EOT	
1 0 0 1 0				•		%		
1 0 0 1 1		&	—		Ø			
1 0 1 0 0				$		,	.	
1 0 1 0 1		I	R		Z			
1 0 1 1 0		H	Q		Y			
1 0 1 1 1	7							
1 1 0 0 0				(NA)		(NA)	(NA)	
1 1 0 0 1		F	O		W			
1 1 0 1 0		E	N		V			
1 1 0 1 1	4							
1 1 1 0 0		C	L		T			
1 1 1 0 1	2							
1 1 1 1 0	1							
1 1 1 1 1								

Figure 8.2 IBM data transceiver code. TPH/TGH = telephone/telegraph. SOC/EOC = start or end of card; EOT = end of transmission; (NA) = Valid but not assigned; $\overset{+}{0}$ = plus zero; $\overline{0}$ = minus zero; "Lozenge" (special symbol). Transmission order; bit X → bit 1.

a common language between computing, input–output, and terminal equipments of the field data equipment family.

To explain parity and parity checks a little more clearly, let us look at some examples. Consider a seven-level code with an extra parity bit. By system convention, even parity has been established. Suppose that a character is transmitted as 1111111. There are seven marks, so to maintain even parity we would need an even

Bit Number											
			→ 0	0	0	0	1	1	1	1	
			→ 0	0	1	1	0	0	1	1	
			→ 0	1	0	1	0	1	0	1	
			Column								
$b_7\,b_6\,b_5\,b_4\,b_3\,b_2\,b_1$	Row		0	1	2	3	4	5	6	7	
0 0 0 0	0		NUL	DLE	SP	Ø	@	P	`	p	
0 0 0 1	1		SOH	DC1	!	1	A	Q	a	q	
0 0 1 0	2		STX	DC2	"	2	B	R	b	r	
0 0 1 1	3		ETX	DC3	#	3	C	S	c	s	
0 1 0 0	4		EOT	DC4	$	4	D	T	d	t	
0 1 0 1	5		ENQ	NAK	%	5	E	U	e	u	
0 1 1 0	6		ACK	SYN	&	6	F	V	f	v	
0 1 1 1	7		BEL	ETB	'	7	G	W	g	w	
1 0 0 0	8		BS	CAN	(	8	H	X	h	x	
1 0 0 1	9		HT	EM	)	9	I	Y	i	y	
1 0 1 0	10		LF	SS	*	:	J	Z	j	z	
1 0 1 1	11		VT	ESC	+	;	K	[	k	{	
1 1 0 0	12		FF	FS	,	<	L	\	l	|	
1 1 0 1	13		CR	GS	–	=	M	]	m	}	
1 1 1 0	14		SO	RS	.	>	N	∧	n	~	
1 1 1 1	15		SI	US	/	?	O	–	o	DEL	

Figure 8.3 American Standard code for information interchange.

Notes

1. Columns 2, 3, 4, and 5 indicate the printable characters in the U.S. Defense Communication System automatic digital network (DCS Autodin).

2. Columns 6 and 7 fold over into columns 4 and 5, respectively, except DEL.

GENERAL DEFINITIONS

Communication control. A function character intended to control or facilitate transmission of information over communication networks.

Format effector. A functional character which controls the layout or positioning of information in printing or display devices.

Information separator. A character used to separate and qualify information in a logical sense. There is a group of four such characters, which are to be used in a hierarchical order. In order rank, highest to lowest, they appear as follows: FS-file separator; GS-group separator; RS-record separator; US-unit separator.

SPECIFIC CHARACTERS

NUL. The all-zeros character.

SOH (*start of heading*). A communication control character used at the beginning of a sequence is referred to as the *heading*. An STX character has the effect of terminating a heading.

STX (*start of text*). A communication control character which precedes a sequence of characters that is to be treated as an entity and thus transmitted through to the ultimate destination. Such a sequence is referred to as *text*. STX may be used to terminate a sequence of characters started by SOH.

ETX (*end of text*). A communication control character used to terminate a sequence of characters started with STX and transmitted as an entity.

EOT (*end of transmission*). A communication control character used to indicate the conclusion of a transmission, which may have contained one or more texts and any associated headings.

ENQ (*enquiry*). A communication control character used in data communication systems as a request for a response from a remote station.

ACK (*acknowledge*). A communication control character transmitted by a receiver as an affirmative response to a sender.

BEL. A character used when there is a need to call for human attention. It may trigger an alarm or other attention devices.

BS (*backspace*). A format effector which controls the movement of the printing position one printing space backward on the same printing line.

HT (*horizontal tabulation*). A format effector which controls the movement of the printing position to the next in a series of predetermined positions along the printing line.

LF (*line feed*). A format effector which controls the movement of the printing position to the next printing line.

VT (*vertical tabulation*). A format effector which controls the movement of the printing position to the next in a series of predetermined printing lines.

FF (*form feed*). A format effector which controls the movement of the printing position to the first predetermined printing line on the next form or page.

CR (*carriage return*). A format effector which controls the movement of the printing position to the first printing position on the same printing line.

SO (*shift out*). A control character indicating that the code combinations which follow shall be interpreted as outside of the character set of the standard code table until a shift in (SI) character(s) is (are) reached.

SI (*shift in*). A control character indicating that the code combinations which follow shall be interpreted according to the standard code table.

DLE (*data link escape*). A communication control character which will change the meaning of a limited number of contiguously following characters. It is used exclusively to provide supplementary controls in data communication networks. DLE is usually terminated by a shift in character(s).

DC1, DC2, DC3, DC4 (*device controls*). Characters for the control of ancillary devices associated with data processing or telecommunication systems, especially for switching devices on or off.

NAK (*negative acknowledgment*). A communication control character transmitted by a receiver as a negative response to the sender.

SYN (*synchronous idle*). A communication control character used by a synchronous transmission system in the absence of any other character to provide a signal from which synchronism may be achieved or retained.

ETB (*end of transmission block*). A communication control character used to indicate the end of a block of data for communication purposes.

CAN (*cancel*). A control character used to indicate that the data with which it is sent is in error or is to be disregarded.

EM (*end of medium*). A control character associated with the sent data which may be used to identify the physical end of the medium, or the end of the used or wanted portion of information recorded on a medium.

SS (*start of special sequence*). A control character used to indicate the start of a variable-length sequence of characters which have special significance or which are to receive special handling. SS is usually terminated by a shift in (SI) character(s).

ESC (*escape*). A control character intended to provide code extension (supplementary characters) in general information interchange. The escape character itself is a prefix affecting the

360

interpretation of a limited number of contiguously following characters. ESC is usually terminated by a shift in (SI) character(s).

DEL (*delete*). This character is used primarily to erase or obliterate erroneous or unwanted characters in perforated tape.

SP (*space*). Normally a nonprinting graphic character used to separate words. It is also a format effector which controls the movement of the printing position, one printing position forward.

Diamond. A noncoded graphic which is printed by a printing device to denote the sensing of an error when such an indication is required.

Heart. A noncoded graphic which may be printed by a printing device in lieu of the symbols for the control characters shown in columns 0 and 1.

$b_7\ b_6\ b_5\ b_4\ b_3\ b_2\ b_1$ Row	Column 0	1	2	3	4	5	6	7
0 0 0 0 0	NUL	(TC$_r$)DLE	SP	0	(@) ③	P	' ④	p
0 0 0 1 1	(TC$_1$)SOH	DC$_1$	!	1	A	Q	a	q
0 0 1 0 2	(TC$_2$)STX	DC$_2$	" ⑥	2	B	R	b	r
0 0 1 1 3	(TC$_3$)ETX	DC$_3$	£ ③ ⑦	3	C	S	c	s
0 1 0 0 4	(TC$_4$)EOT	DC$_4$	$ ③ ⑦	4	D	T	d	t
0 1 0 1 5	(TC$_5$)ENQ	(TC$_8$)NAK	%	5	E	U	e	u
0 1 1 0 6	(TC$_6$)ACK	(TC$_9$)SYN	&	6	F	V	f	v
0 1 1 1 7	BEL	(TC$_{10}$)ETB	' ⑥	7	G	W	g	w
1 0 0 0 8	FE$_0$(BS)	CAN	(	8	H	X	h	x
1 0 0 1 9	FE$_1$(HT)	EM	)	9	I	Y	i	y
1 0 1 0 10	FE$_2$(LF)①	SUB	*	: ⑧	J	Z	j	z
1 0 1 1 11	FE$_3$(VT)	ESC	+	; ⑧	K	([) ③	k	③
1 1 0 0 12	FE$_4$(FF)	IS$_4$(FS)	,	<	L	⑨	l	③
1 1 0 1 13	FE$_5$(CR)①	IS$_3$(GS)	—	=	M	(]) ③	m.	③
1 1 1 0 14	SO	IS$_2$(RS)	.	>	N	∧ ④ ⑥	n	– ④ ⑥
1 1 1 1 15	SI	IS$_1$(US)	/	?	O	—	o	DEL

Figure 8.4 CCITT no. 5 code for information interchange. (See CCITT Rec. V.3.)

Notes

1. The controls CR and LF are intended for printer equipment which requires separate combinations to return the carriage and to feed a line.

For equipment which uses a single control for a combined carriage return and line feed operation, the function FE$_2$ will have the meaning of *new line* (NL).

These substitutions must be in agreement between the sender and the recipient of the data.

The use of this function NL is not allowed for international transmission on general switched telecommunication networks (telegraph and telephone networks).

2. For international information interchange, $ and £ symbols do not designate the currency of a given country. The use of these symbols combined with other graphic symbols to designate national currencies may be the subject to other recommendations.

3. Reserved for national use. These positions are intended primarily for alphabetic extensions. If they are not required for that purpose, they may be used for symbols, and a recommended choice is shown in parentheses in some cases.

4. Positions 5/14, 6/0, and 7/14 of the 7-bit set table normally are provided for the diacritical signs *circumflex*, *grave accent*, and *overline*. However, these positions may be used for other graphical symbols when it is necessary to have 8, 9, or 10 positions for national use.

5. For international information interchange, position 7/14 is used for the graphical symbol – (overline), the graphical representation of which may vary according to national use to represent ~ (tilde) or another diacritical sign provided that there is no risk of confusion with another graphical symbol included in the table.

6. The graphics in positions 2/2, 2/7, 5/14 have respectively the significance of *quotation mark*, *apostrophe*, and *upwards arrow*; however, these characters take on the significance of the diacritical signs *diaeresis*, *acute accent*, and *circumflex accent* when they precede or follow the backspace character.

7. For international information interchange, position 2/3 of the 7-bit code table has the significance of the symbol £, and position 2/4 has the significance of the symbol $.

By agreement between the countries concerned where there is no requirement for the symbol £, the symbol *number sign* (#) may be used in position 2/3. Likewise, where there is no requirement for the symbol $, the symbol *currency sign* ( ) may be used in position 2/4.

8. If 10 and 11 as single characters are needed (for example, for Sterling currency subdivision), they should take the place of *colon* (:) and *semicolon* (;), respectively. These substitutions require agreement between the sender and the recipient of the data. On the general telecommunication networks, the characters *colon* and *semicolon* are the only ones authorized for international transmission.

number of marks. Thus an eighth bit is added and must be a mark (1). Look at another bit pattern, 1011111. Here there are six marks, even; then the eighth (parity) bit must be a space. Still another example would be 0001000. To get even parity, a mark must be added on transmission, and the character transmitted would be 00010001, maintaining even parity. Suppose that, owing to some sort of signal interference, one signal element was changed on reception. No matter which element was changed, the receiver would indicate an error because we would no longer have even parity. If two elements were changed, though, the error could be masked. This would happen in the case of even or odd parity if two marks were substituted for two spaces or vice versa at any element location in the character.

The IBM data transceiver code (see Figure 8.2) is used for the transfer of digital data information recorded on perforated cards. It is an eight-bit code providing a total of 256 mark–space combinations. Only those combinations or patterns having a "fixed count" of four 1's (marks) and four 0's (spaces) are made available for assignment as characters. Thus only 70 bit patterns satisfy the fixed count condition, and the remaining 186 combinations are invalid. The parity (fixed count) or error checking advantage is obvious. Of the 70 valid characters, 54 are assigned to alphanumerics and a limited number of punctuation signs and other symbols. In addition, the code includes special bit patterns assigned to control functions peculiar to the transmission of cards such as "start card" and "end card."

The ASCII (see Figure 8.3) is the latest effort on the part of the U.S. industry and common carrier systems, backed by the American National Standards Institute, to produce a universal common language code. ASCII is a seven-unit code with all 128 combinations available for assignment. Here again the 128 bit patterns are divided into two groups of 64. One of the groups is assigned to a subset of graphic printing characters. The second subset of 64 is assigned to control characters. An eighth bit is added to each character for parity check. ASCII is widely used in North America and has received considerable acceptance in Europe and Latin America.

CCITT Rec. V.3 offers a seven-level code as an international standard for information interchange. It is not intended as a substitute for CCITT ITA no. 2 code. CCITT no. 5, or the new alphabet no. 5, as the seven-level code is more commonly referred to, is basically intended for data transmission.

Although CCITT no. 5 is considered a seven-level code, CCITT Rec. V.4 advises that an eighth bit may be added for parity. Under certain circumstances odd parity is recommended; on others, even parity.

Figure 8.4 shows the CCITT no. 5 code. b_1 is the first signal element in serial

Bits xxxx 4567	Hex Row	Digit Punches	0 (00·00)	1 (00·01)	2 (00·10)	3 (00·11)	4 (01·00)	5 (01·01)	6 (01·10)	7 (01·11)	8 (10·00)	9 (10·01)	A (10·10)	B (10·11)	C (11·00)	D (11·01)	E (11·10)	F (11·11)	
0000	0	8-1	•1 NUL	•2 DLE	•3	•4	•5 SP	•6 &	•7 -	•8					•9 {	•10 }	•11 \	•12 0	8-1
0001	1	1	SOH	DC1			•13 /				a	j	¬		A	J	•14	1	1
0010	2	2	STX	DC2		SYN					b	k	s		B	K	S	2	2
0011	3	3	ETX	DC3							c	l	t		C	L	T	3	3
0100	4	4									d	m	u		D	M	U	4	4
0101	5	5	HT		LF						e	n	v		E	N	V	5	5
0110	6	6			BS	ETB					f	o	w		F	O	W	6	6
0111	7	7	DEL		ESC	EOT					g	p	x		G	P	X	7	7
1000	8	8			CAN						h	q	y		H	Q	Y	8	8
1001	9	8-1			EM					`	i	r	z		I	R	Z	9	9
1010	A	8-2					[	]	•15 ¦	:									8-2
1011	B	8-3	VT				.	$	,	ǂ									8-3
1100	C	8-4	FF	FS		DC4	<	*	%	@									8-4
1101	D	8-5	CR	GS	ENQ	NAK	(	)	_	'									8-5
1110	E	8-6	SO	RS	ACK		+	;	>	=									8-6
1111	F	8-7	SI	US	BEL	SUB	!	^	?	"									8-7

Card Hole Patterns
•1, 12-0-9-8-1   •4, 12-11-0-9-8-1   •8, 12-11-0   •12, 0
•2, 12-11-9-8-1   •5, No Punches   •9, 12-0   •13, 0-1
•3, 11-0-9-8-1   •6, 12   •10, 11-0   •14, 11-0-9-1
•7, 11   •11, 0-8-2   •15, 12-11

Figure 8.5 Extended binary decimal interchange code (EBCDIC), as shown for punched card application. (Source: MIL-STD-188C.)

ZONES	12				12	12		12
		11				11	11	11
			0		0		0	0
DIGIT PUNCH ROWS	&	-	0	SP	{	|	}	11/10
1	A	J	/	1	a	j	~	13/9
2	B	K	S	2	b	k	s	13/10
3	C	L	T	3	c	l	t	13/11
4	D	M	U	4	d	m	u	13/12
5	E	N	V	5	e	n	v	13/13
6	F	O	W	6	f	o	w	13/14
7	G	P	X	7	g	p	x	13/15
8	H	Q	Y	8	h	q	y	14/0
9	I	R	Z	9	i	r	z	14/1
8-2	[	]	\	:	12/4	12/11	13/2	14/2
8-3	.	$	,	#	12/5	12/12	13/3	14/3
8-4	<	*	%	@	12/6	12/13	13/4	14/4
8-5	(	)	_	'	12/7	12/14	13/5	14/5
8-6	+	;	>	=	12/8	12/15	13/6	14/6
8-7	!	^	?	"	12/9	13/0	13/7	14/7
8-1	10/8	11/1	11/9	~	12/3	12/10	13/1	13/8
9-1	SOH	DC1	8/1	9/1	10/0	10/9	9/15	11/11
9-2	STX	DC2	8/2	SYN	10/1	10/10	11/2	11/12
9-3	ETX	DC3	8/3	9/3	10/2	10/11	11/3	11/13
9-4	9/12	9/13	8/4	9/4	10/3	10/12	11/4	11/14
9-5	HT	8/5	LF	9/5	10/4	10/13	11/5	11/15
9-6	8/6	BS	ETB	9/6	10/5	10/14	11/6	12/0
9-7	DEL	8/7	ESC	EOT	10/6	10/15	11/7	12/1
9-8	9/7	CAN	8/8	9/8	10/7	11/0	11/8	12/2
9-8-1	8/13	EM	8/9	9/9	NUL	DLE	8/0	9/0
9-8-2	8/14	9/2	8/10	9/10	14/8	14/14	15/4	15/10
9-8-3	VT	8/15	8/11	9/11	14/9	14/15	15/5	15/11
9-8-4	FF	FS	8/12	DC4	14/10	15/0	15/6	15/12
9-8-5	CR	GS	ENQ	NAK	14/11	15/1	15/7	15/13
9-8-6	SO	RS	ACK	9/14	14/12	15/2	15/8	15/14
9-8-7	SI	US	BEL	SUB	14/13	15/3	15/9	15/15

NOTE:

A card code position that has not been assigned a corresponding ASCII Code character is designated with the corresponding, though not yet assigned, column/row of the ASCII Codes.

Figure 8.6 Hollerith punched card code. *Note.* A card code position that has not been assigned a corresponding ASCII code character is designated with the corresponding, though not yet assigned, column/row of the ASCII codes. (Source: MIL-STD-188-100.)

transmission and b_7 is the last element of a character. Like the ASCII, CCITT no. 5 does not normally need to shift out (i.e., uppercase, lowercase as in CCITT ITA no. 2). However, like ASCII, it is provided with an escape, 1101100. Eight footnotes explaining peculiar character usage are provided. Few differences exist between the ASCII and the CCITT no. 5 codes.

The reader's attention is called to what are known as computable codes, such as ASCII and CCITT no. 5 codes. Computable codes have the letters of the alphabet plus all other characters and graphics assigned values in continuous binary sequence.

Thus these codes are in the native binary language of today's common digital computers. The CCITT ITA no. 2 is not, and when used with a computer, it often requires special processing.

The Extended Binary Coded Decimal Interchange Code (EBCDIC) is similar to the ASCII, but it is a true eight-bit code. The eighth bit is used as an added bit to "extend" the code, providing 256 distinct code combinations for assignment. Figure 8.5 illustrates the EBCDIC code.

The Hollerith code was specifically designed for use with perforated (punched) cards. It has attained wide acceptance in the business machine and computer fields. Hollerith is a 12-unit character code in that a character is represented on a card by one or more holes perforated in one column having 12 potential hole positions. It is most commonly used with the standard 80-column punched cards.

The theoretical capacity of a 12-unit binary code is very great, and by our definition, using all hole patterns available, it is 2^{12} or 4096 bit combinations. In the modern version of the Hollerith code only 64 of these, none using more than three holes, are assigned to graphic characters.

Because of its unwieldiness, the Hollerith code is seldom used directly for transmission. Most often it is converted to one of the more conventional transmission codes such as the ASCII, CCITT no. 5, BCD interchange, or EBCDIC. Figure 8.6 shows the Hollerith code as extended to cover ASCII equivalents.

8.3.3 Hexadecimal Representation and the BCD Code

The hexadecimal system is a numeric representation in the number base 16. The number base uses 0 through 9 as in the decimal base, and the letters A through F to represent the decimal numbers 10 through 15. The hexadecimal numbers can be translated to the binary base as follows:

Hexadecimal	Binary	Hexadecimal	Binary
0	0000	8	1000
1	0001	9	1001
2	0010	A	1010
3	0011	B	1011
4	0100	C	1100
5	0101	D	1101
6	0110	E	1110
7	0111	F	1111

Two examples of the hexadecimal notation are:

Number Base 10	Number Base 16
21	15
64	40

The BCD is a compromise code assigning 4-bit binary numbers to the digits between 0 and 9. The BCD equivalents to decimal digits appear as follows:

Decimal Digit	BCD Digit	Decimal Digit	BCD Digit
0	1010	5	0101
1	0001	6	0110
2	0010	7	0111
3	0011	8	1000
4	0100	9	1001

To cite some examples, consider the number 16; it is broken down into 1 and 6. Thus its BCD equivalent is 0001 0110. If it were written in straight binary notation, it would appear as 10000. The number 25 in BCD combines the digits 2 and 5 above as 0010 0101.

8.4   ERROR DETECTION AND ERROR CORRECTION

8.4.1   Introduction

In the transmission of data the most important goal in design is to minimize the error rate. Error rate may be defined as the ratio of the number of bits incorrectly received to the total number of bits transmitted. On many data circuits the design objective is an error rate no poorer than one error in 1×10^5 and on telegraph circuits, one error in 1×10^4.

One method to minimize the error rate is to provide a "perfect" transmission channel, one that will introduce no errors in the transmitted information by the receiver. The engineer designing a data transmission system can never achieve that perfect channel. Besides improvement of the channel transmission parameters themselves, the error rate can be reduced by forms of systematic redundancy. In old-time Morse on a bad circuit words often were sent twice; this is redundancy in its simplest form. Of course, it took twice as long to send a message. This is not very economical if useful words per minute received is compared to channel occupancy.

This brings up the point of channel efficiency. Redundancy can be increased such that the error rate could approach zero. Meanwhile the information transfer or *throughput* across the channel also approaches zero. Thus unsystematic redundancy is wasteful and merely lowers the rate of useful communication. Maximum efficiency or throughput could be obtained in a digital transmission system if all redundancy and other code elements, such as start and stop elements, were removed from the code, and, in addition, if advantage were taken of the statistical phenomenon of our written language by making high-usage letters, such as E, T, and A, short in code length and low-usage letters, such as Q and X, longer.

8.4.2 Throughput

The *throughput* of a data channel is the expression of how much data are put through. In other words, throughput is an expression of channel efficiency. The term gives a measure of *useful* data put through the data communication link. These data are directly useful to the computer or data terminal equipment (DTE).

Therefore on a specific circuit, throughput varies with the raw data rate; is related to the error rate and the type of error encountered (whether burst or random); and varies with the type of error detection and correction system used, the message handling time, and the block length from which we must subtract the "nonuseful" bits such as overhead bits. Among overhead bits we have parity bits, flags, cyclic redundancy checks, and so forth.

8.4.3 The Nature of Errors

In data/telegraph transmission an error is a bit that is incorrectly received. For instance, a 1 is transmitted in a particular time slot and the element received in that slot is interpreted as a 0. Bit errors occur either as single random errors or as bursts of error. In fact we can say that every transmission channel will experience some random errors, but on a number of channels burst errors may predominate. For instance, lightning or other forms of impulse noise often cause bursts of errors, where many contiguous bits show a very high number of bits in error. The IEEE defines error burst as "a group of bits in which two successive bits are always separated by less than a given number of correct bits" (Ref. 37).

8.4.4 Error Detection and Correction Defined

The data transmission engineer differentiates between error detection and error correction. Error detection identifies that a symbol has been received in error. As discussed above, parity is primarily used for error detection. Parity bits, of course, add redundancy and thus decrease channel efficiency or throughput.

Error correction corrects the detected error. Basically there are two types of error correction techniques: forward-acting (FEC) and two-way error correction [automatic repeat request (ARQ)]. The latter technique uses a return channel (backward channel). When an error is detected, the receiver signals this fact to the transmitter over the backward channel, and the block* of information containing the error is transmitted again. FEC utilizes a type of coding that permits a limited number of errors to be corrected at the receiving end by means of special coding and software (or hardware) implemented at both ends of a circuit.

ERROR DETECTION

There are various arrangements or techniques available for the detection of errors. All error detection methods involve some form of redundancy, those additional

*A block is a group of digits or data characters transmitted as a unit over which a coding procedure is usually applied for synchronization and error control purposes.

bits or sequences that can inform the system of the presence of error or errors. Parity, discussed above, was character parity, and its weaknesses were presented. Commonly the data transmission engineer refers to such parity as *vertical redundancy checking* (VRC). The term *vertical* comes from the way characters are arranged on paper tape (i.e., hole positions).

Another form of error detection utilizes longitudinal redundancy checking (LRC), which is used in block transmission where a data message consists of one or more blocks. Remember that a block is a specific group of digits or data characters sent as a "package" (not to be confused with *packet*). In such circumstances an LRC character, often called a block check character (BCC), is appended at the end of each block. The BCC verifies the total number of 1's and 0's in the columns of the block (vertically). The receiving end sums the 1's (or the 0's) in the block, depending on the parity convention for the system. If that sum does not correspond to the BCC, an error (or errors) exists in the block. The LRC ameliorates much of the problem of undetected errors which could slip through with VRC if used alone. The LRC method is not foolproof, however, as it uses the same thinking as VRC. Suppose that errors occur such that two 1's are replaced by two 0's in the second and third bit positions of characters 1 and 3 in a certain block. In this case the BCC would read correctly at the receive end and the VRC would pass over the errors as well. A system using both LRC and VRC is obviously more immune to undetected errors than either system implemented alone. A more effective method of error detection is cyclic redundancy check (CRC), which is based on a cyclic code and is used in block transmission with a BCC. In this case the transmitted BCC represents the remainder of a division of the message block by a "generating polynomial."

Mathematically, a message block can be treated as a function such as:

$$a_n X^n + a_{n-1} x^{n-1} + a_{n-2} X^{n-2} + \cdots + a_1 X + a_0$$

where coefficients a are set to represent a binary number. Consider the binary number 11011, which is represented by the polynomial

$$\begin{array}{ccccc} 1 & 1 & 0 & 1 & 1 \\ a_4 & a_3 & a_2 & a_1 & a_0 \end{array}$$

and then becomes

$$X^4 + X^3 + X + 1$$

or consider another example,

$$\begin{array}{ccccc} 0 & 1 & 1 & 0 & 1 \\ a_4 & a_3 & a_2 & a_1 & a_0 \end{array}$$

which then becomes

$$X^3 + X^2 + 1$$

The CRC character used as the BCC is the remainder of a data polynomial divided by the generating polynomial, as mentioned.

More specifically, if a data polynomial $D(X)$ is divided by the generating polynomial $G(X)$, the result is a quotient polynomial $R(X)$, or

$$\frac{D(X)}{G(X)} = Q(X) + \frac{R(X)}{G(X)}$$

The CRC character in most applications is 16 bits in length or two 8-bit bytes. At present there are three standard generating polynomials commonly in use:

CRC-16 (ANSI) $X^{16} + X^{15} + X^2 + 1$

CRC (CCITT) $X^{16} + X^{12} + X^5 + 1$

CRC-12 $X^{12} + X^{11} + X^3 + X^2 + X + 1$

Of course, again, if the computed BCC at the receive end differs from the BCC received from the transmit end, there would be an error (or errors) in the received block.

In (Ref. 25) it is stated that CRC-12 provides error detection of bursts of up to 12 bits in length. Additionally, 99.955% of error bursts up to 16 bits in length can be detected. CRC-16 provides detection of bursts up to 16 bits in length, and 99.955% of error bursts greater than 16 bits can be detected.

FORWARD-ACTING ERROR CORRECTION (FEC)

FEC uses certain binary codes that are designed to be self-correcting for errors introduced by the intervening transmission media. In this form of error correction the receiving station has the ability to reconstitute messages containing errors.

FEC uses *channel encoding*, whereas the encoding covered previously in this chapter is generically called *source encoding*. The channel encoding used in FEC can be broken down into two broad categories, block codes and convolutional codes. These are discussed at considerable length in Section 15.7.

Key to this type of coding is the modulo 2 adder. Modulo 2 addition is denoted by the symbol $\oplus$. It is binary addition without the "carry", or $1 + 1 = 0$, and we do *not* carry the 1. Summing 10011 and 11001 in modulo 2, we get 01010.

A measure of the error detection and correction capability of a code is given by the Hamming distance. The distance is the minimum number of digits in which two encoded words differ. For example, to detect E digits in error, a code of a minimum Hamming distance of $(E + 1)$ is required. To correct E errors, a code must display a minimum Hamming distance of $(2E + 1)$. A code with a minimum Hamming distance of 4 can correct a single error *and* detect two digits in error.

ERROR CORRECTION WITH FEEDBACK CHANNEL

Two-way or feedback error correction is used very widely today on data and some telegraph circuits. Such a form of error correction is called ARQ, which derives from the old Morse and telegraph signal, "automatic repeat request."

In most modern data systems block transmission is used, and the block is a convenient length of characters sent as an entity. There are two aspects in that "convenience" of length. One relates to the material that is being sent. For instance, the standard "IBM" card has 80 columns. With 8 bits per column, a block of 8 X 80 or

640 bits would be desirable as data text so that we could transmit an IBM card in each block. In fact, one such operating system, Autodin, bases block length on that criterion, with blocks 672 bits long. The remaining bits, those in excess of 640, are overhead and check bits.

The second aspect of block length is the trade-off for optimum length between length and error rate, or the number of block repeats that may be expected on a particular circuit. Longer blocks tend to amortize overhead bits better but are inefficient regarding throughput when an error rate is high. Under these conditions long blocks tend to tie up a circuit with longer retransmission periods.

ARQ, as we know, is based on the block transmission concept. When a receiving station detects an error, it requests a repeat of the block in question from the transmitting station. The request is made on a feedback channel, which may be a channel especially dedicated for that purpose or may be the return side of a full-duplex link. If the channel is "specially dedicated," it generally is a slow-speed channel, often 75 bps, whereas the forward channel may be 2400 bps or better.

8.5   THE DC NATURE OF DATA TRANSMISSION

8.5.1   Loops

Binary data is transmitted on a dc loop. More correctly the binary data end instrument delivers to the line and receives from the line one or several dc loops. In its most basic form a dc loop consists of a switch, a dc voltage source, and a termination. A pair of wires interconnects the switch and termination. The voltage source

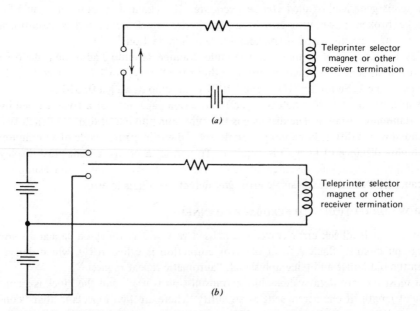

Figure 8.7 Simplified diagram of a dc loop. (*a*) With neutral keying. (*b*) With polar keying.

in data and telegraph work is called the battery, although the device is usually electronic, deriving the dc voltage from an ac power line source. The battery is placed in the line such that it provides voltage(s) consistent with the type of transmission desired. A simplified dc loop is shown in Figure 8.7.

8.5.2 Neutral and Polar DC Transmission Systems

Nearly all dc data and telegraph systems functioning today are operated in either a neutral or a polar mode. The words *neutral* and *polar* describe the manner in which the battery is applied to the dc loop. On a neutral loop, following the mark–space convention in Table 8.1, the battery is applied during marking (1) conditions and is switched off during spacing (0). Current therefore flows in the loop when a mark is sent and the loop is closed. Spacing is indicated on the loop by a condition of no current. Thus we have the two conditions for binary transmission, an open loop (no current flowing) and a closed loop (current flows). Keep in mind that we could reverse this, change the convention (Table 8.1), and, say, assign spacing to a condition of current flowing or closed loop and marking to a condition of no current or open loop. This is sometimes done in practice and is called changing the sense. Either way, a neutral loop is a dc loop circuit where one binary condition is represented by the presence of voltage, flow of current, and the other by the absence of voltage/current. Figure 8.7a shows a simplified neutral loop.

Polar transmission approaches the problem a little differently. Two batteries are provided. One is called negative battery and the other positive. During a condition of marking, a positive battery is applied to the loop, following the convention of Table 8.1, and a negative battery is applied to the loop during the spacing condition. In a polar loop, current is always flowing. For a mark or binary 1 it flows in one direction and for a space or binary 0 it flows in the opposite direction. Figure 8.7b shows a simplified polar loop.

8.6 BINARY TRANSMISSION AND THE CONCEPT OF TIME

8.6.1 Introduction

Time and timing are most important factors in digital transmission. For this discussion consider a binary end instrument sending out in series a continuous run of marks and spaces. Those readers who have some familiarity with Morse will recall that the spaces between dots and dashes told the operator where letters ended and where words ended. With the sending device or transmitter delivering a continuous series of characters to the line, each consisting of 5, 6, 7, 8, or 9 elements (bits) to the character, let the receiving device start its print cycle when the transmitter starts sending. If the receiver is perfectly in step with the transmitter, ordinarily one could expect good printed copy and few, if any, errors at the receiving end.

It is obvious that when signals are generated by one machine and received by another, the speed of the receiving machine must be the same or very close to that of the transmitting machine. When the receiver is a motor-driven device, timing stability and accuracy are dependent on the accuracy and stability of the speed of rotation of the motors used.

Most simple data/telegraph receivers sample at the presumed center of the signal element. It follows, therefore, that whenever a receiving device accumulates a timing error of more than 50% of the period of one bit, it will print in error.

The need for some sort of synchronization is shown in Figure 8.8a. A 5-unit code is employed, and three characters transmitted sequentially are shown. Sampling points are shown in the figure as vertical arrows. Receiver timing begins when the first pulse is received. If there is a 5% timing difference between transmitter and receiver, the first sampling at the receiver will be 5% away from the center of the transmitted pulse. At the end of the tenth pulse or signal element the receiver may sample in error. The eleventh signal element will indeed be sampled in error, and all subsequent elements are errors. If the timing error between transmitting machine and receiving machine is 2%, the cumulative error in timing would cause the receiving device to print all characters in error after the twenty-fifth bit.

8.6.2 Asynchronous and Synchronous Transmission

In the earlier days of printing telegraphy, start–stop transmission or asynchronous operation was developed to overcome the problem of synchronism. Here timing starts at the beginning of a character and stops at the end. Two signal elements are added to each character to signal the receiving device that a character has begun and ended.

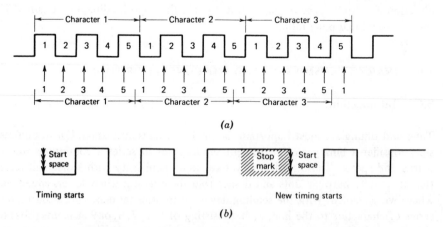

(a)

(b)

Figure 8.8 (*a*) 5-unit synchronous bit stream with timing error. (*b*) 5-unit start and stop stream of bits with a 1.5-unit stop element.

As an example consider a five-element code, such as CCITT ITA no. 2 (see Figure 8.1). In the front of a character an element is added called a start space and at the end of each character a stop mark is inserted. Send the letter A in Figure 8.1. The receiving device starts its timing sequence on receiving element no. 1, a space or 0, then a 11000 is received; the A is selected, then the stop mark is received, and the timing sequence stops. On such an operation an accumulation of timing errors can take place only inside each character.

Suppose that the receiving device is again 5% slower or faster than its transmitting counterpart; now the fifth information element will be no more than 30% displaced in time from the transmitted pulse and well inside the 50% or halfway point for correct sampling to take place.

In start–stop transmission the information signal elements are each of the same duration, which is the duration or pulse width of the start element. The stop element has an indefinite length or pulse width beyond a certain minimum.

If a steady series of characters are sent, then the stop element is always the same width or has the same number of unit intervals. Consider the transmission of two A's, $0110001011000111111 \rightarrow 11111$. The start space (0) starts the timing sequence for six additional elements, which are the five code elements in the letter A and the stop mark. Timing starts again on the mark-to-space transition between the stop mark of the first A and the start space of the second. Sampling is carried out at pulse center for most asynchronous systems. One will note that at the end of the second A a continuous series of marks is sent. Thus the signal is a continuation of the stop element or just a continuous mark. It is the mark-to-space transition of the start element that tells the receiving device to start timing a character.

Minimum lengths of stop elements vary. The example discussed above shows a stop element of 1-unit interval duration (1 bit). Some are 1.42-unit intervals, others are of 1.5- and 2-unit interval duration. The proper semantics of data/telegraph transmission would describe the code of the previous paragraph as a 5-unit start–stop code with a 1-unit stop element.

A primary objective in the design of telegraph and data systems is to minimize errors received or to minimize the error rate. There are two prime causes of errors. These are noise and improper timing relationships. With start–stop systems a character begins with a mark-to-space transition at the beginning of the start space. 1.5-unit intervals later the timing causes the receiving device to sample the first information element which simply is a mark or space decision. The receiver continues to sample at 1-bit intervals until the stop mark is received. In start–stop systems the last information bit is most susceptible to cumulative timing errors. Figure 8.8b is an example of a 5-unit start–stop bit stream with a 1.5-unit stop element.

Another problem in start–stop systems is that of mutilation of the start element. Once this happens the receiver starts a timing sequence on the next mark-to-space transition it sees and thence continues to print in error until, by chance, it cycles back properly on a proper start element.

Synchronous data/telegraph systems do not have start and stop elements but consist of a continuous stream of information elements or bits as shown in Figure 8.8*a*. The cumulative timing problems eliminated in asynchronous (start–stop) systems are present in synchronous systems. Codes used on synchronous systems are often 7-unit codes with an extra unit added for parity, such as the ASCII or CCITT no. 5 codes. Timing errors tend to be eliminated by virtue of knowing the exact rate at which the bits of information are transmitted.

If a timing error of 1% were to exist between transmitter and receiver, not more than 100 bits could be transmitted until the synchronous receiving device was 100% apart in timing from the transmitter and up to all bits received were in error. Even if timing accuracy were improved to 0.05%, the correct timing relationship between transmitter and receiver would exist for only the first 2000 bits transmitted. It follows, that no timing error at all can be permitted to accumulate, since anything but absolute accuracy in timing would cause eventual malfunctioning. In practice the receiver is provided with an accurate clock which is corrected by small adjustments as explained below.

8.6.3 Timing

All currently used data transmission systems are synchronized in some manner. Start–stop synchronization has been discussed. Fully synchronous transmission systems all have timing generators or clocks to maintain stability. The transmitting device and its companion receiver at the far end of the data circuit must be a timing system. In normal practice the transmitter is the master clock of the system. The receiver also has a clock which in every case is corrected by one means or another to its transmitter equivalent at the far end.

Another important timing factor which must also be considered is the time it takes a signal to travel from the transmitter to the receiver. This is called propagation time. With velocities of propagation as low as 20,000 mi/s, consider a circuit 200 mi in length. The propagation time would then be 200/20,000 s or 10 ms, which is the time duration of one bit at a data rate of 100 bps. Thus the receiver in this case must delay its clock by 10 ms to be in step with its incoming signal.

Temperature and other variations in the medium may affect this delay. One can also expect variations in the transmitter master clock as well as other time distortions due to the medium.

There are basically three methods of overcoming these problems. One is to provide a separate synchronizing circuit to slave the receiver to the transmitter's master clock. This wastes bandwidth by expending a voice channel or subcarrier just for timing. A second method, which was used fairly widely up to several years ago, was to add a special synchronizing pulse for groupings of information pulses, usually for each character. This method was similar to start–stop synchronization and lost its appeal largely owing to the wasted information capacity for synchronizing. The most prevalent system in use today is one that uses transition timing. With this type of timing the receiving device is automatically adjusted to the signaling rate of the transmitter, and adjustment is made at the receiver by sampling the transitions of the incoming pulses. This offers many advantages, most important of which is that it automatically compensates for variations in propagation time. With this type of synchronization the receiver determines the average repetition rate and phase of the incoming signal transition and adjusts its own clock accordingly.

In digital transmission the concept of a transition is very important. The transition is what really carries the information. In binary systems the space-to-mark and mark-to-space transitions (or lack of transitions) placed in a time reference contain the information. Decision circuits regenerate and retime in sophisticated systems and care only *if* a transition has taken place. Timing cares *when* it takes place. Timing circuits must have memory in case a long series of marks or spaces is received. These will be periods of no transition, but they carry meaningful information. Likewise, the memory must maintain timing for reasonable periods in case of circuit outage. Keep in mind that synchronism pertains to both frequency and phase and that the usual error in high-stability systems is a phase error (i.e., the leading edges of the received pulses are slightly advanced or retarded from the equivalent clock pulses of the receiving device).

High-stability systems once synchronized need only a small amount of correction in timing (phase). Modem internal timing systems may be as stable as 1×10^{-8} or greater at both the transmitter and the receiver. Before a significant error condition can build up owing to a time rate difference at 2400 bps, the accumulated time difference between transmitter and receiver must exceed approximately 2×10^{-4} s. This figure neglects phase. Once the transmitter and receiver are synchronized and the circuit is shut down, then the clock on each end must drift apart by at least 2×10^{-4} s before significant errors take place. Again this means that the leading edge of the receiver clock equivalent timing pulse is 2×10^{-4} in advance or retarded from the leading edge of the received pulse from the distant end. Often an idling signal is sent on synchronous data circuits during conditions of no traffic to maintain timing. Other high-stability systems need to resynchronize only once a day.

Bear in mind that we are considering dedicated circuits only, not switched synchronous data. The problems of synchronization of switched data immediately come to light. Two such problems are the following:

- No two master clocks are in perfect phase synchronization.
- The propagation time on any two paths may not be the same.

Thus such circuits will need a time interval for synchronization at each switching event before traffic can be passed.

To sum up, synchronous data systems use high-stability clocks, and the clock at the receiving device is undergoing constant but minuscule corrections to maintain an in-step condition with the received pulse train from the distant transmitter by looking at mark–space and space–mark transitions.

8.6.4 Distortion

It has been shown that the key factor in data transmission is timing. The signal must be either a mark or a space, but that alone is not sufficient. The marks and spaces (or 1's and 0's) must be in a meaningful sequence based on a time reference.

In the broadest sense distortion may be defined as any deviation of a signal in any parameter, such as time, amplitude, or wave shape, from that of the ideal signal. For data and telegraph binary transmission, distortion is defined as a displacement in time of a signal transition from the time which the receiver expects to be correct. In other words the receiving device must make a decision whether a received signal element is a mark or a space. It makes the decision during the sampling interval which is usually at the center of where the received pulse or bit should be. Thus it is necessary for the transitions to occur between sampling times and preferably halfway between them. Any displacement of the transition instants is called distortion. The degree of distortion a data signal suffers as it traverses the transmission medium is a major contributor in determining the error rate that can be realized.

Telegraph and data distortion is broken down into two basic types, systematic and fortuitous. Systematic distortion is repetitious and is broken down into bias distortion, cyclic distortion, and end distortion (more common in start–stop systems). Fortuitous distortion is random in nature and may be defined as a distortion type in which the displacement of the transition from the time interval in which it should occur is not the same for every element. Distortion caused by noise spikes in the medium or other transients may be included in this category. Characteristic distortion is caused by transients in the modulation process which appear in the demodulated signal.

Figure 8.9 shows some examples of distortion. Figure 8.9*a* is an example of a binary signal without distortion, and Figure 8.9*b* shows the sampling instants, which occur ideally in the center of the pulse to be sampled. From this we can see that the displacement tolerance is nearly 50%. This means that the point of sample could be displaced by up to 50% of a pulse width and still record the mark or space

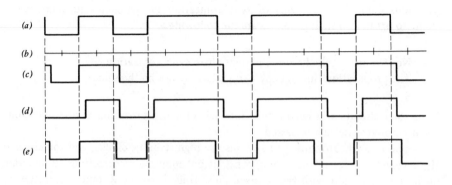

Figure 8.9 Three typical distorted data signals.

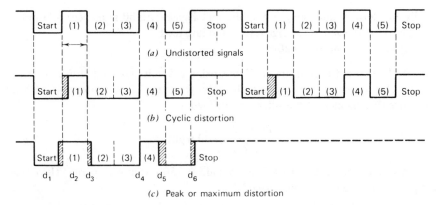

(a) Undistorted signals

(b) Cyclic distortion

(c) Peak or maximum distortion

Figure 8.10 Distorted telegraph signals illustrating cyclic and peak distortion. The peak distortion in the character appears at transition d_5.

condition present without error. However, the sampling interval does require a finite amount of time so that in actual practice the displacement permissible is somewhat less than 50%. Figure 8.9c and d shows bias distortion. An example of spacing bias is shown in Figure 8.9c, where all the spacing impulses are lengthened at the expense of the marking impulses. Figure 8.9d shows the reverse of this; the marking impulses are elongated at the expense of the spaces. This latter is called marking bias. Figure 8.9e shows fortuitous distortion, which is a random type of distortion. In this case the displacement of the signal element is not the same as the time interval in which it should occur for every element.

Figure 8.10 shows distortion that is more typical of start and stop transmission. Figure 8.10a is an undistorted start and stop signal. Figure 8.10b shows cyclic or repetitive distortion typical of mechanical transmitters. In this type of distortion the marking elements may increase in length for a period of time, and then the spacing elements will increase in length. Figure 8.10c shows peak distortion. Identifying the type of distortion present on a signal often gives a clue to the source or cause of distortion. Distortion measurement equipment measures the displacement of the mark-to-space transition from the ideal of a digital signal. If a transition occurs too near to the sampling point, the signal element is liable to be in error. Standards dealing with distortion are published by the Electronic Industries Association (Ref. 26, 27).

8.6.5 Bits, Baud, and Words per Minute

There is much confusion among transmission engineers in handling some of the semantics and simple arithmetic used in data and telegraph transmission.

The bit has been defined previously. Now the term *words per minute* will be introduced. A word in our telegraph and data language consists of six characters, usually five letters, numbers, or graphics and a space. All bit sequences transmitted must be counted, such as carriage return and line feed.

Let us look at some arithmetic:

1. A channel is transmitting at 75 bps using a 5-unit start and stop code with a 1.5-unit stop element. Thus for each character there are 7.5 unit intervals (7.5 bits). Therefore the channel is transmitting at 100 wpm:

$$\frac{75 \times 60}{6 \times 7.5} = 100 \text{ wpm}$$

2. A system transmits in CCITT no. 5 code at 1500 wpm with parity. How many bits per second are being transmitted?

$$1500 \times 8 \times \tfrac{6}{60} = 1200 \text{ bps}$$

The baud is the unit of modulation rate. In binary transmission systems baud and bit per second are synonymous. Thus a modem in a binary system transmitting to the line 110 bps has a modulation rate of 110 baud. In multilevel or M-ary systems the number of bauds is indicative of the number of transitions per second. The baud is more meaningful to the transmission engineers concerned with the line side of a modem. This concept will be discussed more at length further on.

8.7 DATA INTERFACE

A data link permits communication between a data terminal or computer and another data terminal or computer. A terminal may or may not have data processing capability. One with such a capability is called a "smart" terminal; without any such capability it is a "dumb" terminal. Doll (Ref. 29) defines an *end terminal* as

> an equipment comprising the ultimate source or sink, or both, in a digital data transmission system. It may include features designed to initiate or react to various end-to-end control procedures such as carriage return or line feed on a keyboard printer, but does not include features for the execution or control of communication procedures.

A data link requires some form of transmission media. To connect a computer to a cluster of access terminals down the hall requires little more than multipair cable. In this case the procedure is essentially electrical and little problem is presented to the transmission system engineer. However, extend the distance, and the transmission problems become more interesting.

Often we find ourselves with terminals across town from the central processing unit (CPU) (the computer), in another town, or hundreds or even thousands of miles away. The most at hand transmission medium in this case may well be the telephone network. The data signal to be transmitted must now be conditioned to meet the characteristics of the transmission medium. Let us assume in further

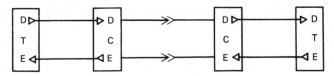

▷   Interface generator
▷   Interface load
≫   Telecommunication channel

Figure 8.11 Digital interface circuit illustrating DTE, DCE, generator, and load. DTE = data terminal equipment; DCE = data communication equipment.

argument that it is a telephone wire pair. The device that conditions or converts the data to make it compatible with the telephone network is the modem.

To keep terminology consistent with the industry, we call the data terminal with its input–output device or CPU the DTE (data terminal equipment) and the communication device (modem) the DCE (data communication equipment). A simplified data link illustrating this terminology is shown in Figure 8.11.

Since its beginnings in the 1950s, the world of data communications has been faced with many makes and varieties of computers and data terminals. Modem types also proliferated, although the Western Electric Company set a number of modem standards, many of which still exist today. These standards were for the line side of the circuit, between DCE and DCE in Figure 8.11. On the data terminal side of the modem, that is, the interface between DTE and DCE, there was poor definition and it took a great deal of care and engineering to assure interworking of these two devices on each end of the data link.

To bring order to the chaos, the U.S. Electronic Industries Association (EIA) established a number of standards for the DTE–DCE interface. The CCITT followed shortly thereafter to issue standards for this important area of data communications as well. Many of the CCITT standards were similar to (but not exactly identical to) their EIA counterparts for the data interface. U.S. military and U.S. federal standards must also be considered if one is to work in those environments.

At present the leading DTE–DCE interface standard is EIA RS-232C (Ref. 23). CCITT Rec. V.24 is its counterpart for that international agency. These two standards define an electromechanical interface, and, with several related standards, define signal levels, conditions, and polarity at each interface connection. The interface, in effect, is a 25-pin plug/socket. Each of the utilized 25 interface pins can be placed in one of four categories based on the dedicated function the particular pin performs:

1. Electrical ground
2. Data (interchange)
3. Control
4. Clock/timing

Table 8.2   EIA RS-232C Pin Assignments with CCITT Rec. V. 24 Equivalents

Pin Number	Circuit Mnemonic	Description	Source	Type	Equivalent CCITT Rec. V. 24
1	AA	Protective ground (chassis)	Ground	Ground	101
2	BA	Transmit data from terminal	Terminal	Data	103
3	BB	Receive data from terminal	Modem	Data	104
4	CA	Request to send (terminal on line)	Terminal	Control	105
5	CB	Clear to send (modem response to CA)	Modem	Control	106
6	CC	Data set ready (telephone function)	Modem	Control	107
7	AB	Signal ground (signal common return)	Ground	Ground	102
8	CF	Carrier detect (received line signal detect)	Modem	Control	109
9	–	(Positive test voltage/data set testing)	–	–	–
10	–	(Negative test voltage/data set testing)	–	–	–
11	–	Unassigned	–	–	–
12	SCF	Secondary carrier detect	Modem	Control	122
13	SCB	Secondary clear to send	Modem	Control	121
14	SBA	Secondary transmit data	Terminal	Data	118
15	DB	Transmit clock positive edge = signal element transition	Modem	Timing	114
16	SBB	Secondary receive data	Modem	Data	119
17	DD	Receive clock negative edge = signal element center	Modem	Timing	115
18	–	Unassigned			–
19	SCA	Secondary request to send	Terminal	Control	120
20	CD	Data Terminal ready (terminal on line)	Terminal	Control	108
21	CG	Signal quality detect (off for receive error)	Modem	Control	110
22	CE	Ring indicator	Modem	Control	125
23	CH/CI	Data signal rate selector	Terminal/modem	Control	111/112
24	DA	Transmit clock negative edge = signal element center	Terminal	Timing	113
25	–	Unassigned			–

Terminal = DTE; Modem = DCE.

Table 8.2 gives RS-232C pin assignments, circuit number, and function for each of the 25 pins with CCITT Rec. V.24 equivalence.

The following are the pertinent CCITT recommendations:

V.10, "Electrical Characteristics for Unbalanced Double-Current Interchange Circuits for General Use with Integrated Circuit Equipment in the Field of Data Communications"

Table 8.3   EIA RS-449 Interchange Circuits

Circuit Mnemonic	Circuit Name	Circuit Direction	Circuit Type
SG	Signal ground	—	Common
SC	Send common	To DCE	
RC	Receive common	From DCE	
IS	Terminal in service	To DCE	Control
IC	Incoming call	From DCE	
TR	Terminal ready	To DCE	
DM	Data mode	From DCE	
SD	Send data	To DCE	Primary channel
RD	Receive data	From DCE	data
TT	Terminal timing	To DCE	Primary channel
ST	Send timing	From DCE	timing
RT	Receive timing	From DCE	
RS	Request to send	To DCE	Primary channel
CS	Clear to send	From DCE	control
RR	Receiver ready	From DCE	
SQ	Signal quality	From DCE	
NS	New signal	To DCE	
SF	Select frequency	To DCE	
SR	Signal rate selector	To DCE	
SI	Signal rate indicator	From DCE	
SSD	Secondary send data	To DCE	Secondary channel data
SRD	Secondary receive data	From DCE	nel data
SRS	Secondary request to send	To DCE	Secondary
SCS	Secondary clear to send	From DCE	channel
SRR	Secondary receiver ready	From DCE	control
LL	Local loopback	To DCE	Control
RL	Remote loopback	To DCE	
TM	Test mode	From DCE	
SS	Select standby	To DCE	Control
SB	Standby indicator	From DCE	

Table 8.4 EIA RS-449 Equivalencies

EIA RS-449		RS-232C		CCITT Rec. V. 24	
SG	Signal ground	AB	Signal Ground	102	Signal ground
SC	Send common			102a	DTE common
RC	Receive common			102b	DCE common
IS	Terminal in service				
IC	Incoming call	CE	Ring indicator	125	Calling indicator
TR	Terminal ready	CD	Data terminal ready	108/2	Data terminal ready
DM	Data mode	CC	Data set ready	107	Data set ready
SD	Send data	BA	Transmitted data	103	Transmitted data
RD	Receive data	BB	Received data	104	Received data
TT	Terminal timing	DA	Transmitter signal element timing (DTE source)	113	Transmitter signal element timing (DTE source)
ST	Send timing	DB	Transmitter signal element timing (DCE source)	114	Transmitter signal element timing (DCE source)
RT	Receive timing	DD	Receiver signal element timing	115	Receiver signal element timing (DCE source)
RS	Request to send	CA	Request to send	105	Request to send
CS	Clear to send	CB	Clear to send	106	Ready for sending
RR	Receiver ready	CF	Received line signal detector	109	Data channel received line signal detector
SQ	Signal quality	CG	Signal Quality Detector	110	Data signal quality detector
NS	New signal				
SF	Select frequency			126	Select transmit frequency
SR	Signaling rate selector	CH	Data signal rate selector (DTE source)	111	Data signaling rate selector (DTE source)

		CI	Data signal rate selector (DCE source)	112	Data signaling rate selector (DCE source)
SI	Signaling rate indicator	CI	Data signal rate selector (DCE source)	112	Data signaling rate selector (DCE source)
SSD	Secondary send data	SBA	Secondary transmitted data	118	Transmitted backward channel data
SRD	Secondary receive data	SBB	Secondary received data	119	Received backward channel data
SRS	Secondary request to send	SCA	Secondary request to send	120	Transmit backward channel line signal
SCS	Secondary clear to send	SCB	Secondary clear to send	121	Backward channel ready
SRR	Secondary receiver ready	SCF	Secondary received line signal detector	122	Backward channel received line signal detector
LL	Local loopback			141	Local loopback
RL	Remote loopback			140	Remote loopback
TM	Test mode			142	Test indicator
SS	Select standby			116	Select standby
SB	Standby indicator			117	Standby indicator

V.11, "Electrical Characteristics for Balanced Double-Current Interchange Circuits for General Use with Integrated Circuit Equipment in the Field of Data Communications"

V.24, "List of Definitions for Interchange Circuits between Data-Terminal and Data Circuit-Terminating Equipment"

V.28, "Electrical Characteristics for Unbalanced Double-Current Interchange Circuits"

V.31, "Electrical Characteristics for Single-Current Interchange Circuits Controlled by Contact Closure"

Note. Where CCITT refers to "double-current" we use polar keying or polar transmission; for "single-current" we use neutral keying or neutral transmission.

The relevant U.S. military standard is MIL-STD-188-114, "Electrical Characteristics of Digital Interface Circuits."

EIA RS-232C will probably remain in force for some time, as well as its CCITT counterpart. However, during the mid-1980s EIA RS-232C will be phased out slowly and replaced by EIA RS-449, supplemented by EIA RS-422 and RS-423. With a few additional provisions for interoperability, equipment conforming to EIA RS-449 can interoperate with equipment designed to EIA RS-232C. Essentially EIA RS-449 specifies the functional and mechanical characteristics of the interface between DTE and DCE, whereas EIA RS-422 and 423 specify the electrical characteristics of digital interface circuits, the former dealing with balanced voltage interface, and the latter, unbalanced.

EIA RS-232C defines a single 25-pin plug socket and EIA RS-449 a 37-pin interface. Table 8.3 lists interchange circuits and their mnemonics relating to EIA RS-449 and Table 8.4 is an equivalency table: EIA RS-449/RS-232C and CCITT Rec. V. 24.

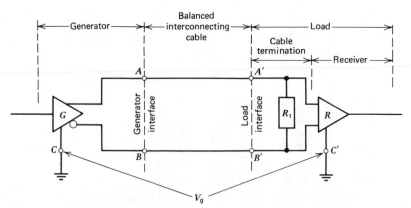

Figure 8.12 Balanced digital interface circuit, EIA RS-422. A, B = generator interface; A', B' = load interface; C = generator circuit ground; C' = load circuit ground; R = optional cable termination resistance; V_g = ground potential difference. (Courtesy EIA.)

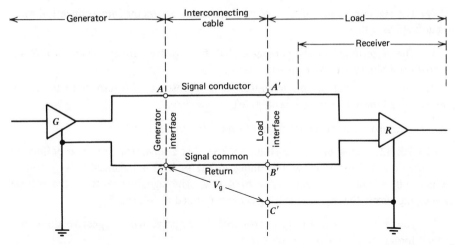

Figure 8.13 Unbalanced digital interface circuit, RS-423. A, C = generator interface; A', B' = load interface; C' = load circuit ground; C = generator circuit ground; V_g = ground potential difference. (Courtesy EIA.)

Figures 8.12 and 8.13 illustrate the digital interface circuits for EIA RS-422 and RS-423, respectively. EIA RS-422, as mentioned, specifies a balanced interface and EIA RS-423 an unbalanced interface. As we would expect, the unbalanced circuit uses a common return (ground) which is shown in Figure 8.13. In either figure the load may be considered to be one or more receivers. Also see how the generator and load are configured in Figure 8.11.

The following presents some basic guidelines for the application of balanced and unbalanced electrical connections DTE–DCE. EIA states in RS-422.

While the balanced interface is intended for use at the higher modulation rates, it may, in preference to the unbalanced interface circuit, generally be required where any of the following conditions prevail:

- The interconnecting cable (i.e., DTE–DCE) is too long for effective unbalanced operation.
- The interconnecting cable is exposed to extraneous noise sources that may cause an unwanted voltage in excess of plus or minus 1 V measured between the signal conductor and the circuit common at the load end of the cable with a 50-Ω resistor substituted for the generator.
- It is necessary to minimize interference with other signals.
- Inversion of signals may be required, e.g., PLUS MARK to MINUS MARK, may be obtained by inverting the cable pair.

RS-422 states, in essence (Figure 8.12), that a generator, as defined in the standard, results in a low-impedance (100 Ω or less) balanced voltage source which will produce a differential voltage applied to the interconnecting cable in the range of

2–V. The signaling sense of the voltages appearing across the interconnection cable is defined as follows:

1. The A terminal of the generator shall be negative with respect to the B terminal for a binary 1 (MARK or OFF state).

2. The A terminal of the generator shall be positive with respect to the B terminal for a binary 0 (SPACE or ON state).

Note. Compare this signaling convention to that in Table 8.1.

EIA RS-423 states, in essence (Figure 8.13), that a generator circuit, as defined in the standard results in a low-impedance (50 Ω or less) unbalanced voltage source which will produce a voltage range of 4–6 V. The signaling sense of the voltage appearing across the interconnecting cable is defined as follows:

1. The A terminal of the generator shall be negative with respect to the C terminal for a binary 1 (MARK or OFF state).

2. The A terminal of the generator shall be positive with respect to the C terminal for a binary 0 (SPACE or ON state).

The test load termination for a balanced circuit is two 50-Ω resistors in series between the two signal leads (A and B in Figure 8.12). The test lead termination for the unbalanced circuit is 450 Ω between the generator output terminal and "common."

For the balanced or unbalanced case load characteristics during real operating conditions result in a differential receiver having a high input impedance (greater than 4kΩ), a small input threshold transition region between −0.2 and +0.2 V, and an allowance for an internal bias voltage not to exceed 3 V in magnitude.

EIA RS-422 and RS-423 provide guidance on maximum cable length DTE–DCE. For the case of balanced signal lines (RS-422): up to 90 kbps, 4000 ft; to 1 Mbps, 380 ft; and to 10 Mbps, 40 ft. For the unbalanced signal lines (RS-423): up to 900 bps, 4000 ft; to 10 kbps, 380 ft; to 100 kbps, 40 ft. EIA states that these lengths are on the conservative side.

The following provides a cross reference of interface standards (Ref. 32):

EIA RS-423A	U.S. Fed. Std. 1030A	CCITT V.10 (X. 26)
EIA RS-422A	U.S. Fed. Std. 1020A	CCITT V.11 (X. 27)
EIA RS-449	U.S. Fed. Std. 1031	CCITT V.24/V.10/V.11, ISO 4902
RS-232C		CCITT V.24/V.28, ISO 2110

8.8 DATA INPUT–OUTPUT DEVICES

The following is meant to give the reader a broad-brush familiarity with data subscriber equipment, which is more often referred to as input–output devices. Such equipments convert user information (data or messages) into electrical signals or vice versa. Many refer to the input–output devices as the human interface. Electrically a data subscriber terminal consists of the end instrument, the DTE, and a device called a modem or communication set. The DTE has been discussed in Section 8.7. Modems are described in Section 8.13. The data source is the input device and the data sink is the output device.

Data input–output devices handle paper tape, punched cards, magnetic tape, drums, disks, visual displays, and printed page copy. Input devices* may be broken down into the following categories:

- Keyboard sending units
- Card readers
- Paper tape readers
- Magnetic tape, disk, and drum readers
- Optical character readers (OCR)

Output devices* are as follows:

- Printers (paper, hard copy)
- Card punches
- Paper tape punches
- Magnetic tape recorders, magnetic cores and disks
- Visual display units (VDU) (cathode ray tubes and plasma)

Further, these devices may be used as on-line devices or off line. Off-line devices are not connected directly to the communication system, but serve as auxiliary equipment.

Off-line devices are used for tape or card preparation for eventual transmission. In this case a keyboard is connected to a card punch for card preparation. Also the keyboard may be connected to a paper tape punch.

Once tape or cards are prepared, they then are handled by on-line equipment, either tape readers or card readers. Intermediate equipment or line buffers supply timing, storage, and serial–parallel conversion to the input–output devices. The following table provides equivalence between telegraph and data terminology of input–output devices.

Data	Telegraph
Keyboard	Keyboard
Tape reader	Transmitter-distributor[†]
Printer	Teleprinter
Tape punch	Perforator, reperforator
Visual displays	—

*Compatible with the data rates under discussion.
[†]The distributor performs the parallel-to-serial equivalent conversion of data transmission.

8.9 DIGITAL TRANSMISSION ON AN ANALOG CHANNEL

8.9.1 Introduction

There are two fundamental approaches to the practical problem of data transmission. The first approach is often to design and construct a complete new network expressly for the purpose of data transmission. The second approach is to adapt the many existing telephone facilities for data transmission. The paragraphs that follow deal with the second approach.

Transmission facilities designed to handle voice traffic have characteristics which make it difficult to transmit dc binary digits or bit streams. To permit the transmission of data over voice facilities (i.e., the telephone network), it is necessary to convert the dc data into a signal within the VF range. The equipment which performs the necessary conversion to the signal is generally termed a data modem. Modem is an acronym for modulator–demodulator.

8.9.2 Modulation–Demodulation Schemes

A modem modulates and demodulates. The types of modulation used by present-day modems may be one or combinations of the following:

- Amplitude modulation (AM), double sideband (DSB)
- Amplitude modulation (AM), vestigial sideband (VSB)
- Frequency shift modulation (FSK)
- Phase shift modulation (PSK)

AMPLITUDE MODULATION, DOUBLE SIDEBAND

With this modulation technique binary states are represented by the presence, or absence of an audio tone or carrier. More often it is referred to as on-off telegraphy. For data rates up to 1200 bps, one such system uses a carrier frequency centered at 1600 Hz. For binary transmission AM has significant disadvantages which include (1) susceptibility to sudden gain change, and (2) inefficiency in modulation and spectrum utilization, particularly at higher modulation rates (see CCITT Rec. R. 70).

AMPLITUDE MODULATION, VESTIGIAL SIDEBAND

An improvement in the AM-DSB technique results from the removal of one of the information-carrying sidebands. Since the essential information is present in each of the sidebands, there is no loss of content in the process. The carrier frequency must be preserved to recover the dc component of the information envelope. Therefore digital systems of this type use VSB modulation in which one sideband, a portion of the carrier and a "vestige" of the other sideband, is retained. This is accomplished by producing a DSB signal and filtering out the unwanted sideband components. As a result, the signal takes only about three fourths of the bandwidth required for a DSB system. Typical VSB data modems are operable up to 2400 bps in a telephone channel. Data rates up to 4800 bps are achieved using multilevel (M-ary) techniques. The carrier frequency is usually located between 2200 and 2700 Hz.

FREQUENCY SHIFT MODULATION

A large number of data transmission systems utilize frequency shift modulation, commonly called frequency shift keying (FSK). The two binary states are represented by two different frequencies and are detected by using two frequency tuned sections, one tuned to each of the two bit frequencies. The demodulated signals are then integrated over the duration of a bit, and upon the result a binary decision is based.

Digital transmission using frequency shift modulation (FSK) has the following advantages. (1) The implementation is not much more complex than an AM system. (2) Since the received signals can be amplified and limited at the receiver, a simple limiting amplifier can be used, whereas the AM system requires sophisticated AGC in order to operate over a wide level range. Another advantage is that FSK can show a 3- or 4-dB improvement over AM in most types of noise environment, particularly at distortion threshold (i.e., at the point where the distortion is such that good printing is about to cease). As the frequency shift becomes greater, the advantage over AM improves in a noisy environment.

Another advantage of FSK is its immunity from the effects of nonselective level variations even when extremely rapid. Thus it is used almost exclusively on worldwide HF radio transmission where rapid fades are a common occurrence. In the United States it has nearly universal application for the transmission of data at the lower data rates (i.e., 1200 bps and below).

PHASE MODULATION

For systems using higher data rates, phase modulation (PSK) becomes more attractive. Various forms are used such as two phase, relative phase, and quadrature phase. A two-phase system uses one phase of the carrier frequency for one binary state and the other phase for the other binary state. The two phases are $180°$ apart and are detected by a synchronous detector using a reference signal at the receiver which is of known phase with respect to the incoming signal. This known signal is at the same frequency as the incoming signal carrier and is arranged to be in phase with one of the binary signals.

In the relative phase system, a binary 1 is represented by sending a signal burst of the same phase as that of the previous signal burst sent. A binary 0 is represented by a signal burst of a phase opposite to that of the previous signal transmitted. The signals are demodulated at the receiver by integrating and storing each signal burst of one bit period for comparison in phase with the next signal burst.

In the quadrature phase system, two binary channels (2 bits) are phase multiplexed onto one tone by placing them in phase quadrature as shown in the sketch below. An extension of this technique places two binary channels on each of several tones spaced across the voice channel of a typical telephone circuit.

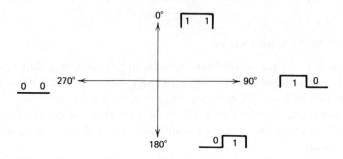

Some of the advantages of phase modulation are as follows:

- All available power is utilized for intelligence conveyance.
- The demodulation scheme has good noise rejection capability.
- The system yields a smaller noise bandwidth.

A disadvantage of such a system is the complexity of equipment required.

8.9.3   Critical Parameters

The effect of the various telephone circuit parameters on the capability of a circuit to transmit data is a most important consideration. The following discussion is to familiarize the reader with the problems most likely to be encountered in the transmission of data over analog circuits (e.g., the telephone network) and to make some generalizations in some cases, which can be used to help in planning the implementation of data systems.

DELAY DISTORTION

Delay distortion "constitutes the most limiting impairment to data transmission, particularly over telephone voice channels" (Ref. 4). When specifying delay distortion, the terms *envelope delay distortion* and *group delay* are often used. The IEEE standard dictionary (Ref. 37) states that "envelope delay is often defined the same as group delay, that is, the rate of change, with angular frequency, of the phase shift between two points in a network" (see Section 1.9.4).

The problem is that in a band-limited analog system, such as the typical telephone voice channel, not all frequency components of the input signal will propagate to the receiving end in exactly the same elapsed time, particularly on loaded cable

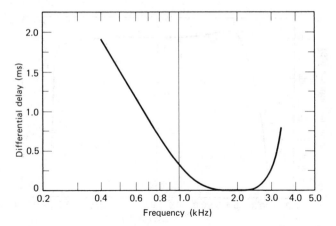

Figure 8.14 Typical differential delay across a voice channel. FDM equipment back to back.

circuits and carrier systems. In carrier systems it is the cumulative effect of the many filters used in the FDM equipment. On long-haul circuits the magnitude of delay distortion is generally dependent on the number of carrier modulation stages that the circuit must traverse rather than the length of the circuit. Figure 8.14 shows a typical frequency–delay response curve in milliseconds of a voice channel due to FDM equipment only. For the voice channel (or any symmetrical passband for that matter), delay increases toward band edge and is minimum around the center portion of the passband.

In essence, therefore, we are dealing with the phase linearity of a circuit. If the phase–frequency relationship over the passband is not linear, distortion will occur in the transmitted signal. This distortion is best measured by a parameter called *envelope delay distortion* (EDD). Mathematically, envelope delay is the derivative of the phase shift with respect to frequency. The maximum difference in the derivative over any frequency interval is called EDD. Therefore EDD is always a difference between the envelope delay at one frequency and that at another frequency of interest in a passband. The EDD unit of measurement is milliseconds or microseconds.

When transmitting data, the shorter the pulse (or symbol) width (in the case of binary systems this would be the width of 1 bit), the more critical the EDD constraints become. As a rule of thumb, delay distortion in the passband should be below the period of 1 bit (or symbol).

AMPLITUDE RESPONSE (ATTENUATION DISTORTION)

Another parameter which seriously affects the transmission of data and which can place very definite limits on the modulation rate is that of amplitude response. Ideally all frequencies across the passband of the channel of interest should suffer the same attenuation. Place a −10-dBm signal at any frequency between 300 and 3400

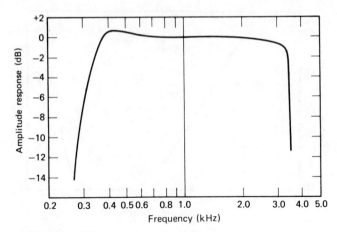

Figure 8.15 Typical amplitude versus frequency response across a voice channel. Channel modulator, demodulator back to back, FDM equipment.

Hz, and the output at the receiving end of the channel may be −23 dBm, for example, at any and all frequencies in the band; we would then describe a fully flat channel. Such a channel has the same loss or gain at any frequency within the band. This type of channel is ideal but would be unachievable in a real, working system. In Rec. G.132, CCITT recommends no more than 9 dB of amplitude distortion relative to 800 Hz between 400 and 3000 Hz. This figure of 9 dB describes the maximum variation that may be expected from the reference level at 800 Hz. This variation of amplitude response often is called attenuation distortion. A conditioned channel, such as a Bell System C-4 channel, will maintain a response of −2 to +3 dB from 500 to 3000 Hz and −2 to +6 dB from 300 to 3200 Hz. Channel conditioning is discussed in Section 8.12.

Considering tandem operation, the deterioration of amplitude response is arithmetically summed when sections are added. This is particularly true at band edge in view of channel unit transformers and filters which account for the upper and lower cutoff characteristics.

Amplitude response is also discussed in Section 1.9.3. Figure 8.15 illustrates a typical example of amplitude response across FDM carrier equipment (see Chapter 3) connected back to back at the voice channel input–output.

Tables 8.5 and 8.6 give the ATT requirements for leased lines covering attenuation distortion and EDD.

NOISE

Another important consideration in the transmission of data is that of noise. All extraneous elements appearing at the voice channel output which were not due to the input signal are considered to be noise. For convenience noise is broken down into four categories:

Table 8.5 ATT Requirements for Two-Point or Multipoint Channel Attenuation Distortion

	Frequency band (Hz)	Attenuation[a] (dB)
Basic requirements	500–2500	−2 to +8
	300–3000	−3 to +12
C1 conditioning	1000–2400	−1 to +3
	300–2700	−2 to +6
	2700–3000	−3 to +12
C2 conditioning	500–2800	−1 to +3
	300–3000	−2 to +6
C4 conditioning	500–3000	−2 to +3
	300–3200	−2 to +6
C5 conditioning	500–2800	−0.5 to +1.5
	300–3000	−3 to +3

Source: From (Ref. 30).
[a] Relative to 1000 Hz.

Table 8.6 ATT Requirements for Two-Point or Multipoint Channel Envelope Delay Distortion

	Frequency Band (Hz)	EDD (μs)[a]
C1 conditioning	800–2600	1750
	1000–2400	1000
C2 conditioning	1000–2600	500
	600–2600	1500
	500–2800	3000
C4 conditioning	1000–2600	300
	800–2800	500
	600–3000	1500
	500–3000	3000
C5 conditioning	1000–2600	100
	600–2600	300
	500–2800	600

Source: From (Ref. 30).
[a] Maximum inband envelope delay difference.

- Thermal
- Crosstalk
- Intermodulation
- Impulse

Thermal noise, often called resistance noise, white noise, or Johnson noise, is of a Gaussian nature or fully random. Any system or circuit operating at a temperature above absolute zero inherently will display thermal noise. It is caused by the random motions of discrete electrons in the conduction path.

Crosstalk is a form of noise caused by unwanted coupling from one signal path into another. It may be due to direct inductive or capacitive coupling between conductors or between radio antennas (see Section 1.9.6).

IM noise is another form of unwanted coupling usually caused by signals mixing in nonlinear elements of a system. Carrier and radio systems are highly susceptible to IM noise, particularly when overloaded (see Section 1.9.6).

Impluse noise is a primary source of errors in the transmission of data over telephone networks. It is sporadic and may occur in bursts or dicrete impulses called *hits*. Some types of impulse noise are natural, such as that from lightning. However, man-made impulse noise is ever-increasing, such as from automobile ignition systems or power lines. Impulse noise may be of high level in telephone switching centers due to dialing, supervision, and switching impulses which may be induced or otherwise coupled into the data transmission channel.

For our discussion of data transmission, only two forms of noise will be considered: random or Gaussian noise and impulse noise. Random noise measured with a typical transmission measuring set appears to have a relatively constant value. However, the instantaneous value of the noise fluctuates over a wide range of amplitude levels. If the instantaneous noise voltage is of the same magnitude as the received signal, the receiving detection equipment may yield an improper interpretation of the received signal and an error or errors will occur. For a proper analytical approach to the data transmission problem, it is necessary to assume a type of noise which has an amplitude distribution that follows some predictable pattern. White noise or random noise has a Gaussian distribution and is considered representative of the noise encountered on the analog telephone channel (i.e., the voice channel). From the probability distribution curve of Gaussian noise shown in Figure 8.16 we can make some accurate predictions. It may be noted from this curve that the probability of occurrence of noise peaks which have amplitudes 12.5 dB above the rms level is 1 in 10^5. Hence if we wish to ensure an error rate of 10^{-5} in a particular system using binary polar modulation, the rms noise should be at least 12.5 dB below the signal level (Ref. 4, p. 114; Ref. 5, p. 6). This simple analysis is valid for the type of modulation used, assuming that no other factors are degrading the operation of the system and that a cosine shaped receiving filter is used. If we were to interject EDD, for example, into the system, we could translate the degradation into an equivalent signal-to-noise ratio improvement necessary to restore the desired error rate. For example, if the delay distortion were the equiv-

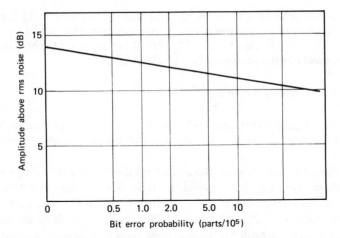

Figure 8.16 Probability of bit error in Gaussian noise, binary polar transmission.

alent of one pulse width, the signal-to-noise ratio improvement required for the same error rate would be about 5 dB, or the required signal-to-noise ratio would now be 17.5 dB.

For reasons that will be discussed later, let us assume that the signal level is -10 dBm at the zero transmission level point of the system. Then the rms noise measured at the same point would be -27.5 dBm to retain the error rate of 1 in 10^5.

In order for the above figure to have any significance, it must be related to the actual noise found in a channel. CCITT recommends no more than 50,000 pW of noise psophometrically weighted on an international connection made up of six circuits in a chain. However, CCITT states (Recs. G.142A and 142D) that for data transmission at as high a modulation rate as possible without significant error rate, a reasonable circuit objective for maximum random noise would be -40 dBm0p for leased circuits (impulse noise not included) and -36 dBm0p for switched circuits without compandors. This figure obviously appears quite favorable when compared to the -27 dBm0 (-29.5 dBm0p) required in the example above. However, other factors which will be developed later will consume much of the noise margin that appears available.

Whereas random noise has an rms value when we measure level, impulse noise is another matter entirely. It is measured as the number of "hits" or "spikes" per interval of time over a certain threshold. In other words it is a measurement of the recurrence rate of noise peaks over a specified level. The word *rate* should not mislead the reader. The recurrence is not uniform per unit time, as the word *rate* may indicate, but we can consider a sampling and convert it to an average.

ATT (Ref. 30) states the following:

The impulse noise objective is specified in terms of the rate of occurrence of the impulse voltages above a specified magnitude. The objective is expressed as the threshold in dBrnc0 at which no more than 15 impulses in 15 minutes are measured by an impulse counter with a maximum counting rate of 7 counts per second. The overall objective of 71 dBrnc0 implies a 6-dB signal-to-impulse noise threshold in the presence of a −13-dBm0 signal.

CCITT states (Rec. Q.45) that

in any four-wire international exchange the busy hour impulsive noise counts should not exceed 5 counts in 5 minutes at a threshold level of −35 dBm0.

Remember that random noise has a Gaussian distribution and will produce peaks at 12.5 dB over the rms value (unweighted) 0.001% of the time on a data bit stream for an equivalent error rate of 1×10^{-5}. It should be noted that some references use 12 dB, some 12.5 dB, and others 13 dB. The 12.5 dB above the rms random noise floor should establish the impulse noise threshold for measurement purposes. We should assume in a well-designed data transmission system traversing the telephone network, that the signal-to-noise ratio of the data signal will be well in excess of 12.5 dB. Thus impulse noise may well be the major contributor to degrade the error rate.

Care must be taken when measuring impulse noise. A transient such as an impulse noise spike on a band-limited system (which our telephone network most certainly is) tends to cause "ringing." Here the initial impulse noise spike causes what we might call a main bang or principal spike followed by damped subsidiary spikes. If we are not careful, these subsidiary spikes, that ringing effect, may also be counted as individual hits in our impulse noise count total. To avoid this false counting, impulse noise meters have a built-in dead time after each count. It is a kind of damping. The Bell System, for example, specifies a 150-ms dead time after each count. This limits the counting capability of the meter to no more than 6 or 7 counts per minute (cpm).

In this damping or dead time period, missed (real) impulse noise hits may seem to be a problem. For example, the Bell System suggests that the average improved (increased) sensitivity to measure "all" hits is only 0.9 dB, with a standard deviation of 0.76 dB (Ref. 24).

The period of measurement is also important. How long should the impulse noise measurement set remain connected to a line under test to give an accurate count? It appears empirically that 30 min is sufficient. However, a good estimate of error can be made and corrected for if that period of time is reduced to 5 min. This is done by reducing the threshold (on paper) of the measuring set. From (Ref. 24) the standard deviation for a 5-min period is about 2.2 dB. Thus 95% of all 5-min measurements will be within ±3.6 dB of a 30-min measurement period.

To clarify this, remember that impulse noise distributions are log normal and impulse noise level distributions are normal. With this in mind we can relate count distributions, which can be measured readily, to level distributions. The mean of the level distribution is the threshold value of which the impulse noise level meter was set to record the count distribution (in dBm, dBmp, dBrnC, or whatever unit). The set has a count associated with that threshold which is simply the median of the observed count distribution. The sigma σ_1 standard deviation of the impulse noise level distribution is estimated by the expression

$$\sigma_1 = m\sigma_D$$

where m = inverse slope of the peak amplitude distribution in decibels per decade of counts and average 7.0 dB and σ_D = standard deviation of the log normal count distribution, which is the square root of the $\log_{10}$ of the ratio of the average number of counts to the median count, or

$$\sigma_D = \sqrt{\log_{10} \frac{\text{average count}}{\text{median count}}}$$

where the median is not equal to zero. For instance, if we measured 10 cpm at a given threshold, the 1-cpm threshold would be 7 dB above the 10-cpm threshold.

When an unduly high error rate has been traced to impulse noise, there are some methods for improving conditions. Noisy areas may be bypassed, repeaters may be added near the noise source to improve the signal-to-impulse noise ratio, or in special cases pulse smearing techniques may be used. This latter approach uses two delay distortion networks which complement each other such that the net delay distortion is zero. By installing the networks at opposite ends of the circuit, impulse noise passes through only one network and is therefore smeared because of the delay distortion. The signal is unaffected because it passes through both networks.

MULTIPATH DISTORTION

This is a phenomenon of radio transmission which results from the difference in the time of arrival of signals which have traveled along different paths in going from a transmitter to a receiver. The strongest signal usually is the first to arrive at the receiver, having taken the shortest path. This is followed by signals of lesser amplitude which arrive later, with the result that the end of a transmitted pulse appears to be elongated or stretched. In the case of longer delay times and shorter pulse widths it results in unwanted second pulse or noise spikes. Pulse elongation is most common, and the modulation rate, therefore, must be limited to something less than an equivalent of one-half of the elongation or "smear."

For tropospheric scatter circuits, representative figures are from 0.05 to 0.5 μs, and where multiple reflections are involved, up to 10 μs. On HF radio circuits delays of 1–3 ms are observed frequently. However, maxima of 6, 12, and even 40 ms also have been observed.

An HF circuit with a 3-ms multipath delay must limit its modulation rate to an equivalent of 6-ms pulse width on any one carrier or subcarrier; or $1/6 \times 10^{-3}$ = 1000/6 = 166 baud.

Multipath delay on LOS microwave paths is not considered to be serious enough to limit the modulation rates covered in this discussion.

LEVELS AND LEVEL VARIATIONS

The design signal levels of telephone networks traversing FDM carrier systems are determined by average talker levels, average channel occupancy, permissible overloads during busy hours, and so on. Applying constant amplitude digital data tone(s) over such an equipment at 0 dBm0 on each channel would result in severe overload and IM within the system.

Loading does not affect hard wire systems except by increasing crosstalk. However, once the data signal enters carrier multiplex (voice) equipment, levels must be considered carefully, and the resulting levels most probably have more impact on the final signal-to-noise ratio at the far end than anything else. CCITT (Recs. G.151C, H51, H23, and V2) recommends a level of −10 dBm0 in some cases, and −13 dBm0 when the proportion of nonspeech circuits on an international carrier circuit exceeds 10 or 20%. For multichannel telegraphy the composite level is −8.7 dBm0, or for 24 channels each channel would be adjusted for −22.5 dBm0. Even this loading may be too heavy if a high proportion of the voice channels are loaded with data. Depending on the design of the carrier equipment, cutbacks to −13 dBm0 or less may be advisable.

In a properly designed transmission system the standard deviation of the variation in level should not exceed 1.0 dB per circuit. However, data communication equipment should be able to withstand level variations in excess of 4 dB.

FREQUENCY TRANSLATION ERRORS

Total end-to-end frequency translation errors on a voice channel being used for data or telegraph transmission must be limited to 2 Hz (CCITT Rec. G.135). This is an end-to-end requirement. Frequency translations occur mostly owing to carrier equipment modulation and demodulation steps. FDM carrier equipment widely uses SSB suppressed carrier techniques. Nearly every case of error can be traced to errors in frequency translation (we refer here to deriving the group, supergroup, mastergroup, and its reverse process; see also Chapter 3) and carrier reinsertion frequency offset, the frequency error being exactly equal to the error in translation and offset or the sum of several such errors. Frequency locked (e.g., synchronized) or high-stability master carrier generators (1×10^{-7} or 1×10^{-8}, depending on the system), with all derived frequency sources slaved to the master source, usually are employed to maintain the required stability.

Although 2 Hz seems to be a very rigid specification, when added to the possible back-to-back error of the modems themselves, the error becomes more appreciable. Much of the trouble arises with modems that employ sharply tuned filters. This is particularly true of telegraph equipment. But for the more general case, high-speed data modems can be designed to withstand greater carrier shifts than those that will be encountered over good telephone circuits.

PHASE JITTER

The unwanted change in phase or frequency of a transmitted signal due to modulation by another signal during transmission is defined as phase *jitter*. If a simple sinusoid is frequency or phase modulated during transmission, the received signal will have sidebands. The amplitude of these sidebands compared to the received signal is a measure of the phase jitter imparted to it during transmission.

Phase jitter is measured in degrees of variation peak to peak for each hertz of transmitted signal. Phase jitter shows up as unwanted variations in zero crossings of a received signal. It is the zero crossings that data modems use to distinguish marks and spaces. Thus the higher the data rate, the more jitter can affect the error rate on the receive bit stream.

The greatest cause of phase jitter in the telephone network is FDM carrier equipment, where it shows up as undesired incidental phase modulation. Modern FDM equipment derives all translation frequencies from one master frequency source by multiplying and dividing its output. To maintain stability, phase lock techniques are used. Thus the low jitter content of the master oscillator may be multiplied many times. It follows, then, that we can expect more phase jitter in the voice channels occupying the higher baseband frequencies.

Jitter most commonly appears on long-haul systems at rates related to the power line frequency (e.g., 60 Hz and its harmonics and submultiples, or 50 Hz and its harmonics of submultiples) or derived from 20-Hz ringing frequency. Modulation components that we define as jitter usually occur close to the carrier from about 0 to about ±300 Hz maximum.

8.10 CHANNEL CAPACITY

A leased or switched voice channel represents a financial investment. The goal of the system engineer is to derive as much benefit as possible from the money invested. For the case of digital transmission this is done by maximizing the information transfer across the system. This subsection discusses how much information in bits can be transmitted, relating information to bandwidth, signal-to-noise ratio, and error rate. An empirical discussion of these matters is carried out in Section 8.11.

First, looking at very basic information theory, Shannon stated in his bandwidth paper (Ref. 13) that if the input information rate to a band-limited channel is less than C (bps), a code exists for which the error rate approaches zero as the message length becomes infinite. Conversely, if the input rate exceeds C, the error rate cannot be reduced below some finite positive number.

The usual voice channel approximates to a Gaussian band-limited channel (GBLC) with additive Gaussian noise. For such a channel consider a signal wave of a mean power of S watts applied at the input of an ideal low-pass filter having a bandwidth of W hertz and containing an internal source of mean Gaussian noise with a mean power of N watts uniformly distributed over the passband. The capacity in bits per second is given by

$$C = W \log_2 \left(1 + \frac{S}{N}\right)$$

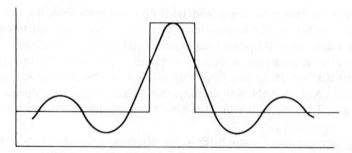

Figure 8.17   Pulse response through a Gaussian band-limited channel (GBLC).

Applying Shannon's "capacity" formula (GBLC) to some everyday voice channel criteria, $W = 3000$ Hz and $S/N = 1023$, then

$$C = 30,000 \text{ bps}$$

(Remember that bits per second and baud are interchangeable in binary systems.)

Neither S/N or W is an unreasonable value. Seldom, however, can we achieve a modulation rate greater than 3000 baud. The big question in advanced design is how to increase the data rate and keep the error rate reasonable.

One important item that Shannon's formula did not take into consideration is intersymbol interference. A major problem of a pulse in a band-limited channel is that the pulse tends not to die out immediately, and a subsequent pulse is interfered with by "tails" from the preceding pulse. This is shown in Figure 8.17.

Nyquist provided another approach to the data rate problem, this time using intersymbol interference (the tails in Figure 8.17) as a limit (Ref. 12). This resulted in the definition of the so-called Nyquist rate, which is $2W$ elements per second. W is the bandwidth in hertz of a band-limited channel as shown in Figure 8.17. In binary transmission we are limited to $2W$ bps. If we let $W = 3000$ Hz, the maximum data rate attainable is 6000 bps. Some refer to this as "the Nyquist 2-bit rule."

The key here is that we have restricted ourselves to binary transmission and we are limited to $2W$ bps no matter how much we increase the signal-to-noise ratio. The Shannon GBLC equation indicates that we should be able to increase the information rate indefinitely by increasing the signal-to-noise ratio. The way to attain a higher value of C is to replace the binary transmission system with a multi-level system, often termed an M-ary transmission system with $M > 2$. An M-ary channel can pass $2W \log_2 M$ bps with an acceptable error rate. This is done at the expense of the signal-to-noise ratio. As M increases (as the number of levels increases), so must the signal-to-noise ratio increase to maintain a fixed error rate.

8.11   VOICE CHANNEL DATA MODEMS VERSUS CRITICAL DESIGN PARAMETERS

The critical parameters that affect data transmission have been discussed. They are amplitude–frequency response (sometimes called amplitude distortion), EDD, and noise. Now we relate these parameters to the design of data modems to establish some general limits or "boundaries" for equipment of this type. The discussion that follows purposely avoids HF radio considerations.

As stated earlier in the coverage of envelope delay distortion, it is desirable to keep the transmitted pulse (bit) length equal to or greater than the residual differential EDD. Since about 1.0 ms is assumed to be a reasonable residual delay after equalization (conditioning), the pulse length should then be no less than approximately 1 ms. This corresponds to a modulation rate of 1000 pulses per second (binary). In the interest of standardization (CCITT Rec. V.22), this figure is modified to 1200 bps.

The next consideration is the usable bandwidth required for the transmission of 1200 bps. This figure is approximately 1800 Hz, using modulation methods such as PSK, FSK, or DSB-AM, and somewhat less for VSB-AM. Since delay distortion of a typical voice channel is at its minimum between 1700 and 1900 Hz, the required band when centered about these points extends from 800 to 2600 Hz or from 1000 to 2800 Hz. From the previous discussion we can see from Figure 8.14 that the EDD requirement is met easily over the range of 800–2800 Hz.

Bandwidth limits modulation rate (as will be discussed below). However, the modulation rate in bauds and the data rate in bits per second may not necessarily be the same. This is a very important concept.

Suppose a modulator looked at the incoming serial bit stream 2 bits at a time rather than the conventional 1 bit at a time. Now let four amplitudes of a pulse be used to define each of the four possible combinations of two consecutive bits, such that

$$A_1 = 00$$

$$A_2 = 01$$

$$A_3 = 11$$

$$A_4 = 10$$

where A_1, A_2, A_3, and A_4 represent the four pulse amplitudes. This form of treating two bits at a time is called di-bit coding (see Section 8.9.2).

Similarly, we could let eight pulse levels cover all the possible combinations of three consecutive bits so that with a modulation rate of 1200 baud it is possible to transmit information at a rate of 3600 bps. Rather than vary amplitude to four or eight levels, phase can be varied. A four-phase system (PSK) could be coded as follows:

$$F_1 = 0° = 00$$

$$F_2 = 90° = 01$$

$$F_3 = 180° = 11$$

$$F_4 = 270° = 10$$

Again, with a four-phase system using di-bit coding, a tone with a modulation rate of 1200 baud PSK can be transmitting 2400 bps. An eight-phase PSK system at 1200 baud could produce 3600 bps of information transfer. Obviously this process cannot be extended indefinitely. The limitation comes from channel noise. Each time the number of levels or phases is increased, it is necessary to increase the signal-to-noise ratio to maintain a given error rate. Consider the case of a signal voltage S and a noise voltage N. The maximum number of increments of signal (amplitude) that can be discerned is S/N (since N is the smallest discernible increment). Add to this the no-signal case, and the number of discernible levels now becomes $S/N + 1$ or $S + N/N$, where S and N are expressed as power. The number of levels becomes the square root of this expression. This formula shows that in going from two to four levels or from four to eight levels, approximately 6-dB noise penalty is incurred each time we double the number of levels.

A similar analysis is carried out for the multiphase case, the penalty in going from four phases to eight phases is 3 dB, for example. See Chapter 15.

Sufficient background has been developed to appraise the data modem for the voice channel. Now consider a data modem for a data rate of 2400 bps. By using QPSK, as described above, 2400 bps is transmitted with a modulation rate of 1200 baud. Assume that the modem uses differential phase detection wherein the detector decisions are based on the change in phase between the last transition and the preceding one.

Assume the bandwidth to be present for the data modem under consideration (for most telephone networks the minimum bandwidth discussed for the sample case is indeed present—1800 Hz). It is now possible to determine if the noise requirements can be satisfied. Figure 8.17 shows that a 12.5-dB signal-to-noise ratio (Gaussian noise) is required to maintain an error rate of 1×10^{-5} for a binary polar (AM) system. As is well established, PSK systems have about 3-dB improvement. In this case only a 9.5-dB signal-to-noise ratio would be needed, all other factors held constant (no other contributing factors).

Assume the input from the line to be -10 dBm0 in order to satisfy loading conditions. To maintain the proper signal-to-noise ratio, the channel noise must be down to -19.5 dBm0.

To improve the modulation rate without the expense of increased bandwidth, QPSK (four-phase) is used. Allow a 3-dB general noise degradation factor, bringing the required noise level down to -22.5 dBm0.

Consider now the effects of EDD. It has been found that for a four-phase differential system, this degradation will amount to 6 dB if the permissible delay distortion is one pulse length. This impairment brings the noise requirement down to -28.5 dBm0 of average noise power in the voice channel. Allow 1 dB for frequency translation error or other factors, and the noise requirement is now down to -29.5 dBm0.

If the transmit level were -13 dBm0 instead of -10 dBm0, the numbers for noise must be adjusted another 3 dB such that it is now down to -32.5 dBm0 (19.5-dB signal-to-noise ratio). Thus it can be seen that to achieve a certain error rate for a given modulation rate, several modulation schemes should be considered. It is safe to say that in the majority of these schemes the noise requirement will fall somewhere between -25 and -40 dBm0. This is safely inside the CCITT figure of -43 dBm (see Section 8.9.3, noise). More discussion on this matter may be found in (Ref. 4).

8.12 CIRCUIT CONDITIONING

Of the critical circuit parameters mentioned in Section 8.9.3, two that have severe deleterious effects on data transmission can be reduced to tolerable limits by circuit conditioning. These two are amplitude–frequency response (distortion) and EDD.

Another name for circuit conditioning is equalization. There are several methods of performing equalization. The most common is to use one or several networks in tandem. Such networks tend to flatten response. In the case of amplitude, they add attenuation increasingly toward channel center and less toward its edges. The overall effect is one of making the amplitude response flatter. The delay equalizer operates fairly much in the same manner. Delay increases toward channel edges parabolically from the center. Delay is added in the center much like an inverted parabola, with less and less delay added as the band edge is approached. Thus the delay response is flattened at some small cost to absolute delay which, in most data systems, has no effect. However, care must be taken with the effect of a delay equalizer on an amplitude equalizer and, conversely, the amplitude equalizer on the delay equalizer. Their design and adjustment must be such that the flattening of the channel for one parameter does not entirely distort the channel for the other.

Another type of equalizer is the transversal type of filter. It is useful where it is necessary to select among or to adjust several attenuation (amplitude) and phase characteristics. The basis of the filter is a tapped delay line to which the input is presented. The output is taken from a summing network which adds or sums the outputs of the taps. Such a filter is adjusted to the desired response (equalization of both phase and amplitude) by adjusting the tap contributions.

If the characteristics of a line are known, another method of equalization is predistortion of the output signal of the data set. Some devices use a shift register and a summing network. If the equalization needs to be varied, then a feedback circuit from the receiver to the transmitter would be required to control the shift register. Such a type of active predistortion is valid for binary transmission only.

A major drawback of all the equalizers discussed (with the exception of the last with a feedback circuit) is that they are useful only on dedicated or leased circuits where the circuit characteristics are known and remain fixed. Obviously a switched circuit would require a variable automatic equalizer, or conditioning would be required on every circuit in the switched system that would be transmitting data.

Circuits are usually equalized on the receiving end. This is called postequalization. Equalizers must be balanced and must present the proper impedance to the line.

Administrations* may choose to condition (equalize) trunks and attempt to elimi-nate the need to equalize station lines; the economy of considerably fewer equalizers is obvious. In addition, each circuit that would possibly carry high-speed data in the system would have to be equalized, and the equalization must be good enough that any possible combination will meet the overall requirements. If equalization require-ments become greater (i.e., parameters more stringent), then consideration may have to be given to the restriction of the maximum number of circuits (trunks) in tandem.

Conditioning to meet amplitude–frequency response requirements is less exacting on the overall system than envelope delay. Equalization for envelope delay and its associated measurements are time consuming and expensive. Envelope delay in general is arithmetically cumulative. If there is a requirement of overall EDD of 1 ms for a circuit between 1000 and 2600 Hz, then in three links in tandem, each link must be better than 333 μs between the same frequency limits. For four links in tandem, each link would have to be 250 μs or better. In practice accumulation of delay distortion is not entirely arithmetical, resulting in a loosening of requirements by about 10%. Delay distortion tends to be inversely proportional to the velocity of propagation. Loaded cables display greater delay distortion than nonloaded cables. Likewise, with sharp filters a greater delay is experienced for frequencies approach-ing band edge than for filters with a more gradual roll-off.

In carrier multiplex systems channel banks contribute more to the overall EDD than any other part of the system. Because channels 1 and 12 of the standard CCITT modulation plan, those nearest the group band edge, suffer additional delay distor-tion owing to the effects of group and, in some cases, supergroup filters, the system engineer should allocate channels for data transmission near group and supergroup centers. On long-haul critical data systems the data channels should be allocated to throughgroups and throughsupergroups, minimizing as much as possible the steps of demodulation back to voice frequencies (channel demodulation).

Automatic equalization for both amplitude and delay shows promise, particularly for switched data systems. Such devices are self-adaptive and require a short adapta-tion period after switching, on the order of 1–2 s or less. This can be carried out during synchronization. Not only is the modem clock being "averaged" for the new circuit on transmission of a synchronous idle signal, but the self-adaptive equalizer adjusts for optimum equalization as well. The major drawback of adaptive equalizers is their expense.

8.13 PRACTICAL MODEM APPLICATIONS

8.13.1 Voice Frequency Carrier Telegraph (VFCT)

Narrow shifted FSK transmission of digital data goes under several common names. These are VFTG and VFCT. VFTG stands for VF telegraph and VFCT for VF carrier telegraph.

*Telephone companies.

In practice VFCT techniques handle data rates up to 1200 bps by a simple application of FSK modulation. The voice channel is divided into segments or frequency bounded zones or bands. Each segment represents a data or telegraph channel, each with a frequency shifted subcarrier.

For proper end-to-end system interface it is convenient to use standardized modulation plans, particularly on international circuits. In order for the far-end demodulator to operate with the near-end modulator, it must be tuned to the same center frequency and accept the same shift. Center frequency is the frequency in the center of the passband of the modulator–demodulator. The shift is the number of hertz that the center frequency is shifted up and down in frequency for the mark and space condition. From Table 8.1, by convention, the mark condition is the center frequency shifted downward, and the space upward. For modulation rates below 80 baud bandpasses have either 170-* or 120-Hz bandwidths with frequency shifts of ±42.5 or ±30 Hz, respectively. CCITT recommends (Rec. R.31) the 120-Hz channels for operating at 50 baud and below. However, some administrations operate these channels at higher modulation rates. Figure 8.18 shows graphically the partial modulation plan for 120-, 170-, and 240-Hz spacing. The 240-Hz channel is recommended by the CCITT for 100-baud operation with ±60 Hz frequency shift.

The number of tone telegraph or data channels that can be accommodated on a voice channel depends for one thing on the usable voice channel bandwidth. For HF radio with a voice channel limit on the order of 3 kHz, 16 channels may be accommodated using 170-Hz spacing (170 Hz between center frequencies). 24 VFCT channels may be accommodated between 390 and 3210 Hz with 120-Hz

*CCITT Rec. R.39.

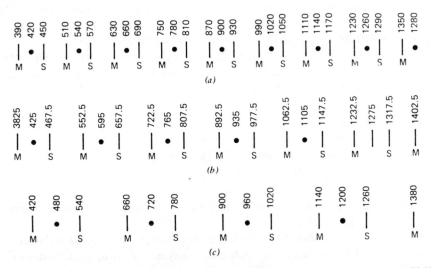

Figure 8.18 CCITT channel frequencies—VFCT. (*a*) 120-Hz spacing, ± 30-Hz shift. (*b*) 170-Hz spacing, ±42.5-Hz shift. (*c*) 240-Hz spacing, ± 60-Hz shift. Partial modulation plan shown.

spacing, or 12 channels with 240-Hz spacing. This can easily meet standard tele-phone FDM carrier channels of 300–3400 Hz.

Some administrations use a combination of voice and telegraph/data simultane-ously on a telephone channel. This technique is commonly referred to a "voice plus" or S + D (speech plus derived). There are two approaches to this technique. The first is recommended by CCITT and is used widely by INTELSAT orderwires. It places five telegraph channels (channels 20–24) above a restricted voice band with a roofing filter near 2500 Hz. Speech occupies a band between 300 and 2500 Hz. Above 2500 Hz appear up to five 50-baud telegraph channels.

The second approach removes a slot from the center of the voice channel into which up to two telegraph channels may be inserted. The slot is a 500-Hz band centered on 1275 Hz.

However, some administrations use a slot for telegraphy of frequencies 1680 and 1860 Hz by either AM or FM (FSK) (see CCITT Rec. R.43).

The use of S + D should be avoided on trunks in large networks because it causes degradation to speech and also precludes the use of the channel for higher speed data. In addition the telegraph channels should be removed before going into two-wire telephone service (i.e., at the hybrid or term set); otherwise service drops to half-duplex on telegraph.

8.13.2 Medium-Data-Rate Modems

In normal practice FSK is used for the transmission of data rates up to 1200 bps. The 120-Hz channel is nominally modified as in Figure 8.18c such that one 240-Hz channel replaces two 120-Hz channels. Administrations use the 240-Hz channel for modulation rates up to 150 baud. The same process can continue using 480-Hz channels for 300 baud FSK, and 960-Hz channels for 600 baud. CCITT Rec. V.23 specifies 600/1200-baud operation in the nominal 4-kHz voice band.

CCITT Recs. V.21, 23, 26, 27, and 29 (VIII-1) recommend the following. Rec. V.21, which refers to "200-baud modem standardized for use in the general switched telephone networks," recommends

- Frequency shift ±100 Hz
- Center frequency of channel 1, 1080 Hz
- Center frequency of channel 2, 1750 Hz
- In each case space (0) is the higher frequency

It also provides for a disabling tone on echo suppressors, a very important con-sideration on long circuits.

Rec. V.23 recommends 600/1200-baud modem standardized for use in the general switched telephone network for application to synchronous or asynchronous sys-tems. Provision is made for an optional backward channel for error control.

For the forward channel the following modulation rates and characteristic frequencies are presented:

	F_0	F_Z	F_A
Mode 1, up to 600 baud	1500 Hz	1300 Hz	1700 Hz
Mode 2, up to 1200 baud	1700 Hz	1300 Hz	2100 Hz

The backward channel for error control is capable of modulation rates up to 75 baud. Its mark and space frequencies are

F_Z	F_A
390 Hz	450 Hz

Refer to Table 8.1 for the mark-space convention (F_Z = mark or binary 1, F_A = space or binary 0).

CCITT has tried to achieve a universality, recommending a modem that can be used nearly anywhere in the world "in the general switched telephone network." It considers worst case conditions of amplitude–frequency response and EDD. The following are applicable CCITT recommendations dealing with data modems:

V.26, "2400 bps Modem Standardized for Use on 4-Wire Leased Telephone-Type Circuits"

1. Carrier frequency 1800 Hz ±1 Hz, 4-phase modulation, synchronous mode of operation

2. Dibit coding as follows:

	Phase Change (°)	
Dibit	Alternative A	Alternative B
00	0	+45
01	+90	+135
11	+180	+225
10	+270	+315

3. Data signaling rate 2400 bps ±0.01%
4. Modulation rate 1200 baud ±0.01%
5. Maximum frequency error at receiver ±7 Hz [allowing ±1 Hz for modulator (transmitter) error]
6. Backward channel 75 bps (see Rec. V.23)

V.26bis, "2400/1200 bps Modem for Use in the General Switched Network"

Generally the same as Rec. V.26. At 1200 bps line operation, differential PSK is recommended where

$$0 = +90°$$

$$1 = +270°$$

V.27, "4800 bps Modem with Manual Equalizer for Use on Leased Telephone-Type Circuits"

The principal characteristics of this modem are:

   1. Full-duplex or half-duplex operation
   2. Differential 8-phase modulation, synchronous mode of operation
   3. Possibility of backward (supervisory) channel with modulation rates to 75 baud in each direction of transmission
   4. Inclusion of a manually adjustable equalizer
   5. Carrier frequency 1800 Hz ±1 Hz
   6. Tribit coding as follows:

Tribit Values	Phase Change (°)
001	0
000	45
010	90
011	135
111	180
110	225
100	270
101	315

   7. Frequency tolerance same as Recs. V.26 and V.26bis
   8. Line signal characteristics: a 50% raised cosine energy spectrum is equally divided between transmitter and receiver
   9. A self-synchronizing scrambler/descrambler having a generating polynomial $1 + X^{-6} + X^{-7}$ with additional guards against repeating patterns of 1, 2, 3, 4, 6, 9, and 12 bits shall be included in the modem

V.27bis, "4800 bps Modem with Automatic Equalizer Standardized for Use on Leased Telephone-Type Circuits"

The principal characteristics of this modem are the same as for that of Rec. V.27 with the following exceptions:

   1. Includes fall-back rate of 2400 bps with V.26 Alternative A characteristics.
   2. Automatic equalizer with two types of turn-on sequences and a turn-off

Table 8.7 Phase Encoding

Q_2	Q_3	Q_4	Phase Change ($^\circ$)
0	0	1	0
0	0	0	45
0	1	0	90
0	1	1	135
1	1	1	180
1	1	0	225
1	0	0	270
1	0	1	315

sequence. The first type of turn-on sequence is short in duration for comparatively good circuits, meeting CCITT Rec. M.1020, and a second is a longer sequence for relatively poor circuits, below the Rec. M.1020 standard. The sequences, which are transmitted on-line prior to traffic, provide for equalizer conditioning and serve to set descrambler synchronization into proper operation.

V.27ter, "4800/2400 bps Modem Standardized for Use in the General Switched Telephone Network"

Similar to Rec. V.27bis, except for the turn-on sequence which, in this case, includes the capability to protect against talker echo.

V.29, "9600 bps Modem Standardized for Use on Leased Telephone-Type Circuits"

The principal characteristics of this modem are:

1. Includes a fallback to rates of 7200 and 4800 bps
2. Full-duplex and half-duplex operation
3. Combined amplitude and phase modulation with synchronous mode operation
4. Automatic equalizer
5. Optional inclusion of a multiplexer for combining data rates of 7200, 4800, and 2400 bps

Table 8.8 Amplitude–Phase Relationships

Absolute Phase	Q_1	Relative Signal Element Amplitude
$0^\circ, 90^\circ, 180^\circ, 270^\circ$	0	3
	1	5
$45^\circ, 135^\circ, 225^\circ, 315^\circ$	0	$\sqrt{2}$
	1	$3\sqrt{2}$

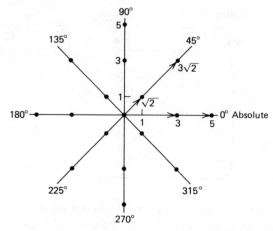

Figure 8.19   Signal space diagram, 9600-bps operation (CCITT Rec. V.29).

6. Line signal 1700 Hz ±1 Hz

7. Signal space coding. At 9600 bps the scrambled data stream to be transmitted is divided into groups of four consecutive data bits (quadbits). The first bit Q_1 in time of each quadbit is used to determine the signal element amplitude to be transmitted. The second Q_2, third Q_3, and fourth Q_4 bits are encoded as a phase change relative to the phase of the immediately preceding element (Table 8.7). The phase encoding is identical to that of Rec. V.27.

The relative amplitude of the transmitted signal element is determined by the first bit Q_1 of the quadbit and the absolute phase of the signal element (Table 8.8). The absolute phase is initially established by the synchronizing signal. The four possible signaling elements Q_1, Q_2, Q_3, and Q_4 represent 32 phase–amplitude possibilities ($2^4 = 32$) which are shown in Figure 8.19.

8.13.3 High-Data-Rate Modems

Section 8.10 put a limit on the data rate in a bandwidth of 3000 Hz. The Nyquist limit was 6000 bits for binary transmission. It was also noted that a practical limit for this bandwidth is about 3000 baud without automatic equalization. For the binary case this would be equivalent to 1 b/Hz (3000 baud ≈ 3000 bps). For the quaternary case a data rate of 6000 bps may be reached (3000 baud ≈ 6000 bps), or 2 b/Hz. However, most telephone lines do not have 3000 Hz of usable bandwidth available.

Consider the following modems, their required bandwidths, and their modulation schemes, which permit an improved data rate on a telephone channel (Table 8.9).

Table 8.9   Characteristics of Some High-Data-Rate-Modems

Data Rate (bps)	Modulation Rate (baud)	Modulation	Bits per Hertz	Bandwidth Required (Hz)
2400 synchronous	1200	Differential four-phase	2	1200
4800 synchronous	1600	Differential four-phase	3	1600
3600 synchronous	1200	Differential four-phase, two-level (combined PSK-AM)	3	1200
2400 synchronous	800	Differential eight-phase	3	800
9600 synchronous	4800	Differential two-phase, two-level	2	2400[a]

[a] Uses automatic equalizer.

8.14 SERIAL-TO-PARALLEL CONVERSION FOR TRANSMISSION ON IMPAIRED MEDIA

Often the transmission medium, in most cases the voice channel, cannot support a high data rate even with conditioning. The impairments may be due to poor amplitude–frequency response, EDD, or excessive impulse noise.

One step that may be taken in these circumstances is to convert the high-speed serial bit stream at the dc level (e.g., demodulated) to a number of lower speed parallel bit streams. One technique widely used on HF radio systems is to divide a 2400-bit serial stream into 16 parallel streams, each carrying 150 bps. If each of the slower streams is di-bit coded (2 bits at a time, discussed in Section 8.11) and applied to a QPSK tone modulator, the modulation rate on each subchannel is reduced in this case to 75 baud. The equivalent period for a di-bit interval is $\frac{1}{75}$ s or 13 ms.

There are two obvious advantages of this technique. First, each subchannel has a comparatively small bandwidth and thus looks at a small and tolerable segment of the total delay across the channel. The impairment of EDD is less on slower speed channels. Second, there is less of a chance of a noise burst or hit of impulse noise to smear the subchannel signal beyond recognition. If the duration of the noise burst is less than half the pulse width, the data pulse can be regenerated, and the pulse will not be in error. The longer one can make the pulse width, the less chance there is of disturbance from impulse noise. In this case the interval or pulse width has had an equivalent lengthening by a factor of 32.

8.15 PARALLEL-TO-SERIAL CONVERSION FOR IMPROVED ECONOMY OF CIRCUIT USAGE

Long high-quality (conditioned) toll telephone circuits are costly to lease or are a costly investment. The user is often faced with a large number of slow-speed circuits (50–300 bps) which originate in one general geographic location, with a general destination to another common geographic location. If we assume that these are 75-bps circuits (100 wpm), which are commonly encountered in practice, only 18–24 can be transmitted on a high-grade telephone channel by conventional VFTG techniques (see Section 8.13.1).

Circuit economy can be affected using a data/telegraph TDM. A typical application of this type is illustrated in Figure 8.20. It shows one direction of transmission only. Here incoming slow-speed VFCT channels are converted to equivalent dc bit streams. Up to 32 of these bit streams serve as input to a TDM in the application that is illustrated in the figure. The output of the TDM is a 2400-bit synchronous series bit stream. This output is fed to a conventional 2400-bit modem. At the far end the 2400-bps serial stream is demodulated to dc and fed to the equivalent demultiplexer. The demultiplexer breaks the serial stream back down to the original 75-bps circuits. Figure 8.20 illustrates the concept. It does not show clocking or other interconnect circuitry. By use of a TDM a savings of up to 2 : 1 can be effected.

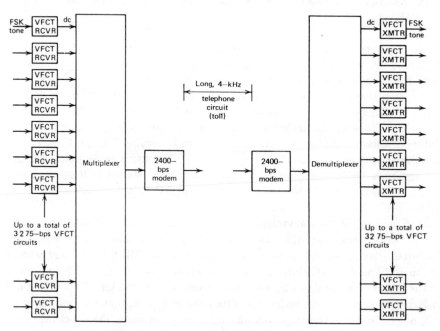

Figure 8.20 Typical application of parallel-to-serial conversion. VFCT = voice frequency carrier telegraph; RCVR = receiver (converter); XMTR = transmitter (keyer); bps = bits per second; FSK = frequency shift keying.

Whereas by conventional VFCT means that only about 18 75-bps circuits can be transmitted on a good telephone channel, by means of the multiplexer up to 32 such circuits can now be transmitted on the same channel.

TDMs are also available with line data rates of 4800 and 9600 bps, accepting inputs for multiplexing from 75 bps up to 2400 bps for the 4800-bps line rate and up to 4800 bps for the 9600-bps line rate.

The statistical TDM is an interesting variation of the conventional TDM and is particularly useful where the input port data traffic is "bursty" (i.e., where data messages are relatively short and where there are periods where there is no traffic being transmitted). The conventional TDM creates a permanently dedicated time slot or subchannel for each input port in the sharing group. A statistical TDM, by contrast, dynamically allocates the subchannels or time slots on a statistical basis to increase efficiency by providing time slots only for ports actively transmitting data. In effect these devices not only multiplex, but concentrate as well. Of course it will be appreciated that in the case of a true statistical TDM the number of 75-bps input ports shown in Figure 8.20 can exceed 32, whereas for the conventional TDM, 32 is the limit.

The amount by which the statistical TDM can exceed the limiting number of input ports (whatever the data rate) depends largely on its storage capabilities as well as on the nature and intensity of the data traffic appearing at the input ports.

Data multiplexers frame traffic when placing the signal on line, and we must then expect overhead bits in this case. Many frames provide error control and other housekeeping duties in accordance with particular protocols. A protocol is a set of rules or conventions that govern the operation of a data communication system. A protocol involves some or all of the following topics:

1. *Framing.* Frame makeup; or block makeup, message or packet.

2. *Error control.*

3. *Sequence control.* The numbering of messages (blocks, packets) to keep order in sequence, to eliminate duplication, to maintain a record of proper identification of data messages, especially with some of the more complex ARQ systems.

4. *Transparency.* Transparency of communication links, link control equipment, multiplexers, concentrators, modems, and so on to any bit pattern the user wishes to transmit, even though these patterns resemble control characters or "prohibited bit sequences" such as long series of 1's or 0's.

5. *Line control.* Determination, in the case of half-duplex or multipoint lines, of which station is going to transmit and which station(s) is (are) going to receive.

6. *Idle patterns* to maintain network synchronization.

7. *Time-out control.* The procedures to follow if message (block or packet) flow ceases entirely.

8. *Startup control.* Getting a network into operation initially or after some period of remaining idle for one reason or another.

9. *Sign-off control.* Under normal conditions the process of ending a communication or transaction before starting the next transaction or message exchange.

There are numerous protocols now available to the data system designer. Among these are IBM's BISYNC and SDLC, ISO's HDLC, CCITT X.25, ANSI ADCCP, and DEC DDCMP. For more information on these protocols, (Refs. 25, 29, and 31) may be consulted.

8.16 MODEMS FOR APPLICATION TO CHANNEL BANDWIDTHS IN EXCESS OF 4 kHz

Well-equalized (conditioned) telephone circuits are handling data rates in excess of 7500 bps. For data rates in excess of 9600 bps, the use of 48-kHz channels becomes attractive. 48 kHz is the bandwidth of the standard CCITT 12-channel group (e.g., 60–108 kHz). Group equalization in a 36-kHz portion of the band can bring EDD (group delay) down to 50 μs. From previous reasoning, it can be assumed that the channel will support a 20-kbaud modulation rate. It should also be noted that 50 μs is considerably greater than the multipath delay encountered on tropospheric scatter circuits (10 μs). Hence operation will be possible over most wideband media—coaxial cable, tropospheric scatter, satellite, and radiolink.

The industry is now standardizing on data rates of 75×2^n, where n is any whole number. Thus a 48-kHz channel can support 19.2 kbaud of two-level FSK or PSK or 38.4 kbps of four-level AM or PSK.

Taking the above discussion into account, as well as some of the other design rules set forth in the chapter, the following specifications are suggested for a 48-kHz data modem:

Modulation rate	19.2 kbaud
Data rate	38.4 kbps
Input–output to the line	60–108 kHz
Envelope delay distortion	
within 200 μs	64–104 kHz
within 50 μs	66–102 kHz
Transmit level to the line	+1 dBm0
(derived from -10 dBm0 + $10 \log n$,	
where n = number of channels, = 12)	
Noise (equivalent white noise)	-20 dBm0 with flat weighting
Quaternary differential phase-shift modulation	
(with data input di-bit coded)	

8.17 DATA TRANSMISSION SYSTEM—FUNCTIONAL BLOCK DIAGRAM

Figure 8.21 is a simplified functional block diagram showing a typical data transmission system end to end. The diagram is meant to be representative and con-

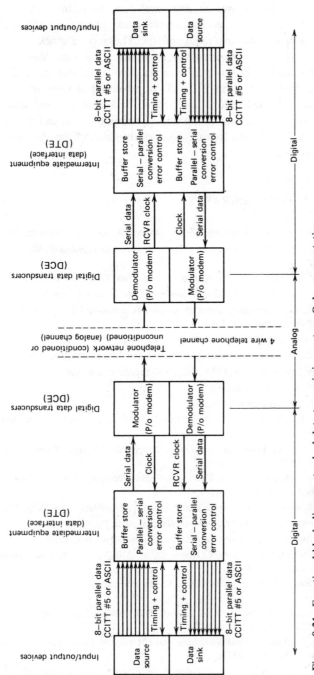

Figure 8.21 Functional block diagram standard data transmission system. Only representative electrical connections are shown.

415

ceptual. The system shown uses an eight-level code operating in a synchronous mode. Sections 8.3.2 and 8.6.2 cover eight-level codes and synchronous transmission, respectively, as well as clocking requirements. The receive clock in this case is corrected by averaging transitions on the incoming bit stream. Master clocks are contained in the modems. Timing problems are covered in Section 8.6.3. The data interface is discussed in Section 8.7 and input–output devices in Section 8.8. Analog (telephone) channel criteria are covered in Section 8.9 and subsequently. The diagram shows the functions of the various building blocks and how the higher speed more sophisticated data transmission system differs from the more conventional telegraph system.

REFERENCES AND BIBLIOGRAPHY

1. *Reference Data for Radio Engineers*, 6th ed., Howard W. Sams, Indianapolis, IN, 1977.
2. D. H. Hamsher Ed., *Communication System Engineering Handbook*, McGraw-Hill, New York, 1967.
3. *Transmission Systems for Communications*, 4th ed., Bell Telephone Laboratories.
4. W. R. Bennett and J. R. Davey, *Data Transmission*, McGraw-Hill, New York, 1965.
5. *Data Transmission, Parameters and Capabilities*, International Telephone and Telegraph Co., Federal Laboratories, Oct. 1961.
6. *Understanding Telegraph Distortion*, Stelma, Inc., Stamford, CT, 1962.
7. J. M. Weir, "Digital Data Communication Techniques," *Proc. IRE*, Jan. 1961.
8. CCITT Orange Books, Geneva, 1976, Vol. III, G recommendations.
9. *Ibid.*, Vol. VII, R recommendations.
10. *Ibid.*, Vol. VIII, V recommendations.
11. R. W. Lucky, J. Salz, and E. J. Weldon, *Principles of Data Communication*, McGraw-Hill, New York, 1968.
12. H. Nyquist, "Certain Topics in Telegraph Transmission Theory," *Trans. AIEE*, Vol. 47, 617–644, Apr. 1928.
13. C. E. Shannon, "A Mathematical Theory of Communication," *Bell Sys. Tech. J.*, Vol. 27, 379–423, July 1948; 623–656, Oct. 1948.
14. J. Martin, *Systems Analysis for Data Transmission*, Prentice-Hall, New York, 1972.
15. W. P. Davenport, *Modern Data Communication*, Hayden, New York, 1971.
16. S. Goldman, *Information Theory*, Dover Publications, New York, 1968.
17. *DCS Autodin Interface and Control Criteria*, DCAC 370-D-175-1, Defense Communication Agency, Washington, DC, 1965.
18. M. P. Ristenbatt, "Alternatives in Digital Communications," *Proc. IEEE*, Vol. 61, June 1973.
19. *Notes on Transmission Engineering*, 3rd ed., U.S. Independent Telephone Association, New York, 1971.
20. C. L. Cuccia, "Subnanosecond Switching and Ultra-Speed Data Communications," *Data and Communications*, Nov. 1971.
21. D. R. Doll, "Controlling Data Transmission Errors," *Data Dynamics*, July 1971.
22. E. N. Gilbert, *Information Theory after 18 Years*, Bell Telephone Monograph, Bell Telephone Laboratories.

23. EIA RS-232C, "Interface between Data Terminal Equipment and Data Communication Equipment Employing Serial Binary Data Interchange," Electronic Industries Association, Washington, DC, Aug. 1969.

24. *Bell System Technical Reference, Transmission Parameters Affecting Voiceband Data Transmission–Description of Parameters*, Publ. 41008, American Telephone and Telegraph Co., New York, July 1974.

25. J. E. McNamara, "Technical Aspects of Data Communication," Digital Equipment Corp., Maynard, MA, 1977.

26. EIA RS-363, "Standard for Specifying Signal Quality for Transmitting and Receiving Data Processing Terminal Equipments Using Serial Data Transmission at the Interface with Nonsynchronous Data Communication Equipment," Electronic Industries Association, Washington, DC, May 1969.

27. EIA RS-404, "Standard for Start–Stop Signal Quality between Data Terminal Equipment and Nonsynchronous Data Communication Equipment," Electronic Industries Association, Washington, DC, Mar. 1973.

28. EIA RS-422 "Electrical Characteristics of Balanced Voltage Digital Interface Circuits," Electronic Industries Association, Washington, DC, Apr. 1975.

29. D. R. Doll, *Data Communications*, Wiley-Interscience, New York, 1978.

30. *Telecommunication Transmission Engineering*, Vol. 2, 2nd ed., American Telephone and Telegraph Co., New York, 1977.

31. R. L. Freeman, *Telecommunication System Engineering*, Wiley-Interscience, New York, 1980.

32. H. C. Folts, "A Powerful Standard Replaces the Old Interface Standby," *Data Communications*, May 1980.

33. *Proceedings of the IEEE*, Special Issue on Data Networks and Protocols, Vol. 66, Nov. 1978.

34. EIA RS-423, "Electrical Characteristics of Unbalanced Voltage Digital Interface Circuits," Electronic Industries Association, Washington, DC, Apr. 1975.

35. EIA RS-449, "General Purpose 37-Position and 9-Position Interface for Data Terminal Equipment and Data Circuit-Terminating Equipment Employing Serial Binary Data Interchange," Electronic Industries Association, Washington, DC, Nov. 1977.

36. Bell System Technical Reference, *Data Communications Using Voiceband Private Line Channels October 1973*, Publ. 41004, American Telephone and Telegraph Co., New York, Oct. 1973.

37. IEEE Std. 100-1977, "Dictionary of Electrical and Electronics Terms," 2nd ed., IEEE, New York, 1977.

9 | COAXIAL CABLE SYSTEMS

9.1 INTRODUCTION

A coaxial cable is simply a transmission line consisting of an unbalanced pair made up of an inner conductor surrounded by a grounded outer conductor, which is held in a concentric configuration by a dielectric. The dielectric can be of many different types, such as solid "poly" (polyethylene or polyvinyl chloride), foam, Spirafil, air, or gas. In the case of air/gas dielectric, the center conductor is kept in place by spacers or disks.

Systems have been designed to use coaxial cable as a transmission medium with a capability of transmitting an FDM configuration ranging from 120 to 13,200 voice channels. Community antenna television (CATV) systems use single cables for transmitted bandwidths on the order of 300 MHz.

FDM was developed originally as a means to increase the voice channel capacity of wire systems. At a later date the same techniques were applied to radio. Then for a time, the 20 years after World War II, radio systems became the primary means for transmitting long-haul toll telephone traffic. Lately coaxial cable has been making a strong comeback in this area.

One advantage of coaxial cable systems is reduced noise accumulation when compared to radiolinks. For point-to-point multichannel telephony the FDM line frequency (see Chapter 3) configuration can be applied directly to the cable without further modulation steps as required in radiolinks, thus substantially reducing system noise.

In most cases radiolinks will prove more economical then coaxial cable. Nevertheless, owing to the congestion of centimetric radio wave systems (radiolinks) (see Chapter 10), coaxial cable is making a new debut. Coaxial cable should be considered in lieu of radiolinks using the following general guidelines:

- In areas of heavy microwave (including radiolink) RFI.
- On high-density routes where it may be more economical than radiolinks. (Think here of a system that will require 5000 circuits at the end of 10 years.)
- On long national or international backbone routes where the system designer is concerned with noise accumulation.

418

Coaxial cable systems may be attractive for the transmission of TV or other video applications. Some activity has been noted in the joint use of TV and FDM telephone channels on the same conductor. Another advantage in some circumstances is that system maintenance costs may prove to be less than for equal-capacity radiolinks.

One deterrent to the implementation of coaxial cable systems, as with any cable installation, is the problem of getting the right-of-way for installation, and its subsequent maintenance (gaining access), particularly in urban areas. Another consideration is the possibility of damage to the cable once it is installed. Construction crews may unintentionally dig up or cut the cable. For more details on the choice, whether coaxial cables or radiolinks, refer to Section 9.12.

9.2 BASIC CONSTRUCTION DESIGNS

Each coaxial line is called a *tube*. A pair of these tubes is required for full-duplex long-haul application. One exception is the CCITT small-bore coaxial cable system where 120 voice channels, both "go" and "return," are accommodated in one tube. For long-haul systems more than one tube is included in a sheath. In the

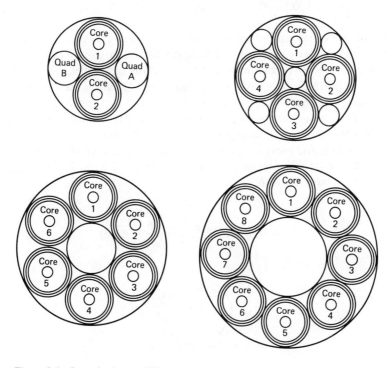

Figure 9.1 Some basic coaxial lay-ups.

same sheath filler pairs or quads are included, sometimes placed in the interstices, depending on the size and lay-up of the cable. The pairs and quads are used for orderwire and control purposes as well as for local communication. Some typical cable lay-ups are shown in Figure 9.1. Coaxial cable is usually placed at a depth of 90-120 cm, depending on frost penetration, along the right-of-way. Tractor-drawn trenchers or plows are used to open the ditch where the cable is placed, using fully automated procedures.

Cable repeaters are spaced uniformly along the route. Secondary or "dependent" repeaters are often buried. Primary power feeding or "main" repeaters are installed in surface housing. Cable lengths are factory cut so that the splice occurs right at repeater locations.

9.3   CABLE CHARACTERISTICS

For long-haul transmission, standard cable sizes are as follows:

(inches)	(mm)
0.047/0.174	1.2/4.4 (small diameter)
0.104/0.375	2.6/9.5

The fractions express the outside diameter of the inner conductor over the inside diameter of the outer conductor. For instance, for the large-bore cable the outside diameter of the inner conductor is 0.104 in. and the inside diameter of the outer conductor is 0.375 in. This is shown in Figure 9.2. As can be seen from equation (9.1) relating to Figure 9.2, the ratio of the diameters of the inner and outer conductors has an important bearing on attenuation. If we can achieve a ratio of $b/a = 3.6$, a minimum attenuation per unit length will result.

For an air dielectric cable pair, $\epsilon = 1.0$, the outside diameter of the inner conductor $= 2a$, the inside diameter of the outer conductor $= 2b$ and $f =$ frequency. The attenuation constant α is then

$$\alpha = 2.12 \times 10^{-5} \frac{\sqrt{f\left(\frac{1}{a} + \frac{1}{b}\right)}}{\log b/a} (\text{dB/mi}) \qquad (9.1)$$

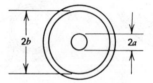

Figure 9.2   Basic electrical characteristics of coaxial cable.

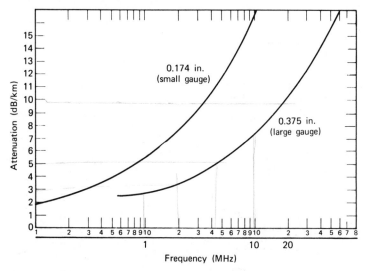

Figure 9.3 Attenuation–frequency response per kilometer of coaxial cable.

where a = radius of inner conductor and b = radius of outer conductor.

The characteristic impedance is

$$Z_0 = \left(\frac{138}{\sqrt{\epsilon}}\right) \log\left(\frac{b}{a}\right) = 138 \log\left(\frac{b}{a}\right) \text{ (in air)} \tag{9.2}$$

The characteristic impedance of coaxial cable is $Z_0 = 138 \log(b/a)$ for an air dielectric. If $b/a = 3.6$, then $Z_0 = 77\ \Omega$. Using dielectric other than air reduces the characteristic impedance. If we use the disks mentioned above to support the center conductor, the impedance lowers to $75\ \Omega$.

Figure 9.3 is a curve giving the attenuation per unit length in decibels versus frequency for the two most common types of coaxial cable discussed in this chapter. Attenuation increases rapidly as a function of frequency and is a function of the square root of frequency as shown in Figure 9.2. The transmission system engineer is basically interested in how much bandwidth is available to transmit an FDM line frequency configuration (Chapter 3). For instance, the 0.375-in. cable has an attenuation of about 5.8 dB/mi at 2.5 MHz and the 0.174-in. cable, 12.8 dB/mi. At 5 MHz the 0.174-in. cable has about 19 dB/mi and the 0.375-in. cable, 10 dB/mi. Attenuation is specified for the highest frequency of interest.

Coaxial cable can transmit signals down to dc, but in practice, frequencies below 60 kHz are not used because of difficulties of equalization and shielding. Some engineers lift the lower limit to 312 kHz. The HF limit of the system is a function of the type and spacing of repeaters as well as cable dimensions and the dielectric constant of the insulating material. It will be appreciated from Figure 9.3 that the gain frequency characteristics of the cable follow a root frequency law, and equalization and preemphasis should be designed accordingly.

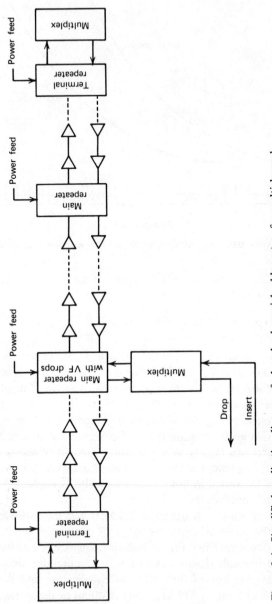

Figure 9.4 Simplified application diagram of a long-haul coaxial cable system for multichannel telephony.

9.4 SYSTEM DESIGN

Figure 9.4 is a simplified application diagram of a coaxial cable system in long-haul point-to-point multichannel telephone service. To summarize system operation, an FDM line frequency (Chapter 3) is applied to the coaxial cable system via a line terminal unit. Dependent repeaters are spaced uniformly along the length of the cable system. These repeaters are fed power from the cable itself. In the ITT design (Ref. 1) the dependent repeater has a plug-in automatic level control unit. In temperate zones, where cable laying is sufficient and where diurnal and seasonal temperature variations are within the "normal" (a seasonal swing of $\pm 10°C$), a plug-in level control (regulating) unit is incorporated in every fourth dependent amplifier (see Figure 9.5). We use the word "dependent" for the dependent repeater for two reasons. It depends on a terminal or main repeater for power and it provides to the terminal or main repeater fault information.

Let us examine Figures 9.4 and 9.5 at length. Assume that we are dealing with a nominal 12-MHz system on a 0.375-in. (9.5-mm) cable. Up to 2700 voice channels can be transmitted. To accomplish this, two tubes are required, one in each direction. Most lay-ups, as shown in Figure 9.1, have more than two tubes. Consider Figure 9.4 from left to right. Voice channels in a four-wire configuration connect

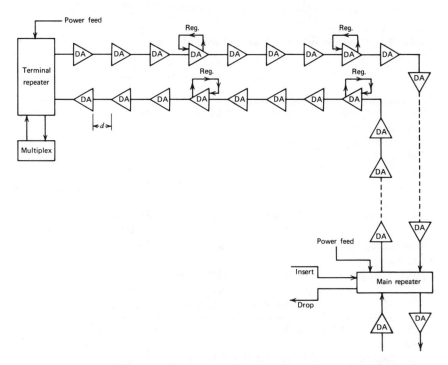

Figure 9.5 Detail of application diagram. DA = dependent amplifier (repeater); Reg. = regulation circuitry; d = distance between repeaters.

with the multiplex equipment in both the "go" and the "return" directions. The output of the multiplex equipment is the line frequency (baseband) to be fed to the cable. Various line frequency configurations are shown in Figures 3.11, 3.18, and 3.19. The line signal is fed to the terminal repeater, which performs the following functions:

- Combines the line control pilots with the multiplex line frequency.
- Provides preemphasis to the transmitted signal, distorting the output signal such that the higher frequencies get more gain than the lower frequencies, as shown in Figure 9.3.
- Equalizes the incoming wideband signal.
- Feeds power to dependent repeaters.

The output of the terminal repeater is a preemphasized signal with required pilots along with power feed. In the ITT design this is a dc voltage up to 650 V with a stabilized current of 110 mA. A main (terminal) repeater feeds, in this design, up to 15 dependent repeaters in each direction. Thus a maximum of 30 dependent repeaters appear in a chain for every main or terminal repeater. Other functions of a main repeater are to equalize the wideband signal and to provide access for drop and insert of telephone channels by means of throughgroup filters.

Table 9.1 Characteristic of L Coaxial Cable Systems

| Item | L System Identifier | | | |
	L1	L3	L4	L5
Maximum design line length	4000 mi	1000 mi	4000 mi	4000 mi
Number of 4-kHz FDM VF channels	600	1860	3600	10,800[a]
TV NTSC	Yes	Yes plus 600 VF	No	Not stated
Line frequency	60–2788 kHz	312–8284 kHz	564–17,548 kHz	1590–68,780 kHz
Nominal repeater spacing	8 mi	4 mi	2 mi	1 mi
Power feed points	160 mi or every 20 repeaters	160 mi or every 42 repeaters	160 mi or every 80 repeaters	75 mi or every 75 repeaters

[a] L5E = 13,200 VF channels.

Notes
1. Cable type of all L systems, 0.375 in.
2. Number of VF channnles expressed per pair of tubes, one tube "go" and one tube "return."

Table 9.2 Characteristics of CCITT Specified Coaxial Cable Systems (Large Diameter Cable)

Item	Nominal Top Modulation Frequency				
	2.6 MHz	4 MHz	6 MHz	12 MHz	60 MHz
CCITT Rec.	G.337A	G.338	G.337B	G.332	G.333
Repeater type	Tube	Tube	Tube	Transistor	Transistor
Video capability	No	Yes	Yes	Yes	Not stated
Video + FDM capability	No	No	No	Yes	Not stated
Nominal repeater spacing	6 mi / 9 km	6 mi / 9 km	6 mi / 9 km	3 mi / 4.5 km	1 mi / 1.55 km
Main line reg. pilot	2604 kHz	4092 kHz	See CCITT Rec. J.72	12,435 kHz	12,435 /4287 kHz
Auxiliary reg. pilot(s)		308, 60 kHz	See Rec. J.72	4287, 308 kHz	61,160, 40,920 and 22,372 kHz

Note. Cable type for all systems, 0.104 /0.375 in. = 2.6 /9.5 mm.

Figure 9.5 is a blowup of a section of Figure 9.4 showing each fourth repeater with its automatic level regulation circuitry. The distance d between DA repeaters is 4.5 km or 2.8 mi for a nominal 12-MHz system (0.375-in. cable). Amplifiers have gain adjustments of ± 6 dB, equivalent to varying repeater spacings of ± 570 m (1870 ft).

As can be seen from the above, the design of coaxial cable systems for both long-haul multichannel telephone service as well as CATV systems has become, to a degree, a "cookbook" design. Basically, system design involves the following:

- Repeater spacing as a function of cable type and bandwidth
- Regulation of signal level
- Temperature effects on regulation
- Equalization
- Cable impedance irregularities
- Fault location or the so-called supervision
- Power feed

Other factors are, of course, the right-of-way for the cable route with access for maintenance and the laying of the cable. With these factors in mind, Tables 9.1 and 9.2 review the basic parameters of the Bell System approach (Table 9.1) and the CCITT approach (Table 9.2).

For the 0.375-in. coaxial cable systems practical noise accumulation is less than 1 pWp/km, whereas radiolinks allocate 3 pWp/km. These are good guideline numbers to remember for gross system considerations. Noise in coaxial cable systems derives from the active devices in the line (e.g., the repeaters and terminal equipment, both line conditioning and multiplex). Noise design of these devices is a trade-off between thermal and IM noise. IM noise is the principal limiting parameter forcing the designer to install more repeaters per unit length with less gain per repeater.

Refer to Chapter 3 for CCITT recommended FDM line frequency configurations, in particular Figures 3.18 and 3.19 valid for 12-MHz systems. CCITT pilot frequencies and system levels are covered in Section 9.7.

9.5 REPEATER DESIGN—AN ECONOMIC TRADE-OFF FROM OPTIMUM

9.5.1 General

Consider a coaxial cable system 100 km long using 0.375-in. cable capable of transmitting up to 2700 VF channels in an FDM/SSB configuration (12 MHz). At 12 MHz cable attenuation per kilometer is approximately 8.3 dB (from Figure 9.3). The total loss at 12 MHz for the 100-km cable section is $8.3 \times 100 = 830$ dB. Thus one approach the system design engineer might take would be to install a 830-dB amplifier at the front end of the 100-km section. This approach is rejected out of hand. Another approach would be to install a 415-dB amplifier at the front end and another at the 50-km point. Suppose that the signal level was -15 dBm composite at the originating end. Thus -15 dBm + 415 dB = +400 dBm or +370 dBW. Remember that +60 dBW is equivalent to 1 MW; otherwise we would have an amplifier with an output of 10^{37} W or 10^{31} MW. Still another approach is to have 10 amplifiers with 83-dB gain, each spaced at 10-km intervals. Another would be 20 amplifiers or $830/20 = 41.5$ dB each; or 30 amplifiers at $830/30 = 27.67$ dB, each spaced at 3.33-km intervals. As we shall see later, the latter approach begins to reach an optimum from a noise standpoint, keeping in mind that the upper limit for noise accumulation is 3 pWp/km. The gain most usually encountered in coaxial cable amplifiers is 30–35 dB.

If we remain with the 3-pWp/km criterion, in nearly all cases radiolinks (Chapter 5) will be installed because of their economic advantage. Assuming 10 full-duplex RF channels per radio system at 1800 VF channels per RF channel, the radiolink can transmit 18,000 full-duplex channels, and do it probably more cheaply on an installed cost basis. On the other hand, if we can show noise accumulation less on coaxial cable systems, these systems will prove in at some number of channels less than 18,000 if the reduced cumulative noise is included as an economic factor. There are other considerations, such as maintenance and reliability, but let us discuss noise further.

Suppose that we design our coaxial cable systems for no more than 1 pWp/km. Most long-haul coaxial cable systems being installed today meet this figure. However, we will use the CCITT figure of 3 pWp/km in some of the examples that follow.

In Chapter 1, noise for this discussion consists of two major components, namely,

- Thermal noise (white noise)
- IM noise

Coaxial cable amplifier design, to reach a goal of 1 pWp/km of noise accumulation, must walk a "tightrope" between thermal and IM noise. It is also very sensitive to overload with its consequent impact on IM noise.

The purpose of the abbreviated and highly simplified discussion in this section is to give the transmission system engineer some appreciation of coaxial repeater design. For a deeper analysis, the reader should refer to (Ref. 2, chaps. 12–16; Ref. 3, chaps. 3–7, and Ref. 15, chap. 12).

9.5.2 Thermal Noise

From Section 1.12 the thermal noise threshold P_n may be calculated for an active two-port device as follows:

$$P_n = -174 \text{ dBm/Hz} + \text{NF} + 10 \log B_w \qquad (9.3)$$

where B_w = bandwidth (Hz) and NF_{dB} = noise figure of the amplifier.

Restating equation (9.3) for a voice channel with a nominal B_w of 3000 Hz, we can then have

$$P_n = -139 \text{ dBm} + \text{NF} \qquad (\text{dBm/3 kHz}) \qquad (9.4)$$

Assume a coaxial cable system with identical repeaters, each with gain G_r, spaced at equal intervals along a uniform cable section. Here G_r exactly equals the loss of the intervening cable between repeaters. The noise output of the first repeater is $P_n + G_r$ (in dBm). For N repeaters in cascade, the total noise (thermal) output of the Nth repeater is

$$P_n + G_r + 10 \log N \qquad (\text{dBm}) \qquad (9.5)$$

An important assumption all along is that the input–output impedance of the repeaters just equals the cable impedance Z_0.

9.5.3 Overload and Margin

The exercise of this section is to develop an expression for system noise and discuss methods of reducing it. In Section 9.5.2 we developed a term for the thermal noise for a string of cascaded amplifiers $(P_n + G_r + 10 \log N)$. The next step is to establish 0 dBm as a reference, or more realistically -2.5 dBmp, because we are dealing with a voice channel nominally 3 kHz wide and we want it weighted psophometrically (see Section 1.9.6). Now we can establish a formula for a total thermal noise level as measured at the end of a coaxial cable system with N amplifiers in cascade:

$$P_t = P_n + G_r + 10 \log N - 2.5 \text{ dBmp} \qquad (9.6)$$

As before we assume that all the amplifiers are identical and spaced at equal intervals and that the gain of each is G_r, which is exactly equal to the loss of the intervening cable between each amplifier.

Examining equation (9.6) we see that the operating level is high. The next step is to establish an operating level that should never be exceeded and call it L. A margin to that level must also be established to take into account instability of the amplifiers caused by aging effects, poor maintenance, temperature variations, misalignment, and so forth. The margin to the maximum operating level point is M_g. All units are in decibels. A more realistic equation can now be written for total thermal noise including a suitable margin:

$$P_t = P_n + G_r + 10 \log N - 2.5 \text{ dBmp} + L + M_g \qquad (9.7)$$

These levels are shown graphically in Figure 9.6.

A number of interesting relationships can be developed if we consider a hypothetical example. CCITT permits 3-pWp/km noise accumulation (CCITT Rec. G.222). Allow 2 pWp of that figure to be attributed to thermal noise. If we were to build a system 100 km long, we could then accumulate 200 pWp of the thermal noise. Now set 200 pWp equal to P_t in equation (9.7). First convert 200 pWp to dBmp (-67 dBmp). Thus

$$-67 \text{ dBmp} = P_n + G_r + 10 \log N - 2.5 \text{ dBmp} + L + M_g$$

$$P_r + G_r + 10 \log N + L + M_g = -64.5 \text{ dBmp}$$

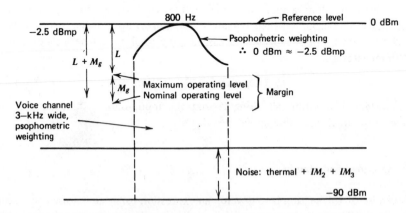

Figure 9.6 Graphic representation of reference level, signal levels, and noise levels in a coaxial cable system (*Note.* levels are not drawn to scale.)

Let us assign some numbers to the equation which are somewhat reasonable. To the 100-km system install 20 repeaters at equal intervals. Cable loss is 5 dB/km, or 500 dB total loss at the highest operating frequency. Thus repeater gain G_r is 25 dB, with $N = 20$. Let

$$L + M = 15 \text{ dB}$$

From Section 9.5.2,

$$P_n = -139 \text{ dBm} + NF$$

Thus

$$-139 + NF + 25 + 10 \log 20 + 15 = -64.5 \text{ dBmp}$$

$$NF = 21 \text{ dB or less}$$

This is a NF that is fairly easy to meet.

Let us examine this exercise a little more closely and see whether we cannot derive some important relationships which can offer the system and amplifier design engineer some useful guidance.

1. By doubling the length of the system, system noise increases 3 dB, or by doubling the number of amplifiers, G_r being held constant, system noise doubles (i.e., $10 \log 2N$).

2. By making the terms L and M_g smaller, or in other words, increasing the maximum operating level, reducing the margin, system thermal noise improves on a decibel for decibel basis.

3. Of course by reducing NF, system noise may also be reduced. But suppose that NF turned out to be very small in the calculations, a figure that could not be met or would imply excessive expense. Then we would have to turn to other terms in the equation, such as reducing terms G_r, L, and M_g. However, there is little room to maneuver with the latter two, 15 dB in the example. That leaves us with G_r. Of course, reducing G_r is at the cost of increasing the number of amplifiers (or increasing the size of the cable to reduce attenuation, etc.). As we reduce G_r, the term $10 \log N$ increases because we are increasing the number of amplifiers N. The trade-off between the term $10 \log N$ and G_r occurs where G_r is between 8 and 9 dB.

Another interesting relationship is that of the attenuation of the cable. It will be noted that the loss in the cable is approximately inversely proportional to the cable diameter. As an example, let us assume that the loss of a cable section between repeaters is 40 dB. By increasing the cable diameter 25%, the loss of the cable section becomes $40/(1 + 0.25) = 32$ dB. In our example above, by increasing the cable diameter, repeater gain may be decreased with the consequent improvement in system noise (thermal). *Note.* The above examples are given as exercises and may not necessarily be practicable owing to economic constraints.

9.5.4 IM Noise

The second type of noise to be considered in coaxial cable system and repeater design is IM noise. IM noise on a multichannel FDM system may be approximated by a Gaussian distribution (see Section 3.4.4) and consists of second; third; and higher order IM products. Included in these products, in the wide-band systems we cover here, are second and third harmonics. IM products (e.g., IM noise) are a function of the nonlinearity of active devices* (see Section 1.9.6).

To follow our argument on IM noise in coaxial cable repeater design, the reader is asked to accept the following (Ref. 2). If a simple sinusoid wave is introduced at the input port of a cable amplifier, the output of the amplifier could be expressed by an equation with three terms, the first of which is linear, representing the desired amplification. The second and third terms are quadratic and cubic representing the nonlinear behavior of the amplifier (i.e., second- and third-order products). On the basis of this power series, for each 1-dB change of fundamental input to the amplifier, the second harmonic changes 2 dB, and the third harmonic 3 dB. Furthermore, for two waves A and B a second-order sum $(A + B)$ or difference $(A - B)$ is equivalent to the second harmonic of A at the output plus 6 dB. Likewise, the sum of $A + B + C$ would be equivalent to the level of the third harmonic at the output plus 15.6 dB of one of the waves. We consider that all inputs are of equal level. The situation for $2A + B$ would be equivalent to $3A + 9.6$ dB, and so forth. These last three power series may be more clearly expressed when set down as follows, where P_H is the harmonic IM power:

$$P_{H(A \pm B)} = P_{H(2A)} + 6 \,\mathrm{dB} \tag{9.8a}$$

$$P_{H(A \pm B \pm C)} = P_{H(3A)} + 15.6 \,\mathrm{dB} \tag{9.8b}$$

$$P_{H(2A \pm B)} = P_{H(3A)} + 9.6 \,\mathrm{dB} \tag{9.8c}$$

For this discussion let IM_2 and IM_3 express the nonlinearity of a repeater; they are, respectively, the power of the second and third harmonics corresponding to a 0 dBm fundamental (-2.5 dBmp). Adjusted to the maximum operating level L (see Figure 9.6), the second harmonic power P_{2A} is

$$P_{2A} = IM_2 + 10 \log N + L \quad (\mathrm{dBmp}) \tag{9.9}$$

L is assumed to be positive as in our argument in the preceding section.

Now decrease the applied signal level to a repeater by L dB, assuming that the power of the fundamental of a wave at the 0 TLP was L dBmp. It follows that, by decreasing the applied power L dB so that the fundamental of a wave is now 0 dBmp at the 0 TLP, the magnitude of the fundamental is decreased L dB at the input of the first amplifier.

---

*IM products may also be produced in passive devices, but to simplify the argument in this chapter, we have chosen to define IM products as those derived in active devices.

For every decrease of 1 dB in the fundamental, the second harmonic decreases by 2 dB. Therefore the new power of the second harmonic amplitude will be decreased by $2L$ dB, or

$$P_{2A} = IM_2 + 10 \log N + L - 2L \quad \text{(for the system)}$$
$$= IM_2 + 10 \log N - L \quad \text{(dBmp)} \tag{9.10}$$

which is the second harmonic noise power level.

Let us consider the $A + B$ product. For fundamentals of equal magnitudes, such a product in a single repeater will be 6 dB higher than a $2A$ product. It also varies by 2 dB per 1 dB variation in both fundamentals and adds in power addition as a function of the number of repeaters. Thus

$$P_{(A \pm B)} = IM_3 + 10 \log N - L + 6 \, dBmp \tag{9.11}$$

Similarly for the $3A$ condition (e.g., third harmonic) of a wave fundamental A,

$$P_{3A} = IM_3 + 10 \log N - 2L \quad \text{(dBmp)} \tag{9.12}$$

For the $A + B - C$ condition,

$$P_{(A + B - C)} = IM_3 + 20 \log N - 2L + 15.6 \, dBmp \tag{9.13}$$

Again we find $2L$ because third-order products vary 3 dB for every 1-dB change in the fundamental. We use 20 log rather than 10 log, assuming that the products add in phase (i.e., voltage-wise) versus the number of repeaters. The 10 log term represented power addition. For the $2A - B$ condition,

$$P_{2A - B} = IM_3 - 2L + 20 \log N + 9.6 \, dBmp \tag{9.14}$$

9.5.5 Total Noise and Its Allocation

In summary there are three noise components to be considered:

- Thermal noise
- Second-order IM noise
- Third-order IM noise

Let us consider them two at a time. If a system is thermal and second-order IM noise limited, minimum noise is achieved allowing an equal contribution. For the 3-pWp/km case we would assign 1.5 pW to each component.

For thermal and third-order IM noise limited systems, twice the contribution is assigned to thermal noise as to third-order IM noise. Again for the 3-pWp case 2 pWp is assigned to thermal noise and 1 pWp to third-order IM noise.

Expressed in decibels with P_t equal to total noise, the following table expresses these relationships in another manner:

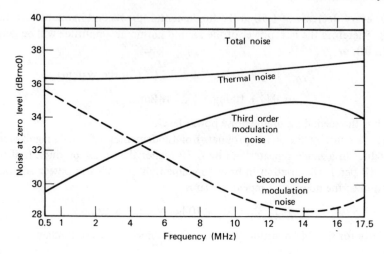

Figure 9.7 Allocation of noise in a practical system. Accumulated noise over 4000 mi of the North American L4 system. (Source: *Bell Sys. Tech. J.*, 830, Apr. 1969; Copyright © 1969 by American Telephone and Telegraph Company.)

	Noise Assigned	
System	to Thermal	to IM
Thermal and second-order limited	$P_t - 3$	$P_t - 3$
Thermal and third-order limited	$P_t - 1.8$	$P_t - 4.8$
Overload limited	P_t	—

The parameter L is established such that these apportionments can be met by adjusting repeater spacing and repeater design. As an example in practice, Figure 9.7 shows noise allocation of the North American L4 system (Ref. 5).

9.6 EQUALIZATION

9.6.1 Introduction

Consider the result of transmitting a signal down a 12-MHz coaxial cable system with the amplitude–frequency response shown in Figure 9.3. The noise per voice channel would vary from an extremely low level for the channels assigned to the very lowest frequency segments of the line frequency (baseband) to extremely high levels for those channels that were assigned to the spectrum near 12 MHz. For long systems there would be every reason to believe that these higher frequencies would be unusable if nothing was done to correct the cable to make the amplitude response more uniform as a function of frequency. Ideally we would wish it to be linear.

Equalization of a cable deals with the means used to assure that the signal-to-noise ratio in each FDM telephone channel is essentially the same no matter what its assignment in the spectrum (see Chapter 3). In the following discussion we consider both fixed and adjustable equalizers.

9.6.2 Fixed Equalizers

The coaxial cable transmission system design engineer has three types of cable equalization available which fall into the category of fixed equalizers:

- Basic equalizers
- Line build-out (LBO) networks
- Design deviation equalizers

The basic equalizer is incorporated in every cable repeater. It is designed to compensate for the variation of the loss frequency characteristic of uniform cable sections. This is done by simply making the fixed gain proportional to the square root of frequency, matching the loss for the nominal length. For the North American L4 system the gain characteristic is shown in Figure 9.8. For the case of 12-MHz cable, the section nominal length would be 4.5 km (CCITT Rec. G.332).

The key word in the preceding paragraph is *uniform*. Unfortunately some cable sections are not uniform in length. It is not economically feasible to build tailor-made repeaters for each nonuniform section. This is what LBO networks are used for. Such devices are another class of fixed equalizer for specific variations of nominal repeater spacing. One way of handling such variations is to have available LBO equalizers for 5, 10, 15%, and so on, of the nominal distance.

The third type of equalizer compensates for design deviations of the nominal characteristics which are standard for dependent repeaters and the actual loss

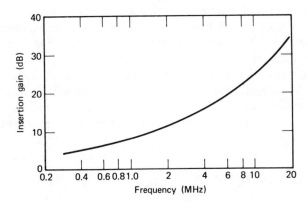

Figure 9.8 Repeater gain characteristic for North American L4 system.

characteristic of the cable system in which the repeaters are to be installed. Such variations are systematic so that the third level of equalization, the design deviation equalizer, is installed one for each 10, 15, or 20 repeaters to compensate for gross design deviations over that group of repeaters.

9.6.3 Variable Equalizers

Figure 9.9a shows the change of loss of cable as a function of temperature variations, and Figure 9.9b shows approximate earth temperature variations with time. Adjustable equalizers are basically concerned with gain frequency variations with time. Besides temperature, variations due to the aging of components may also be a problem. However, this is much less true with transistorized equipment. The cable loss per kilometer shown in Figure 9.3 is the loss at a mean temperature. The $\frac{3}{8}$-in. cable used in the L system application has a mean variation of ±20°F from the nominal (+55°F) buried at a 4-ft depth in the United States. At 20 MHz the loss due to temperature effects is about ±0.38 dB/mi, and it is about ±0.67 dB/mi at 60 MHz (Ref. 2). This loss can be estimated at 0.11%/°F.

The primary purpose of automatic regulation is to compensate for the gain variation due to temperature changes. Such automatic regulation usually is controlled by a pilot tone at the highest cable frequency. For instance, in the ITT cable design for 12 MHz (Ref. 1),

> The pilot controlled system will always apply exact compensation . . . at the pilot frequency of 12,435 kHz, an error may occur at other frequencies. On a single amplifier this error is very small but will add systematically along the route.

Such an error is usually corrected by manually adjustable equalizers.

In the North American L4 system (Ref. 5) the regulation is controlled by a 11,648-kHz line pilot. The gain frequency characteristic is varied to compensate for temperature-associated changes in the loss of a regulated section. Another regulator is controlled by a thermistor buried in the ground near the repeater to monitor ground temperature. This latter regulator provides about half the temperature compensation necessary.

The L4 system also uses additional repeaters called equalizing repeaters, which are spaced up to 54 mi apart. The repeater includes six networks for adjusting the gain frequency characteristics to mop up collective random deviations in the 54-mi section. The equalization of the repeater is done remotely from manned stations while the system is operational.

9.7 LEVEL AND PILOT TONES

Intrasystem levels are fixed by cable and repeater system design. These are the L and M_g established in Section 9.5.

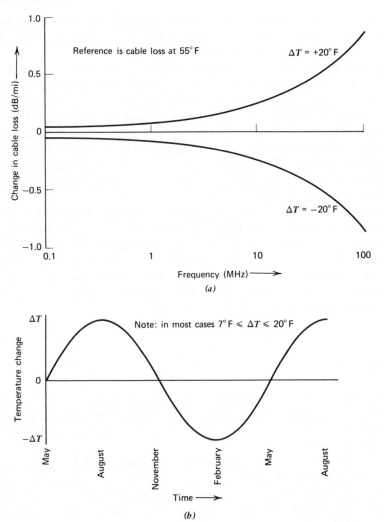

Figure 9.9 (a) Change in loss of 1 mi of $\frac{3}{8}$-in. coaxial cable for $\pm 20°\,$F change in temperature. (b) Approximate earth temperature variations with time at 4-ft depth. (Copyright © 1970 by Bell Telephone Laboratories.)

Modern 12-MHz systems display an overload point of +24 dBm or more.* Remember that

Overload point = equivalent peak power level + relative sending level + margin

The margin is M_g, as in Section 9.5, and L may be related to relative sending level.

M_g can be reduced, depending on how well system regulation is maintained. System pilots, among other functions (covered in Section 9.8), provide a means for

*CCITT Rec. G.223 calls for at least a +20-dBm overload point.

AGC of some or all cable repeaters so as to compensate (partially or entirely; see Section 9.6) for transmission loss deviations due to temperature effects on the system and aging of active components (e.g., repeaters).

Typically pilot levels are -10 dBm0. The level is a compromise, bearing in mind system loading, to minimize the pilots' contribution to IM noise in a system that is multichannel in the frequency domain (FDM). Another factor tending to force the system designer to increase the pilot level is the signal-to-noise ratio of the pilot tone required to effectively actuate level-regulating circuitry. Pilot level adjustment at the injection point usually requires a settability better than 0.1 dB. Internal pilot stability should display a stability improved over the desired cable system level stability. If the system level stability is to be ±1 dB, then the internal pilot stability should be better than ±0.1 dB.

The number of system pilots assigned and their frequencies depend on bandwidth and the specific system design. Commonly 12,435 kHz is used for regulation and 13.5 MHz for supervisory in the L4 system. In the same system an auxiliary pilot is offered at 308 kHz and, as an option, a frequency-comparison pilot at 300 kHz.

The only continuous in-band pilot in the L4 system is located at 11,648 kHz. Supervisory pilot tones are transmitted in the band of 18.50–18.56 MHz. An L multiplex synchronizing pilot is located at 512 kHz.

9.8 SUPERVISORY

The term *supervisory* in coaxial cable system terminology refers to a method of remotely monitoring the repeater condition at some manned location. In the case of the L4 system, 16 pilot tones are brought up on command giving the status of 16 separate buried repeaters.

The ITT method uses a common oscillator frequency (13.5 MHz) and relies on time separation to establish the identity of each repeater being monitored. An interrogation signal is injected at the terminal repeater or other manned station. At the first dependent repeater the signal is filtered off and, after a delay, regenerated and passed on to the next repeater. Simultaneously on the receipt of the regenerated pulse, a switch is closed, connecting a local oscillator signal to the output of the repeater for a short time interval. This local oscillator pulse is transmitted back to the terminal or other manned station. The delay added at each repeater is added to the natural delay of the intervening cable. This added delay allows for a longer return pulse from each repeater, thereby simplifying circuitry. This same interrogating pulse, delayed, regenerated, and then passed on to the next repeater, carries on down the line of dependent repeaters, causing returning "tone bursts" originating from successive amplifiers along the cable route.

The tone burst response pulses are rectified and fed to a counter at the manned station. The resulting count is compared with the expected count, and an alarm is indicated if there is a discrepancy. The faulty amplifier is identified automatically.

Table 9.3 Supervisory System Parameters (ITT)

Response oscillator frequency	13.5 MHz
Response oscillator level	-20 dBm0
Response pulse duration	10 μs
Response pulse delay	250 μs
Interrogation pulse repetition rate	6/s
Interrogation pulse amplitude	$\sim$0.5 V
Interrogation pulse rise time	$\sim$10 μs

Table 9.3 gives basic operating parameters of the system. Such a system can be used for coaxial cable system segments up to 280 km in length.

9.9 POWERING THE SYSTEM

Power feeding of buried repeaters in the ITT system permits the operation of 15 dependent repeaters from each end of a feed point (12-MHz cable). Thus up to 30 dependent repeaters can be supplied power between power feed points. A power feed unit at the power feed point (see Figure 9.4) provides up to 650-V dc voltage between center conductor and ground, using 110-mA stabilized direct current. Power feed points may be as far apart as 140 km (87 mi) on large-diameter cable.

9.10 60-MHz COAXIAL CABLE SYSTEMS

Wideband coaxial cable systems are presently being implemented due to the ever-increasing demand for long-haul toll quality telephone channels. Such systems are designed to carry 10,800* FDM nominal 4-kHz VF channels. The line frequency configuration for such a system, as recommended in CCITT Rec. G.333, is shown in Figure 9.10. To meet long-haul noise objectives the large-diameter cable is recommended (e.g., 2.6/9.5 mm).

When expanding a coaxial cable system, a desirable objective is to use the same repeater locations as with the old cable and add additional repeaters at intervening locations. For instance, if we have 4.5-km spacing for a 12-MHz system and our design shows that we need three times the number of repeaters for an equal-length 60-MHz system, then repeater spacing should be at 1.5-km intervals.

The ITT 12-MHz system uses 4.65-km spacing. Thus its 60-MHz system will use 1.55-km (0.95-mi) spacing with a mean cable temperature of 10°C. The attenuation characteristic of the large-gauge cable is shown in Figure 9.11. This is an extension of Figure 9.3.

Repeater gain for the ITT system is nominally 28.5 dB at 60 MHz and can be

---

*The ATT L5E system is designed for 13,200 VF channels.

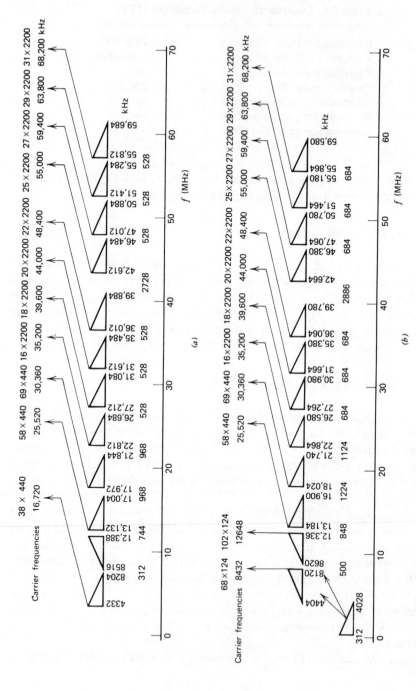

Figure 9.10 Line frequency allocation recommended for 40- and 60-MHz systems on 2.6/9.5-mm coaxial cable pairs. (a) Using plan 1. (b) Using plan 2. (From CCITT Rec. G.333.)

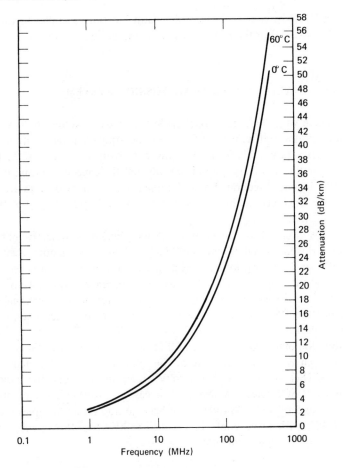

Figure 9.11 Attenuation of large-diameter coaxial cable (0.375 in.).

varied ±1.5 dB. LBO networks allow still greater tolerance. The overload point, following CCITT Rec. 223, is taken at +20 dBm with a transmit level of −18 dBm.

The system pilot frequency is 61.160 MHz for regulation. A second pilot frequency of 4.287 MHz corrects the level of the lower frequency range. Pilot regulation repeaters are installed at from 7 to 10 nonregulated repeaters, with deviation equalization at every twenty-fourth repeater. All repeaters have temperature control (controlled by the buried ambient).

Power feeding is planned at every 100 km (63 mi). Thus 64 repeaters will be fed remotely using constant dc feed over the conductors. Each repeater will tap off about 15 V, requiring 2 W. Thirty-two repeaters at 2 × 15 V each will require 960 V. An additional 120 V dc is required for pilot-regulated repeaters plus one repeater with deviation equalization. Added to this is the 50-V *IR* drop on the cable.

The total feed voltage adds to 1226 V dc. Fault location is similar to that for the 12-MHz ITT system. [For the ITT 60-MHz system, see (Ref. 14).]

9.11   THE L5 COAXIAL CABLE TRANSMISSION SYSTEM

A good example of a 60-MHz coaxial cable transmission system that is presently operational, carrying traffic, is the L5 system operating on a transcontinental route in North America (see Table 9.1). In its present lay-up it consists of 22 tubes, of which 20 are on-line and 2 are spare. Each tube has the capacity to transmit 10,800 VF channels* in one direction. For full-duplex operation, two tubes are required for 10,800 VF channels, or the total system capacity is [(22 – 2)/2] × 10,800 or 108,000 VF channels.

The system is designed for a 40-dBrnC0 (8000-pWp) noise objective in the worst VF channel at the end of a 4000-mi (6400-km) system. This is a noise accumulation of 8000/6400 or 1.25 pWp/km. The system design is such that it is second-order modulation and thermal noise limited. Repeater overload is at the +24-dBm point.

Compared to the L4 system, L5 provides three times the voice channel capacity at three times the L4 spectrum, twice the repeaters with a small deterioration in noise accumulation.

The modulation plan is an extension of that shown in Chapter 3, Figure 3.21. The basis of the plan is the development of the "jumbo group" (JG) made up of six master groups (Bell System FDM hierarchy). Keep in mind that the basic master-group consists of 600 voice channels (in this case) or 10 standard supergroups and occupies the band of 564–3084 kHz. The basic jumbo group occupies the band of 564–17,548 kHz with a level control pilot at 5888 kHz. The three jumbo groups are assigned the following line frequencies:

JG 1   3,124–20,108 kHz
JG 2   22,068–39,056 kHz
JG 3   43,572–60,556 kHz

Equalizing pilots are at 2976, 20,992, and 66,048 kHz as transmitted to the line. There is a temperature pilot at 42,880 kHz.

The basic jumbo group frequency generator is built around an oscillator which has an output of 5.12 MHz. This oscillator has a drift rate of less than 1 part in 10^{10} per day after aging and a short-term stability of better than 1 part in 10^{8} per millisecond. Excessive frequency offset is indicated by an alarm.

Automatic protection of the 10 operating systems is afforded by the line protec-

* 13,200 VF channels in the case of L5E.

tion switching system (LPSS) on a 1 : 10 basis. The maximum length of switching span is 150 mi. Power feeds are at 150-mi intervals, feeding power in both directions. Thus a power span is 75 mi long, or 75 repeaters. Power is 910 mA on each cable, ±1150 V operating against ground.

The basic repeater is a fixed-gain amplifier, spaced at 1-mi intervals. Typically every fifth repeater is a regulating repeater, and this regulation is primarily for temperature compensation.

9.12 COAXIAL CABLE OR RADIOLINK—THE DECISION

9.12.1 General

One major decision that the transmission system engineer often faces is whether to install on a particular point-to-point circuit a radiolink or coaxial cable.

What are the factors that will determine the choice? Most obviously they fall into two categories, technical and economic. Table 9.4 compares the two media from a technical viewpoint. These comparisons can serve as a fundamental guide for making a technical recommendation in the selection of a facility. System mixes may also be of interest. (Refer to Chapter 5 for a discussion of radiolink engineering.)

The discussion that follows is an expansion of some of the points covered in Section 9.1. Table 9.4 summarizes the factors listed below.

9.12.2 Land Acquisition as a Limitation to Coaxial Cable Systems

Acquisition of land detracts more from the attractiveness of the use of coaxial cable than any other consideration; it adds equally to the attractiveness of selecting radiolinks (LOS microwave). With a radiolink system large land areas are jumped and the system engineer is not concerned with what goes on between. One danger which many engineers tend to overlook is that of the chance building of a structure in the path of the radio beam after installation on the routes has been completed.

Cable, on the other hand, must physically traverse the land area that intervenes. Access is necessary after the cable is laid, particularly at repeater locations. This may not be as difficult as it first appears. One method is to follow parallel to public highways, keeping the cable lay on public land. Otherwise, with a good public relations campaign, easement or rights-of-way often are not hard to get.

This leads to another point. The radiolink relay sites are fenced. Cable lays are marked, but the chances of damage by the farmer's plow or construction activity are fairly high with the cable alternative.

Table 9.4 Comparison of Coaxial Cable Versus Radio Link

Item	Cable	Radio
Land acquisition	Requires land easements or right-of-way along entire route and recurring maintenance access later	Repeater site acquisition every 30–50 km with building, tower, access road at each site
Insert and drop	Insert and drop at any repeater. Should be kept to minimum. Land buys, building required at each insert location	Insert and drop at more widely spaced repeater sites
Fading	None aside from temperature variations	Important engineering parameter
Noise accumulation	Less, 1 pWp/km	More, 3 pWp/km
Radio frequency interference (RFI)	None	A major consideration
Limitation on number of carriers or basebands transmitted	None	Strict, band-limited plus RFI ambient limitations
Repeater spacing	1.5, 4.5, 9 km	30–50 km
Comparative cost of repeaters	Considerably lower	Considerably higher
Power considerations	High voltage dc in milli-ampere range	48 V dc static no-break at each site in ampere range
Cost versus traffic load	Full load proves more economical than radiolink	Less load proves more economical than cable
Multiplex	FDM-CCITT	FDM-CCITT
Maintenance and engineering	Lower level, lower cost	Higher level, higher cost
Terrain	Important consideration in cable laying	Can jump over, even take advantage of difficult terrain

9.12.3 Fading

Radiolinks are susceptible to fading. Fades of 40 dB on long hops are not unknown. Overbuilding a radiolink system tends to keep the effect of fades on system noise within specified limits.

On coaxial cable systems signal level variation is mainly a function of temperature variation. Level variations are well maintained by regulators controlled by pilot tones and, in some cases, auxiliary regulators controlled by ground ambient.

9.12.4 Noise Accumulation

Noise accumulation has been discussed in Section 9.4. Either system will serve for long-haul backbone routes and meet the minimum specific noise criteria established by CCITT/CCIR. However, in practice the engineering and installation of a radio system may require more thought and care to meet those noise requirements. Modern coaxial cable systems have a design target of 1 pWp/km of noise accumulation. Applying good design techniques and using IF repeaters, radio systems can meet the 3-pWp criterion. Besides fading, radiolinks, by definition, have more modulation steps and thus are noisier.

9.12.5 Group Delay—Attenuation Distortion

Group delay is less of a problem with radiolinks. Figure 9.3 shows the amplitude response of a cable section before amplitude equalization. The cable plus amplifiers plus amplitude equalizers add to the group delay problem.

It should be noted that for video transmission on cable an additional modulation step is required to translate the video to the higher frequencies and invert the band (see Section 12.8.2). While on radiolinks, video can be transmitted directly without additional translation or inversion besides RF modulation.

9.12.6 Radio Frequency Interference (RFI)

There is no question as to the attractiveness of broadband buried coaxial cable systems over equivalent radiolink systems when the area to be traversed by the transmission medium is one of dense RFI. Usually these areas are built-up metropolitan areas with high industrial/commercial activity. Unfortunately, as a transmission route enters a dense RFI area, land values increase disproportionately, as do construction costs for cable laying. Yet the trade-off is there.

9.12.7 Maximum VF Channel Capacity

In heavily populated areas of highly developed nations frequency assignments are becoming severely limited or unavailable. Although some of the burden on assignment will be removed as the tendency toward usage of the millimeter region of the spectrum is increased, coaxial cable remains the most attractive of the two for high-density FDM configurations.

If it is assumed that there are no RFI of frequency assignment problems, a radiolink can accommodate up to eight carriers in each direction (CCIR Rec. 384-1) with

2700 VF channels per carrier. Thus the maximum capacity of such a system is 8 ×
2700 = 21,600 VF channels.

Assume a 12-MHz coaxial cable system with 22 tubes, 20 operative, that is, 10
"go" and 10 "return." Each coaxial tube has a capacity of 2700 channels. Thus the
maximum capacity is 10 × 2700 = 27,000 VF channels.

It should be noted that the radio system with a full 2700 VF channels may suffer
from some multipath problems. Coaxial cable systems have no similar interference
problems. However, cable impedance must be controlled carefully when splicing
cable sections. Such splices usually are carried out at repeater locations.

Consider now 60-MHz cable systems with 20 active tubes, 10 "go" and 10 "return."
Assume a 10,000-channel capacity per tube; thus 10,000 × 10 = 100,000 VF channels, or the equivalent to five full radiolink systems.

9.12.8 Repeater Spacing

As discussed in Chapter 5, a high average for radiolink repeater spacing is 50 km (30
mi), depending on drop and insert requirements as well as an economic trade-off
between tower height and hop distance. For coaxial cable systems, repeater separation depends on the highest frequency to be transmitted, ranging from 9.0 km for
4-MHz systems to 1.5 km for 60-MHz systems. A radiolink repeater is much more
complex than a cable repeater.

Coaxial cable repeaters are much cheaper than radiolink repeaters, considering
tower, land, and access roads for radiolinks. However, much of this advantage for
coaxial cables is offset because radiolinks require many fewer repeaters. It also
should be kept in mind that a radiolink system is more adaptable to difficult
terrain.

9.12.9 Power Considerations

The 12-MHz ITT coaxial cable system can have power feed points separated by as
much as 140 km (87 mi) using 650 V dc at 150 mA. In a 140-km section of a radiolink route at least four power feed points would be required, one at each repeater
site. About 2 A is required for each transmitter–receiver combination using standard
48 V dc battery, usually with static no-break power. Power also will be required for
tower lights and perhaps for climatizing equipment enclosures.

9.12.10 Engineering and Maintenance

Cable systems are of "cook-book" design. Radiolink systems require a greater
engineering effort prior to and during installation. Likewise, the level of maintenance
of radio systems is higher than that for cable.

9.12.11 Multiplex Modulation Plans

Interworking or tandem working of radio and coaxial cable systems is made easier because both broadband media use the same standard CCITT or L system modulation plans (see Chapter 3).

REFERENCES AND BIBLIOGRAPHY

1. P. J. Howard, M. F. Alarcon, and S. Tronsli, "12-Megahertz Line Equipment," *Electrical Commun.* (ITT), Vol. 48, No. 1/2, 1973.
2. *Transmission Systems for Communications*, 4th ed., Bell Telephone Laboratories, American Telephone and Telegraph Co., New York, 1971.
3. W. A. Rheinfelder, *CATV System Engineering*, TAB Books, Blue Ridge Summit, PA, 1970.
4. P. Norman and P. J. Howard, "Coaxial Cable System for 2700 Circuits," *Electrical Commun.* (ITT), Vol. 42, No. 4, 1967.
5. "The L-4 Coaxial System," *Bell Sys. Tech. J.*, Vol. 48, Apr. 1969.
6. CCITT Orange Books, Geneva, 1976, Vol. III, G recommendations, particularly the G.200 and G.300 series (see Appendix A).
7. J. A. Lawlor, *Coaxial Cable Communication Systems, Management Overview*, Tech. Mem., ITT, New York, Feb. 1972.
8. *Lenkurt Demodulator*, Lenkurt Electric Corp., San Carlos, CA, June 1967, May and June 1970, and May 1971.
9. *Data Handbook for Radio Engineers*, 6th ed., Howard W. Sams, Indianapolis, IN, 1977.
10. F. J. Herr, "The L5 Coaxial System Transmission System Analysis," *IEEE Trans. Commun.*, Feb. 1974.
11. F. C. Kelcourse and T. A. Tarbox, "Design of Repeatered Lines for Long-Haul Coaxial Systems," *IEEE Trans. Commun.*, Feb. 1974.
12. E. H. Angell and M. M. Luniewicz, "Low Noise Ultralinear Line Repeaters for the L5 Coaxial System," *IEEE Trans. Commun.*, Feb. 1974.
13. Y.-S. Cho *et al.*, "Static and Dynamic Equalization of the L5 Repeatered Line," *IEEE Trans. Commun.*, Feb. 1974.
14. L. Becker, "60-Megahertz Line Equipment," *Electrical Commun.* (ITT), Vol. 48, No. 1/2, 1973.
15. *Telecommunications Transmission Engineering*, Vol. 2, 2nd ed., American Telephone and Telegraph Co., New York, 1977.
16. R. L. Freeman, *Telecommunication System Engineering*, Wiley-Interscience, New York, 1980.

10.1 GENERAL

Denominating frequency bands for radio communication is arbitrary. The microwave LOS or centimeter band discussed in Chapter 5 covered the spectrum from 150 MHz (200 cm) to 13 GHz (2.3 cm). For the sake of discussion in this chapter, millimeter waves will encompass that region of the electromagnetic spectrum from 13 GHz (23 mm) to over 100 GHz (3 mm).*

One major concern of the radio transmission engineer is propagation, and this is the primary topic throughout this chapter. Millimeter wave transmission through the atmosphere is more adversely affected by certain propagation properties than its centimeter counterpart. These properties are the absorption and scattering of a wave as it is transmitted through the atmosphere. The result of this phenomenon is the principal reason why the transmission engineer is reluctant to use these frequencies for point-to-point communication. Much emphasis will be given to rainfall attenuation; it should be noted that there was some discussion of fading due to rainfall in Chapter 5.

With all its apparent shortcomings, millimeter wave transmission has been given a second look in the last 15 years for various reasons. Probably the primary reason is the increasing congestion in the centimetric wave bands in certain areas of the world. Another consideration is the need for much greater bandwidths (Chapter 11) to accommodate digital transmission or spread spectrum waveforms. The third factor is that of equipment development; again much of this development is a result of military research and technology. Development of millimeter wave transmission has reached a point equivalent to about where centimetric wave development was in the late 1950s, when that region of the spectrum was opened for wide usage by multichannel, point-to-point transmission.

*The ITU designates 3–30 GHz as centimetric waves and 30–300 GHz as millimetric waves (Ref. 1).

10.2 PROPAGATION

Consider that we are dealing with propagation of millimetric waves through the atmosphere. Therefore, in addition to free-space attenuation, several other factors affecting path loss may have to be considered. As expressed in Ref. 2, these are as follows:

(a) The gaseous contribution of the homogeneous atmosphere due to resonant and nonresonant polarization mechanisms.

(b) The contribution of inhomogeneities in the atmosphere.

(c) The particulate contributions due to rain, fog, mist, and haze (dust, smoke and salt particles in the air).

Figure 10.1 illustrates these properties diagrammatically. Under (a) above we are dealing with the propagation of a wave through the atmosphere under the influence of several molecular resonances, such as water vapor at 22 and 183 GHz, oxygen with lines around 60 GHz, and a single oxygen line at 119 GHz. These points with their relative attenuation are shown in Figure 10.2.

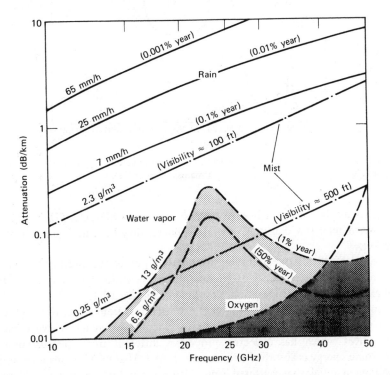

Figure 10.1 Attenuation due to rain, mist, water vapor, and oxygen for a temperate maritime climate at ground level (Ref. 2; Courtesy Institute of Telecommunication Sciences, Office of Telecommunications, U.S. Department of Commerce.)

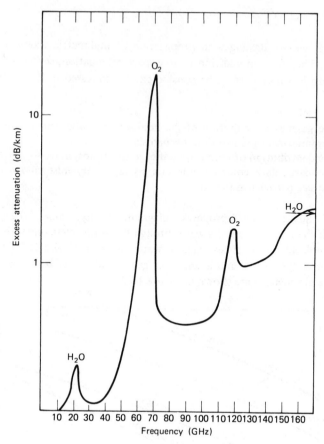

Figure 10.2 Areas of the millimetric spectrum of high attenuation due to oxygen and water vapor. Values are excess attenuation and are estimates (after Ref. 3).

Other gases display resonant lines as well, such as N_2O, SO_2, O_3, NO_2, and NH_3, but because of their low density in the atmosphere, they have negligible effect on propagation.

The major offender is precipitation attenuation [under (b) and (c) above]. It exceeds that of all other sources of attenuation in the atmosphere above 18 GHz. Rainfall and its effect on propagation is covered in Section 10.3.

The immediate concern in this section is the total loss from absorption and scattering. It will be remembered that when an incident electromagnetic wave passes over an object which has dielectric properties different from the surrounding medium, some energy is absorbed and some scattered. That which is absorbed heats the absorbing material; that scattered is quasi-isotropic and relates to the wavelength of the incident wave. The smaller the scatterer, the more isotropic it is in direction with respect to the wavelength of the incident energy.

The ideal for the transmission system engineer would be to establish a formula valid anywhere in the world which would provide path loss in decibels. In free space such a formula is available [equation (5.1)]. This classic formula is

$$\text{Attenuation (dB)} = 32.44 + 20 \log f + 20 \log D$$

where D = hop or path length (km) and f = operating frequency (MHz).

From the strictly simplified engineering point of view it can be seen that milli-metric wave propagation is ideal in free space for communication between satellites, from satellite to deep space probe, and so on. Of course factors f and D must be considered, but no other variables enter the calculation.

For millimetric transmission through the atmosphere the problem takes on con-siderably greater complexity. To the free-space loss formula (with f in GHz) several terms are added:

$$\text{Attenuation (dB)} = 92.45 + 20 \log F_{\text{GHz}} + 20 \log D_{\text{km}} + a + b + c + d + e$$

$$(10.1)$$

where a = loss (dB) due to water vapor
$\quad b$ = loss (dB) due to mist and fog
$\quad c$ = loss (dB) due to oxygen (O_2)
$\quad d$ = sum of the absorption losses due to other gases
$\quad e$ = losses (dB) due to rainfall

Notes

1. a varies with relative humidity, temperature, pressure, and altitude. The trans-mission engineer assumes that the water vapor content is linear with these param-eters and that the atmosphere is homogeneous (actually horizontally homogeneous but vertically stratified). There is a water vapor absorption band at about 22 GHz caused by molecular resonance.

2. c and d are assumed to vary linearly with atmospheric density, thus directly with atmospheric pressure, and are also a function of altitude (e.g., it is assumed that the atmosphere is homogeneous).

3. b and e vary with the density of the rainfall cell or cloud and the size of raindrops or water particles such as fog or mist. In this case the atmosphere is most certainly not homogeneous. (Droplets less than 0.01 cm in diameter are considered mist/fog, more than 0.01 cm, rain). Ordinary fog produces about 0.1 dB/km excess attenuation at 35 GHz, rising to 0.6 dB/km at 75 GHz.

With reference to Figure 10.2, terms a, b, c, and d can be disregarded except in regions indicated in the figure. However, it is just these "forbidden" regions of the electromagnetic spectrum that, in themselves, may offer an advantage under special circumstances. Suppose that one were designing a ship-to-ship secure communication system. Attenuation like that encountered in the region of 60 GHz, the oxygen line area, adds excess attenuation on the order of 15 dB/km to the free-space loss. That

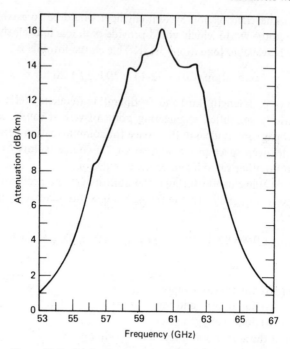

Figure 10.3 Sea-level oxygen attenuation (Ref. 22).

is 15 dB/km of additional security. Such additional attenuation would help ensure that overshoot of a concentrated beam would not be successfully intercepted by an enemy. This shows that good advantage can be taken of even the so-called forbidden regions in millimeter transmission, in this case, the factor c in equation (10.1). Another example is the satellite crosslinks of the U.S. strategic satellite system (StratSat) where the operational frequency band is 59–62 GHz to minimize the effects of earth-based jammers. Crosslinks are those radiolinks interconnecting satellites. As the reader can see from Figure 10.3, these links are centered exactly on the oxygen attenuation band.

Figure 10.2 gives total gaseous absorption in decibels per kilometer as a function of frequency. The comparatively high absorption bands of oxygen and water vapor are patently evident. As can be seen from the figure, certain frequency bands are relatively "open." These openings are often called windows. Three such windows are suggested for point-to-point service as follows:

| | Excess Attenuation by |
Band (GHz)	Atmospheric Absorption (dB/km)
28–42	<0.13
75–95	<0.4
125–140	<1.8

10.3 RAINFALL LOSS

10.3.1 Basic Rainfall Considerations

Of the factors a through e in equation (10.1), factor e, excess attenuation due to rainfall, is the principal one affecting path loss, provided care is used in path analysis in the area of the forbidden regions. For instance, even at 22 GHz (1.35 cm), the water vapor line, excess attenuation accumulates at only 0.15 dB/km, and for a 10-km path only 1.5 dB must be added to the free-space loss to compensate for the water vapor loss. This is negligible when compared to free-space loss itself, such as 128.4 dB for the first kilometer, accumulating thence approximately 6 dB each time the path length is doubled (i.e., add 6 dB for 2 km, 12 dB for 4 km, etc.). Thus a 10-km path would have a free-space loss of 148.4 dB plus 1.5 dB added for water vapor loss (22 GHz), or a total of 149.9 dB.

Rain is another matter. It has been common practice to express path loss due to rain as a function of the precipitation rate. Such a rate depends on the liquid water content and the fall velocity of the drops. The velocity, in turn, depends on raindrop size. Thus our interest in rainfall boils down to drop size and drop size distribution for point rainfall rates. All this information is designed to lead the transmission engineer to fix an excess attenuation due to rainfall on a particular path as a function of time and time distribution. This is a familiar approach, which was used in Chapter 5 for fading.

One source (Ref. 3) suggests limiting path length for millimetric transmission to 1 km. Further on we will consider this aspect for system design. However, such a measure may be overly severe and very costly. Another approach would be to get more rainfall information and its effect on higher frequency transmission systems as a function of wavelength. These should then be applied to good statistics available for the area in which a millimeter wave transmission system is to be installed.

Research carried out so far usually has dealt with rain on a basis of rainfall in millimeters per hour. Often this has been done with rain gauges using collected rain averaging over a day or even periods of days. In millimeter path design such statistics are not sufficient for paths where we desire a propagation reliability well over 99.9% and do not wish to resort to overconservative design procedures (i.e., assign excessive fade margins).

As pointed out in Chapter 5, several weeks of light drizzle will affect the overall long-term propagation reliability much less than several good downpours that are short-lived (20 min duration, for instance). It is simply this downpour activity for which we need statistics. Such downpours are cellular in nature. How big are the cells? What is the rainfall rate in the cell? What are the size of drops and their distribution?

Hogg (Ref. 3) suggests the use of new high-speed rain gauges with outputs readily available for computer analysis. These gauges can provide minute-by-minute analysis of the rate of fall, something lacking with the older type of gauges. Of course it would be desirable to have several years of statistics for a specific path to provide

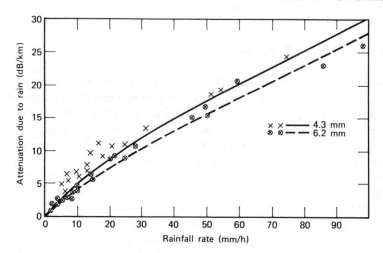

Figure 10.4 Measurements made by Bell Telephone Laboratories of attenuation due to rainfall at wavelengths of 6.2 and 4.3 mm compared with calculated values (Ref. 3; Copyright © 1968 by American Telephone and Telegraph Company.)

the necessary information on fading which will govern system parameters such as repeater spacing, antenna size, diversity, and so forth.

Some such information is now available and is indicative of a great variation of short-term rainfall rates from one geographical location to another. For instance, in one period of measurement it was found that Miami, FL, has maximum rain rates about 20 times greater than those of heavy showers occurring in Oregon, the region of heaviest rainfall in the United States. In Miami a point rainfall rate may even exceed 700 mm/h. The effect of 700 mm/h on two millimetric frequencies can be extrapolated from Figure 10.4. In this figure the rainfall rate in millimeters per hour extends to 100, which at 100 mm/h provides an excess attenuation of from 25 to 30 dB/km.

When identical systems were compared (Ref. 3) at 30 GHz with repeater spacings of 1 km and equal desired signal levels (e.g., a 30-dB signal-to-noise ratio), 140 min of total time below the desired level was obtained at Miami, FL, 13 min at Coweeta, NC, 4 min at Island Beach, NJ, 0.5 min at Bedford, England, and less than 0.5 min at Corvallis, OR. As pointed out, such outages may be improved by increasing transmitter output power, improving receiver NF, increasing antenna size, using diversity, and so on.

One valid method suggested to lengthen repeater sections (space between repeaters) is by the use of path diversity. This is the most effective form of diversity for downpour rainfall fading. Path diversity is the simultaneous transmission of the same information on paths separated by at least 6 km, the idea being that rain cells affecting one path will have a low probability of affecting the other at the same time. A switch would select the better path of the two. Careful phase equalization between the two paths would be required, particularly for the transmission of high-bit-rate data information. (see Section 10.7.3).

10.3.2 Calculation of Excess Path Attenuation Due to Rainfall

When designing millimetric radiolinks (or satellite links), certainly the greatest problem in link engineering is to determine the excess path attenuation due to rainfall. The adjective "excess" is used to denote path attenuation in *excess* of free-space loss [i.e., the terms a, b, c, d, and e in equation (10.1) are in excess of free-space attenuation (loss)].

Before treating the methodology of calculation of excess rain attenuation, we will review some general link engineering information dealing with rain. When discussing rainfall below, all measurements are in millimeters per hour of rain and are point-rainfall measurements. From our previous discussion we know that heavy downpour rain is the most seriously damaging to millimetric propagation. Such rain is cellular in nature and has limited coverage. We must address the question whether the entire hop is in the storm for the whole period of the storm. Light rainfall (less than 2 mm/h), on the other hand, is usually widespread in character, and the path average is the same as the local value. Heavier rain occurs in convective storm cells which are typically 2–6 km across and are often imbedded in larger regions measured in tens of kilometers (Ref. 24). Thus for short hops (2–6 km) the path-averaged rainfall rate will be the same as the local rate, but for longer paths it will be reduced by the ratio of the path length on which it is raining to the total path length.

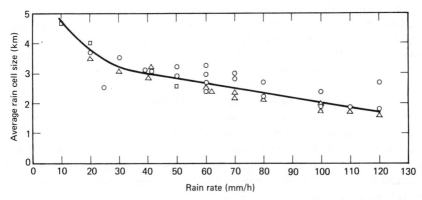

Figure 10.5 Average rain cell size as a function of rain rate, (CCIR Rep. 563-1; Courtesy ITU-CCIR.)
○ Derived from attenuation measurements carried out in Japan (1968, 1969, 1970): climate 2
▲ Derived from attenuation measurements carried out in Malaysia (1971, 1972): climate 1
□ Derived from weather radar observations carried out in Switzerland (1973): climate 3
● Derived from rain gauge observations in France (1977): climate 3
The measurements carried out in Japan and Malaysia are indirect, and are therefore subject to caution.

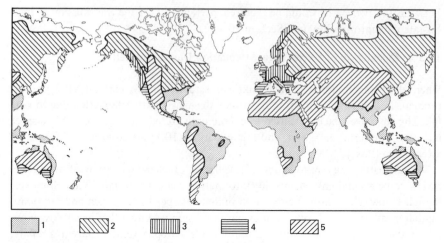

Figure 10.6 Regions corresponding to rainfall rate distributions (CCIR Rep. 563-1; Courtesy ITU–CCIR.)

This concept is further expanded upon by CCIR Rep. 563-1 where rain cell size is related to rainfall rate. This is shown in Figure 10.5. The reference climate zones in the figure are detailed in Figure 10.6. This concept of raincell size is very important, whether engineering a millimeter terrestrial link or a satellite link, particularly when it is of low elevation angle. CCIR Rep. 338-3 is quoted in part below:

> Measurements in the United Kingdom over a period of two years . . . at 11, 20, and 37 GHz on links of 4–22 km in length show that the attenuation due to rain and multipath, which is exceeded for 0.01% of the time and less, increased rapidly with path length up to 10 km, but a further increase up to 22 km produced only a small additional effect . . .

It is also interesting to note (CCIR Rep. 338-3) that horizontally polarized millimeter radio signals are attenuated considerably more by rain than are vertically polarized signals. In France it was found that the ratio of vertical to horizontal polarization attenuation at 13 GHz averages 0.75. In Italy the difference in attenuation increased as the attenuation increased and reached values of about 15 dB when an attenuation of 40 dB was measured in the vertical polarization, which substantially agrees with the French data.

One of the most accepted methods of dealing with excess path attenuation A due to rainfall is an empirical procedure based on the approximate relation between A and the rain rate R,

$$A = aR^b \tag{10.2}$$

where a and b are functions of frequency f and the rain temperature T.

Allowing a rain temperature of $0°C$, as suggested in (Ref. 25), Table 10.1 gives

Table 10.1 Functions of $a(f)$ and $b(f)$.

Sources of Fitted Curve Points	30 GHz		40 GHz		50 GHz		60 GHz		70 GHz	
	$a(f)$	$b(f)$	$a(f)$	$b(f)$	$a(f)$	$b(f)$	$a(f)$	$b(f)$	$a(f)$	$b(f)$
De Bettencourt										
Fitted curve	0.25	0.95	0.4	0.88	0.6	0.82	0.82	0.77	1.1	0.73
Data range	0.3–0.5	0.8–1.0	0.18–0.3	1–1.1					0.8–1.3	0.65–0.78
Atlas	0.16	1.05	0.30	0.98			0.64	0.87		
Olsen[a]										
LP drop size	0.16	1.06	0.31	0.98	0.49	0.91	0.66	0.85	0.80	0.81
MP drop size	0.19	1.04	0.36	0.97	0.58	0.91	0.80	0.85	1.0	0.81
JT drop size	0.27	0.82	0.45	0.76	0.63	0.71	0.80	0.68	0.83	0.66
Crane	0.16	1.07	0.30	1.03	0.46	0.98	0.60	0.94	0.73	0.89
Selected average[b]	0.16		0.30		0.47		0.63		0.77	

Source: From (Ref. 22).

[a] LP = Laws–Parsons; MP = Marshall–Palmer; JT = Joss thunderstorm.

[b] Average of Atlas, Olsen LP, and Crane values.

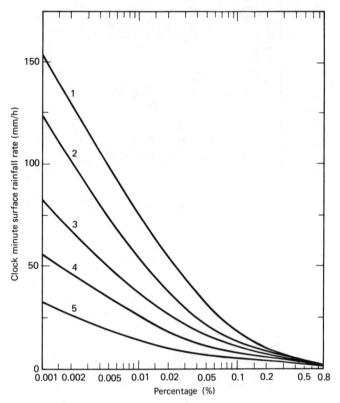

Figure 10.7 Percentage of an average year for which the rainfall rate is exceeded for the five rain climates in Figure 10.6. (CCIR Rep. 563-1; Courtesy ITU–CCIR.)

guidance on values of a and b as functions of frequency. The Laws–Parsons raindrop size seems to provide the more accurate results (Ref. 25).

As an example consider a 10-km path at 30 GHz in West Germany. From Table 10.1 use Laws–Parsons values $a = 0.16$ and $b = 1.06$. Turn now to Figure 10.7, where rainfall rates R are given for the five climatic regions shown in Figure 10.6. This is region 3, and assume we design for 99.99% path reliability (due to rainfall) and $R = 35$ mm/h. Then $A = 0.16 \times 35^{1.06} = 6.93$ dB/km. From Figure 10.5 we see that the rain cell size is about 3.2 km in extension. Thus the excess attenuation due to rainfall would be 3.2 × 6.93, or 22.2 dB. This value is then added to the free-space loss. It will be noted that the 6.93-dB/km value fairly well agrees with that derived from Figure 10.8, a nomogram which may be used for estimating excess attenuation due to rain.

For a treatment of excess rainfall attenuation on a satellite link operating in the millimeter spectrum we recommend using the rain model developed by Crane (Refs. 23, 24, 26). There is considerable similarity in approach between the Crane model and the CCIR approach. Crane uses eight climatic regions, whereas CCIR uses five.

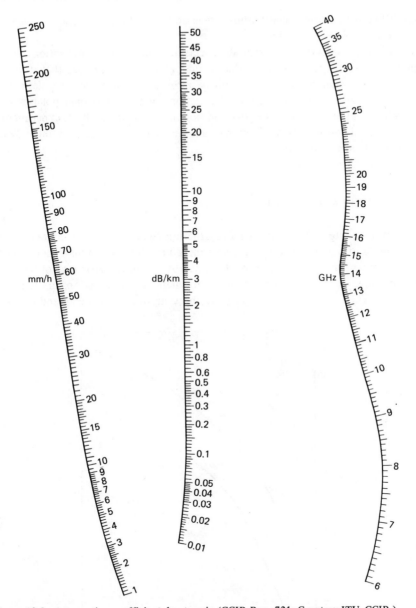

Figure 10.8 Attenuation coefficient due to rain (CCIR Rep. 721; Courtesy ITU–CCIR.)
Raindrop size distributions (Laws and Parsons, 1943)
Terminal velocity of raindrops (Gunn and Kinzer, 1949)
Index of refraction of water at 18°C (Ray, 1972)

Both utilize the $A = aR^b$ relationship. Crane adds a latitude relationship as well as an elevation angle relationship.

A further consideration on rainfall effects on satellite communications is the depolarization of a received signal due to rain. (Ref. 27) points out that it would appear that circular polarization would be more attractive to avoid polarization tracking and also to produce equal attenuations in the orthogonal polarizations (i.e., left-hand polarized versus right-hand polarized). However, both experimental and theoretical considerations show that rain-induced depolarization is much worse for circularly polarized waves.

10.4 SCINTILLATION FADING

Scintillation fading is short-term signal fluctuation. Short term in this context means in the period of 1 min. We might observe from 20 to 50 scintillations or fluctuations in such a time frame. They are brought about by localized departures of temperature, pressure, and homogeneity of the atmosphere causing variations in the refractive index. These inhomogeneities degrade the transmission of broadband informa-

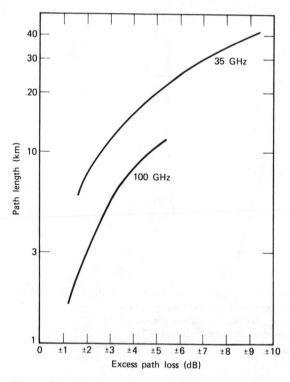

Figure 10.9 Scintillation fading values for 35 and 100 GHz (Ref. 6).

tion by limiting the useful gain at the receiving station because the coherent phase front will disperse, and short-term refractivity fluctuations limit the useful signal bandwidth that can be transmitted because of the time delays introduced.

Scintillation fading is expressed in decibels about the median signal level. It is a function of frequency and path length. Figure 10.9 gives values of scintillation fading for 35 and 100 GHz. Admittedly for most applications these values are on the conservative side.

10.5 PRACTICAL MILLIMETER WAVE SYSTEMS

For point-to-point multichannel communications the same radiolink principles will be used as set forth in Chapter 5, namely, heterodyne receivers, some form of angle modulation, low-level outputs, and parabolic antennas. Wider bandwidths will be used, the 70-MHz standard IF being replaced by a higher frequency. Both 300 and 700 MHz are now being employed.

The millimeter wave region shows great promise for the use of digital transmission techniques utilizing a number of the advantages that the region provides. Besides its being a region that is wide open for development and that provides greater bandwidths, the greater path attenuation and even the absorption attenuation can be put to advantage. This greater loss can reduce mutual interference problems of overshoot, side lobe radiation, and a general reduction of the effectiveness of RFI. This is another reason why millimeter wave links can be installed today in large metropolitan or built-up industrial areas where there are heavily congested conventional centimeter wave transmission systems. Once away from the built-up areas, the system can interface at IF with conventional centimeter systems in areas where RFI problems are fewer in the more popular frequency bands.

Millimeter wave antennas will get more gain for less cross-sectional area, beamwidths will be sharper, and with the improved operation of millimeter transmission systems regarding RFI in general, more of these systems will be able to operate "in harmony" in a given geographical area than their centimeter counterparts. Some CATV radiolinks are now operating at 18 GHz.

The obvious problem in long-haul point-to-point radio systems operating in the conventional analog mode is the accumulation of noise. As proposed in Section 10.3, millimeter wave hops will be much shorter. Thus many more repeaters will be required for a given route than for comparable centimeter radiolinks. To avoid excessive noise buildup, one answer is to use digital transmission techniques. PCM is proposed using M-ary (multilevel) modulation such as differential QPSK or eight-level PSK. The regeneration implicit at each repeater obviates much of the noise problem. This transmission mode will be further discussed below.

Millimeter radio is a galloping technology. Consider the following performance parameters for the noise temperature of 20-GHz low-noise amplifiers with similar bandwidths in 1978 versus 1985 (forecast) (Ref. 28):

| | Performance (K) | |
Amplifier	1978	1985 (forecast)
Parametric (uncooled)	160	100
Parametric (cooled)	100	70
Mixer (uncooled)	530	300
Mixer (cooled)	230	140
GaAs FET (uncooled)	630	300
GaAs FET (cooled)	360	120

FET amplifiers are now demonstrating bandwidths of better than 10%. At 20 GHz 2-GHz bandwidth systems are now possible.

At 44 GHz 250-W TWT amplifiers are now in production. GaAs FET power amplifiers at 30 GHz are now available with up to 0.5-W power outputs. IMPATT devices display 25 watts output at 30 GHz. The U.S. Institute of Telecommunication Sciences in Boulder, CO, is now operating an experimental link at 95 GHz, and other similar experiments are reported from Europe.

An example calculation of a 10-km (6-mi) path follows. Free-space loss may be taken from Table 10.2. We assume an operating frequency of 40 GHz. Given this frequency and path length, the free-space path loss is 144.5 dB. To this figure we must add 10×0.13 dB for gas absorption (see Figure 10.2 and Section 10.2). Assume a 100-MHz bandwidth B_{IF}, and we will leave aside for the present the type of modulation used. Allow a 10-dB NF for the receiver. With this information, compute the receiver noise threshold:

$$\text{Noise threshold (dBW)} = -204 + 10 + 80$$

$$= -114 \text{ dBW}$$

$$\text{Path loss} = \text{free-space loss} + \text{excess loss (absorption)}$$

$$= 144.5 + 1.3 \text{ dB}$$

$$= 145.8 \text{ dB}$$

Assign 100 mW as the output from the transmitter, or -10 dBW. Allow 2-dB total line losses for both ends of the link by locating the transmitter–receiver butt-on with the antenna feed. We can picture link calculations better from the model in Figure 10.10. The imaginary isotropic antennas allow us no gain and we find that we are 43 dB short to reach receiver threshold. Thus at points A and B parabolic antennas may be installed with 43/2 dB, or 21.5 dB each (see Table 10.2).

The transmission design engineer must now assign a margin to the noise threshold. The margin will depend on the type of modulation used. With FM, for example, the advantage gained with the trade-off of thermal noise versus bandwidth is achieved only when the carrier-to-noise ratio is 10 dB. Thus, for FM at least, 10 dB must be included to the gain in the model. It is to this figure that we add a margin for propagation conditions or, more properly, for rainfall and scintillation.

Table 10.2   Parabolic Reflector Gain Versus 10-km Path Loss (Eff. 54%)

Frequency (GHz)	Diameter (m)	Gain (dB)	10-km Free-Space Loss (dB)
20	0.5	37.7	138.5
	1.0	43.7	
	1.5	47.2	
	2.0	49.7	
30	0.5	41	141.9
	1.0	47	
	1.5	50.6	
	2.0	53	
40	0.5	43.7	144.5
	1.0	49.7	
	1.5	53.2	
	2.0	55.7	
50	0.5	45.7	146.5
	1.0	51.7	
	1.5	55.2	
	2.0	57.7	
60	0.5	47.2	148.1
	1.0	53.2	
	1.5	56.7	
	2.0	59.2	
70	0.5	48.6	149.3
	1.0	54.6	
	1.5	58.1	
	2.0	60.6	
80	0.5	49.7	150.5
	1.0	55.7	
	1.5	59.2	
	2.0	61.7	
90	0.5	50.8	151.5
	1.0	56.8	
	1.5	60.3	
	2.0	62.8	
100	0.5	51.7	152.5
	1.0	57.7	
	1.5	61.2	
	2.0	63.7	

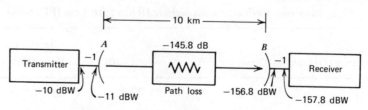

Figure 10.10 Millimeter wave link model.

For excess attenuation due to rainfall we turn back to Section 10.3.2. For the example let us assume that we are in region 2 (Figure 10.6) and that the path must have a propagation reliability of 99.95%. From Table 10.1 the values of a and b (40-GHz column, LP) are 0.31 and 0.98, respectively. Apply formula (10.2):

$$A = aR^b$$

When R for region 2 (0.05%) is 20 mm/h (from Figure 10.7), then

$$A_{(0.05)} = 0.31(20)^{0.98}$$

$$= 5.8 \text{ dB/km}$$

The rain cell size is about 3.8 km (Figure 10.5), so the excess attenuation due to rainfall on this path is 5.8 × 3.8, or about 22 dB. From Figure 10.8 the margin for scintillation fading is approximately 3 dB. Thus the total path attenuation we should use is 145.8 dB + 22 db + 3 dB. Turning to Table 10.2 with 1.5-m antennas, we find the antenna gain at 40 GHz to be 53.2 dB providing a total antenna gain at each end (i.e., transmit and receive) of 106.4 dB, and the level into the receiver front end would be -76.4 dBW. Assume that the modulation is conventional FM. Then the FM improvement carrier-to-noise level would be -104 dBW, permitting a total margin to this level of 27.6 dB, 2.6 dB in excess of the necessary margin of 25 dB for rainfall and scintillation fading.

Continue the exercise by extending the propagation reliability to 99.99% (or an outage probability due to rainfall of 0.01%). Now Figure 10.7 for region 2 shows the rainfall rate R at about 55 mm/h; Figure 10.5 gives a rain cell size of 2.8. km. Then

$$A_{(0.01)} = 0.31 (55)^{0.98}$$

$$= 15.7 \text{ dB/km}$$

For this example excess attenuation due to rainfall is 2.8 × 15.7 or 44.1 dB. Previously the figure was 22 dB; in this case we will need 44.1 - 22 = 22.1 dB additional path gain. The natural tendency of the link design engineer would be to increase the antenna reflector diameter. This would require an antenna on the order of 6 m in diameter. Such an antenna is certainly technically feasible, but may not be the most cost-effective solution with out present technology for two reasons:

- As wavelength is reduced, reflector surface accuracy requirements become more stringent.
- Beamwidths consistent with the antenna gains now encountered place stricter constraints on tower design for twist and sway.

Other alternatives may well be worth considering:

- Reduce propagation reliability.
- Increase power output of the transmitter. (*Note.* If we were to increase it to 1 W, we would only produce 10-dB additional gain.)
- Reduce bandwidth (*Note.* If B_{IF} were reduced to 10 MHz, we would only achieve an equivalent gain of 10 dB.)
- Use path diversity.
- Possibly use another form of modulation, particularly if we are transmitting digital information, for example, QPSK.
- Reduce link length (hop distance). For instance, reducing link length to 7 km would satisfy requirements.

It should be noted regarding the above analysis that excess attenuation due to rainfall as calculated here assumes the raincell size according to Figure 10.5. We emphasize the caution statement provided by CCIR under that figure.

10.6 THE SHORT HOP CONCEPT

As has been suggested earlier in this chapter, one way to ameliorate the excess attenuation problem due to rainfall is to significantly reduce hop length, for instance, to 1 km. An average repeater spacing in the centimetric region discussed in Chapter 5 was 50 km (30 mi), and 50 such repeaters might be an average number required on a 2500-km path. Over the same distance with the short hop concept 2500 repeaters would be required. For analog transmission this number of repeaters in tandem would make accumulated noise intolerable. However, using digital modulation techniques with regeneration at each repeater, the signal delivered at the output of the 2500-km system would be as good as the signal injected at the input.*

This concept fits well with the proposed all-digital network. The digital format would be PCM or possibly DM. A 100-Mbps system, which is well within today's technology, would serve as a vehicle for the transmission of about 1400 nominal 4-kHz analog VF channels.

For centimetric systems 100-m high towers are fairly common and a major cost item. For millimetric wave systems operating with this short hop concept (e.g., 1 km) towers will be proportionately lower (20–30 m), less expensive, and less

---

*Leaving aside jitter, or assuming that the jitter problem has been solved.

unsightly. These stubby towers or poles would indeed require very good twist and sway specifications. The entire radio installation would be colocated with the antennas reducing transmission line losses to small amounts.

Consider a 1-km path using 100-GHz equipment. The free-space path loss is 132.5 dB plus an absorption loss factor to be added of about 0.8 dB (Figure 10.2). This gives a total path loss of 133.3 dB. Assume B_{IF} to equal 100 MHz (sufficient bandwidth to permit the transmission of 100 Mb/s allowing 1 b/Hz). Use a 10-dB noise figure for the receiver, and its noise threshold is calculated to be -118 dBW.

At the transmit end of the path, the transmitter power output is 10 mW (-20 dBW). With zero antenna gains, the input to the receiver is -20 dBW (+) -133 dBW = -153 dBW with low line losses. If the receiver operating point is placed 30 dB above noise threshold (equivalent to -183.3 dBW), the sum of the antenna gains turns out to be 65.3. Therefore each antenna would require a gain of 65.3/2 or 32.7 dB. From Table 10.2, a 0.5-m antenna provides a 51.7-dB gain. Thus we end up with more than 38 dB over and above requirements, which can be allotted to excess rainfall attenuation and scintillation.

We have been emphasizing propagation and building in a fade margin sufficient to provide a highly reliable system. One important point has been put aside which only lately is taking on its proper priority. This is equipment reliability (including reliable power). On some centimetric systems the yearly outages due to poor equipment reliability exceed outages due to propagation by as much as 10:1.

In this respect, with approximately 1-km repeater spacing versus 50 km average for centimetric systems, the number of repeaters for a given section utilizing this short hop concept is multiplied 50 times. It certainly follows that equipment and power reliabilities must be multiplied manyfold. Likewise, maintenance must be reduced. This, too, is within our grasp with present-day solid-state and strip-line techniques. Another important feature is low-level transmitter power outputs which permit smaller, more compact no-break power sources with sealed batteries floating on-line. Thermoelectric generation may be feasible as a backup during long line voltage outages. Purposely we used 10 mW (-20 dBW) as the transmitter output power in our example above.

The approach to reliability for such a system would need to be similar to that used on present undersea multichannel cable systems.

10.7 EARTH–SPACE COMMUNICATION

Another obvious candidate for propagated millimeter wave transmission would be earth–space communications and, in particular, satellite communications. Nearly all the problems associated with the propagation of millimeter waves come from our atmosphere, such as rain, fog, and gas absorption (see Chapter 7).

Consider the free-space loss figures and some parabolic antenna gains related to synchronous satellites given in Table 10.3. The tendency would be to reduce the

Table 10.3 Parabolic Antenna Gains for Synchronous Satellites

Frequency (GHz)	Free-Space Loss (dB)	15-m Diameter Antenna Gain[a] (dB)	20-m Diameter[b] Antenna Gain[a] (dB)
20	210.5	67.5	70.0
30	214.0	71.0	72.7
40	216.5	73.0	76.0

[a]With 65% efficiency.
[b]Surface accuracy requirements severely limit the use of large-diameter dishes in the high millimeter region.

diameter of the earth station antenna to keep the accuracy of the surface within reasonable bounds, and generally reduce overall cost. By doing so, of course, antenna gain suffers. From Chapter 7 we see that the nominal 30-m "standard" antenna has a gain in excess of 60 dB at 4 GHz, and the free-space loss to a synchronous satellite at that frequency is about 196 dB. Unless we use a 30-m antenna here as well, it would seem that the "standard" earth station criteria may not be met.

There are some legislative restrictions and a number of other items in our favor that could reduce these "standard" requirements.

1. By legislation a millimeter frequency band could be assigned for exclusive use by satellite–earth communication, and thereby the +32-dBW limit of satellite EIRP now in effect for the centimetric range would be increased. For instance, in the band 17.7-22 GHz a 13-dB increase in power flux density is permitted (Ref. 29).

2. The "standard" earth station is specified to meet its minimum G/T at $10°$ elevation angles, where sky noise is the greatest. This angle should be increased to 15 or $20°$. The operating area of the satellite would be reduced somewhat as a consequence, but the trade-off in cost and technology should be worthwhile. Likewise, it is just at these low look angles, where the millimeter beam must travel through the most atmosphere, where it will suffer the most outage due to rain and the most gaseous absorption.

Rainfall continues to be the largest deterrent. Rainfall margins of 35 dB or more in the millimetric frequency range are not uncommon. Remember that such margins for "standard" earth stations operating in the 4-GHz band are on the order of 4-6 dB. Path diversity should reduce this figure significantly if separation between receiving sites is 5 mi (8 km) or greater. This leads into a discussion of the INTELSAT standard C earth stations, which are designed to operate in the lower millimetric frequency region, as well as a review of excess noise considerations and path diversity, which are covered in the following sections.

10.7.1 INTELSAT Standard C Earth Stations

INTELSAT V satellites provide uplink operation at 14.0–14.5 GHz and downlink operation in the bands of 10.95–11.2 GHz and 11.45–11.7 GHz. These satellites operate with the standard C earth stations. We review standard C earth stations here rather than in Chapter 7 because of the influence of rainfall on standard C operation.

Reference 30 is quoted in part regarding G/T of the standard C earth station:

> To ensure the best utilization of the space segment, the aim is to achieve for the receiving system a gain to noise temperature ratio (G/T) that is sufficient to ensure that CCIR performance criteria are met. This requires a consideration of long-term rainfall data and the associated attenuation and sky noise temperature data at each earth station. Considering the form in which propagation information is available, it is more convenient to express CCIR monthly noise criteria in terms of percentage-of-a-year relationships which are chosen to be equivalent to CCIR values.
>
> Annual noise criteria are given in terms of "nominal" performance requirements associated with 90% of the time in a year and degraded performance requirements associated with a small percentage of the time in a year.

For any frequency in the bands 10.95–11.20 and 11.45–11.70 GHz, the value of G/T_i must not be less than the values given below for the various criteria.

- The value of $G/T_1 - L_1$ shall be no less than $A + 20 \log_{10} (f/11.2)$ dB/K for all but P_1 % of the time.*
- The value of $G/T_2 - L_2$ shall be no less than $B + 20 \log_{10} (f/11.2)$ dB/K for all but P_2 % of the time.*
- In the case of diversity earth stations the quantity that shall comply with the above specifications is the one which, at any given time, is equal to the value of $G/T - L$ of the operating terminal.

The following notes apply to this specification (Ref. 30):

1. The term G denotes the receive antenna gain referred to the input of the receiving amplifier at the frequency of interest.

2. The term L_i ($i = 1, 2$) is the predicted attenuation relative to a clear sky, at the frequency of interest, exceeded for no more than P_i ($i = 1, 2$) percent of the time,* along the path to the satellite(s) with which operation is desired. The value of L_i shall be that predicted on the basis of the statistical distribution of mean attenuation within periods of the order of 1 min.

3. The term T_i ($i = 1, 2$) is the receiving system noise temperature, including

---

*The time period to which this percentage applies shall be that period for which statistics are available, preferably a minimum of 5 years.

noise contributions from the atmosphere, referred to the input of the low-noise amplifier at the frequency of interest, when the attenuation L_i prevails.

4. The terms G/T_i and L_i are assumed to be given in decibel notation; f is the frequency of interest expressed in gigahertz.

5. The following values for A, B, P_1 and P_2 shall apply:

INTELSAT V Coverage in Which
Earth Station is Located

Parameter	West spot	East Spot
A	39.0 dB	39.0 dB
B	29.5 dB	32.5 dB
P_1	10.0%	10.0%
P_2	0.017%	0.017%

6. An applicant seeking approval for access for a standard C 11-GHz receiving earth station to the INTELSAT V space segment shall submit to INTELSAT information on earth station receive main beam antenna gain, clear-sky predicted system noise temperature, the use of diversity where applicable, and the rain model or relevant propagation data used for the derivation of T_i and L_i.

POLARIZATION

In the 14/11-GHz frequency band the INTELSAT V satellite polarization is linear where the beam generating the east coverage is orthogonal to the beam generating the west coverage. The transmission and reception polarizations for a given coverage will be orthogonal. It shall be possible to match the spacecraft polarization angle within 1° under clear sky conditions. The spacecraft polarization angle for each beam will be known prior to launch and made available to earth station owners. It is assumed that earth stations will have an elliptical polarization pattern which will meet the axial ratio specified (30.0-dB polarization discrimination).

RF BANDWIDTH AND BEAM SWITCHING FOR INTELSAT V SATELLITES

The RF bandwidth of the INTELSAT V is divided into segments of 34, 36, 41, 72, 77, and 241 MHz, depending upon the frequency band and beam connections employed to meet traffic requirements. INTELSAT V has the capability to operate in both the 6/4-GHz and the 14/11-GHz frequency bands and to interconnect these bands so that standard A (6/4-GHz) and standard C (14/11-GHz) earth stations can communicate.

A transponder numbering system from 1 to 12 is used to designate frequency slots having a bandwidth of approximately 36 MHz. When a transponder with a bandwidth larger than 36 MHz is used, a multiple number is employed (e.g., 1-2).

In the 14/11-GHz band INTELSAT V transponders have the center frequencies shown below:

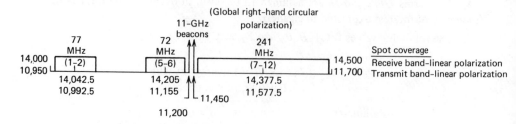

TOTAL RECEIVE POWER FLUX DENSITY

The maximum receive power flux density that may be expected at an earth station is −116 dBW/m² within the nominal coverage areas [refer to (Ref. 30)]. It should be noted that the INTELSAT V spot beam pointing directions may be changed from time to time to meet traffic requirements. Presently these spot beams cover the northeastern United States and Canada (W) and Europe (E) for the Atlantic

Table 10.4 EIRP For FDM/FM Carriers and Uplink Margins

Bandwidth Unit (MHz)	EIRP Regular Carrier[a] (dBW)	EIRP High-Density Carrier (dBW)
East up (14–4-GHz cross strapping)		
2.5	+68.8–77.2	+79.5
10.0	+73.8–80.5	+83.1
25.0	+80.2–88.4	+91.8
East up (14–11-GHz) direct links		
2.5	+67.5–75.9	+78.2
10.0	+72.5–80.2	+82.8
25.0	+78.6–85.7	+90.5
West up (14–4-GHz cross strapping)		
2.5	+65.7–74.1	+76.4
10.0	+71.7–78.4	+81.0
25.0	+78.1–85.2	+88.7
West up (14–11 GHz) direct links		
2.5	+64.4–72.8	+75.1
10.0	+70.4–76.6	+79.7
25.0	+75.5–84.0	+87.4

[a] EIRP varies in this range depending on the number of voice channels to be transmitted.

**Table 10.5   Atlantic Region 14-GHz Uplink Path Margins for
Power Control Determination**

	Uplink Path Margin M (dB)	Criterion K [a] (%)
West up		
4-GHz links	6.0	0.017
11-GHz links (80 MHz)	6.5	0.010
11-GHz links (240 MHz)	12.5	0.010
East up		
4-GHz links	6.0	0.017
11-GHz links (80 MHz)	8.5	0.010
11-GHz links (240 MHz)	9.0	0.010

[a] Based on up-link rain criterion for this percentage of the time in one year.

satellite and Europe (W) and Japan (E) for the Indian Ocean satellite. Table 10.4 shows EIRP sample values from (Ref. 30). The following is quoted from (Ref. 30) to provide some insight on how uplink rain margins are handled:

> In order to meet the short-term CCIR performance criterion it is mandatory that means be provided to prevent the power flux density at the satellite from falling by more than M dB below the nominal clear-sky value for more than K percent of the time in a year. The values of M and K are given in Table 10.5.

DIVERSITY

Where measured statistics indicate the use of space diversity (which we have been calling path diversity), it will be necessary to duplicate the 14/11-GHz antenna and RF system at a suitable distance from the main site and provide interconnection by means of an appropriate link. In some cases an economic trade-off may show it less expensive to duplicate only the receiving facilities and provide an additional power control range for the transmitter at the main site. In either case it will be necessary for the diversity earth station to meet the mandatory characteristics of a standard C antenna.

It should be noted that the CCIR hypothetical reference circuit contained in CCIR Rec. 352-2 includes the diversity interconnection links to the diversity switching point and any additional modulation–demodulation equipment required. This means that system noise budgets must include this additional link. It is also necessary to account for weather effects when a microwave interconnection link is employed.

Another element of importance is associated with the differential time delay between the diversity signals as they arrive at the switching point. Minimizing this time delay before switching or combining will reduce the requirements for built-in memory to avoid loss of data. The circuit interruption during switching should be such that errors are not introduced into telephone signaling circuits.

INTELSAT V TRANSLATION FREQUENCIES

The following are the applicable INTELSAT V translation frequencies with respective cross strapping.

Cross Connection (GHz)	Translation (MHz) Frequency
6 up to 10.95–11.2 down	5025
6 up to 11.45–11.7 down	5275
14 up to 4 down	10,300
14 up to 10.95–11.2 down	3050
14 up to 11.45–11.7 down	2800

Note. Section 10.7.1 has been excerpted in large part from (Ref. 30). (Courtesy INTELSAT.)

10.7.2   Excess Sky Noise Temperature Due to Precipitation and Clouds

An additional source of degradation on a millimeter-wave downlink is the increase in noise temperature caused by precipitation and clouds. This will result in an increased system margin requirement. The increase is dependent on the clear-sky noise temperature and can be expressed as follows:

$$M = 10 \log \frac{T_{sys} + T_{sky}}{T_{sys}} \qquad (10.3)$$

where M = increase in margin requirement, T_{sys} = clear-sky system noise temperature, and

$$T_{sky} = [1 - 10^{-A/10}] T_r$$

with T_r = physical temperature of the precipitation (when no other information available, use 273 K), A = excess attenuation over clear-sky conditions (see Section 10.3.2). (Ref. 34).

Thus given a system noise temperature of 250 K (a reasonable value for the lower millimetric region), the increase in margin requirement would approach about 3 dB at high values of attenuation and would be less at lower values.

10.7.3   Notes on Path Diversity

Figure 10.11 provides guidance on the amount of site separation required for path diversity. It was taken from (Ref. 31) and is based on the Hodge rainfall model. Of course the idea of path diversity is that, due to the cellular nature of heavy rainfall,

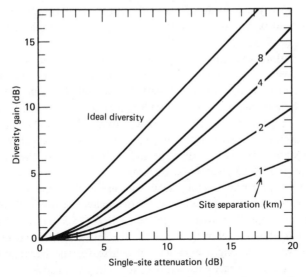

Figure 10.11 Diversity improvement for various site separations (Ref. 31).

the diversity site is separated from the main site sufficiently that both sites will not suffer the high attenuation of a major rain cell. From Figure 10.5 we can reason that minimum separation should be 4 or 5 km, depending on the rainfall region involved.

10.8 LIMITATIONS OF REFLECTOR DIAMETER OF EHF PARABOLIC ANTENNAS

There is a limit to antenna gain for large parabolic antennas as the operational frequency is increased. Two major factors affect the gain. First, deviations of the surface of the reflector from the ideal surface tend to defocus the antenna beam and produce a concomitant loss in gain. Second, beam pointing errors can result in significant gain reductions in the desired direction because of the narrowness of the antenna ray beam formed by the reflector.

It can be shown that there is an approximate linear relationship between surface reflector tolerance ϵ and the antenna diameter D. It can also be stated that the maximum surface accuracy currently achievable decreases as the reflector diameter increases. The upshot of the argument is that we just cannot continue to increase antenna gain by building larger dishes and/or decreasing wavelength (e.g., increasing frequency). As we know, the gain of a parabolic reflector can be related to the diameter of the reflector expressed in wavelengths. Given an ϵ/D ratio of 10^{-3}, the maximum achievable reflector is about 100 wavelengths in diameter, and increasing the reflector diameter, D by more than 100 wavelengths shows a gain reduction rather than a gain increase. If $\epsilon/D = 10^{-4}$, 1000 wavelengths is achievable, and for 10^{-5}, about 10,000 wavelengths. Commonly available antennas today have ϵ/D ratios on the order of 10^{-4}. Thus at 44 GHz we are limited to a reflector diameter of about 17 ft (5 m) and a gain of about 61 dBi. Specially built custom antennas can achieve a ratio of 5×10^{-5}, permitting 35-ft (11-m) antennas with a gain of 67 dBi at 44 GHz. (Ref. 32).

With earth stations operating in the extremely-high frequency (EHF) band, as beamwidth decreases, pointing error increases. For a given tracking accuracy, the achievable gain of an antenna in the direction of the satellite is limited to a specific maximum as the antenna aperture diameter is increased. Antennas have a tracking error which is proportional to the half-power (3-dB) beamwidth of the antenna. The amount of tracking error (resulting in pointing loss) we can expect from a specific earth station depends on the type and sophistication of the tracking system used (see Chapter 7). The greater the tracking error (from any source), the greater the pointing loss. Currently tracking accuracies are in the range of 0.02° for large sophisticated antenna systems down to 0.1° for smaller less sophisticated terminals. At 44 GHz antenna aperture sizes are limited to 46 ft (14.5 m) for the more sophisticated terminals, down to 9 ft (2.8 m) for the less sophisticated terminals with presently achievable tracking accuracies. Of course for the larger antennas [e.g., 46 ft (14.5 m)] we have already passed the gain limit given above.

REFERENCES AND BIBLIOGRAPHY

1. *Reference Data for Radio Engineers,* 6th ed., Howard W. Sams, Indianapolis, IN, 1977.
2. H. J. Liebe, "Atmospheric Propagation Properties in the 10 to 75-GHz Region: A Survey and Recommendations," ESSA Tech. Rep. ERL 130-ITS 91, Boulder, CO 1969.
3. D. C. Hogg, "Millimeter-Wave Propagation through the Atmosphere," *Science,* 1968.
4. J. J. Taub, "The Future of Millimeter Waves," *Microwave J.,* Nov. 1973.
5. B. R. Dean and E. J. Dutton, *Radio Meteorology,* Dover, New York, 1968.
6. J. A. Lane, "Scintillation and Absorption Fading on Line-of-Sight Links at 35 and 100 GHz," IEE Conference on Tropospheric Scatter Propagation, London, Oct. 1968.
7. M. C. Thompson *et al.,* "Phase and Fading Characteristics in the 10 to 40-GHz Band," Institute of Telecommunication Sciences, Boulder, CO, Oct. 1972.
8. G. E. Weibel and H. O. Dressel, "Propagation Studies in Millimeter-Wave Link Systems," *Proc. IEEE,* Apr. 1967.
9. F. Dale, "The Reach for Higher Frequencies," *Telecommunications,* July 1970.
10. E. M. Hickin, "18 GHz Propagation," IEE Conference on Tropospheric Scatter Propagation, London, Oct. 1968.
11. R. E. Skerjanec and C. A. Samson, "Rain Attenuation Study for 15 GHz Relay Design," Rep. FAA-RD-70-21, Institute of Telecommunication Sciences, Boulder, CO, May 1970.
12. R. E. Skerjanec and C. A. Samson, "Microwave Link Performance Measurements at 8 and 14 GHz," Rep. FAA-RD-72-115, Institute of Telecommunication Sciences, Boulder, CO, Oct. 1972.
13. F. A. Benson, *Millimetre and Submillimetre Waves,* Iliffe, London, 1969.
14. D. E. Setzer, "Computed Transmission Through Rain at Microwave and Visible Frequencies," *Bell Sys. Tech. J.,* 1973, Oct. 1970.
15. IEE Conference on Millimeter Wave Propagation (Conference Papers), London, 1972.

16. "Bibliography on Propagation Effects 10 GHz to 1000 THz," Telecommunications and Engineering Rep. 30, Office of Telecommunications, Boulder, CO, Mar. 1972.

17. L. C. Tillotson, "Use of Frequencies above 10 GHz for Common Carrier Applications," *Bell Sys. Tech. J.*, 1563, July–Aug. 1969.

18. C. L. Ruthroff *et al.,* "Short Hop Radio System Experiment," *Bell Sys. Tech. J.,* 1577, July–Aug. 1969.

19. A. E. Freeny and J. D. Gabbe, "A Statistical Description of Intense Rainfall," *Bell Sys. Tech. J.,* 1789, July–Aug. 1969.

20. R. G. Medhurst, "Rainfall Attenuation of Centimeter Waves: Comparison of Theory and Experiment," *IEEE Trans. Ant. Prop.,* Vol. **AP-13,** 500, July 1965.

21. D. B. Hodge, "The Characteristics of Millimeter Wavelength Satellite-to-Ground Space Diversity Links," IEE Conference No. 98, London, April 1973.

22. W. Sollfrey, "Nomograms for the Calculation of Propagation Effects on Tactical Millimeter-Wave Radio Links," CORADCOM Report 77-0142-1 The RAND Corp., June 1979.

23. N. E. Feldman, "Rain Attenuation over Earth–Satellite Paths," Science Applications, Inc., El Segundo, CA, 1979.

24. R. K. Crane, "Prediction of the Effects of Rain on Satellite Communications Systems," *Proc. IEEE,* Vol. 65, 456–474, Mar 1977.

25. R. Olsen, D. Rogers, and D. Hodge, "The aR^b Relation in the Calculation of Rain Attenuation," *IEEE Trans. Ant. Prop.,* Vol. AP-26, 318–329, Mar. 1978. .

26. R. K. Crane, "Rain Attenuation Prediction," CCIR Doc. F5/003, 1978.

27. D. C. Hogg and T. Chu, "The Role of Rain in Satellite Communications," *Proc. IEEE,* 1308–1331, Sep. 1975.

28. "18 and 30 GHz Fixed Service Communication Satellite System Study," NAS-3-21367, Hughes Aircraft Co., 1979.

29. R. G. Gould and Y. F. Lum, *Communication Satellite Systems: An Overview of the Technology,* IEEE Press, New York, 1975.

30. "Standard C Performance Characteristics of Earth Stations in the INTELSAT V System (14 and 11 GHz Frequency Bands)," BG-28-72E M/6/77, INTELSAT, Washington, DC, 1977.

31. "Concepts for 18/30 GHz Satellite Communication System Study," Vol. 1, Final Rep. NAS3-21362, Ford Aerospace & Communications Corp., Western Development Labs., Palo Alto, CA, Nov. 1979.

32. W. C. Cummings, P. C. Jain, and L. J. Richardi, "Fundamental Performance Characteristics that Influence EHF MILSATCOM Systems," *IEEE Trans Commun.,* Vol. COM-27, Oct. 1979.

33. "Recommendations and Reports of the CCIR 1978," XIV Plenary Assembly, Kyoto, 1978, Vols. V and IX (Geneva, 1978).

34. "A Propagation Effects Handbook for Satellite Systems Design," NASA Report, ORI TR-1679 NTIS N80-25520, March 1980.

DIGITAL TRANSMISSION SYSTEMS—PCM AND DELTA MODULATION

11.1 WHAT IS PCM?

Pulse code modulation (PCM) is a method of modulation in which a continuous analog wave is transmitted in an equivalent digital mode. The cornerstone of an explanation of the functioning of PCM is the sampling theorem, which states (Ref. 1, Section 23):

> If a band-limited signal is sampled at regular intervals of time and at a rate equal to or higher than twice the highest significant signal frequency, then the sample contains all the information of the original signal. The original signal may then be reconstructed by use of a low-pass filter.

As an example of the sampling theorem, the nominal 4-kHz channel would be sampled at a rate of 8000 samples per second (i.e., 4000 × 2).

To develop a PCM signal from one or several analog signals, three processing steps are required: sampling, quantization, and coding. The result is a serial binary signal or bit stream, which may or may not be applied to the line without additional modulation steps. At this point a short review of Chapter 8 may be in order so that terminology such as mark, space, regeneration, and information bandwidth [Shannon (Ref. 21) and Nyquist (Ref. 20)] will not be unfamiliar to the reader.

One major advantage of digital transmission is that signals may be regenerated at intermediate points on links involved in transmission. The price for this advantage is the increased bandwidth required for PCM. Practical systems require 16 times the bandwidth of their analog counterpart (e.g., a 4-kHz analog voice channel requires 16 × 4 or 64 kHz when transmitted by PCM). Regeneration of a digital signal is simplified and particularly effective when the transmitted line signal is binary, whether neutral, polar, or bipolar. An example of bipolar transmission is shown in Figure 11.1.

Binary transmission tolerates considerably higher noise levels (i.e., degraded signal-to-noise ratios) when compared to its analog counterpart (i.e., FDM, Chapter

474

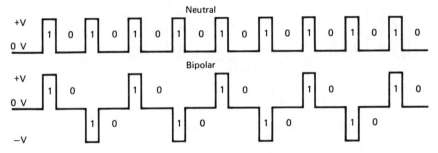

Figure 11.1 Neutral versus bipolar bit streams. (*a*) Alternate 1's and 0's transmitted in a neutral mode. (*b*) Equivalent in a bipolar mode.

3). This plus the regeneration capability is a great step forward in transmission engineering. The regeneration that takes place at each repeater by definition recreates a new digital signal. Therefore noise, as we know it, does not accumulate. However, there is an equivalent to noise in PCM systems which is generated in the modulation–demodulation processes. This is called quantizing distortion and can be equated in annoyance to the listener with thermal noise. Regarding thermal/IM noise, let us compare a 2500-km conventional analog circuit using FDM multiplex over cable or radio with an equivalent PCM system over either medium:

	FDM/Radio/Cable	PCM/Radio/Cable
Multiplex	2,500 pWp	130 pWp equivalent
Radio/cable	7,500 pWp	0 pWp
Total	10,000 pWp	130 pWp equivalent*

Error rate is another important factor (see Chapter 8). If we can maintain an end-to-end error rate on the digital portion of the system of 1×10^{-5}, intelligibility will not be degraded. A third factor is important in PCM cable applications. This is crosstalk spilling from one PCM system to another or from the send path to the receive path inside the same cable sheath.

The purpose of this chapter is to provide a background of the problems involved in PCM and its transmission, including the several PCM formats now in use. Practical aspects are stressed later in the chapter, such as the design of interexchange trunks (junctions) and the prove-in distance,† compared with other forms of multiplex or VF cable media. Long-distance (toll) transmission via PCM is also discussed, and reference should then be made to Chapters 14 and 15. Finally a second method of digital modulation is briefly described, namely, delta modulation.

---

*Not dependent on system length, see Section 11.2.6.
†The point at which a PCM transmission system becomes an economically viable alternative when applied to an analog telephone network.

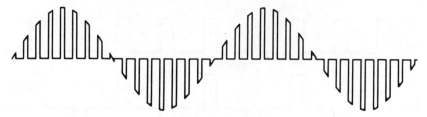

Figure 11.2   PAM wave as a result of sampling a single sinusoid.

11.2   DEVELOPMENT OF A PCM SIGNAL

11.2.1   Sampling

Consider the sampling theorem given above. If we now sample the standard CCITT voice channel, 300–3400 Hz (a bandwidth of 3100 Hz), at a rate of 8000 samples per second, we will have complied with the theorem and we can expect to recover all the information in the original analog signal. Therefore a sample is taken every 1/8000 s, or every 125 μs. These are key parameters for our future argument.

Another example may be a 15-kHz program channel. Here the lowest sampling rate would be 30,000 times per second. Samples would be taken at 1/30,000-s intervals or every 33.3 μs.

11.2.2   The PAM Wave

With at least one exception (i.e., SPADE, Section 7.13) practical PCM systems involve TDM. Sampling in these cases does not involve just one voice channel, but several. In practice, one system to be discussed samples 24 voice channels in sequence; another one samples 32 channels. The result of the multiple sampling is a pulse amplitude modulation (PAM) wave. A simplified PAM wave is shown in Figure 11.2, in this case a single sinusoid. A simplified diagram of the processing involved to derive a multiplexed PAM wave is shown in Figure 11.3.

If the nominal 4-kHz voice channel must be sampled 8000 times per second and a group of 24 such voice channels are to be sampled sequentially to interleave them, forming a PAM multiplexed wave, this could be done by gating. Open the gate for a 5.2-μs (125/24) period for each voice channel to be sampled successively from channel 1 through channel 24. This full sequence must be done in a 125-μs period $(1 \times 10^6/8000)$. We call this 125-μs period a *frame*, and inside the frame all 24 channels are successively sampled once.

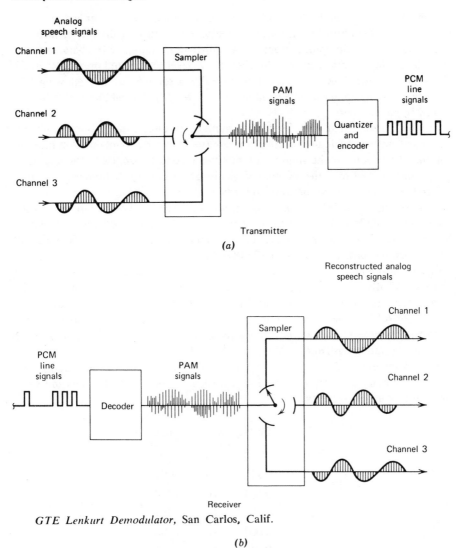

Figure 11.3 Simplified analogy of the formation of a PAM wave and PCM bit stream. (a) Transmitter. (b) Receiver. (Courtesy GTE Lenkurt Inc., San Carlos, CA.)

11.2.3 Quantization

The next step, in the process of forming a PCM serial bit stream, is to assign a binary code to each sample as it is presented to the coder.

Remember from Chapter 8 the discussion of code lengths, or what is more properly called coding *level*. For instance, a binary code with four discrete elements (a four-level code) could code 2^4 separate and distinct meanings or 16 characters, not enough for the 26 letters in our alphabet; a five-level code would provide 2^5 or 32 characters or meanings. The ASCII is basically a seven-level code allowing 128 discrete meanings for each code combination ($2^7 = 128$). An eight-level code would yield 256 possibilities.

Another concept that must be kept in mind as the discussion leads into coding is that bandwidth is related to information rate (more exactly to modulation rate) or, for this discussion, to the number of bits per second transmitted. The goal is to keep some control over the amount of bandwidth necessary. It follows, then, that the coding length (number of levels) must be limited.

As it stands, an infinite number of amplitude levels are being presented to the coder on the PAM highway. If the excursion of the PAM wave is between 0 and +1 V, the reader should ask how many discrete values there are between 0 and 1. All values must be considered, even 0.0176487892 V.

The intensity range of voice signals over an analog telephone channel is on the order of 50 dB (see Section 3.13). The 0–1-V range of the PAM highway at the coder input may represent that 50-dB range. Further, it is obvious that the coder cannot provide a code of infinite length (e.g., an infinite number of coded levels) to satisfy every level in the 50-dB range (or a range from −1 to +1 V). The key is to assign discrete levels from −1 V through 0 to +1 V (50-dB range).

The assignment of discrete values to the PAM samples is called *quantization*. To cite an example, consider Figure 11.4. Between −1 and +1 V 16 quantum steps exist and are coded as follows:

Step	Code	Step	Code
0	0000	8	1000
1	0001	9	1001
2	0010	10	1010
3	0011	11	1011
4	0100	12	1100
5	0101	13	1101
6	0110	14	1110
7	0111	15	1111

Examination of Figure 11.4 shows that step 12 is used twice. Neither time it is used is it the true value of the impinging sinusoid. It is a rounded-off value. These rounded-off values are shown with a dashed line which follows the general outline of the sinusoid. The horizontal dashed lines show the points where the quantum changes to the next higher or next lower level if the sinusoid curve is above or below that value. Take step 14, for example. The curve, dropping from its maximum, is

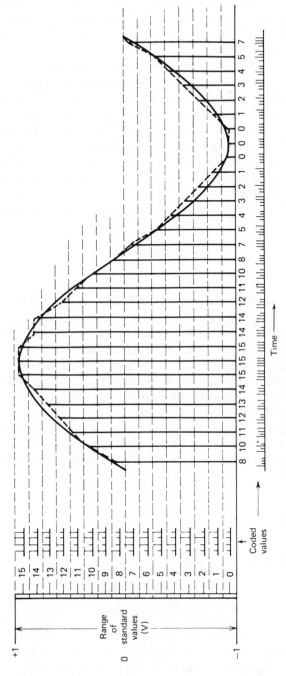

Figure 11.4 Quantization and resulting coding using 16 quantizing steps.

479

given two values of 14 consecutively. For the first, the curve is above 14, and for the second, below. That error, in the case of 14, for instance from the quantum value to the true value is called *quantizing distortion*. This distortion is the major source of imperfection in PCM systems.

In Figure 11.4, maintaining the −1–0–+1-V relationship, let us double the number of quantum steps from 16 to 32. What improvement would we achieve in quantization distortion? First determine the step increment in millivolts in each case. In the first case the total range of 2000 mV would be divided into 16 steps, or 125 mV per step. The second case would have 2000/32, or 62.5 mV per step. For the 16-step case, the worst quantizing error (distortion) would occur when the input to be quantized was at the half-step level, or, in this case, 125/2 or 62.5 mV above or below the nearest quantizing step. For the 32-step case the worst quantizing error (distortion) would again be at the half-step level, or 62.5/2 or 31.25 mV. Thus the improvement in decibels for doubling the number of quantizing steps is:

$$20 \log \left(\frac{62.5}{31.25} \right) = 20 \log 2$$

$$\approx 6 \text{ dB}$$

This is valid for linear quantization only (see Section 11.2.6). Thus increasing the number of quantizing steps for a fixed range of input values reduces quantizing distortion accordingly. Experiments have shown that if 2048 uniform quantizing steps are provided, sufficient voice signal quality is achieved.

For 2048 quantizing steps a coder will be required to code the 2048 discrete meanings (steps). Reviewing Chapter 8, we find that a binary code with 2048 separate characters or meanings (one for each quantum step) requires an 11-element code, or $2^n = 2048$. Thus $n = 11$.

With a sampling rate of 8000 samples per second per voice channel, the binary information rate per voice channel will be 88,000 bps. Consider that equivalent bandwidth is a function of information rate; the desirability of reducing this figure is therefore obvious.

11.2.4 Coding

Practical PCM systems use seven- and eight-level binary codes, or

$$2^7 = 128 \text{ quantum steps}$$

$$2^8 = 256 \text{ quantum steps}$$

Two methods are used to reduce the quantum steps to 128 or 256 without sacrificing fidelity. These are nonuniform quantizing steps and companding prior to quantizing, followed by uniform quantizing. Keep in mind that the primary concern of digital transmission using PCM techniques is to transmit speech, as distinct from digital transmission covered in Chapter 8, which dealt with the transmission of data and message information. Unlike data transmission, in speech transmission there is a much greater likelihood of encountering signals of small amplitudes than those of large amplitudes.

A secondary, but equally important, aspect is that coded signals are designed to convey maximum information considering that all quantum steps (meanings, characters) will have an equally probable occurrence.

(We obliquely referred to this inefficiency in Chapter 8 because practical data codes assume equiprobability. When dealing with a pure number system with complete random selection, this equiprobability does hold true. Elsewhere, particularly in practical application, it does not. One of the worst offenders is our written language. Compare the probability of occurrence of the letter "e" in written text with that of "y" or "q.") To get around this problem larger quantum steps are used for the larger amplitude portion of the signal, and finer steps for signals with low amplitudes.

The two methods of reducing the total number of quantum steps can now be labeled more precisely:

- Nonuniform quantizing performed in the coding process.
- Companding (compression) before the signals enter the coder, which now performs uniform quantizing on the resulting signal before coding. At the receive end, expansion is carried out after decoding.

An example of nonuniform quantizing could be derived from Figure 11.4 by changing the step assignment. For instance, 20 steps may be assigned between 0.0 and +0.1 V (another 20 between 0.0 and –0.1 V, etc.), 15 between 0.1 and 0.2 V, 10 between 0.2 and 0.35 V, 8 between 0.35 and 0.5 V, 7 between 0.5 and 0.75 V, and 4 between 0.75 and 1.0 V.

Most practical PCM systems use companding to give finer granularity (more steps) to the smaller amplitude signals. This is instantaneous companding compared to syllabic companding described in Section 3.13. Compression imparts more gain to lower amplitude signals. The compression and later expansion functions are logarithmic and follow one of two laws, the A-law or the μ-law. The curve for the A-law may be plotted from the following formula:

$$Y = \frac{AX}{1 + \log A}, \qquad 0 \le v \le \frac{V}{A}$$

$$= \frac{1 + \log (AX)}{1 + \log A}, \qquad \frac{V}{A} \le v \le V$$

where $A = 87.6$. The curve for the μ-law may be plotted from the following formula:*

*For A-law and μ-law formulas use natural logarithms.

$$Y = \frac{\log(1 + \mu X)}{\log(1 + \mu)}$$

where $\mu = 100$ for the original North American TI system and 255 for later North American (D2) systems and the CCITT 24-channel system (CCITT Rec. G.733). In these formulas:

$$X = \frac{v}{V}$$

$$Y = \frac{i}{B}$$

where v = instantaneous input voltage, V = maximum input voltage for which peak limitation is absent, i = number of the quantization step starting from the center of the range, and B = number of quantization steps on each side of the center of the range (CCITT Rec. G.711).

A common expression used in dealing with the "quality" of a PCM signal is the signal-to-distortion ratio, expressed in decibels. Parameters A and μ determine the range over which the signal-to-distortion ratio is comparatively constant. This is the dynamic range. Using a μ of 100 can provide a dynamic range of 40 dB of relative linearity in the signal-to-distortion ratio.

In actual PCM systems the companding circuitry does not provide an exact replica of the logarithmic curves shown. The circuitry produces approximate equivalents using a segmented curve, each segment being linear. The more segments the curve

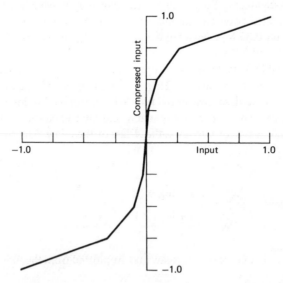

Figure 11.5 Seven-segment linear approximation of the logarithmic curve for the μ-law ($\mu = 100$) (Ref. 2; Copyright © 1970 by Bell Telephone Laboratories.)

has, the more it approaches the true logarithmic curve desired. Such a segmented curve is shown in Figure 11.5.

If the μ-law were implemented using a seven(eight)-segment linear approximate equivalent, it would appear as shown in Figure 11.5. Thus upon coding the first three coded digits would indicate the segment number (e.g., $2^3 = 8$). Of the seven-digit code, the remaining four digits would divide each segment in 16 equal parts to further identify the exact quantum step (e.g., $2^4 = 16$).

For small signals* the companding improvement is approximately

$$A\text{-law} \qquad 24 \text{ dB}$$
$$\mu\text{-law} \qquad 30 \text{ dB}$$

using a seven-level code.

Coding in PCM systems utilizes a straightforward binary coding. Two good examples of this coding are shown in Figures 11.7 and 11.10.

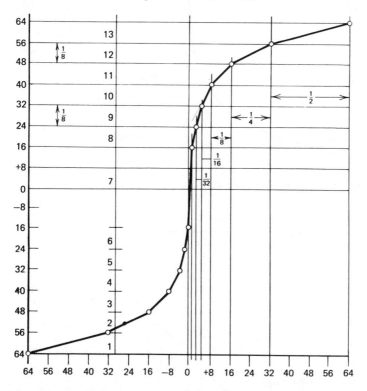

Figure 11.6 A 13-segment approximation of the A-law curve as used on the 24-channel STC system (PSC-24B). The abscissa represents quantized signal levels. Note that there are many more companding values at the lower signal levels than at the higher signal levels.

*Low-level signals

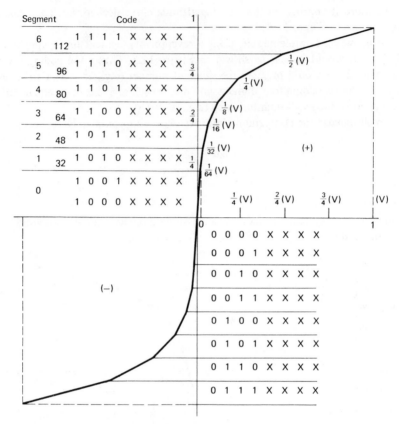

Figure 11.7 Quantization and coding used in the CEPT 30 + 2 PCM system.

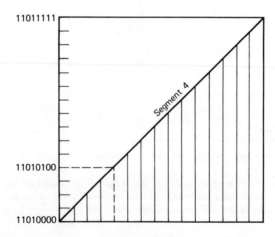

Figure 11.8 CEPT 30 + 2 PCM system, coding of segment 4 (positive).

484

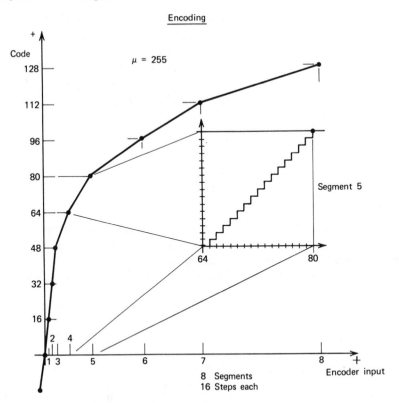

Figure 11.9 Positive portion of the segmented approximation of the μ-law quantizing curve used in the North American DS1 PCM channelizing equipment. (Courtesy ITT Telecommunications, Raleigh, NC.)

The coding process is closely connected to quantizing. In practical systems, whether using A-law or μ-law, quantizing uses segmented equivalents as discussed above and shown in Figure 11.5. Such segmenting is a handy aid to coding. Consider Figure 11.6, which shows the segmenting used on a 24-channel PCM system (A = law) developed by Standard Telephone and Cables, U.K. (STC). Here there are seven linear approximations (segments) above the origin and seven below, providing a 13-segment equivalent of the A-law. It is 13, not 14 (i.e., 7 + 7), because the segments passing through the origin are collinear and are counted as one, not two segments. In this system six bits identify the specific quantum level, and a seventh bit identifies whether it is positive (above the origin) or negative (below the origin). The maximum negative step is assigned 0000000, the maximum positive, 1111111. Obviously we are dealing with a seven-level code providing identification of 128 quantum steps, 64 above the origin and 64 below.

The 30 + 2 PCM system also uses a 13-segment approximation of the A-law, where A = 87.6. The 13 segments (14) lead us to an eight-level code. The coding for this system is shown in Figure 11.7. Again, if the first code element (bit) is 1, it indicates a positive value (e.g., the quantum step is located above the origin). The following three elements (bits) identify the segment, there being seven segments above the seven segments below the origin (horizontal axis).

As an example consider the fourth positive segment given as 1101XXXX in Figure 11.7. The first 1 indicates that it is above the horizontal axis (e.g., it is positive). The next three elements indicate the fourth step, or

$$0\text{—}1000 \text{ and } 1001$$
$$1\text{—}1010$$
$$2\text{—}1011$$
$$3\text{—}1100$$
$$\rightarrow 4\text{—}1101$$
$$5\text{—}1110 \text{ etc.}$$

Figure 11.8 shows a "blowup" of the uniform quantizing and subsequent straight-forward binary coding of step 4; it illustrates final segment coding, which is uniform, providing 16 ($2^4 = 16$) coded quantum steps.

The North American DS1 PCM system uses a 15-segment approximation of the

Code Level		Digit Number							
		1	2	3	4	5	6	7	8
255	(Peak positive level)	1	0	0	0	0	0	0	0
239		1	0	0	1	0	0	0	0
223		1	0	1	0	0	0	0	0
207		1	0	1	1	0	0	0	0
191		1	1	0	0	0	0	0	0
175		1	1	0	1	0	0	0	0
159		1	1	1	0	0	0	0	0
143		1	1	1	1	0	0	0	0
127	(Center levels)	1	1	1	1	1	1	1	1
126	(Nominal zero)	0	1	1	1	1	1	1	1
111		0	1	1	1	0	0	0	0
95		0	1	1	0	0	0	0	0
79		0	1	0	1	0	0	0	0
63		0	1	0	0	0	0	0	0
47		0	0	1	1	0	0	0	0
31		0	0	1	0	0	0	0	0
15		0	0	0	1	0	0	0	0
2		0	0	0	0	0	0	1	1
1		0	0	0	0	0	0	1	0
0	(Peak negative level)	0	0	0	0	0	0	1*	0

* One digit added to ensure that timing content of transmitted pattern is maintained.

Figure 11.10 Eight-level coding of the North American (ATT) DS1 PCM system. Note that actually there are really only 255 quantizing steps because steps 0 and 1 use the same bit sequence, thus avoiding a code sequence with no transitions (i.e., 0's only) (Ref. 6).

logarithmic μ-law. Again, there are actually 16 segments. The segments cutting the origin are collinear and counted as 1. The quantization in the DS1 system is shown in Figure 11.9 for the positive portion of the curve. Segment 5 representing quantizing steps 64–80 is shown blown up in the figure. Figure 11.10 shows the DS1 coding. As can be seen again, the first code element, whether a 1 or a 0, indicates if the quantum step is above or below the horizontal axis. The next three elements identify the segment, and the last four elements (bits) identify the actual quantum level inside that segment.

11.2.5 The Concept of Frame

As shown in Figure 11.3, PCM multiplexing is carried out in the sampling process, sampling several sources sequentially. These sources may be nominal 4-kHz voice channels or other information sources, possibly data or video. The final result of the sampling and subsequent quantization and coding is a series of pulses, a serial bit stream which requires some indication or identification of the beginning of a scanning sequence. This identification tells the far-end receiver when each full sampling sequence starts and ends; it times the receiver. Such identification is called *framing*. A full sequence or cycle of samples is called a frame in PCM terminology.

CCITT Rec. G.702 defines a frame as

A set of consecutive digit time slots in which the position of each digit time slot can be identified by reference to a frame alignment signal. The frame alignment signal does not necessarily occur, in whole or in part, in each frame.

Consider the framing structure of several practical PCM systems. The ATT D1* system is a 24-channel PCM system using a seven-level code (e.g., $2^7 = 128$ quantizing steps). To every seven bits representing a coded quantum step, one bit is added for signaling. To the full sequence one bit is added, called a framing bit. Thus a D1-frame consists of

$$(7 + 1) \times 24 + 1 = 193 \text{ bits}$$

making up a full sequence or frame. By definition 8000 frames are transmitted, so the bit rate is

$$193 \times 8000 = 1,544,000 \text{ bps}$$

The CEPT† 30 + 2 system is a 32-channel system where 30 channels transmit speech derived from incoming telephone trunks and the remaining two channels transmit signaling and synchronization information. Each channel is allotted a time slot, and we can speak of time slots 0–31 as follows:

*Now called D1A.
†CEPT = Conférénce Européenne des Postes et Télécommunications.

Time Slot	Type of Information
0	Synchronizing (framing)
1–15	Speech
16	Signaling
17–31	Speech

In time slot 0 a synchronizing code or word is transmitted every second frame, occupying digits 2–8 as follows:

$$0011011$$

In those frames without the synchronizing word, the second bit of time slot 0 is frozen at a 1 so that in these frames the synchronizing word cannot be imitated. The remaining bits of time slot 0 can be used for the transmission of supervisory information signals.

The current North American basic 24-channel PCM system, typified by the ATT D1D channel bank, varies compared with the older D1A system described above in

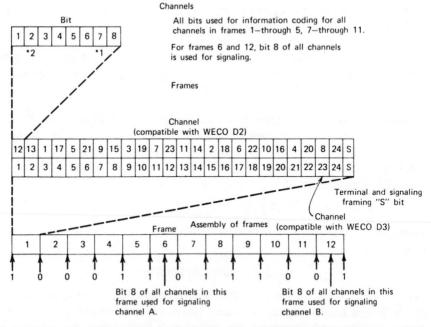

Figure 11.11 Frame structure of the North American (ATT) D1D (DS1) PCM system for the channel bank. Note the bit "robbing" technique used on each sixth frame to provide signaling information. *Notes.* (1) If bits 1–6 and 8 are 0, then bit 7 is transmitted as 1. (2) Bit 2 is transmitted as 0 on all channels for transmission if end-to-end alarm. (3) Composite pattern 000110111001, etc. (Courtesy ITT Telecommunications, Raleigh, NC.)

that all eight bits of a channel word are used in five out of six frames. In the remaining frame digit 1 is used for signaling. To accommodate these changes, it is necessary to change the framing format so that the specific frames containing signaling information can be identified. By using eight bits instead of seven for each channel word, allowing 256 amplitude values to be represented instead of 128 (in five out of six frames), quantizing noise is reduced. The companding characteristic is also different from its older D1A counterpart. D1D uses $\mu = 255$, whereas with D1A $\mu = 100$. This change made a significant difference in the signal-to-noise ratio over a wide range of input signals.

The D1D frame has a similar makeup as the D1A frame in that

$$8 \times 24 + 1 = 193 \text{ bits per frame}$$

producing a line data rate of $193 \times 8000 = 1.544$ Mbps. The frame structure is shown in Figure 11.11. Note that signaling is provided by "robbing" bit 8 from every channel in every sixth frame. For all other frames all bits are used to transmit information coding.

Framing and basic timing should be distinguished. Framing ensures that the PCM receiver is aligned regarding the beginning (and end) of a sequence or frame; timing refers to the synchronization of the receiver clock, specifically, that it is in step with its companion (far-end) transmit clock. Timing at the receiver is corrected via the incoming mark–space (and space–mark) transitions. It is important, then, that long periods without transitions do not occur. This point is discussed later in reference to line codes and digit inversion.

11.2.6 Quantizing Distortion

Quantizing distortion has been defined as the difference between the signal waveform as presented to the PCM multiplex (codec*) and its equivalent quantized value. Quantizing distortion produces a signal-to-distortion ratio S/D given by (Ref. 2, p. 573)

$$\frac{S}{D} = 6n + 1.8 \text{ dB} \qquad \text{(for uniform quantizing)}$$

where n = number of bits used to express a quantizing level. This bit grouping is often referred to as a PCM word. For instance, the ATT D1A system uses a 7-bit code word to express a level, and the 30 + 2 and D1D systems use essentially eight bits.

With a 7-bit code word (uniform quantizing),

---

*The term codec, meaning coder–decoder, is analogous to modem in analog circuits.

$$\frac{S}{D} = 6 \times 7 + 1.8 = 43.8 \text{ dB}$$

This demonstrates the linear relationship between the number of digits per sample and the signal-to-distortion ratio in decibels. Each added digit increases the signal-to-distortion ratio by 6 dB. Practical signal-to-distortion values range on the order of 33–38 dB* for average talker levels.

11.2.7   Idle Channel Noise

An idle PCM channel can be excited by the idle Gaussian noise and crosstalk present on the input analog channel. A decision threshold may be set which would control idle noise if it remains constant. With a constant-level input there will be no change in code word output, but any change of amplitude will cause a corresponding change in code word, and the effect of such noise may be an annoyance to the telephone listener.

One important overall PCM design decision to control idle channel noise is the selection of either the μ or the A values of the logarithmic quantizing curve used. The higher the values of these constants, the more finely granulated are the steps (quantizing steps finer) near the zero signal point. This tends to reduce idle channel noise. Care must also be taken to ensure that hum is minimized at the inputs of voice channels to the PCM equipment (codec).

11.3   OPERATION OF A PCM CODEC

PCM is four-wire. Voice channel inputs and outputs to and from a PCM multiplex are on a four-wire basis. The term *codec* is used to describe a unit of equipment carrying out the function of PCM multiplex and demultiplex and stands for coder-decoder, even though the equipment carries out more functions than just coding and decoding. A block diagram of a codec is shown in Figure 11.12.

A codec accepts 24 or 30 voice channels, depending on the system used, and digitizes and multiplexes the information. It delivers 1.544 Mbps to the line for the ATT DS1 channelizing equipment and 2.048 Mbps for the 30 + 2 (European) channelizing equipment. On the decoder side it accepts a serial bit stream at one or the other line modulation rate, demultiplexes the digital information, and performs digital-to-analog conversion. Output to the analog telephone network is 24 or 30 nominal 4-kHz voice channels. Figure 11.12 illustrates the processing of a single analog voice channel through a codec. The voice channel to be transmitted is passed through a 3.4-kHz low-pass filter. The output of the filter is fed to a sampling circuit. The sample of each channel of a set of n channels (n usually equals 24 or

---

*Using eight-level coding.

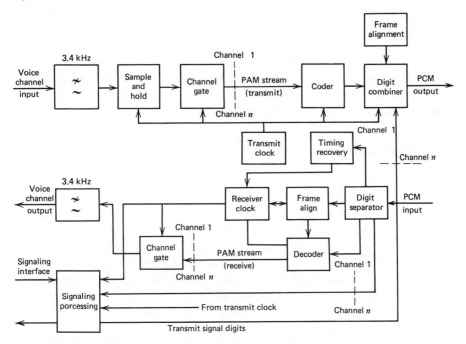

Figure 11.12 Simplified functional block diagram of a typical PCM codec.

30) is released in turn to the PAM highway. The release of samples is under control of a channel gating pulse derived from the transmit clock. The input to the actual coder is the PAM highway. The coder accepts a sample of each channel in sequence and then generates the appropriate signal character (channel word) corresponding to each channel presented. The coder output is the basic PCM signal that is fed to the digit combiner where framing alignment signals are inserted in the appropriate time slots, as well as the necessary supervisory signaling digits corresponding to each channel (European approach). The output of the digit combiner is the serial PCM bit stream fed to the line. As mentioned above, supervisory signaling is carried out somewhat differently in the North American (ATT) approach by robbing one bit in frame 6 and in frame 12 for this purpose.

On the receive side the codec accepts the serial PCM bit stream at the line rate at the digit separator. Here the signal is regenerated and split four ways to carry out the following processing functions: (1) timing recovery, (2) decoding, (3) frame alignment, and (4) signaling (supervisory). The timing recovery keeps the receive clock in synchronism with the far-end transmit clock. The receive clock provides the necessary gating pulses for the receive side of the PCM codec. The frame alignment circuit senses the presence of the frame-alignment signal at the correct time interval, thus providing the receive terminal with frame alignment. The decoder, under control of the receive clock, decodes the code character (channel word) signals corresponding to each channel. The output of the decoder are the reconstituted pulses making up a PAM highway. The channel gate accepts the PAM high-way, gating the *n*-channel PAM highway in sequence under control of the receive clock. The output of the channel gate is fed in turn to each channel filter, thus enabling the reconstituted analog voice signal to reach each appropriate voice path. Gating pulses extract signaling information in the signaling processor and apply this information to each of the reconstituted voice channels with the supervisory signaling interface as required by the analog telephone system in question.

11.4 PRACTICAL APPLICATION

11.4.1 General

PCM has found widest application in expanding interoffice trunks (junctions) which have reached exhaust* or will reach exhaust in the near future. An inter-office trunk is one pair of a circuit group connecting two switching points (ex-changes). Figure 11.13 sketches the interoffice trunk concept. Depending on the particular application, at some point where distance d is exceeded, it will be more economical to install PCM on the existing VF cable plant than to rip up streets and add more VF cable pairs. For the planning engineer, the distance d, where PCM becomes an economic alternative, is called the prove-in distance. d may vary from 8 to 16 km (5–10 mi), depending on the location and other circumstances. For distances less than d, additional VF cable pairs should be used for expanding the plant.

The general rule for measuring the expansion capacity of a given VF cable is as follows:

- For ATT DS1 channelizing equipment, two VF pairs will carry 24 PCM channels.
- For CEPT 30 + 2 system as configured by ITT, two VF pairs plus a phantom pair will carry 30 PCM speech channels.

*Exhaust is an outside plant term meaning that the useful pairs of a cable have been used up (assigned) from a planning point of view.

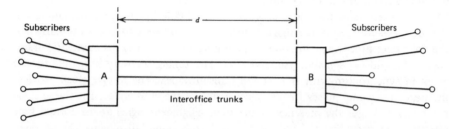

Figure 11.13 Simplified application diagram of PCM as applied to interoffice (interexchange) plant. *A* and *B* are switching centers.

All pairs in a VF cable may not necessarily be usable for PCM transmission. One restriction is brought about by the possibility of excessive crosstalk between PCM carrying pairs. The effect of high crosstalk levels is to introduce digital errors in the PCM bit stream. The error rate may be related on a statistical basis to crosstalk, which in turn is dependent on the characteristics of the cable and the number of PCM carrying pairs.

One method to reduce crosstalk and thereby increase VF pair usage is to turn to two-cable working, rather than have the "go" and "return" PCM cable pairs in the same cable.

Another item that can limit cable pair usage is the incompatibility of FDM and PCM carrier systems in the same cable. On the cable pairs that will be used for PCM, the following should be taken into consideration:

- All load coils must be removed.
- Build-out networks, bridged taps must also be removed.
- No crosses, grounds, splits, high-resistance splices, nor moisture permitted.

The frequency response of the pair should be measured out to 1 MHz and taken into consideration as far out as 2.5 MHz. Insulation should be checked with a megger. A pulse reflection test using a radar test set is also recommended. Such a test will indicate open circuits, short circuits, and high-impedance mismatches. A resistance test and balance test using a Wheatstone bridge may also be in order. Some special PCM test sets are available, such as the Lenkurt Electric 91100 PCM cable test set using pseudorandom PCM test signals and the conventional digital test eye pattern.

11.4.2   Practical System Block Diagram

A block diagram showing the elemental blocks of a PCM transmission link used to expand installed VF cable capacity is shown in Figure 11.14. Most telephone administrations (companies) distinguish between the terminal area of a PCM system and the repeatered line. The term *span* comes into play here. A span line is composed of a number of repeater sections permanently connected in tandem at repeater apparatus cases mounted in manholes or on pole lines along the span. A span is defined as the group of span lines which extend between two office (switching center) repeater points.

A typical span is shown in Figure 11.14. The spacing between regenerative repeaters is important. Section 11.4.1 mentioned the necessity of removing load coils from those trunk (junction) cable pairs which are to be used for PCM transmission. It is at these load points that the PCM regenerative repeaters are installed. On a VF line with H-type loading (see Section 2.8.4), spacing between load points is normally about 6000 ft (1830 m). It will be remembered from Chapter 2 that the first load coil out from the exchange on a trunk pair is at half-distance, or 3000 ft

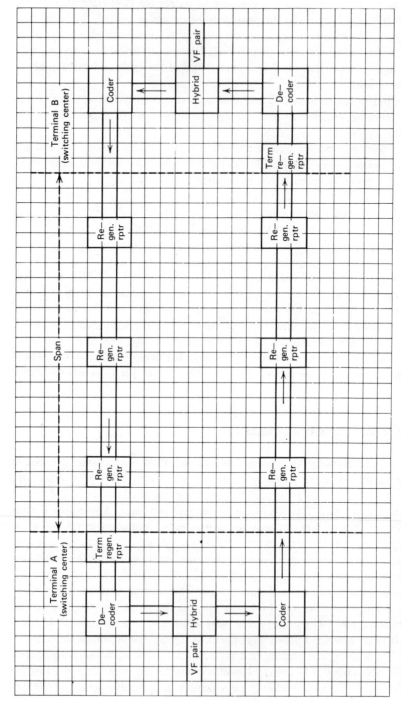

Figure 11.14 Simplified functional block diagram of a PCM link used to expand capacity of an existing VF cable. For simplicity, interface with only one VF pair is shown. Note the spacing between repeaters in the span line.

494

(915 m). This is provident, for a regenerative repeater must also be installed at this point. This spacing is shown in Figure 11.14 (1 space = 1000 ft). The purpose of installing a repeater at this location is to increase the pulse level before entering the environment of an exchange area where the levels of impulse noise may be quite high. High levels of impulse noise induced into the system may cause significant increases in the digital error rate of the incoming PCM bit streams, particularly when the bit stream is of a comparatively low level.

Commonly the PCM pulse amplitude output of a regenerative repeater is on the order of 3 V. Likewise, 3 V is the voltage on the PCM line crossconnect field at the exchange (terminal area).

A guideline used by Bell Telephone Manufacturing Company (Belgium) is that the maximum distance separating regenerative repeaters is that corresponding to a cable pair attenuation of 36 dB at the maximum expected temperature at 1024 kHz. This frequency is equivalent to the half-bit rate for the CEPT systems (e.g., 2048 kbs). Actually repeater design permits operation on lines with attenuations anywhere from 4 to 36 dB, allowing considerable leeway in placing repeater points. Table 11.1 gives some other practical repeater spacing parameters for the CEPT–ITT–BTM 30 + 2 system.

The maximum distance is limited by the maximum number of repeaters, which in this case is a function of power feeding and supervisory considerations. For instance, the fault location system can handle up to a maximum of 18 tandem repeaters for the BTM (ITT) configuration.

Power for the BTM system is via a constant-current feeding arrangement over a phantom pair serving both the "go" and the related "return" repeaters, providing up to 150 V dc at the power feed point. The voltage drop per regenerative repeater is 5.1 V. Thus for a "go" and "return" repeater configuration the drop is 10.2 V.

As an example, let us determine the maximum number of regenerative repeaters in tandem which may be fed from one power feed point by this system using 0.8-mm diameter pairs with a 3-V voltage drop in an 1830-m spacing between adjacent repeaters:

$$\frac{150}{(10.2 + 3)} = 11$$

Table 11.1   Line Parameters for ITT–BTM PCM

Pair Diameter (mm)	Loop Attenuation at 1 MHz (dB/km)	Loop Resistance (Ω/km)	Voltage Drop (V/km)	Maximum Distance[a] (km)	Total Number of Repeaters	Maximum System Distance (km)
0.9	12	60	1.5	3	18	54
0.6	16	100	2.6	2.25	16	36

[a]Between adjacent repeaters.

Assuming power fed from both ends and an 1800-m "dead" section in the middle, the maximum distance between power feed points is approximately

$$(2 \times 11 + 1) \ 1.8 \ \text{km} = 41.4 \ \text{km}$$

Fault tracing for the North American (ATT) T1 system is carried out by means of monitoring the framing signal, the 193rd bit (Section 11.2.5). The framing signal (amplified) normally holds a relay closed when the system is operative. With loss of the framing signal, the relay opens, actuating alarms. By this means a faulty system is identified, isolated, and dropped from "traffic."

To locate a defective regenerator on the BTM (Belgium)–CEPT system, traffic is removed from the system, and a special pattern generator is connected to the line. The pattern generator transmits a digital pattern with the same bit rate as the 30 + 2 PCM signal, but the test pattern can be varied to contain selected low frequency spectral elements. Each regenerator on the repeatered line is equipped with a special audio filter, each with a distinctive passband. Up to 18 different filters may be provided in a system. The filter is bridged across the output of the regenerator, sampling the output pattern. The output of the filter is amplified and transformer-coupled to a fault transmission pair, which is normally common to all PCM systems on the route, span, or section.

To determine which regenerator is faulty, the special test pattern is tuned over the spectrum of interest. As the pattern is tuned through the frequency of the distinct filter of each operative repeater, a return signal will derive from the fault transmission pair at a minimum specified level. Defective repeaters will be identified by the absence of a return signal or a return level under specification. The distinctive spectral content of the return signal is indicative of the regenerator undergoing test.

11.4.3   The Line Code

PCM signals as transmitted to the cable are in the bipolar mode (biternary), as shown in Figure 11.1. The marks or 1's have only a 50% duty cycle. There are several advantages to this mode of transmission:

- No dc return is required; thus transformer coupling can be used on the line.
- The power spectrum of the transmitted signal is centered at a frequency equivalent to half the bit rate.

It will be noted in bipolar transmission that the 0's are coded as absence of pulses, and the 1's are alternately coded as positive and negative pulses, with the alternation taking place at every occurrence of a 1. This mode of transmission is also called alternate mark inversion (AMI).

One drawback to straightforward AMI transmission is that when a long string of 0's is transmitted (e.g., no transitions), a timing problem may come about because repeaters and decoders have no way of extracting timing without transitions. The problem can be alleviated by forbidding long strings of 0's. Codes have been developed which are bipolar but with N zeros substitution; they are called BNZS codes. For instance, a B6ZS code substitutes a particular signal for a string of six 0's.

Another such code is the HDB3 code (high density binary 3), where the 3 indicates that it substitutes for binary formations with more than three consecutive 0's. With HDB3 the second and third zeros of the string are transmitted unchanged. The fourth 0 is transmitted to the line with the same polarity as the previous mark sent, which is a "violation" of the AMI concept. The first 0 may or may not be modified to a 1 to assure that the successive violations are of opposite polarity.

11.4.4 Signal-to-Gaussian-Noise Ratio on PCM Repeatered Lines

As we mentioned earlier, noise accumulation on PCM systems is not an important consideration. This does not mean that Gaussian noise (nor crosstalk, impulse noise) is not important. Indeed, it does affect the error performance, expressed as error rate (see Chapter 8). The error rate, from one point of view, is cumulative. A decision in error, whether 1 or 0, made anywhere in the digital system, is not recoverable.* Thus such an incorrect decision made by one regenerative repeater adds to the existing error rate on the line, and errors taking place in subsequent repeaters further down the line add in a cumulative manner, tending to deteriorate the received signal.

In a purely binary transmission system, if a 20-dB signal-to-noise ratio is maintained, the system operates nearly error free. In this respect, consider Table 11.2.

As discussed in Section 11.4.3, PCM, in practice, is transmitted on-line with alternate mark inversion. The marks have a 50% duty cycle, permitting energy concentration at a frequency of half the transmitted bit rate. Thus it is advisable to add 1 or 2 dB to the values shown in Table 11.2 to achieve a desired error rate on a practical system.

Table 11.2 Error Rate of a Binary Transmission System Versus Signal-to-RMS Noise Ratio

Error Rate	S/N (dB)	Error Rate	S/N (dB)
10^{-2}	13.5	10^{-7}	20.3
10^{-3}	16	10^{-8}	21
10^{-4}	17.5	10^{-9}	21.6
10^{-5}	18.7	10^{-10}	22
10^{-6}	19.6	10^{-11}	22.2

*Unless some special form of coding is used to correct the errors (see Chapters 8 and 15).

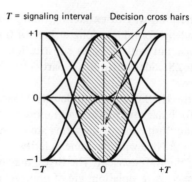

Figure 11.15   Sketch of an eye pattern.

11.4.5   The Eye Pattern

The "eye" pattern provides a convenient method of checking the quality of a digital transmission line. A sketch of a typical eye pattern is shown in Figure 11.15. Any oscilloscope can produce a suitable eye pattern, provided it has the proper rise time, which most quality oscilloscopes now available on the market have. The oscilloscope should either terminate or bridge the repeatered line or output of a terminal repeater. The display on the oscilloscope contains all the incoming bipolar pulses superimposed on one another.

Eye patterns are indicative of decision levels. The greater the eye opens, the better defined is the decision (whether 1 or 0 in the case of PCM). The opening is often referred to as the decision area (crosshatched in Figure 11.15). Degradations reduce the area. Eye patterns are often measured off in the vertical, giving a relative measure of the margin of decision.

Amplitude degradations shrink the eye in the vertical. Among amplitude degradations can be included echos, intersymbol interference, and decision threshold uncertainties.

Horizontal shrinkage of the eye pattern is indicative of timing degradations (i.e., jitter and decision time misalignment).

Noise is the other degradation to be considered. Usually noise may be expressed in terms of some improvement in the signal-to-noise ratio to bring the operating system into the bounds of some desired objective (see Table 11.2, for example). This ratio may be expressed as $20 \times \log$* of the ideal eye opening (in the vertical as read on the oscilloscope's vertical scale) to the degraded reading.

11.5   HIGHER ORDER PCM MULTIPLEX SYSTEMS

In Chapter 3 an FDM multiplex hierarchy was developed based on the 12-channel group. Five such groups were formed into a supergroup, thence the mastergroup

---

*Oscilloscopes are commonly used to measure voltage. Thus we can measure the degraded opening in voltage units and compare it to the full-scale perfect opening in the same units. A ratio is developed, and we take 20 log that ratio to determine the signal-to-noise ratio.

Table 11.3 PCM Multiplex Hierarchy Comparison

System Type	Level				
	1	2	3	4	5
North American ATT type	1	2	3	4	
Number of voice channels	24	96	672	4032	
Line bit rate (Mbps)	1.544	6.312	44.736	274.176	
Japan					
Number of voice channels	24	96	480	1440.0	5760.0
Line bit rate (Mbps)	1.544	6.312	32.064	97.728	400.352
Europe					
Number of voice channels	30	120	480	1920.0	7680.0
Line bit rate (Mbps)	2.048	8.448	34.368	139.264	560.0

and the supermastergroup. Likewise, in PCM a hierarchy of multiplex is developed based on the 24- or 30-channel group, which is called level 1. Subsequent levels are then developed (i.e., levels 2, 3, 4, and in one system, level 5). Table 11.3 summarizes and compares these multiplex levels for the North American system, Japan, and Europe (based on CCITT). The North American PCM hierarchy is shown in Figure 11.16, giving respective DS line rates and multiplex nomenclature. Regarding this nomenclature, we see from the figure that M12 accepts 1-level input, delivering 2-level to the line. It actually accepts four DS1 inputs deriving a DS2 output (6.312 Mbps). M13 accepts 1-level inputs, delivering 3-level to the line. In this case 28 DS1 inputs form one DS3 output (44.736 Mbps), the M34 takes six DS3 inputs

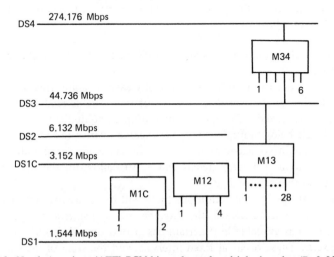

Figure 11.16 North American (ATT) PCM hierarchy and multiplexing plan (Ref. 22).

(level 3) to form one DS4 line rate (274.176 Mbps). DS1C is a special case where two 1.544-Mbps DS1 rates are multiplexed to form a 48-channel group with a line rate of 3.152 Mbps.

By simple multiplication we can see that the higher order line rate is a multiple of the lower input rate plus some number of bits. The DS1C is an example. Here the line rate is 3152 kbps, which is 2×1544 kbps + 64 kbps. The additional 64 kbps are used for multiplex synchronization and framing. Multiplex (and demultiplex) timing is very important, as one might imagine. The two DS1 signal inputs are each 1.544 Mbps plus and minus some tolerance (actually specified as ±130 ppm). The two input signals must be made alike in repetition rate and a rate suitable for multiplexing. This is done by *pulse stuffing.* In this process time slots are added to each signal in sufficient quantity to make the signal operate at a precise rate controlled by a common clock circuit in the multiplex. Pulses are inserted (or stuffed) into these time slots but carry no information. Thus it is necessary to code the signal in such a manner that these noninformation bits can be recognized and removed at the receiving terminal (demultiplex).

Of course, in the above example we are dealing with two DS1 sources which are physically separate and controlled by different clocks. If the sources were colocated (no difference in arrival due to delay) and controlled by a common clock, bit stuffing would not be necessary. However, this is academic because a common digital network multiplex rate at a specific level is a requirement, whether or not the input line rates are from independent tributaries.

Consider the more general case using CCITT terminology. If we wish to multiplex several lower level PCM bit streams deriving from separate tributaries into a single PCM bit stream at a higher level, a process of *justification* is required (called *bit stuffing* above). CCITT defines justification (Rec. G.702)

> as a process of changing the rate of a digital signal in a controlled manner so that it can accord with a rate different from its own inherent rate usually without loss of information.

Positive justification (as above) adds or stuffs digits; negative justification deletes or slips digits.

In the case of positive justification, normally each separate tributary bit stream is read into a store at its own data rate t, but the store is read out at a rate corresponding to T/n, where T = rate of the multiplex equipment and n = number of tributary signals being multiplexed. T/n is selected relative to t so that $T/n > t$ with a sufficient margin to accommodate the difference in relative data rates of the multiplex and input tributary signals and also to allow for the addition of frame alignment and other service digits.

Under normal operational conditions there will be variations between T/n and t. To provide for these variations, the sequence of time slots at the output of each tributary store has available in it certain designated time slots known as *justifiable digit time slots.* These occur at fixed intervals, and the state of the store fill deter-

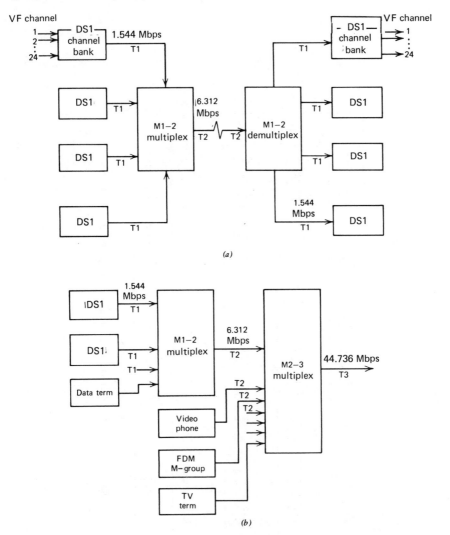

Figure 11.17 (a) Development of the 96-channel DS2 (ATT) system by multiplexing four 24-channel DS1 channel bank outputs. (b) Development of higher order PCM (ATT plan). [From (Ref. 2).]

mines whether or not the justifiable digit time slot has information written into it from the store. If the store is filling, the time slot is used, if not, the time slot is ignored. By this means a degree of "elasticity" is acquired which enables the relative timing difference to be absorbed.

Figure 11.17a is a simplified functional block diagram of the ATT M12 multiplex, and Figure 11.17b of a higher level multiplex scheme.

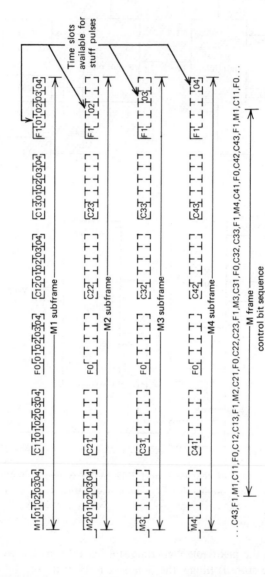

Figure 11.18 Organization of the DS2 signal bit stream (Ref. 22; Copyright © 1977 by American Telephone and Telegraph Co.)

11.5.1 Second-Level Frame Structures

The organization of the DS2 bit stream is shown in Figure 11.18. DS2 is transmitted on line at 6.312 Mbps (equivalent of 96 VF channels) and consists of four DS1 bit streams at 1.544 Mbps multiplexed, plus synchronization, framing, and stuff (justification) bits. All of the control information for the far-end demultiplexer is carried within an 1176-bit frame which is divided into four 294-bit subframes. The control-bit word, disbursed throughout the frame, begins with an M bit, as shown in Figure 11.18. The four M bits are transmitted as $011X$; the fourth bit X may be a 1 or a 0. This bit may be used as an alarm where 0 indicates an alarm condition and 1 no alarm. The 011 sequence for the first three M bits identifies (formats) the frame.

Within each of the four subframes two other control sequences are used. Each control bit is followed by a 48-bit block of information of which 12 bits are taken

Table 11.4 8448 kbps Multiplexing Frame Structure

Frame Structure	Bit Number
	Set I
Frame alignment signal (1111010000)	1–10
Alarm indication to the remote digital multiplex equipment	11
Bit reserved for national use	12
Bits from tributaries	13–212
	Set II
Justification control bits $C_{j1}{}^a$	1–4
Bits from tributaries	5–212
	Set III
Justification control bits $C_{j2}{}^a$	1–4
Bits from tributaries	5–212
	Set IV
Justification control bits $C_{j3}{}^a$	1–4
Bits from tributaries available for justification	5–8
Bits from tributaries	9–212
Tributary bit rate	2048 kbps
Number of tributaries	4
Frame length	848 bits
Bits per tributary	206 bits
Maximum justification rate per tributary	10 kbps
Nominal justification ratio	0.424

Source: CCITT Rec. G.742 (Courtesy ITU–CCITT.)

$^a C_{ji}$ indicates the ith justification control bit of the jth tributary.

from each of the four DS1 input signals. These are interleaved sequentially in the 48-bit block. The first bit in the third and sixth blocks is designated an *F* bit. The *F* bits are the sequence 0101 and are used to identify the location of the control bit sequences and the start of each information block.

The stuff (justification) control bits are transmitted at the beginning of each of the 48-bit blocks numbered 2, 4, and 5 within each subframe. These control bits are designated C in Figure 11.18. When a sequence is 000, no stuff pulse is present; when 111, a stuff pulse is added in the stuff position. The stuff bit positions are all assigned to the sixth 48-bit block of each subframe. In subframe 1 the stuff bit is the first bit after the F1 bit; for subframe 2 it is the second bit after the F1 bit, and so on through the fourth subframe. The nominal stuffing rate is 1796 bps for each DS1 signal input; the maximum is 5367 bps.

The output of the M12 multiplexer is in the B6ZS format.

The basic European second-order multiplex using positive justification is described in CCITT Rec. G.742. It accepts four tributary inputs, each from the standard 30 + 2 PCM channel bank which has a nominal line data rate of 2.048 Mbps. The output of the multiplex is a nominal 8.448 Mbps on-line, an equivalent of 4 X 30 or 120 VF channels. The multiplex frame structure is described in Table 11.4. CCITT Rec. G.742 recommends cyclic bit interleaving in the tributary numbering order and positive justification. Justification should be distributed using C_{jn} bits ($n = 1$, 2, 3). Positive justification is indicated by the signal 111 (stuffing), no stuffing by 000. Two bits per frame are available as service digits. Bit eleven of set I is used to transmit an alarm indication to the remote multiplex equipment. Faults indicated by service digits may be power supply failure, loss of tributary input (2.048 Mbps), loss of incoming 8.448-Mbps line signal, and loss of frame alignments.

11.5.2 Line Rates and Codes

Table 11.5 summarizes the North American DS series of PCM multiplex line rates, tolerances of these rates, and line codes. CCITT Rec. G.703 deals with line inter-

Table 11.5   ATT DS Series Line Rates, Tolerances, and Line Codes (Format)

Signal	Repetition Rate (Mbps)	Tolerance (ppm)[a]	Format	Duty Cycle (%)
DS0	0.064	[b]	Bipolar	100
DS1	1.544	±130	Bipolar	50
DS1C	3.152	±30	Bipolar	50
DS2	6.312	±30	B6ZS	50
DS3	44.736	±20	B3ZS	50
DS4	274.176	±10	Polar	100

[a]Parts per million.
[b]Expressed in terms of slip rate.

Table 11.6 Summary of 30 + 2 Related Line Rates and Codes

Level	Line Data Rate (Mbps)	Tolerance (ppm)	Code	Mark Peak Voltage (V)
1	2.048	±50	HDB3	2.37 or 3[a]
2	8.448	±30	HDB3	2.37
3	34.368	±20	HDB3	1.0
4	139.264	±15	CMI[b]	1 ± 0.1 V[c]

[a]2.37 V on coaxial pair; 3 V on symmetric wire pair.
[b]Coded mark inversion.
[c]Peak-to-peak voltage.

faces. For the 30 + 2 PCM and higher order multiplex derived therefrom, Table 11.6 provides a similar summary as Table 11.5.

11.6 TRANSMISSION OF DATA USING PCM, AS EXEMPLIFIED BY THE ATT DDS

The ATT digital data system (DDS) provides duplex point-to-point and multipoint private line digital data transmission at a number of synchronous data signal rates. This system is based on the standard 1.544–Mbps DS1 PCM line rate, where individual bit streams have data rates which are submultiples of that line rate (i.e., based on 64 kbps). However, pulse slots are reserved for identification in the demultiplexing of individual user bit streams as well as for certain status and control signals and to ensure that sufficient line pulses are transmitted for receive clock recovery and pulse regeneration. The maximum data rate available to a subscriber to the system is 56 kbps, some 87.5% of the 64-kbps theoretical maximum.

The 1.544-Mbps line signal as applied to DDS service consists of 24 sequential 8-bit words (i.e., channel time slots) plus one additional framing bit. This entire sequence is repeated 8000 times a second. Note that again we have $(192 + 1)8000 = 1.544$ Mbps, where the value 192 is 8×24 (see Section 11.2.5). Thus the line rate of a DDS facility is compatible with the DS1 (T1) PCM line rate and offers the advantage of allowing a mix of voice (PCM) and data where the full dedication of a DS1 facility to data transmission would be inefficient.

ATT calls the basic 8-bit word a *byte*. One bit of each 8-bit word is reserved for network control and for stuffing to meet nominal line bit rate requirements. This control bit is called a C-bit. With the C-bit removed we see where the standard channel bit rate is derived, namely, 56 kbps or 8000×7. Three subrate or submultiple data rates are also available: 2.4, 4.8, and 9.6 kbps. However, when these rates are implemented, an additional bit must be robbed from the basic byte to establish flag patterns to route each subrate channel to its proper demultiplexer port. This allows only 48 kbps out of the original 64 kbps for the transmission of user data. The 48-kbps composite total may be divided down to five 9.6-kbps channels, or ten 4.8

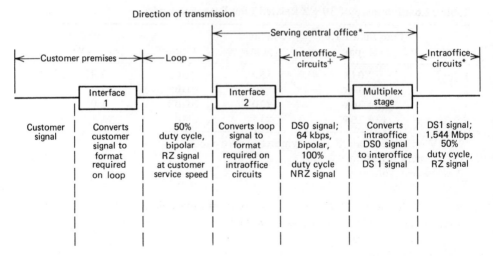

Figure 11.19 Subhierarchy of DDS signals. *Note.* Inverse processing must be provided for the opposite direction of transmission. Four-wire transmission is used throughout. *exchange; †PCM trunk. (Copyright © 1977 by American Telephone and Telegraph Company.)

kbps channels, or twenty 2.4 kbps channels. The subhierarchy of DDS signals is shown in Figure 11.19.

11.7 LONG-DISTANCE (TOLL) TRANSMISSION BY PCM

11.7.1 General

PCM, with its capability of regeneration, essentially eliminating the accumulation of noise as a signal traverses its transmission media, would appear to be the choice for toll transmission or backbone long-haul routes. This has not been the case. One must consider the disadvantages of PCM as well. Most important is the competition with FDM systems, the L5 system, for instance (Table 3.3). ATT's L5 system provides 10,800 VF channel capacity over long-haul coaxial cable media. The required bandwidth for this capacity on the cable is 60 MHz. To transmit the same number of channels by PCM would require on the order of 16 times the bandwidth.

Keep in mind the relationship briefly covered in Section 11.1 wherein this 16-multiple concept is shown: a 4-kHz voice channel requires an equivalent PCM bandwidth of 64 kHz, assuming 1-Hz bandwidth per PCM bit transmitted.

Thus a 10,800 VF (4-kHz) channel would require, if transmitted by PCM, about 691.2 MHz. The available bandwidth is still at a premium, whether by wire cable or radio. PCM at the DS1 line rate on the T1 repeatered line is capable of up to 200-mi (320-km) routes. In this case regenerative repeaters are spaced about 1 statute mi (1.6 km) apart. The transmission medium is a wire pair for each direction in standard multipair telephone cable. Powering points are at 36-mi (58-km) intervals. The ATT T2 system designed for the DS2 line rate (equivalent to 96 VF channels) may be used for distances up to 500 mi (800 km) in length. T2 requires special low-capacitance wire pairs in separate cables for opposite directions of transmission, crosstalk being a major system design consideration. Nominal repeater spacing is 15,000 ft. (4570 m).

The ATT T4M system is designed for the DS4 line rate, 274.176 Mbps, equivalent to 4032 VF channels, over 0.375-in. (9.5-mm) coaxial cable. T4M repeaters are spaced up to 5700 ft (1738 m) apart, and systems can work up to 500 mi (800 km) in length.

PCM on fiber optic cables, on both local trunk and toll (long-distance) routes, is now being implemented. With fiber optics, equivalent bandwidth is provided more economically, this being the major restraint covered above. The implementation of digital transmission on glass fiber cable is being spurred forward by the ever-increasing number of digital exchanges being installed requiring good-quality wideband landline trunks. One such high-capacity long-distance route in the United States is ATT's eastern corridor route from Boston via New York City to Washington, DC. [see Chapter 14 (FT3)]. Transcontinental PCM optical fiber systems will be implemented shortly. For one reason, optical fiber can provide the bandwidth relatively cheaply; another reason is that fewer repeaters are required per 100 km (60 mi) easing the jitter build-up problem.

11.7.2 Jitter

There is one other important limitation of present-day technology on using PCM as a vehicle for long-haul transmission. This is jitter, more particularly, timing jitter.

A general definition of jitter is "the movement of zero crossings of a signal (digital or analog) from their expected time of occurrence." In Chapter 8 it was called unwanted phase modulation or incidental FM. Such jitter or phase jitter affected the decision process or the zero crossing in a digital data modem. Much of this sort of jitter can be traced to the intervening FDM equipment between one end of a data circuit and the other.

PCM has no intervening FDM equipment, and jitter in PCM systems takes on different characteristics. However, essentially the effect is the same—uncertainty in a decision circuit as to when a zero crossing (transition) took place, or the shifting of a zero crossing from its proper location. In PCM it is more proper to refer to jitter as timing jitter.

The primary source of timing jitter is the regenerative repeater. In the repeatered line jitter may be systematic or nonsystematic. Systematic jitter may be caused by offset pulses (i.e., where the pulse peak does not coincide with regenerator timing peaks, or transitions are offset), intersymbol interference (dependent on specific pulse patterns), and local clock threshold offset. Nonsystematic jitter may be traced to timing variations from repeater to repeater and to crosstalk.

In long chains of regenerative repeaters, systematic jitter is predominant and cumulative, increasing in rms value as $N^{1/2}$, where N = number of repeaters in the chain. Jitter is also proportional to a repeater's timing filter bandwidth. Increasing the Q of these filters tends to reduce the jitter of the regenerated signal, but it also increases the error rate due to sampling the incoming signal at nonoptimum times.

The principal effect of jitter on the resulting analog signal after decoding is to distort the signal. The analog signal derives from a PAM pulse train which is then passed through a low-pass filter. Jitter displaces the PAM pulses from their proper location, showing up as undesired pulse position modulation (PPM).

Because jitter varies with the number of repeaters in tandem, it is one of the major restricting parameters of long-haul high-bit-rate PCM systems. Jitter can be reduced in future systems by using elastic store at each regenerative repeater (costly) and high-Q phase-locked loops.

11.8   DELTA MODULATION (DM)

11.8.1   Basic DM

DM is another method of transmitting an audio (analog) signal in a digital format. It is quite different from PCM in that coding is carried out before multiplexing and the code is far more elemental, actually coding at only 1 bit at a time.

The DM code is a one-element code and differential in nature. Of course we mean here that comparison is always made to the prior condition. A 1 is transmitted to the line if the incoming signal at the sampling instant is greater than the immediate previous sampling instant; it is a 0 if it is of smaller amplitude. With DM the derivative of the analog input is transmitted rather than the instantaneous amplitude as in PCM. This is achieved by integrating the digitally encoded signal and comparing it with the analog input to decide which of the two has the larger amplitude. The polarity of the next binary digit placed on line is either plus or minus, to reduce the amplitude of the two waveforms [i.e., analog input and integrated digital output (previous digit)]. We thus see the delta encoder basically as a feedback circuit, as shown in Figure 11.20.

Let's see how this feedback concept is applied to the delta encoder. Figure 11.21 illustrates the application. The switch is the double NAND gate and flip-flop. The comparator is the amplifier in Figure 11.20, and the feedback network is the integrating network.

The basic delta decoder consists of a current source, integrating network, amplifier, and low-pass filter. Figure 11.22 illustrates a simplified delta decoder.

We have seen that the digital output signal of the delta coder is indicative of the slope of the analog input signal (its derivative is the slope)—a 1 for positive slope

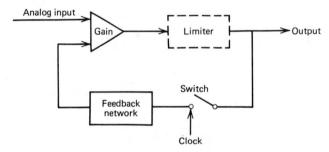

Figure 11.20   Basic electronic feedback circuit used in DM.

and 0 for negative slope. But the 1 and 0 give no idea of an instantaneous or even semiinstantaneous slope. This leads to the basic weakness found in the development of the DM system, namely, poor dynamic range or poor dynamic response, given a satisfactory signal-to-quantizing noise ratio. For delta circuits this limit is about 26 dB. A number greater than 26 dB is generally satisfactory, and a number numerically less than 26 dB is unsatisfactory. The reader is cautioned not to numerically equate quantizing noise in PCM to quantizing noise in DM, although the concept is the same.

One method used to improve the dynamic range of a DM system is by using two integrator circuits (we showed just one in Figure 11.21). This is called *double*

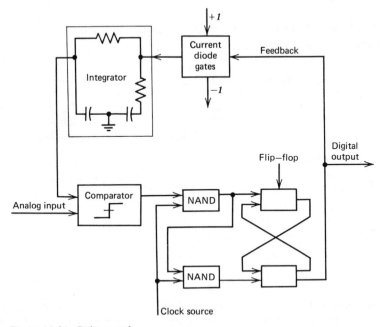

Figure 11.21   Delta encoder.

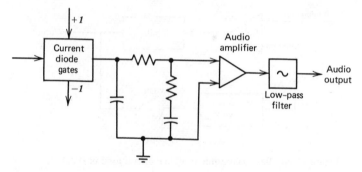

Figure 11.22 Delta decoder.

integration. Companding provides further improvement. Table 11.7 compares several 56-kbps digital systems with a 3-kHz bandwidth input analog signal regarding dynamic range.

As seen from Table 11.7, companded DM has properties equal to PCM. One reason we can take advantage of the good coding nature of companded DM is that voice signals are predictable, unlike band-limited random signals. The predictability can be based on knowledge of the speech spectrum or on the autocorrelation function. Thus delta coders can be designed on the principle of prediction.

The advantages of DM over PCM are as follows:

- Multiplexing is carried out by simple digital multiplexers, whereas PCM interleaves analog samples.

Table 11.7 Dynamic Range—Digital Modulation Systems Compared

System	Maximum Signal-to-Quantizing Noise Ratio (dB)	Dynamic Range for Minimum Signal-to-Quantizing Noise Ratio of 26 dB (dB)
Basic DM, single integration	34	8
Basic DM, double integration	44	18
Companded DM, single integration	34	15
Companded DM, double integration	44	31
7-digit companded PCM	30 dB UVR[a]	At S/N of 31 dB

[a]Useful volume range

- It is essentially more economical for small numbers of channels.
- It has few varieties of building blocks.
- It is less complex, thus giving improved reliability.
- Intelligibility is maintained with a BER down to 10^{-2}.

The disadvantages remain for the dynamic range and for the multiplexing of many channels.

DM has found wide application in military communications where low-bit-rate digital systems are required. Both the U.S. and NATO forces have fielded large quantities of delta multiplexers and switches based on 16 and 32 kbps. DM is also used in the commercial telephony world on thin-line telephone systems, for rural subscribers, and in certain satellite DAMA applications, such as Canada's TeleSat.

11.8.2 CVSD—A Subset of DM

Continuous variable-slope DM (CVSD) is a DM scheme that improves the basic weakness of DM, namely, small dynamic range over which the noise level is constant. CVSD is used by the U.S. military forces Tri-Tac communications system and by NATO's Eurocon with 16- and 32-kbps digital data line rates.

A CVSD coder is shown in Figure 11.23, a decoder in Figure 11.24, and the CVSD waveforms in Figure 11.25. CVSD circuitry provides increased dynamic range capability by adjusting the gain of the integrator. For a given clock frequency and input bandwidth, the additional circuitry increases the delta modulator's dynamic range (i.e., up to 50 dB of range). External to the basic delta modulator is an algorithm which monitors the past few outputs of the delta modulator in a

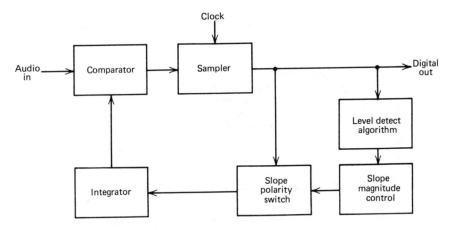

Figure 11.23 A CVSD encoder.

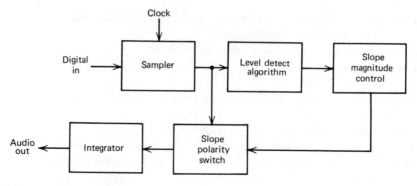

Figure 11.24   A CVSD decoder.

simple shift register. The register is usually 3–4 bits long, depending on the application. A CVSD algorithm monitors the contents of the shift register and indicates
whether it contains all 1's or all 0's. This condition is called coincidence. When it
occurs, it indicates that the gain of the integrator is too small. In this particular
design the coincidence changes a single-pole low-pass filter. The voltage output
of this syllabic filter controls the integrator gain through a pulse amplitude modulator whose other input is the sign bit or up/down control.

The algorithm provides a means of measuring the average power or level of the
input signal. The purpose of the algorithm is to control the gain of the integrator
and to increase the dynamic range. The algorithm is repeated in the receiver to
recover level data in the decoding process. Because the algorithm only operates
on past serial data, it changes the nature of the bit stream without changing the
channel bit rate.

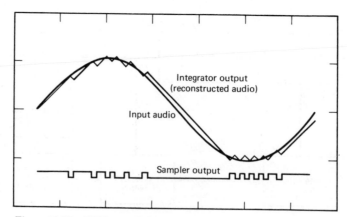

Figure 11.25   CVSD waveforms.

The effect of the algorithm is to compand the input signal. If a CVSD encoder is fed into a basic delta demodulator, the output of that demodulator will reflect the shape of the input signal, but all of the output will be at an equal level. Thus the algorithm at the output is needed to restore the level variations.

REFERENCES AND BIBLIOGRAPHY

1. *Data Handbook for Radio Engineers*, 6th ed., Howard W. Sams, Indianopolis IN, 1977. Section 23.

2. *Transmission Systems for Communications*, 4th ed., Bell Telephone Laboratories, American Telephone and Telegraph Co., New York, 1971.

3. *Lenkurt Demodulator*, Lenkurt Electric Corp., San Carlos, CA, Nov. 1966, Mar. Apr. 1968, Sept, 1971, Dec. 1971, Jan. 1972, Oct. 1973.

4. Seminar on Pulse Code Modulation Transmission, Standard Telephone and Cables Limited, Basildon, England, June 1967.

5. "PCM—System Application—30 + 2*TS*," BTM/ITT, Antwerp, Belgium.

6. *Operations and Maintenance Manual for T324 PCM Cable Carrier System*, ITT Telecommunications, Raleigh, NC, Apr. 1973.

7. *Principles of Modems* (edited draft), Communication Systems, Inc., Falls Church, VA (limited circulation).

8. K. W. Catermole, *Principles of Pulse Code Modulation*, Iliffe, London, 1969.

9. W. C. Sain, "Pulse Code Modulation Systems in North America," *Electrical Commun.* (ITT), Vol. 48, No. 1/2, 1973.

10. J. V. Marten and E. Brading, "30-Channel Pulse Code Modulation System," *Electrical Commun.*, (ITT), Vol. 48, No. 1/2, 1973.

11. K. E. Fultz and D. B. Penick, "The T1 Carrier System," *Bell Sys. Tech. J.*, Vol. 44, Sept. 1965.

12. R. B. Moore, "T2 Digital Line System," *Proceedings IEEE International Conference on Communications*, Seattle WA., June 12, 1973.

13. J. R. Davis, "T2 Repeater and Equalization," *Proceedings IEEE International Conference on Communications*, Seattle, WA, June 12, 1973.

14. E. Cookson and C. Volkland, "Taking the Mystery out of Phase Jitter Measurement," *Telephony*, Sept. 25, 1972.

15. "Phase Jitter and Its Measurement," CCITT Study Group IV, Question 3/IV, 17–28, Jan. 1972.

16. L. Katzschner *et al.*, "An Experimental Local PCM-Switching System," *IEEE Trans. Commun.*, Oct. 1973.

17. A. E. Pinet, "Telecommunication Integrated Network," *IEEE Trans. Commun.*, Aug. 1973.

18. G. C. Hartley *et al.*, *Techniques of Pulse-Code Modulation in Communication Networks*, IEE Monograph Series, Cambridge University Press, 1967.

19. P. Bylanski and D. G. W. Ingram, "Digital Transmission Systems," IEE Telecommunications Series 4, Peter Peregrinus Ltd., Stevenage, Herts, England, 1976.

20. H. Nyquist, "Certain Topics in Telegraph Transmission Theory," *Trans. AIEE*, Vol. 47, 617–644, Apr. 1928.

21. C. E. Shannon, "A Mathematical Theory of Communication," *Bell Sys. Tech. J.* Vol. 27, 623–656, 1948.

22. *Telecommunication Transmission Engineering,* Vol. 2, 2nd ed., American Telephone and Telegraph Co., New York, 1977.

23. G. H. Bennett, "PCM and Digital Transmission," Marconi Instruments, St. Albans, Herts, England, 1978.

24. H. Akima, "Noise Power Due to Digital Errors in a PCM Telephone Channel," OT Rep. 78–139, NTIS PB 277–447, U.S. Department of Commerce, Office of Telecommunications, Jan. 1978.

25. R. L. Freeman, *Telecommunication System Engineering,* Wiley Interscience, New York, 1980.

26. H. R. Schindler, "Delta Modulation," *IEEE Spectrum,* Oct. 1970.

27. CCITT Orange Books, Geneva, 1976, Vol. III, G recommendations; in particular the G. 700 series.

12 VIDEO TRANSMISSION

12.1 GENERAL

This chapter provides the basic essentials for designing point-to-point video transmission systems. To understand the video problem the transmission system engineer must first have an appreciation of video and how the standard TV video signal is developed. The discussion that follows provides an explanation of the "what and why" of video. There follows a review of critical video transmission parameters, black and white and color transmission standards, video program channel transmission, and the transmission of video over specific media. Finally there is a brief discussion of basic tests of video point-to-point facilities. TV broadcast problems are covered only where they specifically interact with point-to-point transmission.

12.2 AN APPRECIATION OF VIDEO TRANSMISSION

A video transmission system must deal with four factors when transmitting images of moving objects:

- A perception of the distribution of luminance or simply the distribution of light and shade
- A perception of depth or a three-dimensioned perspective
- A perception of motion relating to the first two factors above
- A perception of color (hues and tints)

Monochrome TV deals with the first three factors. Color TV includes all four factors.

A video transmission system must convert these three (or four) factors into electrical equivalents. The first three factors are integrated to an equivalent electric current or voltage whose amplitude is varied with time. Essentially, at any one moment it must integrate luminance from a scene in the three dimensions (i.e., width, height, and depth) as a function of time. And time itself is still another variable, for the scene is changing in time.

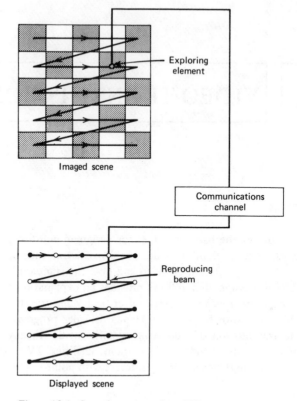

Figure 12.1 Scanning process from TV camera to receiver display.

The process of integration of visual intelligence is carried out by *scanning*. The horizontal detail of a scene is transmitted continuously and the vertical detail discontinuously. The vertical dimension is assigned discrete values which become the fundamental limiting factor in a video transmission system.

The scanning process consists of taking a horizontal strip across the image on which discrete square elements are scanned from left to right. When the right-hand end is reached, another, lower, horizontal strip is explored, and so on until the whole image has been scanned. Luminance values are translated on each scanning interval into voltage and current variations and are transmitted over the system. The concept of scanning by this means is shown in Figure 12.1.

The National Television Systems Committee (U.S.) (NTSC) practice divides an image into 525 horizontal scanning lines. It is the number of scanning lines that determines vertical detail or resolution of a picture.

When discussing picture resolution, the aspect ratio is the width-to-height ratio of the video image. The aspect ratio used almost universally is 4 : 3. In other words, a TV image 12 in. wide would necessarily be 9 in. high. Thus an image divided into 525 (491) vertical elements would then have 700 (652) horizontal elements to maintain an aspect ratio of 4:3. The numbers in parentheses represent the practical maximum active lines and elements. Therefore the total number of elements approaches something on the order of 250,000. We reach this number because, in practice, vertical detail reproduced is 64–87% of the active scanning lines. A good halftone engraving may have as many as 14,400 elements per square inch compared to approximately 3000 elements per square inch for a 9 by 12 in. TV image.

Motion is another variable factor that must be transmitted. The sensation of continuous motion in standard TV video practice is transmitted to the viewer by a successive display of still pictures at a regular rate similar to the method used in motion pictures. The regular rate of display is called the *frame rate.* A frame rate of 25 frames per second will give the viewer a sense of motion, but on the other hand he will be disturbed by luminance flicker (bloom and decay), or the sensation that still pictures are "flicking" on screen one after the other. To avoid any sort of luminance flicker sensation, the image is divided into two closely interwoven (interleaving) parts, and each part is presented in succession at a rate of 60 frames per second, even though *complete* pictures are still built up at a 30 frame-per-second rate. It should be noted that interleaving improves resolution as well as improving apparent persistence of the CRT by tending to reinforce the scanning spots. It has been found convenient to equate flicker frequency to power line frequency. Thus in North American practice, where power line frequency is 60 Hz, the flicker is 60 frames per second. In Europe it is 50 frames per second to correspond to the 50-Hz line frequency used there.

Following North American practice, some other important parameters derive from the previous paragraphs.

1. A field period is 1/60 s. This is the time that is required to scan a full picture on every horizontal line.

2. The second scan covers the lines not scanned on the first period, offset one-half horizontal line.

3. Thus 1/30 s is required to scan all lines on a complete picture.

4. The transit time of exploring and reproducing scanning elements or spots along each scanning line is $1/15,750$ s (525 lines in $1/30$ s) = 63.5 μs.

5. Consider that about 16% of the 63.5 μs is consumed in flyback and synchronization. Therefore only about 53.3 μs are left per line of picture to transmit information.

What will be the bandwidth necessary to transmit images so described? Consider the worst case where each scanning line is made up of alternate black and white squares, each the size of the scanning element. There would be 652 such elements. Scan the picture, and a square wave will result with a positive-going square for white and a negative for black. If we let a pair of adjacent square waves be equivalent to a sinusoid (see Figure 12.2), then the baseband required to transmit the image will have an upper cutoff of about 6.36 MHz, allowing for no degradation in the intervening transmission system. The lower limit will be a dc or zero frequency.

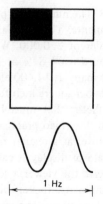

Figure 12.2   Development of a sinusoid wave from the scan of adjacent squares.

12.3   THE COMPOSITE SIGNAL

The word *composite* is confusing in the TV industry. On one hand, composite may mean the combination of the full video signal plus the audio subcarrier; the meaning here is narrower. Composite in this case deals with the transmission of video information as well as the necessary synchronizing information.

Consider Figure 12.3. An image made up of two black squares is scanned. The total time for the line is 63.5 μs, of which 53.3 μs is available for the transmission of actual video information and 10.2 μs is required for synchronization and flyback.

During the retrace time or flyback it is essential that no video information be transmitted. To accomplish this, a blanking pulse is superimposed on the video at the camera. The blanking pulse carries the signal voltage into the reference black region. Beyond this region in amplitude is the "blacker then black" region, which is allocated to the synchronizing pulses. The blanking level (pulse) is shown in Figure 12.3.

The maximum signal excursion of a composite video signal is 1.0 V. This 1.0 V is a video/TV reference and is always taken as a peak-to-peak measurement. The 1.0 V may be reached at maximum synchronizing voltage and is measured between synchronizing "tips."

Of the 1.0 V peak to peak, 0.25 V is allotted for the synchronizing pulses, 0.05 V for the setup, leaving 0.7 V to transmit video information. Thus the video signal varies from 0.7 V for the white through gray tonal region to 0 V for black. The best description of the actual video portion of a composite signal is to call it a succession of rapid nonrepeated transients.

The synchronizing portion of a composite signal is exact and well defined. A TV/video receiver has two separate scanning generators to control the position of the reproducing spot. These generators are called the horizontal and vertical scanning generators. The horizontal one moves the spot in the X or horizontal direction, and the vertical in the Y direction. Both generators control the position of the spot on

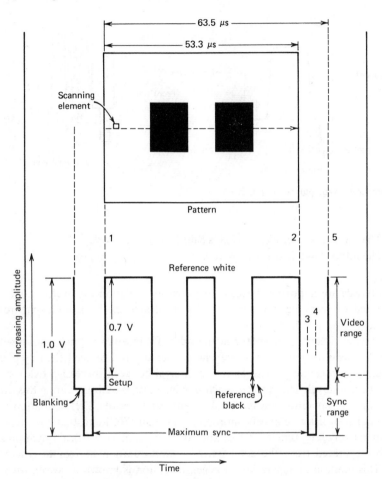

Figure 12.3 Breakdown in time of a scan line.

the receiver and must in turn be controlled from the camera (transmitter) synchronizing generator to keep the receiver in step (synchronization).

The horizontal scanning generator in the video receiver is synchronized with the camera synchronizing generator at the end of each scanning line by means of horizontal synchronizing pulses. These are the synchronizing pulses shown in Figure 12.3 and have the same polarity as the blanking pulses.

When discussing synchronization and blanking, we often refer to certain time intervals. These are discussed as follows:

- The time at the horizontal blanking pulse, 2–5 in Figure 12.3, is called the *horizontal synchronizing interval.*

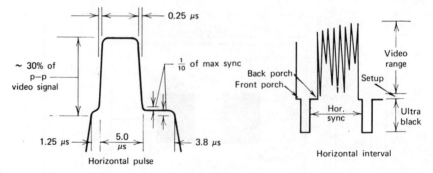

Figure 12.4 Sync pulses and porches.

- The interval 2–3 in Figure 12.3 is called the *front porch*.
- The interval 4–5 is the *back porch*.

The intervals are important because they provide isolation for overshoots of video at the end of scanning lines. Figure 12.4 illustrates the horizontal synchronizing pulses and corresponding porches.

The vertical scanning generator in the video/TV receiver is synchronized with the camera (transmitter) synchronizing generator at the end of each field by means of vertical synchronizing pulses. The time interval between successive fields is called the vertical interval. The vertical synchronizing pulse is built up during this interval. The scanning generators are fed by differentiation circuits. Differentiation for the horizontal scan has a relatively short time constant (RC) and that for the vertical a comparatively long time constant. Thus the long-duration vertical synchronization may be separated from the comparatively short-duration horizontal synchronization. This method of separation of synchronization is known as *waveform separation* and is standard in North America.

In the composite video signal (North American standards) the horizontal synchronization has a repetition rate of 15,750 frames per second, and the vertical synchronization has a repetition rate of 60 frames per second.

12.4 CRITICAL VIDEO TRANSMISSION IMPAIRMENTS

The nominal video baseband is divided into two segments: the HF segment, above 15,750 Hz, and the low-frequency (LF) segment, below 15,750 Hz. Impairments in the HF segment are operative (in general) along the horizontal axis of the received video image. LF impairments are generally operative along the vertical image axis.

Impairment	Cause

HF segment

Impairment	Cause
Undistorted echos	Cyclic gain and phase deviation throughout the passband
Distorted echos	Nonlinear gain and phase deviations, particularly in the higher end of the band
HF cutoff effects, ringing	1. Limited bandpass distorts, shows picture transitions, causing overshoot and undershoot 2. This type of distortion (ringing) may also show up on test pattern from lack of even energy distribution and reduced resolution
Porch distortion (poor reproduction of porches)	Poor attenuation and phase distortion
Porch displacement	Zero wander, dc-restored devices such as clamper circuits
Smearing (blurring of the vertical edges of objects)	1. Coarse variations in attenuation and phase 2. Quadrature distortion

LF segment

Impairment	Cause
Dc suppression, zero wander, distorted image	Lack of clamping
LF roll-off (gradual shading from top to bottom)	Poor clamping, deterioration of coupling networks
Streaking	Phase and attenuation distortion, usually from transmission medium

Nonlinear distortion

Impairment	Cause
Nonlinearity in the extreme negative region, resulting in horizontal striations or streaking	Compression of synchronizing pulses

Impairments due to noise

Noise, in this case, may be considered an undesirable visual sensation. Noise is considered to consist of three types: single frequency, random, and impulse.

Impairment	Cause
Unwanted pattern in received picture	Single-frequency noise
Pattern in alternate fields	Single-frequency noise as an integral

Impairment	Cause

	multiple of field frequency (50 or 60 Hz)
Horizontal or vertical bars It should be noted that the LF region is very sensitive to single-frequency noise.	Single-frequency noise
Picture graininess, "snow"	As random noise increases, graininess of picture increases

Note. System design objective of 47 dB signal-to-weighted noise ratio or 44 dB signal-to-flat noise ratio, based on 4.2-MHz video bandwidth, meets Television Allocation Study Organization (TASO) rating of excellent picture. Peak noise should be 37 dB below peak video. Signal-to-noise ratios are taken at peak synchronizing tips to rms noise

| Noise "hits," momentary loss of synchronization, momentary rolling, momentary masking of picture | Impulse noise |

Note. Bell System limit at receiver input -20 dB reference to 1 V peak-to-peak signal level point, 1 hit/min. Large amplitudes of impulse noise often masked in black.

| Weak, extraneous image superimposed on main image Nonsynchronization of two images causes violent horizontal motion. Effect is most noticeable in line and field synchronization intervals | Strong crosstalk |

Note. Limiting loss in crosstalk coupling path should be 58 dB or greater for equal signal levels, design objective 61 dB.
Crosstalk for video may be defined as the coupling between two TV channels

The preceding sections present a short explanation of the mechanics of video transmission for a video camera directly connected to a receiving device for display. The primary concern of the chapter, however, is to describe and discuss the problems of point-to-point video transmission. Often the entity responsible for the point-to-point transport of TV programs is not the same entity that originated the image transmitted, except in the case of that link directly connecting the studio to a local transmitter (STL links). A great deal of the "why" has now been covered. The remaining parts of the chapter discuss the video transmission problem (the "how") on a point-to-point basis. The medium employed for this purpose may be radiolink, satellite link, coaxial cable, fiber optics cable, or specially conditioned wire pairs.

12.5 CRITICAL VIDEO PARAMETERS

12.5.1 General

Raw video baseband transmission requires excellent frequency response, particularly from dc to 15 kHz and extending to 4.2 MHz for North American systems and to 5 MHz for European systems. Equalization is extremely important. Few point-to-point circuits are transmitted at baseband because transformers are used for line coupling which deteriorate low frequency response and make phase equalization very difficult.

To avoid low-frequency deterioration, cable circuits transmitting video have resorted to the use of carrier techniques and frequency inversion using VSB modulation. However, if raw video baseband is transmitted, care must be taken in preserving its dc component.

12.5.2 Transmission Standard—Level

Standard power levels have developed from what is roughly considered to be the input level to an ordinary TV receiver for a noise-free image. This is 1 mV across 75 Ω. With this as a reference TV levels are given in dBmV. For RF and carrier systems carrying video the measurement refers to rms voltage. For raw video it is 0.707 of instantaneous peak voltage, usually taken on synchronizing tips.

The signal-to-noise ratio is normally expressed for video transmission as follows:

$$\frac{S}{N} = \frac{\text{peak signal (dBmV)}}{\text{rms noise (dBmV)}}$$

TASO picture ratings (4-MHz bandwidth) are related to the signal-to-noise ratio (RF) as follows:

1. Excellent (no perceptible snow)   45 dB
2. Fine (snow just perceptible)   35 dB
3. Passable (snow definitively perceptible but not objectionable)   29 dB
4. Marginal (snow somewhat objectionable)   25 dB

12.5.3 Other Parameters

For black and white video systems there are four critical transmission parameters:

1. Amplitude–frequency response
2. EDD (group delay)
3. Transient response
4. Noise (thermal, IM, crosstalk, and impulse)

Color transmission requires consideration of two additional parameters:

5. Differential gain
6. Differential phase

Descriptions of amplitude–frequency response and EDD may be found in Sections 1.9.3 and 1.9.4, respectively. Because video transmission involves such wide bandwidths compared to the voice channel and because of the very nature of video itself, both delay and amplitude requirements are much more stringent.

Transient response is the ability of a system to "follow" sudden, impulsive changes in signal waveform. It usually can be said that if the amplitude–frequency and envelope delay characteristics are kept within design limits, the transient response will be sufficiently good.

Noise is described in Section 1.9.6. Differential gain is the variation in the gain of the transmission system as the video signal level varies (i.e., as it traverses the ex-

Table 12.1 Critical Parameter Limits

	Input to TV Receiver	Point-to-Point Radio Transmission Facility[a] Single Hop
Amplitude–frequency response		±0.1 dB, 1–300 kHz
		±0.4 dB, 0.3–4.3 MHz
EDD		Deferred
Transient response $2T \sin^2$ pulse–bar K		1%
Noise (expressed as S/N)		
Hum	45 dB	46 dB
IM	50 dB	
Crosstalk	50 dB	
Impulse		
Random (thermal + hum + IM)	47 dB	59 dB
Single-frequency interference		60 dB, 1-kHz 4.3
Differential gain		MHz
10% APL[b]		0.5 dB
50% APL		0.3 dB
90% APL		0.5 dB
Differential phase		
10% APL		±1.0°
50% APL		±0.7°
90% APL		±1.0°
Signal polarity in transmission system		Black negative

[a] From EIA RS-250A.
[b] APL = average picture level.

tremes from black to white). Differential phase is any variation in phase of the color subcarrier as a result of a changing luminance level. Ideally variations in the luminance level should produce no changes in either the amplitude or the phase of the color subcarrier. Table 12.1 summarizes critical transmission parameters for video.

12.6 VIDEO TRANSMISSION STANDARDS (*Criteria for Broadcasters*)

The following outlines video transmission standards from the point of view of broadcasters (i.e., as emitted from TV broadcast transmitters). Figure 12.5 illustrates the components of the emitted wave (North American practice).

12.6.1 Basic Standards

Table 12.2 gives a capsule summary of some national standards as taken from CCIR Rep. 308 (Ref. 1).

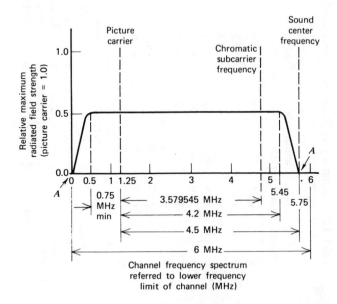

Figure 12.5 RF amplitude characteristics of TV picture transmission. Field strength at points *A* shall not exceed 20 dB below picture carrier. Drawing not to scale.

Table 12.2 Summary of Some Characteristics of TV Systems

System[a]	No. of Lines	Channel Width (MHz)	Vision Width (MHz)	Separation Vision–Sound	VSB	Vision Modulation	Sound Modulation	Sweep Rate	Picture Rate	Field Frame
A	405	5	3	−3.5	0.75	Positive	AM	10.125k	25	50
B	625	7	5	+5.5	0.75	Negative	FM	15.625k	25	50
C	625	7	5	+5.5	0.75	Positive	AM	15.625k	25	50
D	625	8	6	+6.5	0.75	Negative	FM	15.625k	25	50
E	819	14	10	+11.15	2.	Positive	AM	20.475k	25	50
F	819	7	5	+5.5	0.75	Positive	AM	20.475k	25	50
G	625	8	5	+5.5	0.75	Negative	FM	15.625k	25	50
I	625	8	5.5	+6	1.25	Negative	FM	15.625k	25	50
L	625	8	6	+6.5	1.25	Positive	AM	15.625k	25	50
M	525	6	4.2	+4.5	0.75	Negative	FM	15.750k	30	60

Source: CCIR Rep. 308 (Ref. 1). Consult report for complete table.

[a]

Austria B	Denmark B	Italy B	Norway B, G	Switzerland B
Belgium C, F	Finland B	Japan M	Poland D	United Kingdom A, I
Bulgaria D	France E, L	Luxemborg F	Portugal B	United States M
Canada M	Hungary D	Monaco E	Spain E	U.S.S.R B
Czechoslovakia D	Ireland A	Netherlands B	Sweden B	West Germany B, G

UNITED STATES (Ref. 2, sec. 30-13)

Channel width (transmission)	6 MHz
Video	4.2 MHz
Aural	±25 kHz
(see Figure 12.5)	
Picture carrier location	1.25 MHz above lower boundary of channel
Modulation	AM composite picture and synchronizing signal on visual carrier together with FM audio signal on audio carrier
Scanning lines	525 per frame, interlaced 2 : 1
Scanning sequence	Horizontally from left to right, vertically from top to bottom
Horizontal scanning frequency	15,750 Hz for monochrome, or 2/455 × chrominance subcarrier, = 15,734.264 ± 0.044 Hz for NTSC color transmission
Vertical scanning frequency	60 Hz for monochrome, or 2/525 × horizontal scanning frequency for color = 59.94 Hz
Blanking level	Transmitted at 75 ± 25% of peak carrier level
Reference black level	Black level is separated from blanking level by 7.5 ± 2.5% of video range from blanking level to reference white level
Reference white level	Luminance signal of reference white is 12.5 ± 2.5% of peak carrier
Peak-to-peak variation	Total permissible peak-to-peak variation in one frame due to all causes is less than 5%
Polarity of transmission	Negative; a decrease in initial light intensity causes an increase in radiated power
Transmitter brightness response	For monochrome TV, RF output varies in an inverse logarithmic relation to brightness of scene
Aural transmitter power	Maximum radiated power is 20% (minimum 10%) of peak visual transmitter power

BASIC EUROPEAN STANDARD

Channel width (transmission)	7 MHz low band, 8 MHz high band

Video	5 MHz
Aural	±50 kHz

Picture carrier location — 1.25 MHz above lower boundary of channel

Note. VSB transmission is used, similar to North American practice.

Modulation	AM composite picture and synchronizing signal on visual carrier together with FM audio signal on audio carrier
Scanning lines	625 per frame, interlaced 2:1
Scanning sequence	Horizontally from left to right, vertically from top to bottom
Horizontal scanning frequency	15,625 Hz ± 0.1%
Vertical scanning frequency	50 Hz
Blanking level	Transmitted at 75 ± 2.5% of peak carrier level
Reference black level	Black level is separated from blanking by 3–6.5% of peak carrier
Peak white level as a percentage of peak carrier	10–12.5%
Polarity of transmission	Negative; a decrease in initial light intensity causes an increase in radiated power
Aural transmitter power	Maximum radiated power is 20% of peak visual power

SOME VARIANCES (see also Table 12.2)

United Kingdom

Scanning lines	405 per frame, interlaced 2:1 on low band
Horizontal scanning frequency	10,125 Hz low band
Video bandwidth	3 MHz low band
Nominal RF bandwidth	5 MHz low band
Aural transmitter power	Maximum radiated power is 25% of peak visual power
Type of sound modulation	AM, low band
Synchronizing level as % of peak carrier	30%
Blanking level as % of peak carrier	30%

Black level	Same as blanking level, low band
Peak white level as % of peak carrier	100%, low band
Sound carrier relative to vision carrier	3.5 MHz low band

France

Scanning lines	819 per frame
Nominal video bandwidth	10 MHz
Horizontal scanning frequency	20,475 Hz
Nominal RF bandwidth	14 MHz
Sound carrier relative to vision carrier	11.5 MHz
Type of sound modulation	AM
Nominal width of VSB	2 MHz
Synchronizing level as % of peak carrier	30%
Difference between black level and blanking level as % of peak carrier	5%
Peak white level as % of peak carrier	100%
% of effective radiated power of sound compared to vision	25%

Belgium

Same as France except for the following:

Nominal video bandwidth	5 MHz
Nominal RF bandwidth	7 MHz
Sound carrier at	+5.5 MHz
Nominal width of VSB	Same as rest of Europe
Blanking level as % of peak carrier	22.5-27.5 %
Difference between black level and blanking level as % of peak carrier	3-6%

12.6.2 Color Transmission

Three color transmission standards exist:

NTSC	National Television System Committee (North America, Japan)
SECAM	Sequential color and memory (Europe)
PAL	Phase alternation line (Europe)

The systems are similar in that they separate the luminance and chrominance information and transmit the chrominance information in the form of two color difference signals which modulate a color subcarrier transmitted within the video band of the luminance signal. The systems vary in the processing of chrominance information.

In the NTSC system, the color difference signals I and Q amplitude-modulate subcarriers that are displaced in phase by $n/2$, giving a suppressed carrier output. A burst of the subcarrier frequency is transmitted during the horizontal back porch to synchronize the color demodulator.

In the PAL system, the phase of the subcarrier is changed from line to line, which requires the transmission of a switching signal as well as a color burst.

In the SECAM system, the color subcarrier is frequency modulated alternately by the color difference signals. This is accomplished by an electronic line-to-line switch. The switching information is transmitted as a line-switching signal.

12.6.3 Standardized Transmission Parameters* (Point-to-Point TV)

Interconnection at video frequencies:

Impedance	75 Ω unbalanced
Return loss	No less than 24 dB
Input level	1 V peak to peak
Output level	1 V peak to peak
Polarity	Black-to-white transitions, positive going

Interconnection at IF:

Impedance	75 Ω unbalanced
Input level	0.3 V rms
Output level	0.5 V rms
IF up to 1 GHz	35 MHz
IF above 1 GHz	70 MHz

In a hypothetical reference circuit 2500 km long, the signal-to-noise ratios for different systems are as follows:

System (lines)	405	525	625	625	819	819
Video baseband (MHz)	3	4	5	6	5	10
Signal to weighted noise (dB)	50	56	52	57	52	50

*Based on CCIR Recs. 421 and 403.

12.7 METHODS OF PROGRAM CHANNEL TRANSMISSION FOR VIDEO

Composite transmission normally is used on broadcast and CATV distribution. Video and audio carriers are "combined" before being fed to the radiating antenna for broadcast. These audio subcarriers are described in Section 12.6.

For point-to-point TV transmission the audio program channel generally is transmitted separately on coaxial cable, radiolink, and earth station systems. Separate transmission, usually on a separate subcarrier, provides the following advantages:

- Individual channel level control
- Greater control over crosstalk
- Increased guard band between video and audio
- Saves separation at broadcast transmitter
- Leaves TV studio as separate channel
- Permits individual program channel preemphasis

12.8 VIDEO TRANSMISSION OVER COAXIAL CABLE

12.8.1 Early System

Early coaxial cable point-to-point video transmission systems reduced usable bandwidth to about 2800 kHz (L1 system). The 2.8-MHz signal was translated, in the case of the L1 system used in North America, to a line frequency of 200–3111.27 kHz. Carrier modulation techniques are used. In the case of L1, the line frequency mentioned above is a lower sideband occupying between 311.27 and 3111.27 kHz. The VSB occupies 200–311.27 kHz. The modulation process is shown in Figure 12.6. The limited bandwidth of the L1 system reduced picture quality and was usable for monochrome video transmission only.

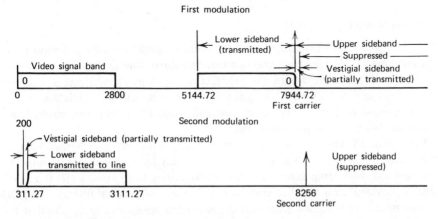

Figure 12.6 Modulation scheme for transmitting TV on the North American L1 carrier system (Ref. 14; Copyright © 1961 by American Telephone and Telegraph Company.)

12.8.2   Modern Broadband Coaxial Cable Systems for Video

THE L3 SYSTEM OF NORTH AMERICA

The L3 carrier system was designed for use on 12-MHz coaxial cable. In most applications when L3 is used for video transmission, it shares the same cable with up to 600 FDM telephone channels. The FDM voice channel segment occupies the band from 564 to 3084 kHz. The video signal occupies the region from 3639 to 8500 kHz. The modulation of the video signal, translating it to the indicated segment of the spectrum, is carried out by VSB methods. The virtual carrier as transmitted on-line is at 4139 kHz and the VSB occupies the space of 3639–4139 kHz. VSB is used to avoid some of the problems encountered with the normally used envelope detection regarding video, such as the production of a spurious envelope wherein video signals which exceed a certain value are inverted. In the L3 system, as in most video transmission systems of this type, homodyne detection is used. Here the demodulator is driven by a locally generated carrier, which is synchronous in phase angle and frequency with the carrier component of the transmitted wave. Homodyne detection also makes possible the necessary suppression of the quadrature distortion associated with VSB transmission. Figure 12.7 illustrates the L3 frequency allocation and modulation processes to develop the line frequency.

To ameliorate somewhat the effects of second harmonic distortion a preemphasis network is used in the transmitting terminal to accentuate the amplitude of the HF components of the signal before transmission. At the receiving terminal a de-emphasis network introduces a complementary frequency characteristic to make the overall transmission characteristic constant with frequency.

Delay and amplitude equalizers are incorporated in the transmit section with an objective of maintaining amplitude in the band of interest to vary no more than ±0.02 dB. Phase shift objectives are on the order of ±0.1°. Mop-up equalizers also are used on receiver terminals. Repeater spacing for the L3 system is 4 mi (6.4 km).

A 12-MHz EUROPEAN SYSTEM

12-MHz coaxial cable systems, if used for video transmission, almost always transmit combined FDM telephone channels with the video. One such system modulates a 5.5-MHz video sideband with a 6.799-MHz carrier. A baseband signal is produced in the band of 6.3–12.3 MHz, using the upper sideband of the modulation process and a vestigial portion of the lower sideband. 1200 FDM telephone channels are transmitted in the lower portion of the band.

The required flatness in envelope delay and amplitude response is maintained in the video transmission band (i.e., 6.3–12.3 MHz) by equalizers built into the coupling and separating filter units for contributions from those units (i.e., each unit is provided with equalizers to flatten response of its own filters). The EDD caused by the coaxial line itself and its associated equipment is equalized at the receiving end.

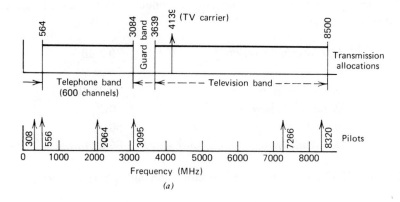

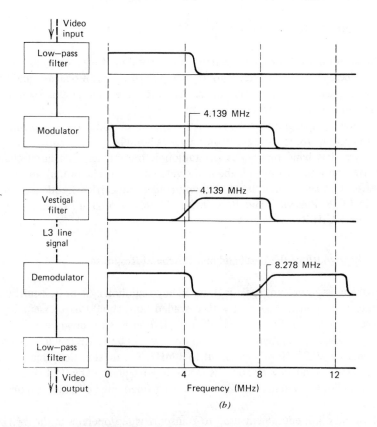

Figure 12.7 (a) Frequency allocations for the North American L3 combined TV–telephone transmission system. (b) L3 TV terminal modulation processes (Ref. 15). (Copyright © 1953 by American Telephone and Telegraph Company.)

The television baseband for a modulator–demodulator back to back is maintained as follows:

Attenuation distortion	0.2 dB
EDD	25 ns
Differential gain	0.05 dB
Differential phase	0.2°
Signal-to-hum ratio	60 dB
Continuous random noise, unweighted	66 dB (down)

CCITT Rec. G.332 covers this type of transmission under "mixed systems."

12.9   TRANSMISSION OF VIDEO OVER RADIOLINKS

12.9.1   General

Telephone administrations increasingly are expanding their offerings to include other services. One such service is to provide point-to-point broadcast-quality video relay on a lease basis over radiolinks. As covered earlier in this section, video transmission requires special consideration.

The following paragraphs summarize the special considerations a planner must take into account for video transmission over radiolinks.

Raw video baseband modulates the radiolink transmitter. The aural channel is transmitted on a subcarrier well above the video portion. The overall subcarriers are themselves frequency modulated. Recommended subcarrier frequencies may be found in CCIR Rec. 402 and Rep. 289-1. Noise on program (aural) channels is discussed in CCIR Rep. 375.

12.9.2   Bandwidth of the Baseband and Baseband Response

One of the most important specifications in any radiolink system transmitting video is frequency response. A system with cascaded hops should have essentially a flat bandpass in each hop. For example, if a single hop is 3 dB down at 6 MHz in the resulting baseband, a system of five such hops would be 15 dB down. A good single hop should be ±0.25 dB or greater out to 8 MHz. The most critical area in the baseband for video frequency response is in the low-frequency area 15 kHz and below. Cascaded radiolink systems used in transmitting video must consider response down to 10 Hz.

Modern radiolink equipment used to transport video operates in the 2-GHz band and above. 525-line video requires a baseband in excess of 4.2 MHz plus available baseband above the video for the aural channel. Desirable characteristics for 525-line video then would be a baseband at least 6 MHz wide. 8 MHz would be required for 625-line TV, assuming that the aural channel would follow the channelization recommended by CCIR Rec. 402.

12.9.3 Preemphasis

Preemphasis–deemphasis characteristics are described in CCIR Rec. 405-1 (also see Figure 5.16).

12.9.4 Differential Gain

Distortion due to differential gain is caused by nonlinear elements of the system such as nonlinearity in a receiver discriminator. One result of poor differential gain is crossmodulation of color signals. A specification of ±0.25 dB of differential gain at full deviation is usually sufficient.

12.9.5 Differential Phase Distortion

The result of this type of distortion is also crossmodulation and often is a modulation of the sound channel by some picture channel element (15-kHz sync, for example). A satisfactory single hop specification for differential phase is ±0.5° per hop at 50% APL.

12.9.6 Signal-to-Noise Ratio

The signal-to-noise ratio for one hop should be 59 dB (video signal) (EIA RS-250-A). For tandem systems 64-dB signal-to-noise ratio is recommended per hop, unweighted. EIA RS-250-A states that the system signal-to-noise ratio may be degraded to 56 dB for multihop systems. The reader should also consult CCIR Rec. 289-1.

12.9.7 Square Wave Tilt

The departure from the horizontal of the top or bottom of a square wave at the field scanning rate shall not exceed ±0.5% (1% total) of the peak-to-peak amplitude. This is specified for one hop. Over a multihop system this figure shall not exceed ±1.0% (2% total) (EIA RS-250-A).

12.9.8 Radiolink Continuity Pilot

For video transmission the continuity pilot is always above the baseband. CCIR recommends an 8.5-MHz pilot. (Refer to CCIR Rec. 401-2 and Table 5.6.)

Table 12.3 Satellite Relay TV Performance Requirements

Paragraph of Part D*	Parameters	Practical Values
3.1	Insertion gain	±0.25 dB
3.1.1 (a)	Insertion gain variation (1 s)	±0.1 dB
3.1.1 (b)	Insertion gain variation (1 h)	±0.25 dB
3.2.1	Continuous random noise	a
3.2.2	Low-frequency noise	—
3.2.3 (a)	Periodic noise (0–1 kHz)	50 dB
3.2.3 (b)	Periodic noise (1 kHz–f_c)^d	55 dB
3.2.4	Impulsive noise	25 dB
3.3 (a)	Crosstalk (undistorted)	—
3.3 (b)	Crosstalk (differentiated)	—
3.4.1.1	Nonlinear distortion (luminance)	b
3.4.1.2	Nonlinear distortion (chrominance)	—
3.4.1.3 (a)	Differential gain x or y	10%
3.4.1.3 (b)	Differential phase x or y	c
3.4.1.4	Chrominance-luminance intermodulation	—
3.4.2.1	Nonlinear distortion (sync/steady state)	+5% −10%
3.4.2.2	Nonlinear distortion (sync/transient)	—
3.5.1.1	Linear waveform distortion (long-time)	—
3.5.1.2	Linear waveform distortion (field-time)	±1%
3.5.1.3	Linear waveform distortion (line-time)	±1%
3.5.1.4 (a)	Linear waveform distortion (short-time: pulse/bar ratio)	96–104%
3.5.1.4 (b)	Linear waveform distortion (short-time: pulse lobes)	—
3.5.3.1	Chrominance-luminance gain inequality	±10%
3.5.3.2	Chrominance-luminance delay inequality	±50 ns
3.5.4.1	Steady state gain/frequency characteristics	+0.5 −1.0 dB^d
3.5.4.2	Steady state delay/frequency characteristics	—

Source: CCIR Rec. 567*. (Courtesy ITU–CCIR.)

^{a}The following values are derived in accordance with CCIR 1974-78a from figures given in the INTELSAT satellites system operations guide:
   525/60 systems, full transponder 53.3 dB
   625/50 systems, full transponder 50.1 dB
   525/60 systems, half transponder 48.7 dB
   625/50 systems, half transponder 47.1 dB
^{b}CCITT Rec. N.62 gives a limit for luminance nonlinearity on a 525-line (International Television Centre) ITC to ITC circuit which includes a satellite section.

536

^{c}Full transponder 3°, half transponder 4°.
^{d}Top video frequencies 5.5 MHz for 625-line systems and 4.2 MHz for 525-line systems.
Note. Table 12.3 is taken from CCIR Rec. 567, annex I to part D. Column 1 provides crossreference to relevant paragraphs in part D of the recommendation.

12.10 TV TRANSMISSION BY SATELLITE RELAY

Table 12.3 provides general guidance on the basic performance requirements for the transmission of broadcast-type TV signals via satellite relay based on CCIR recommendations.

12.11 TRANSMISSION OF VIDEO OVER CONDITIONED PAIRS

12.11.1 General

Broadcast-quality video transmission over conditioned pairs has application to interconnect broadcast facilities and the long-distance transmission system. The broadcaster may lease these facilities from a telephone administration to interconnect a master control point and outlying studios or remote program pick-up points. Normally only the video is transmitted on the conditioned pair. The audio is usually transmitted separately on its own program facilities. The cable is designed for installation in ducts and the repeaters in duct-type cabinets or racks.

12.11.2 Cable Description

A conditioned-pair video transmission system is composed of a shielded wire pair, terminal equipment, and repeaters. The following description is of a system in wide use in North America. Such a description of a typical system will bring forth the advantages and limitations of video transmission over conditioned pairs.

The line facilities consist of 16-gauge polyethylene-insulated pairs generally referred to as PSV.* At 75°F the loss at 4.5 MHz is 3.52 dB/1000 ft and 18.6 dB/mi. The normal slope variation with temperature is approximately 0.1%/°F. The loss at zero frequency is taken as 0 dB.

Because of the effective shielding of the PSV pairs, there is no limitation as to the direction of transmission or the number of circuits obtainable within any given size of cable. Noise considerations cause the requirement that the PSV pairs be separated from the remainder of the cable conductors at building entrances. In this case the shielded video pair is brought to the video equipment under a separate sheath.

*Pair shield video.

The characteristic impedance of the video cable is nearly purely resistive, 124 Ω, at frequencies above 500 kHz. The resistive component increases to 1000 Ω at 60 Hz. The reactive component is about the value of the resistive component at 60 Hz, and drops to nearly zero at the higher frequencies.

12.11.3 Terminal Equipment and Repeaters

A transmitting terminal is provided to match the video output of the line, secure the proper level, and predistort the signal as a first step in conditioning the PSV cable. Equalization is basically one of amplitude. The impedance is from 75 Ω unbalanced to 124 Ω balanced.

Amplitude equalization for this type of transmission facility is such that levels are described as fractions. The level unit is dBV (decibels relative to 1 V). For instance, a voltage level may be expressed as -10/+5 dBV. Such a "fraction" describes the attenuation–frequency characteristic or slope. The numerator refers to a level of zero frequency and the denominator of the fraction refers to the reference HF, in this case 4.5 MHz. Voltages here are peak to peak. Zero frequency may be taken to mean a very low frequency, 30 Hz.

This method of designating level is a useful tool when the amplitude–frequency response and its equalization are of primary concern. At the transmitting end of a PSV link we would expect a zero slope, and the transmitter output would be 1.0 V peak to peak (0 dBV) or 0/0 dBV. To equalize the line, the transmitter must predistort the signal. After equalization the level may be described, assuming 15-dB equalization, as -10/+5 dBV. After traversing 33 dB of cable, the level may then be described as -10/-28 dBV. Repeaters provide both amplification and equalization. It would appear that most systems would require custom design. To simplify engineering and standardize components, PSV systems often are built in blocks. The system is configured with fixed-length repeater sections, plug-in equalizers, and standardized receivers. Repeater spacing is on the order of 4.5 mi (7.2 km).

The receiver provides both gain and equalization. It also includes a clamper to correct for low-frequency distortion. Equalization is usually variable when the block approach is used for residual gain deviations.

12.12 BASIC TESTS FOR VIDEO QUALITY

12.12.1 Window Signal

The window signal when viewed on a picture monitor is a large square or rectangular white area with a black background. The signal is actually a sine-squared pulse. As such it has two normal levels, reference black and reference white. The signal usually is adjusted so that the white area covers one-fourth to one-half the total picture width and one-fourth to one-half the total picture height. This is done in order to locate the maximum energy content of the signal in the lower portion of the frequency band.

A number of useful checks derive from the use of a window signal and a picture monitor. These include the following:

1. *Continuity or level check.* With a window signal of known white level, the peak-to-peak voltage of the signal may be read on a calibrated oscilloscope using a standard roll-off characteristic (i.e., EIA, etc.).

2. *Sync compression or expansion measurements.* Comparison of locally received window signals with that transmitted from the distant end with respect to white level and horizontal synchronization on calibrated oscilloscopes using standard roll-off permits evaluation of linearity characteristics.

3. *Test and adjustment.* Can be made at clamper amplifiers and low-frequency equalizers to minimize streaking by observing the test signal on scopes using the standard roll-off characteristics at both the vertical and the horizontal rates.

4. *Indication of ringing.* With a window signal the presence of ringing may be detected by using properly calibrated wideband oscilloscopes and adjusting the horizontal scales to convenient size. Both amplitude and frequency of ringing may be measured by this method.

12.12.2 Sine-Squared Test Signal

The sine-squared test signal is a pulse type of test signal which permits an evaluation of amplitude–frequency response, transient response, envelope delay, and phase. An indication of the HF amplitude characteristic can be determined by the pulse width and height, and the phase characteristic by the relative symmetry about the pulse axis. However, this test signal finds its principal application in checking transient response and phase delay. The sine-squared signal is far more practical than a square wave test signal to detect overshoot and ringing. The pulse used for checking video systems should have a repetition rate equal to the line frequency, and a duration, at half amplitude, equal to one half the period of the nominal upper cutoff frequency of the system.

12.12.3 Multiburst

This test signal is used for a quick check of gain at a few determined frequencies. A common form of multiburst consists of a burst of peak white (called white flag) which is followed by bursts of six sine wave frequencies from 0.5 to 4.0 MHz (for NTSC systems) plus a horizontal synchronizing pulse. All these signals are transmitted during one-line intervals. The peak white or white flag serves as a reference. For system checks a multiburst signal is applied to the transmit end of a system.

At the receiving point the signal is checked on an oscilloscope. Measurements of

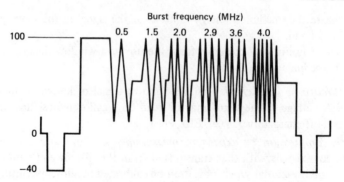

Figure 12.8 Multiburst signal (horizontal frequencies normally used).

peak-to-peak amplitudes of individual bursts are indicative of gain. A multiburst image on an oscilloscope gives a quick check of amplitude–frequency response and changes in setup. Figure 12.8 illustrates a typical NTSC type multiburst signal.

12.12.4 Stair Steps

For the measurement of differential phase and gain a 10-step stair-step signal is often used. Common practice (in the United States) is to superimpose 3.6 MHz on the 10 steps that extend progressively from black to white level. The largest amplitude sine wave block is adjusted on the oscilloscope to 100 standard (i.e., IEEE) divisions and is made a reference block. Then the same 3.6-MHz sine waves from the other steps are measured in relation to the reference block. Any difference in amplitudes of the other blocks represents differential gain. By the use of a color analyzer in conjunction with the above, differential phase may also be measured.

The stair-step signal may be used as a linearity check without the sine wave signal added. The relative height between steps is in direct relation to signal compression or nonlinearity.

12.12.5 Vertical Interval Test Signals (VITS)

The VITS makes use of the vertical retrace interval for the transmission of test signals. In the United States, the FCC specifies the interval for United States use as the last 12 μs of line 17 through line 20 of the vertical blanking interval of each field. For whichever interval boundaries specified, test signals transmitted in the interval may include reference modulation levels, signals designed to check performance of the overall transmission system or its individual components, and cue and control signals related to the operation of TV broadcast stations. These signals are used by broadcasters because, by necessity, they are inserted at the point of origin. Standard test signals are used as described above or with some slight varia-

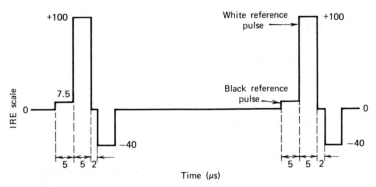

Figure 12.9 Typical vertical-interval reference signal.

tion, such as multiburst, window, and stair step. Some broadcasters use vertical interval reference signals. Figure 12.9 shows one currently in use in the United States.

12.12.6 Test Patterns

Standard test patterns, especially those inserted at a point of program origination, provide a simple means of determining transmission quality. The distant viewer, knowing the exact characteristics of the transmitted image, can readily detect distortion(s). Standard test patterns such as the EIA test pattern used widely in the United States, with a properly adjusted picture monitor, can verify the following:

- Horizontal linearity
- Vertical linearity
- Contrast
- Aspect ratio
- Interlace
- Streaking
- Ringing
- Horizontal and vertical resolution

12.12.7 Color Bars

Color bar test signals are used by broadcasters for the adjustment of their equipment including color monitors. Color bars may also be sent over transmission facilities for test purposes. The color bar also may be used to test color transmission using a black and white monitor by examining gray densities of various bars depending on individual colors. A wideband A-scope horizontal presentation can show whether or not the white reference of the luminance signal and the color information have the proper amplitude relationships. The color bar signal may further be observed on a vector display oscilloscope (chromascope) which allows measurement of absolute amplitude and phase angle values. It also can be used to measure differential phase and gain.

REFERENCES AND BIBLIOGRAPHY

1. CCIR Xth Plenary Assembly, Geneva, 1963, Vol. V.
2. *Reference Data for Radio Engineers*, 6th ed., Howard W. Sams, Indianapolis, IN, 1977.
3. "Fundamentals of Television Transmission," *Bell System Practices*, Section AB 96.100, American Telephone and Telegraph Co., New York, Mar. 1954.
4. "Television Systems Descriptive Information–General Television Signal Analysis," *Bell System Practices*, Section 318-015-100, No. 3, American Telephone and Telegraph Co., New York, Jan. 1963.
5. "Engineering of Local Radio Television Links," *Bell System Practices*, Section R 100.080. American Telephone and Telegraph Co., New York, July 1952.
6. "Television Systems–A2A Video Transmission System Description," *Bell System Practices*, Section 318-200-100, No. 5, American Telephone and Telegraph Co., New York, Feb. 1962.
7. *Transmission Systems for Communications*, 4th ed., Bell Telephone Laboratories, American Telephone and Telegraph Co., New York, 1971, chap. 29.
8. *Lenkurt Demodulator*, Lenkurt Electric Corp., San Carlos, CA, Feb. 1962, Oct., Nov. 1963, Jan. 1965, Mar. 1966, Feb. 1971.
9. D. H. Hamsher, Ed., *Communication System Engineering Handbook*, McGraw-Hill, New York, 1967, chap. 13.
10. J. Herbstreit and H. Pouliquen, "International Standards for Colour Television," (paper), ITU, Geneva, 1967.
11. K. Simons, *Technical Handbook for CATV Systems*, 2nd ed., Jerrold Electronics Corp., Philadelphia, PA, 1966.
12. EIA RS-250-A, Electronic Industries Association, Washington, DC, Feb. 1967.
13. D. Kirk, Jr., *Video Microwave Specifications for System Design*, reprint from *Broadcast Engineering*, Jerrold Electronics Corp., Philadelphia, PA 1966.
14. *Principles of Electricity Applied to Telephone and Telegraph Work*, American Telephone and Telegraph Co., New York, 1961.
15. J. W. Rieke and R. S. Graham, "The L3 Coaxial System Television Terminals," *Bell Sys. Tech. J.*, July 1953.
16. W. von Guttenberg and E. Kugler, "Modulation of TV Signals for Combined Telephone and Television Transmission over Cables," *NTZ-CJ J.*, No. 2, 1965.
17. "Recommendations and Reports of the CCIR 1978," XIV Plenary Assembly, Kyoto, 1978, Vols. IV, IX, and XII.

13 | FACSIMILE COMMUNICATIONS

13.1 APPLICATION

Facsimile is a method of electrical communication of graphic information. It is used to transmit pictorial or printed matter from one location to another with reasonably faithful copy permanently recorded at the receiving end. Facsimile (fax) has been designed primarily to operate over comparatively narrow-band media, lending itself well for transmission over the telephone network and on HF radio.

From the 1920s until sometime after World War II facsimile was used almost exclusively for the transmission of weather maps and wirephoto (news media picture transmission). Today it has much more extensive usage. It is finding ever wider application in such areas as:

- Bank verification of signatures
- The transmission of fingerprints and "mug shots" in the area of law enforcement
- In the commercial world for the delivery of waybills and invoices
- For the production of newspapers and magazines to dispatch news copy from satellite offices or bureaus to main newsrooms and to eliminate duplication of typesetting efforts
- In industry for the transmission of engineering drawings, parts lists, and so forth
- For "electronic mail," particularly as conventional mail service is becoming less efficient and more costly

13.2 ADVANTAGES AND DISADVANTAGES

On first analysis it would appear that conventional data/telegraph methods of transmitting graphic information are faster than facsimile. For instance, a standard printed page may take less than 3 min to transmit over the telephone network using digital data techniques (assuming a throughput of 300 characters per second). The same page may take up to 6 min to transmit by facsimile. However, there is an operational difference which is often overlooked. To the 3 min required to transmit the data page must be added the operator time to keystroke each character, whereas for facsimile the operator just inserts the printed page in the facsimile machine. In addition, operationally facsimile is less prone to error. With the case of digital telegraph or data, the operator can (and often does) cause errors by unknowingly stroking the wrong key. In the case of facsimile no such errors can enter because there is no operator transcription from the original copy. Optical character readers, of course, can eliminate this source of error. Nevertheless, once we enter the domain of computer-to-computer or smart-data-terminal-to-computer operations, such communication should be left to the technique of data transmission discussed in Chapter 8.

13.3 BASIC FACSIMILE OPERATION

13.3.1 General

A facsimile system consists of some method of converting graphic copy on paper to an electrical equivalent signal suitable for transmission on a telephone pair (or other narrowband media), the connection of the pair/telephone circuit and transmission to the desired distant end user, and the recording/printing of the copy on paper by that user. Three basic elements or processes are involved:

- Scanning
- Switching/transmission
- Recording

Basically we are dealing with analog technology. However, lately there is a marked trend toward the use of digital techniques. Figure 13.1 is a simplified block diagram of a facsimile system.

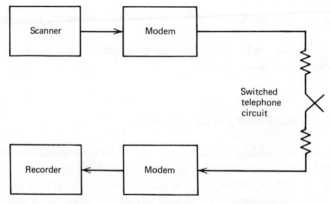

Figure 13.1 Simplified functional diagram of a facsimile system.

The scanner is a photoelectric transducer converting the reflected light, as a document is scanned, to an electrical signal which varies in intensity in accordance with the light intensity. The modem converts or conditions the electrical signal from the scanner, making it compatible for transmission over the telephone network or other media. On the receive side in Figure 13.1 the modem converts or conditions the line signal entering from the telephone network, making it compatible for the facsimile recorder. The recorder prints the received image by various methods which are discussed below. Often the modem is incorporated with the scanner or recorder, as the case may be.

13.3.2 Scanning

The scanner is a photoelectric transducer which produces an electrical analog signal representing the graphic copy to be transmitted. With conventional facsimile, scanning is carried out by one of three basic methods, which are essentially mechanical in variation:

- A spot of light scans a fixed graphic copy.
- The copy moves across a fixed spot of light.
- Both the copy and the scan spot may move, usually at right angles to each other and at different speeds.

There are two common approaches of "lighting" the copy. Both employ the technique of bouncing (reflecting) light off the graphic material to be transmitted. One approach to scanning projects a tiny spot of light onto the surface of the printed copy. The reflection of the spot is then picked up directly by a photoelectric cell or other transducer.

The other approach is the flood projection technique whereby the printed copy is illuminated with difuse light in the area of scan. The reflected light is then optically projected through a very small aperture onto the cathode of a photoelectric transducer.

In modern facsimile systems using electromechanical methods, scanning is either done with the copy attached to a cylinder that rotates, or placed on a flatbed. In the case of cylinder scanning, the cylinder is rotated such as to effect a continuous helical scan of the entire copy in a period of time depending on system constraints, possibly up to 6 min. Flatbed scanners use a feed mechanism. The copy to be transmitted is fed into a slot where a feed mechanism takes over, slowly advancing the copy through the machine. Another type of flatbed scanner uses a flying spot scan where the copy remains at rest and the spot does all the movement.

The electronics of facsimile scanners is based on the photoelectric cell, the photomultiplier tube, or the photodiode. They differ primarily in the relative signal strength of their outputs in relation to scan spot intensity.

The charge-coupled-device (CCD) electronic scanning, which previously had been exclusively in the realm of CRTs, now offers promise in the area of facsimile scanning, making is particularly attractive for direct digital transmission of a facsimile signal.

The electrical output of a conventional scanner may also be digital. In fact the trend is more and more toward digital facsimile systems. In its simplest form of digital transmission, where the signal represents either black or white, a decision circuit may provide a series of marks or 1s for black copy and spaces or 0s for white copy. If, in addition, shades of gray are to be transmitted as well, a level quantizing scheme with subsequent coding is required. This of course in most respects is similar to the quantizing (based on fixed signal thresholds) and coding described in Chapter 11.

13.3.3 Recording

Recording in a facsimile system is the reproduction (e.g., printing) of visual copy of graphic material from an electric signal. Recorders remain electromechanical, just as their scanner counterparts are. And like scanners, recorders are available in cylinder configurations and flatbed. Certain desk-top versions are on the market which are semicylindrical, allowing continuous paper feed. There are four basic electromechanical recording methods in use today:

- Electrolytic
- Electrothermal
- Electropercussive
- Electrostatic

Two types of facsimile recording do not fit into the above categories. One prominent West German manufacturer uses an offset process. The other process involves the use of modulation of a fine spray of ink directed to the surface of plain paper. Yet another process, which is well established, is not electromechanical and is used in the facsimile reproduction of newspapers.

ELECTROLYTIC RECORDING

Electrolytic recording is one of the most popular methods of facsimile recording, yet one of the oldest. It requires a special type of recording paper which is actually an electrolyte-saturated material. When an electric current passes through it, the material tends to discolor. The amount of discoloration or darkness is a function of the current passing through the paper.

To record an image, the electrolytic paper is passed between two electrodes. One is a fixed electrode, a backplate or platen on the machine; the other is a moving stylus. Horizontal lines of varying darkness appear on the paper as the stylus sweeps across the sheet. As each recorded horizontal line is displaced one line width per sweep of the stylus, a "printed" pattern begins to take shape. This pattern is a facsimile of the original pattern transmitted from the distant end scanner.

The more practical helix-blade technique of electrolytic recording is now favored over the stylus-backplate method. The concept is basically the same, except that the two electrodes in this case consist of a special drum containing a helix at the rear and a stationary blade in the front. The drum-helix makes one complete revolution per scanning line. The rotating drum moving the helix carries out the same function as the moving stylus of the more conventional electrolytic recorders.

ELECTROTHERMAL RECORDING (ELECTRORESISTIVE RECORDING)

Electrothermal facsimile recording is misnamed thermal because the recording process gives the appearance of being a "heat" process or burning. More properly it should be called electroresistive. It is similar to the electrolytic process in that the recording paper is interposed between two electrodes (i.e., stylus and backplate). The recorded pattern is made by an electric arc passing through the paper from one electrode to the other. A special recording paper is required, the type varying from equipment to equipment. The paper has a white coating which is decomposed by the current passing through it, the amount of decomposition being a function of the current passing between the electrodes at any moment in time. A major characteristic of electrothermal recording is the high contrast that can be achieved. A true electrothermal process is evolving using specially treated paper and a resistive heat element that responds to rapid temperature changes as a function of signal current.

ELECTROPERCUSSIVE RECORDING

Electropercussive recording is a technique similar to that of recording audio on a record. In this case an amplified facsimile signal is fed to an electromagnetic transducer which actuates a stylus in response to the electrical signal variations of the facsimile signal. When a sheet of carbon paper is interposed between the stylus and a sheet of plain white paper, a carbon copy impression is made on the plain paper by the vibrations of the stylus in accordance with the signal variations. The intensity of the darkness of the copy varies in proportion to the variation in strength of the picture signal. An advantage of this type of recording is that no special recording paper is required. Other names for electropercussive recording are *impact* or *impression* recording. It is also known by the term *pigment transfer*.

ELECTROSTATIC RECORDING

There are essentially two types of electrostatic recording used in facsimile. One type is based on printing the facsimile image from a CRT by means of xerography techniques requiring only plain paper. The other type is a direct copy method requiring a specially coated paper. One type of electrostatic recorder is called a *pin* printer. The advantage of direct electrostatic recording is its exceptional capability of reproducing gray tonal qualities.

13.4 FUNDAMENTAL SYSTEM INTERFACE

13.4.1 General

If we connect a scanner and recorder back to back, will they interoperate correctly? Three requirements must be met to assure this interoperation:

- Phasing compatibility
- Synchronization
- Index of cooperation

13.4.2 Phasing and Synchronization

Proper phasing and synchronization are vital factors in conventional facsimile transmission. Both are time-domain functions. Phasing and synchronization of the far-end facsimile receiver with the near-end transmitting scanner permit the reassembly of picture elements in the same spatial order as when the picture was scanned by the transmitter:

- Phasing assures that the receiving recorder stylus coincides with the transmitter in time and position on the copy at the start of transmission.
- Synchronization keeps the two this way throughout the transmission of a single graphic copy.

In most conventional analog systems phasing is carried out by what may be termed a stop-and-start technique. The receiving end in this case is not really stopped, only retarded until a start-of-stroke recording coincides with the start of scanning stroke at the transmitting end. To indicate start-of-stroke, a phasing signal is generated at the scanner and transmitted on-line. The signal is a pulse of full amplitude equivalent to full black or full white, depending on signal polarity. The pulse duration is from 5 to 7% of line length. The duration of the phasing sequence following CCITT Rec. T.2 is 15 s for the standard 6-min transmission of an (ISO) A4 page size and 6 s for the 3-min transmission of an A4 page.

Synchronization assures that the facsimile recorder remains in step with the transmitting scanner during the transmission period. With conventional analog facsimile systems, synchronization may be carried out by one of the following three methods:

- Both machines are tied to a common ac power source frequency.
- Each machine operates with a stabilized frequency power source or with built-in frequency standards.
- The system operates with the transmission of synchronizing signals during picture transmission.

The first two methods of synchronization are applicable where facsimile scanners and recorders use synchronous drive motors. Of course the first method can only be used where the scanners and recorders derive ac power from the same ac grid. In this case synchronization is simple and automatic after initial phasing.

If the two ends of a facsimile system are not on a common ac grid, such as on numerous international circuits, military systems, air-to-ground, shore-to-ship, and on other mobile and portable operations, then we must resort to one of the two latter methods. With method 2 synchronous drive motors operate from stabilized frequency sources or from frequency standards. In this case frequency stability should be 1×10^{-5} or better. CCITT Rec. T.1 states:

> The speed of transmitters must be maintained as nearly as possible to the nominal speed and in any case within ± 10 parts in 10^6 of nominal speed. The speed of receivers must be adjustable and the range of adjustment should be at least ± 30 parts in 10^6. After regulation, the speeds of the transmitting and receiving sets should not differ by more than 10 parts in 10^6.

Out-of-synchronization conditions cause skewing of the received picture.

Slaving the facsimile recorder to the distant-end scanner essentially avoids the stability problems of the first two methods. There are two ways synchronization can be accomplished. The first uses a synchronization tone, conveniently a multiple of 50 or 60 Hz which can be divided down to drive the recorder synchronous drive motor. The tone is transmitted above or below the picture signal in the VF passband. Effective filtering is then required to separate the picture from the synchronizing signal.

The other method is pulse synchronization, which is continuous during picture transmission. It uses short between-the-line synchronizing pulses which provide a check on the speed of the recorder drive motor. The first method we can denote as operating in the frequency domain and the second in the time domain.

Referring back to CCITT, Rec. T.1 calls for a 1020-Hz synchronizing tone, and Rec. T.3 states:

> During transmission of document information the transmitter should transmit full amplitude carrier during lost time. The phase of the carrier may be reversed at the end of this signal. Both transmitter and receiver should align the end of lost time to this phase reversal with an accuracy of $\pm 2\%$ in this case.

The concept of lost time (flatbed facsimile) or dead sector (cylinder facsimile) is analogous to flyback as used in video transmission (see Section 12.2).

The synchronization of digital facsimile systems is similar to the synchronization of data systems discussed in Chapter 8. There are asynchronous and synchronous systems. There is also a similarity to video transmission in that the scan and recording strokes have to start and end in synchronization. There is also the requirement of picture element synchronization. The element, which is a code word, identifies the quantized level of the picture signal.

Line advance in digital facsimile systems is usually carried out on a stepping basis. This is triggered by a start of line code word. Of course it is similar to carriage return (CR) and line feed (LF) used in data/telegraph printer operation (Chapter 8). Digital facsimile operation varies greatly from manufacturer to manufacturer, depending greatly on the amount of processing involved to reduce redundancy. We find that the normal printed page is 95% white. As a consequence, to reduce transmission time, digital facsimile systems have been developed which are called *white space skipping* systems. White space skipping can be carried out by fairly rudimentary processing.

The distinction between asynchronous and synchronous digital facsimile systems lies in that for asynchronous systems synchronism is only maintained on a line-for-line basis, the line advance acting much like the stop-to-start transition of asynchronous data systems described in Chapter 8. Synchronous digital facsimile systems, on the other hand, maintain synchronization throughout the entire transmission period, similar to the synchronism on a synchronous data link.

13.4.3 Index of Cooperation

On a facsimile circuit where the scanner and recorder have compatible phasing arrangements and synchronization can be maintained, assume that both machines have the same scan rate as measured in lines per minute (LPM). Now they must also have the same index of cooperation for correct interoperability. The index of cooperation is defined by the IEEE (Ref. 3) as follows:

- For rotating systems it is the drum diameter times the lines per unit length.
- The international definition is the product of the scanning or recording line (length) by the scanning or recording lines per unit length divided by π. (That is the product of the line length across the sheet by the lines per inch (LPI) (vertical)).

An older IEEE definition, which is still much in use, is the product of the total line length by the number of lines per unit length. To convert this older standard to the CCITT standard, multiply the IEEE index of cooperation by 0.318. For instance, CCITT Rec. T.1 recommends an index of cooperation of 352, alternatively 264. These convert to IEEE standards of 1105 and 829, respectively. CCITT is now using a term called *factor of cooperation*, which is indeed equivalent to the older IEEE definition of index of cooperation where the conversion factor of 0.318 is valid. The World Meteorological Organization (WMO) specifies indices of cooperation of 576 and 288. These are equivalent to IEEE indices of 1809 and 904, respectively. The U.S. EIA recommends an index of 829 (IEEE) equivalent to a CCITT index of 264.

What happens when the indices of cooperation are not the same? This is shown in Figure 13.2. Figure 13.2*a* and *c* shows examples of transmission between scanner

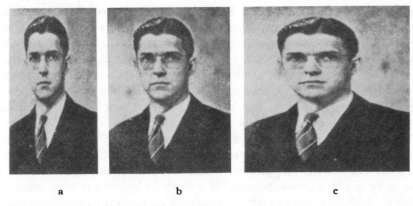

a   b   c

Figure 13.2 Distortion due to different indices of cooperation. (Ref. 5; Courtesy IEEE.)

and recorder with indices of 352 and 264. Figure 13.2*b* is an example where both machines are operating with an index of 264 and the received picture is not distorted. When there is a lengthening distortion in the received copy, such as in Figure 13.2*a*, the scanner index is greater than the recorder index. When this situation is reversed, the distortion appears as a fattening of the received picture, as in Figure 13.2*c*. When the indices of cooperation are the same at each end, it does not necessarily follow that the picture size at the recorder is the same as that of the scanned picture. The size will be the same if the scanner and recorder have the same line lengths.

13.5 FACSIMILE TRANSMISSION

13.5.1 General

The signal output of the conventional scanner operating in the analog mode contains frequency components from subaudio to several thousand hertz. The amplitude variations as a scan spot moves across a scan line consist of rapid transitions in accordance with the white, black, and tonal hues of gray of the line. The transmission problem that the telecommunication engineer faces is to condition or convert this signal to make it compatible with the band-limited media it will traverse (e.g., the telephone network).

13.5.2 Modulation Techniques

Because of the critical subaudio content of a facsimile signal extending down to about 20 Hz, the signal, as is, is not compatible for transmission over the telephone network. To overcome the frequency response problem, simple carrier techniques have been adopted. (See Chapter 3). A common approach is to use VSB modulation with an 1800-Hz carrier. The principal information content is carried in the lower sideband with about 1400 Hz of information bandwidth. The vestige of the upper sideband extends out to about 2300 Hz. If SSB were used where the information bandwidth is greater than half of the carrier frequency, it would give rise to a form of distortion known as the *Kendall effect*. This manifests itself as a fuzziness at the edges of the recorded picture. Thus some Kendall effect must be expected, but the recording quality is still adequate for most applications.

FM is generally more desirable than AM (VSB as described above) for transmission over the switched telephone network and on radio systems, particularly HF. There are two reasons for this. First FM is nearly impervious to noise because most noise is AM in nature. Second, FM is much less sensitive to level variations, which are fairly common on a telephone network, particularly on switched connections. For HF radio using the skywave phenomenen, the receive level is constantly varying.

A common standard for FM transmission of facsimile is to assign 1500 Hz for white and 2300 Hz for black, with a mean frequency of 1900 Hz (CCITT Rec. T.1). Regarding frequency stability, CCITT recommends that "the stability of transmission be such that the frequency corresponding to a given tone does not vary by more than 8 Hz in a period of 1 s and by more than 16 Hz in a period of 15 m." Further, CCITT Rec. T.1 states that the receiver–recorder be capable of operating correctly when the drift of black and white frequencies does not exceed their nominal values by more than ±32 Hz. It is assumed that the frequency deviation varies linearly with the photocell (or equivalent) voltage or, in the case of conversion from AM to FM, with the amplitude of the amplitude-modulated carrier.

CCITT Rec. T.2 has provisionally assigned 1700 Hz as center frequency (mean frequency), and white then corresponds to 1300 Hz and black to 2100 Hz. Rec. T.2 has been specifically established for document transmission, transmitting an A4 page in approximately 6 min.

CCITT Rec. T.3 was prepared for A4 page size document transmission in 3 min. It recommends a 2100-Hz (±10 Hz) carrier frequency using VSB modulation. A white signal is represented by maximum carrier and a black signal by minimum or no carrier. For proper operation the minimum carrier state must be at least 26 dB below the maximum carrier state. Receiver frequency drift should be no more than ±16 Hz. Receivers should be capable of functioning correctly with signal levels as low as – 40 dBm (white signal).

Digital facsimile can operate on any one of the common data rates (see Chapter 8) such as 2400, 4800, 7200, or 9600 bps, and some systems are envisioned with rates as high as 56 kbps. Of course the rate selected is very much dependent on the media to be utilized. Similar data modems are employed for digital facsimile transmission as were discussed in Chapter 8 for digital data transmission.

CCITT Rec. T.30 provides a recommended operating procedure for digital facsimile and is primarily to be used for automatic operation. The introduction to Rec. T.30 states in part:

The binary coded signaling system is based on a high level data link control (HDLC) format developed for data transmission procedures. The basic HDLC structure consists of a number of frames each of which is subdivided into a number of fields. It provides for frame labeling, error checking, and confirmation of correctly received information, and the frames can easily be extended if this should be required in the future.

13.6 CRITICAL TRANSMISSION PARAMETERS

13.6.1 General

The effects of various telephone circuit parameters on the capability of such circuits to transmit facsimile signals are important engineering considerations in system design. Principally the facsimile transmitter output looks into a standard 4-kHz telephone channel.* These circuit parameters are as follows:

- Effective bandwidth
- Amplitude–frequency response
- EDD
- Noise
- Echo
- Level stability
- Phase jitter

Each parameter has been discussed at length previously in this text. However, a brief review is carried out below. Reference paragraphs in other chapters are shown in parentheses.

13.6.2 Effective Bandwidth (Section 1.9.1)

The CCITT voice channel has effective bandwidth limits of 300 and 3400 Hz. An AM facsimile signal requires an information bandwidth of about 1400 Hz. Thus a double-sideband AM signal would occupy twice the frequency, or 2800 Hz of the total 3100 Hz available. As we shall see below, occupying this much of the available bandwidth is impractical due to other constraining factors. Therefore we have had to resort to other types of modulation occupying less bandwidth such as VSB and FM.

---

*Exceptions are noted. For instance, the U.S. Bell System and certain specialized common carriers have special digital offerings, as do some telecommunication administrations in foreign countries where the digital facsimile transmitter looks directly into a digital circuit of fixed data rate and the discussion below is essentially not applicable.

13.6.3 Amplitude–Frequency Response (Sections 1.9.3 and 8.9.3)

CCITT Rec. T.11 refers the facsimile user to CCITT Rec. G.151. It further states that it is desirable to have an amplitude–frequency distortion in the desired band between end user equipments of less than 8.7 dB.

A basic telephone channel in North America between 500 and 2500 Hz referenced to 1004 Hz can expect the worst amplitude–frequency response to be from −2 to +8 dB, or 10 dB of attenuation distortion (Ref. 12).

13.6.4 Envelope Delay Distortion (Sections 1.9.4 and 8.9.3)

This is a most important parameter in facsimile transmission. The reader is advised not to equate EDD with or directly derive it from the more familiar delay distortion. The effect on a facsimile received picture is the same and the cause is the same. Parameters expressed as EDD are preferred simply because it is easier to measure. However, the references used below use delay or delay distortion. Ref. 7 states:

Experience has shown that for effective facsimile transmission within the voice band, delay must be held uniform (±300 μs) for all frequencies within the full frequency range of the transmitted signal.

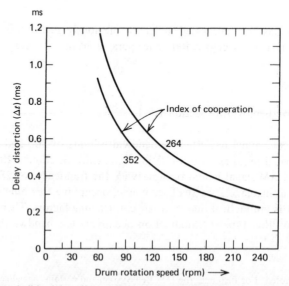

Figure 13.3 Permissible delay distortion in the transmitted frequency band as a function of the phototelegraph transmission speed. *Note.* The scanning spot is assumed to have the same dimensions in both directions (square or circular). (CCITT Rec. T.12; Courtesy ITU–CCITT.)

CCITT Rec. T.12 gives recommended limits of delay distortion as a function of drum rotation speed. This is shown in Figure 13.3.

Excessive delay distortion or EDD between an analog facsimile scanner and a far-end recorder(s) gives rise to received picture distortion, which manifests itself in smears or ghosts not necessarily spread uniformly over the received copy. On digital facsimile circuits it results in intersymbol interference degrading the error rate. If delay distortion becomes very severe, then digital facsimile operation becomes impossible. Automatic delay equalizers are recommended on circuits transmitting 4800 bps and above.

13.6.5 Noise (Sections 1.9.6 and 8.9.3)

On analog facsimile transmission systems the signal-to-noise ratio at the receiver input should be 30 dB or better. On digital facsimile circuits the receiver input signal-to-noise ratio (probably expressed in E_b/N_0) should be such as to meet the system BER requirements.

Impulse noise is momentary, and when severe, it can obliterate certain essential details of a received picture, especially when there are multiple "hits" of some duration. Single-frequency interference, such as from power-line harmonics, is continuous, showing up on received copy as a herring-bone pattern. Crosstalk also gives a herring-bone effect. Whereas the single-frequency interference shows symmetry and continuity in its interference pattern, crosstalk shows up as noncontinuous and nonsymmetrical.

13.6.6 Echo (Section 2.6.4)

On facsimile systems with AM transmission, echo manifests itself with ghosts. On FM systems it shows up as ripple distortion in the received picture.

13.6.7 Level Variation (Sections 1.9.5 and 8.9.3)

A drop in the received level (AM systems) of 1 dB can affect facsimile recording, whereas at least a 3-dB level variation is required to affect voice communication. A level reduction will deteriorate the signal-to-noise ratio and, as a result, reduces tonal response. Level variations affect AM (VSB) facsimile systems much more than FM systems.

13.6.8 Phase Jitter (Section 8.9.3)

Excessive phase jitter shows up on a received picture as a herring-bone pattern similar to that of single-frequency interference. On digital circuits, excessive phase jitter can have much greater deleterious effects (refer to Section 8.9.3).

13.7 END-TO-END QUALITY

The quality of a received facsimile document, map, or picture is a subjective matter, as is the "quality" of a received voice signal. It is difficult to quantify facsimile quality in hard and fast numerical parameters. We do judge picture quality by its resolution, legibility, contrast, and gray-scale reproduction. Each criterion is dicussed below.

13.7.1 Resolution

As in any image communication system, such as radar, TV, and facsimile, resolution is defined as the degree to which adjacent elements of an image are distinguishable as being separate. There is horizontal resolution and vertical resolution. We mean here our ability to define picture elements first horizontally and then vertically.

The resolution of any picture is basically determined by the picture elements per unit area, called *pixels* or *pels* (see Section 12.2). Elements on a scanned facsimile page are a function of LPI (or mm), working vertically down the page, and spot size. Spot size can be related to LPI. Let us assume an LPI of 100. Then the optimum spot size would be the inverse, or $\frac{1}{100}$ (0.25 mm). Thus a square inch would have 100×100, or 10,000 picture elements (pixels or pels). A 10-in. by 10-in. page in this case would have $100 \times 100 \times 100$, or 1×10^6 picture elements. More commonly a practical facsimile system with 96 LPI has a $\frac{1}{200}$ in. spot size, and for an $8\frac{1}{2} \times 11$-in. page the resulting total number of elements (pixels) is about 500,000. The actual number of elements for analog facsimile systems is reduced by about 30% due to the Kell factor or, in this case, $0.70 \times 500{,}000$ or 350,000 pixels.

The Kell factor corrects the ideal picture resolution to the practical. The reason is that scan lines are of finite width. Thus there can be no variation in light intensity values across a scan line. Stated another way, a mark reproduced by a facsimile recorder will always occupy at least the full width of a recording stroke, regardless of the dimensions of the actual mark on the scanned original. On recorded copy there is a certain spreading of line width from the scanned original because effectively the number of elements resolvable per inch is less than the LPI. The amount of reduction is the Kell factor.

The size and shape of the scan aperture are also factors determining resolution in a facsimile system. Just as in photography where the smallest aperture possible for a given set of light conditions produces the "sharpest" picture (i.e., has the best resolution), so in facsimile scanners, the smaller the opening, the sharper the picture. And as in photography, the limiting factor on aperture size is the intensity of the light that will pass through the opening sufficient for system operation.

A round aperture will serve for most facsimile applications but does not give optimum resolution. However, to achieve nearly equal resolution in both the horizontal and the vertical dimensions on a scanned page without unduly degrading fidelity, the scan aperture should have a height somewhat greater than its width.

Table 13.1 Guide to Reflectance

Type of Paper	Reflectivity (%)
White	85
Slight cream tint	83
Deep cream or very light buff	82
Slight sepia cream tint	82
Very light sepia tint	81
Fairly saturated yellow	79
Light yellowish green tint	74
Reddish buff verging on salmon	71
Light blue-green tint	70

Source: Ref. 6 (Courtesy G. F. Stafford, Alden Electronic & Impulse Recording Co., Inc.)

13.7.2 Gray Scale and Contrast

For the transmission of engineering line drawings and document (printed) copy we would want good sharp contrast, but the tonal qualities of gray are relatively insignificant. Of course a lot has to do with the contrast of the copy to be scanned. Where that contrast is weak, the output of the scanner can be increased. But at some point background noise will begin to increase, deteriorating contrast. The amount of black or the blackness of a print is described and measured by density. Whiteness is described by reflectance, and the standard of measurement is the percentage of reflectance. Table 13.1 is taken from Stafford's series of articles on facsimile (Ref. 6). It provides a good guide to reflectivity. With optimum contrast whitest white has 100% reflectance and black nearly 0%, often assigned 1%. When measuring in density units, blackest black is assigned 2.00, and the same white as above, 0.00. Tonal gray is then some number between 0.0 and 2.0 If it had 50% reflectance, it then would have 0.30 density units.

 Contrast is a measure of the faithful reproduction of scanned prints in facsimile communication; resolution is a measure of how well we can distinguish closely spaced objects or identify small items on a print. Up to some point there is a one-to-one trade-off available between contrast and resolution. However, some subjective tests indicate that contrast may affect legibility to a greater extent than resolution.

13.7.3 Legibility

Legibility has a lot to do with the type of copy to be transmitted. Legibility criteria, for instance, would be more severe for photos and drawings than for printed documents. Under certain marginal situations we may not be able to identify certain

letters in a word if they were isolated, but given the entire word or phrase there
would be no trouble reading it.

For proper legibility of a character (i.e., a letter, number, or other common
graphic symbol) at least 10 scan lines are required. Therefore with a 96-LPI facsimile system, minimum letter height on copy to be transmitted must be $\frac{10}{96}$-in.
or about one-tenth of an inch. For word legibility only about half that number of
scan lines is required. A one-eighth-inch high character requires about 80 LPI, and
a word one eighth inch high, 40 LPI. On the other hand, photo transmission,
depending on the original picture quality, requires from 200 to 700 LPI.

13.7.4 Test Charts

Facsimile test charts provide the system engineer and maintenance technician basic
information on the quality of the system regarding contrast, legibility, and resolution. It is an aid to troubleshooting where specific extraneous patterns and types of
chart distortion on the receive side can immediately red flag a problem such as
noise and its type, Kendall effect, poor synchronization or phasing, or incompatible
index of cooperation.

There are three standard charts issued by two agencies: the IEEE and CCITT. The
IEEE chart is identified by a picture of a pretty girl, the chart edition 1 of the
CCITT with a picture of the UNESCO building (Paris), and the edition 2 with a
picture of an Argentinian boy. Any of the three charts will provide the user with
the same basic information. The CCITT charts are shown in Figure 13.4.

13.8 FACSIMILE TRANSMISSION STANDARDS

13.8.1 General

Telecommunication standards are issued to ensure system interoperability. If a
facsimile scanner and distant-end recorder are made by the same manufacturer
and are designed to interoperate, and the intervening transmission medium is provided as specified by the manufacturer, then system compatibility can be expected.

Under many circumstances, particularly where a large community of diverse
users are serviced by a system, we probably cannot expect such ideal pairing arrangements. Typical examples of large diverse communities to be served are wire services
(news), weather/meterological organizations, military forces, and marine and large
corporate systems.

Another factor that has stimulated one form or another of standardization is a
common transmission medium. The international telephone network is the best
example, with constraints on bandwidth, level, noise, amplitude, and delay response.
HF radio is another example.

The following sections review some of the basic standards. Several are what
might be termed domestic U.S. standards, but they are used so widely that they

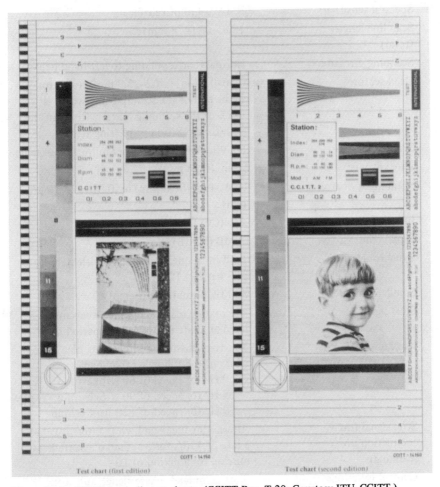

Figure 13.4   CCITT facsimile test charts. (CCITT Rec. T.20; Courtesy ITU–CCITT.)

have taken on worldwide significance. The standard document size varies: for the U.S. civilian market it is $8\frac{1}{2}$ by 11 in., for the U.S. military, 8 by 10 in.; the international paper size is DIN A4 ($8\frac{1}{4}$ by $11\frac{11}{16}$ in). Military machines accept widths up to $8\frac{1}{2}$ in. The standard document transmission speed for a page is 6 min; CCITT has a standard for a 3-min (A4) page.

13.8.2   EIA

EIA has issued several standards dealing with facsimile operation. EIA RS-328, "Message Facsimile Equipment for Operation on Switched Voice Facilities Using Data Communication Terminal Equipment," is summarized below:

Spectral sensitivity	RMA S-4 photo surface
Scan line length (total)	18.85 in. (478 mm)
Scanning direction	Normal (corresponding to left-hand helix)
Dead sector	0.56–0.94 in. (14.2–23.9 mm). (a sector at the end of the scanning line which is 3–5% of the scanning line length; coincides with time position of phasing signal)
Scan speeds	60, 90, or 120 strokes per minute, selectable
Line advance	$\frac{1}{96}$ in. (0.26 mm)
Scanning spot size	0.0104 in. (0.26 mm) by 0.0104 in. (0.26 mm)
Index of cooperation	576 (CCITT definition)
Signal contrast	20 dB (±2 dB)
Synchronization	Built-in frequency standard with stability of 3 parts in 10^6
Standard frequency	300 Hz or multiple thereof
Start signal	Alternating black and white levels modulated at a rate of 300/s for a period of 5 s
Phasing signal	30 s transmission of alternating black and white of the following frequencies: 1.0 Hz for speed of 60 strokes/min 1.5 Hz for speed of 90 strokes/min 2.0 Hz for speed of 120 strokes/min
Stop signal (on completion of scanning)	Alternating black and white interrupted at a rate of 450/s for 5 s, then black transmission for a period of 10 s
Control functions	Start command, phase signal, and stop command
Modulation characteristics	
AM	Maximum amplitude black with carrier frequency 1800 or 2400 Hz (When VSB used, carrier 2400 Hz and upper sideband completely attenuated)
FM	1500 Hz black, 2300 Hz white

Meteorological Receiver, Large Format

Recorded copy size	
Drum recorder	$18\frac{5}{8}$ in. (473 mm) by 12 in. (305 mm) or integral multiples of 12 in. (305 mm)

Scanning rate	180 LPM
Index of cooperation	805–889, eventually to be fixed at 829 to agree with CCITT/CCIR
Total line length	8.5–9.2 in.
Available line length	8.0 in. minimum, eventually to be fixed at 8.5 in.
Phasing time	15 s
Signal sense	Maximum on black
Synchronization	1. 60 Hz (±0.5 Hz), transmitted synchronizing signal
	2. Frequency standard accurate to 1 part in 10^5
LPI	96
Paper speed	$1\frac{7}{8}$ in/min
Phasing pulse	25 ±3 ms. Phasing signal is a black signal with a white pulse once per scanning line, the black-to-white transition occurring at the right-hand edge of an $8\frac{1}{2}$-in. wide sheet

The standard refers to the ANSI IEEE Std 167-1966 (Reaff. 1971), "Test Procedure for Facsimile." Other related EIA standards are: RS-357, "Interface Between Facsimile Terminal Equipment and Voice Frequency Data Communications Terminal Equipment" (June 1968), and RS-373, "Unattended Operation of Facsimile Equipment (as Defined in EIA Standard RS-328" (June 1970).*

13.8.3 U.S. Military Standards

The following excerpts have been taken from the most current U.S. military publication, Mil-Std-188 (with Notices 1, 2, and 3). This will soon be replaced by Mil-Std-188c-260, "Equipment Technical Design Standards for Analog End Instruments and Digital Terminal Equipment." Digital facsimile end instruments will soon be covered in Mil-Std-188-347 on input–output devices (digital).

METEOROLOGICAL EQUIPMENT

Transmitter

Original copy size	
Drum scanner	$18\frac{5}{8}$ in. (473 mm) by 12 in. (305 mm)
Flatbed scanner	$18\frac{5}{8}$ in. (473 mm) by any length (continuous scanners)

---

*These standards may be ordered from the Electronic Industries Association, 2001 Eye St. NW, Washington, DC 20006, U.S.A.

Continuous recorder	400-ft roll wound on 1-in. core (121.9 m on 25.4-mm core), $18\frac{5}{8}$ in. (473 mm) wide
Recorded line length	18.85 in. (478 mm)

The remainder of the specification is essentially the same as that for the transmitter.

Meteorological Receiver, Small Format

General	Continuous recording; able to resolve at least 200 LPI (25.4 mm)
Recorded copy size	$8\frac{1}{2}$ in. (216 mm) wide; length dependent on duration of transmission; 400-ft roll on 1 in. core (121.9 m on 25.4-mm core)
Recording line length	8.64 in. (220 mm)
Recording direction	Normal (corresponding to left-hand helix)
Recording speed	60, 90, or 120 strokes/min, selectable
Index of cooperation	576 and 264 (CCITT definition) (index of 288 will work with specified index of 264 with negligible distortion)
Line advance	1/209.5 in. (0.12 mm) for index of cooperation of 576 or $\frac{1}{96}$ in. (0.26 mm) for index of cooperation of 264, selectable
Recording spot size	1/209.5 by 1/209.5 in. (0.12 by 0.12 mm) for index of cooperation of 576 or $\frac{1}{96}$ by $\frac{1}{96}$ in. (0.26 by 0.26 mm) for index of cooperation of 264, with marking stylus or element being changed for the index employed
Dead sector	Signal transmitted during interval transmitter is scanning dead sector may be blanked if desirable
Input power level	For high signal contrast between -9 and -36 dBm

13.8.4 U.S. National Weather Service

For service over the telephone network, voice channel:

Scan speed	120 LPM
Resolution	48 or 96 LPI
Index of cooperation	576 for 96 LPI and 288 for 48 LPI (CCITT definition)

Transmission/modulation	AM, 2400-Hz carrier
Start signal	Carrier modulated by 300-Hz tone
Phasing signal	Black signal interrupted by 12.5-ms white pulse twice per second prior to recording
Stop signal	Carrier modulated by 450-Hz tone

13.8.5 CCITT Facsimile Recommendations

The CCITT T. recommendations discuss three distinct groups of facsimile equipment as follows:

- *Group 1.* Equipment which uses DSB AM modulation without resorting to special measures to compress bandwidth. Equipment in this group is designed for document transmission of (ISO) A4 size paper at nominally 4 lines per millimeter in about 6 min via a telephone-type circuit. Covered by Rec. T.2.
- *Group 2.* Equipment which uses bandwidth compression techniques to achieve a transmission time for an A4 size document at nominally 4 lines per millimeter via a telephone circuit. Bandwidth compression in this context includes encoding and/or VSB AM but does not include processing to reduce redundancy. Covered by Rec. T.3.
- *Group 3.* Equipment which incorporates a means of reducing redundancy prior to modulation to achieve a transmission time for an A4 size document of 1 min via the telephone network. Bandwidth compression of the line signal may be used. Under study by CCITT Study Group XIV.

REC. T.1, "STANDARDIZATION OF PHOTOTELEGRAPH APPARATUS"

Index of cooperation	352, alternatively 264
Drum scanning	
Drum diameter	66, 70, and 88 mm
Drum factor	2.4 (sending and receiving)
Flatbed scanning, line length	207, 220, and 276 mm, of which 15 mm is not used for effective transmission
Synchronization and phasing	See Section 13.4.2
Modulation	
AM	Carrier frequency 1300 Hz; for systems operating over carrier systems, 1900 Hz is recommended; high amplitude black, low white with at least 30 dB between nominal white and black levels
FM	Mean frequency 1900 Hz; white frequency 1500 Hz; black frequency 2300 Hz

Table 13.2   Common Standards

M	C	D (mm)	L (mm)	P (mm)	F (lines/mm)
264	829	66	207	1/4	4
264	829	70	220	1/3.77	3.77
264	829	88	276	1/3	3
350	1099	70	220	1/5	5
352	1105	66	207	3/16	16/3
352	1105	88	276	1/4	4

Source: CCITT T, recommendations.
Note. The maximum dimensions of the pictures to be transmitted result from the parameters given in the table.

Table 13.2 gives the corresponding values of the index of cooperation M, the factor of cooperation D, the total length L of scanning line, the scanning pitch P, and the scanning density F for equipment in most common use.

Table 13.3 gives the normal and approved alternative combinations of drum rotation speeds or scanning line frequencies and indices of cooperation.

Table 13.3   Standard Drum Rotation Speeds and Scanning Line Frequencies versus Indices of Cooperation

	Drum Rotation Speed (rpm) or Scanning Line Frequency	Index of Cooperation	
		Metallic Circuits	Combined Metallic and Radio Circuits
Normal Conditions	60	352	352
	90		264
Alternatives for use when the phototelegraph apparatus and metallic circuits are suitable	90	264 and 352	
	120	264 and 352	
	150	264	

Notes
1. In the case of transmitters operating on metallic circuits, the index 264 is not intended to be used with an 88-mm drum. In the case of transmitters operating on combined metallic and radio circuits, the index 264 associated with a drum diameter of 88 mm is intended to be used only exceptionally.
2. The provisions given in the table are not intended to require the imposition of such standards on users who use their own equipment for the transmission of pictures over leased circuits. However, the characteristics of the apparatus used should be compatible with the characteristics of the circuits used.

REC. T.2, "STANDARDIZATION OF GROUP 1 FACSIMILE APPARATUS FOR
DOCUMENT TRANSMISSION"

Index of cooperation	264; when a lower vertical resolution is acceptable, 176
Document size	Minimum ISO A4 (210 × 297 mm)
Total scan line length	215 mm (active sector plus dead sector)
Total scan lines per document	1144 for an index of cooperation of 264, and 762 for an index of 176
Scanning density	3.85 lines/mm
Scanning frequency	180 LPM, alternatively 240 LPM with mutual agreement between both ends
Scanning frequency stability	±1 part in 10^5
Phasing and synchronization	See Section 13.4.2
Modulation	
AM	High-level signal is black and carrier frequency should range between 1300 and 1900 Hz depending upon circuit characteristics
FM	Center frequency (provisionally) 1700 Hz, shift ±400 Hz with the higher frequency corresponding to black
Power at the receiver input	Receiver functions correctly with input levels from 0 to - 40 dBm (for AM these correspond to black levels)

REC. T.3, "STANDARDIZATION OF GROUP 2 FACSIMILE APPARATUS FOR
DOCUMENT TRANSMISSION"

Equipment dimensions	
Factor of cooperation	829 ±1%
Total line scanning length	215 mm
Usable scanning line length	200 mm
Document size	Minimum ISO A4 (210 × 297 mm)
Index of cooperation	264
Scanning density	3.85 lines/mm
Number of scanning lines in a document 297 mm long	1145
Scanning frequency	360 LPM, alternatively 300 LPM with mutual agreement between both ends
Phasing and synchronization	See Section 13.4.2

| Modulation, AM | VSB-phase modulation with a carrier at 2100 Hz ±10 Hz; White signal represented by maximum amplitude and black should be at least 26 dB below white level; the phase of the carrier may be reversed after each transition through black |
| Receiver input level | White signal between 0 and −40 dBm |

REC. T.30, "PROCEDURES FOR DOCUMENT FACSIMILE IN THE GENERAL SWITCHED TELEPHONE NETWORK"

This recommendation describes the procedures and signals to be used for facsimile equipments operated over the switched network. Two separate signaling systems are described, a simple system using single-frequency tones and a second binary-coded system which offers a wide range of signals for more complex operational procedures. Automatic operation is implied with the latter, manual and semiautomatic operation with the former.

The binary-coded signaling system is based on a high-level data link control (HDLC) format developed for data transmission procedures. The basic HDLC structure consists of a number of frames, each of which is subdivided into a number of fields. It provides for frame labeling, error checking, and confirmation of correctly received information, and the frames can be easily extended if this should be required in the future.

13.8.6   CCIR Recommendations

In the recommendations below CCIR generally treats facsimile transmission over HF radio circuits and stipulates the use of direct FM or subcarrier FM for these applications.

REC. 343-1, "FACSIMILE TRANSMISSION OF METEROLOGICAL CHARTS OVER RADIO CIRCUITS"

This is the same as CCITT Rec. T.15.

REC. 344-2, "STANDARDIZATION OF PHOTOTELEGRAPH SYSTEMS FOR USE ON COMBINED RADIO AND METALLIC CIRCUITS"

Modulation	See CCITT Rec. T.15
Index of cooperation	352, alternatively 264
Drum speed	60 rpm, alternatively 90/45 rpm

Note. In due course the alternative methods will become obsolete.

13.8.7 World Meteorological Organization (WMO)

The WMO has set forth certain standards for the collection and dissemination of weather data by facsimile (Ref. 10). A partial review of these standards is given below:

Index of cooperation	576 with minimum picture elements of 0.4 mm and 288 for minimum picture elements of 0.7 mm (CCITT definition)
Drum speed	60, 90, 120, or 240 strokes/min (rpm)
Drum diameter	152 mm
Scanning density	4 lines/mm for index of cooperation of 576 and 2 lines/mm for index of 288
Length of drum	55 cm minimum
Stability (scanning speed)	Within 5 parts in 10^6 of its normal value
Modulation	
AM	Maximum carrier amplitude corresponds to black; carrier at 1800 Hz for drum speeds of 60, 90, and 120 rpm; for drum speed of 240 rpm the carrier is at 2600 Hz and VSB must be used.
FM	Center frequency 1900 Hz; black frequency 1500 Hz; white frequency 2300 Hz
Contrast ratio	Between 12 and 25 dB

REFERENCES AND BIBLIOGRAPHY

1. *Reference Data for Radio Engineers*, 6th ed., Howard W. Sams, Indianapolis, IN, 1977.
2. Military Standard, Mil-Std-188-100, U.S. Department of Defense, Washington, DC.
3. IEEE Std. 100–1977, "Dictionary of Electrical and Electronics Terms," 2nd ed, IEEE, New York, 1977.
4. IEEE Std. 168–1956, (Reaff 1971), "Definition of Terms on Facsimile," equivalent to ANSI C16.30-1972, IEEE, New York, 1971.
5. IEEE Std. 167–1966 (Reaff 1971), "Test Procedure for Facsimile," equivalent to ANSI C16.37-1971, IEEE, New York, 1971.
6. G. Stafford, "The Facsimile Series," reprinted from *Communications*.
7. D. M. Costigan, *Electronic Delivery of Documents and Graphics,* Van Nostrand-Reinhold, New York, 1978.
8. CCITT Orange Books, Geneva, 1976, Vol. VII, T recommendation.
9. CCIR Green Books, Kyoto, 1978, Vol. III.

10. *Manual on the Global Telecommunication System*, Vol. I, Part III, Section 7., WMO Publ. 386.

11. EIA RS-328, RS-357, and RS-373, Electronic Industries Association, Washington, DC.

12. Bell System Technical Reference, "Data Communications Using Voiceband Private Line Channels October 1973," Publ. 41004, American Telephone and Telegraph Co., New York, Oct. 1973.

14 | FIBER OPTIC COMMUNICATION LINKS

14.1 OVERVIEW

Glass fiber as a transmission medium holds great promise in this age of information explosion to carry up to several gigabits per second of information over short , medium and eventually long-haul transmission links. A fiber optic link requires far less repeaters for equivalent bit rates than coaxial cable, and its capacity is far greater. It is also lighter, smaller, immune to EMI, and most suitable for digital transmission.

A simplified functional block diagram of a typical fiber optic communication link is shown in Figure 14.1. The optical source in the figure may be a light-emitting diode (LED) or laser diode with outputs in the region of 820–850 nanometers (nm). The transmission medium may be one or more fibers. The optical detector or receiver may be a PIN diode or an avalanche photodiode (APD). The transmission medium may or may not have repeaters, depending on the application. As of this writing, using off-to-shelf devices, a 10-km (6.2-mi) link can carry 100 Mbps on a single fiber without repeaters, operating on a wavelength of 830 nm.

It is more common in the fiber optics industry to speak of optical wavelength rather than frequency. Systems today operate in the region of 800–900 nm or 0.8–0.9 micrometer (μm). This band is shown in Figure 14.2, the radio/optical spectrum from 300 MHz to 1000 terahertz (THz).

14.2 INTRODUCTION TO OPTICAL FIBER AS A TRANSMISSION MEDIUM

The practical propagation of light through an optical fiber might best be explained using ray theory and Snell's law. Simply stated we can say that when light passes from a medium of higher refractive index into a medium of lesser refractive index, the refracted ray is bent away from the normal. For instance, a ray traveling in water and passing into an air region is bent away from the normal to the interface between the two regions. As the angle of incidence becomes more oblique, the refracted ray is bent more until finally the refracted ray emerges at an angle of 90° with respect to the normal and just grazes along the surface. Figure 14.3 shows various incidence

569

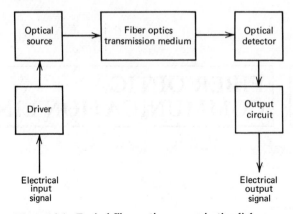

Figure 14.1 Typical fiber optic communication link.

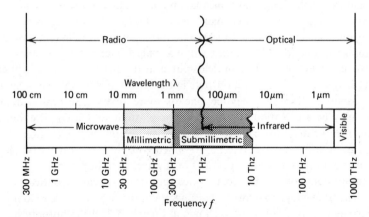

Figure 14.2 Frequency spectrum above 300 MHz.

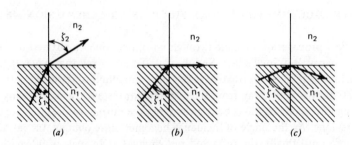

Figure 14.3 Ray paths for several angles of incidence, $n_1 > n_2$.

angles. Figure 14.3*b* illustrates what is called the critical angle, where the refracted ray just grazes along the surface. Figure 14.3*c* is an example of total reflection. This is when the angle of incidence exceeds the critical angle. A glass fiber, when utilized as a medium for the transmission of light, requires total internal reflection.

For effective transmission of light a glass fiber is made up of a glass fiber core which is covered with a jacket called *cladding*. If we let the core have a refractive index of n_1 and the cladding a refractive index of n_2, the structure (core with cladding) will act as a light waveguide when $n_1 > n_2$.

Another property of the fiber for a given wavelength λ is the normalized frequency V, then

$$V = \frac{2\pi a}{\lambda} \sqrt{n_1^2 - n_2^2} \qquad (14.1)$$

where a = the core radius and n_2 for unclad fiber = 1.

In formula (14.1) the term $\sqrt{n_1^2 - n_2^2}$ is called the numerical aperture. In essence the numerical aperture is used to describe the light-gathering ability of a fiber. In fact the amount of optical power accepted by a fiber varies as the square of its numerical aperture. It is also interesting to note that the numerical aperture is independent of any physical dimension of the fiber.

As shown in Figure 14.1, there are three basic elements in an optical fiber transmission system: the source, the fiber link, and the optical detector. Regarding the fiber link, there are two basic design parameters that limit the length of a link without repeaters, or limit the distance between repeaters. These most important parameters are loss, usually expressed in dB/km, and dispersion, which is often expressed in MHz/km. A link may be power limited (loss limited) or it may be dispersion limited.

Dispersion, manifesting itself with intersymbol interference at the far end, is brought about by two factors. One is material dispersion and the other is modal dispersion. Material dispersion is caused by the fact that the refractive index of the material changes with frequency. If the fiber waveguide supports several modes, we have a modal dispersion. Since the different modes have different phase and group velocities, energy in the respective modes arrives at the detector at different times. Consider that most optical sources excite many modes, and if these modes propagate down the fiber waveguide, delay distortion (dispersion) will result. The degree of distortion depends on the amount of energy in the various modes at the detector input.

One way of limiting the number of propagating modes in the fiber is in the design and construction of the fiber waveguide itself. Return again to formula (14.1). The modes propagated can be limited by increasing the radius a and keeping the ratio n_1/n_2 as small as practical, often 1.01 or less.

We can approximate the number of modes N that a fiber can support by applying formula (14.1). If $V = 2.405$, only one mode will propagate (HE_{11}). If V is greater than 2.405, more than one mode will propagate, and when a reasonably large number of modes propagate,

$$N = (\tfrac{1}{2})V^2 \tag{14.2}$$

Dispersion is discussed in more detail later in the chapter.

14.3 TYPES OF FIBER

There are three categories of fiber as distinguished by their modal and physical properties:

- Single mode
- Step index (multimode)
- Graded index (multimode)

Single-mode fiber is designed such that only one mode is propagated. To do this, $V = 2.405$. Such a fiber exhibits no modal dispersion at all. Typically we might encounter a fiber with indices of refraction of $n_1 = 1.48$ and $n_2 = 1.46$. If the optical source wavelength is 820 nm, for single-mode operation the maximum core diameter would be 2.6 μm, a very small diameter indeed.

Step-index fiber is characterized by an abrupt change in refractive index, and graded-index fiber is characterized by a continuous and smooth change in refractive index (i.e., from n_1 to n_2). Figure 14.4 shows the fiber construction and refractive index profile for step-index fiber (Figure 14.4a) and graded-index fiber (Figure 14.4b).

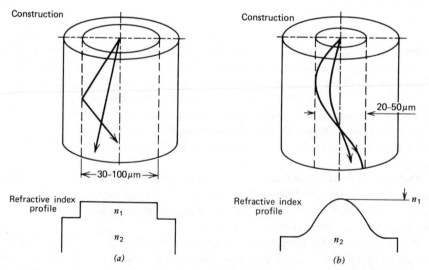

Figure 14.4 Construction and refractive index properties. (a) For step-index fiber. (b) For graded-index fiber.

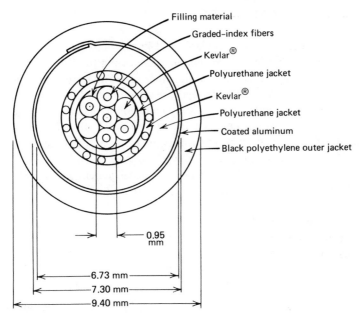

Figure 14.5   Direct-burial optical fiber cable.

Step-index fiber is more economical than graded-index fiber. For step-index fiber the distance–bandwidth product, the measure of dispersion discussed above, is on the order of 10–100 MHz/km. With repeater spacings on the order of 10 km, only a few megahertz of bandwidth is possible.

Graded-index fiber is more expensive than step-index fiber, but it is one alternative for improved distance–bandwidth products. When a laser diode source is used, values of from 400 to 1000 MHz/km are possible. If an LED source is used with its much broader emission spectrum, distance–bandwidth products with graded-index fiber can be achieved up to about 300 MHz/km. Material dispersion in this case is what principally limits the usable bandwidth.

There are two additional criteria for optical fiber that are important in system design. These are minimum bending radius and fiber strength.

Radiation losses at fiber waveguide bends are usually quite small and may be neglected in system design unless the bending radius is smaller than that specified by the manufacturer. Minimum bending radii vary from about 2 to 10 cm, depending on the cable characteristics, or, as a rule of thumb, around 10 times the cable diameter.

Fiber cable strength is also specified by the manufacturer. For example, one manufacturer for a specific cable type specifies a maximum pulling tension of 1780 newtons (N) (400 lb) at 20°C, a maximum permissible compression load of 655 N/cm (375 lb/in.) flat plate, and a maximum permissible impact force of 280 N-cm (160 lb-in.).

Figure 14.5 shows a typical five-fiber cable for direct burial.

14.4 SPLICES AND CONNECTORS

Optical fiber cable is commonly available in 1-km sections. There are two methods of connecting these sections in tandem and connecting the fiber to the source at one end and to the optical detector at the other end. These are by splicing or using special connectors. The objective in either case is to transfer as much light as possible through the coupling. Good splices generally couple more light than connectors.

A good splice can have an insertion loss as low as 0.1 dB, whereas connectors, depending on the type and on how well they are installed, can have insertion losses as low as 0.7 dB and some as high as 3 dB or more.

An optical fiber splice requires highly accurate alignment and an excellent end finish to the fibers. There are three causes of loss at a splice:

- Lateral displacement of fiber axes
- Fiber end separation
- Angular misalignment

Splice loss also varies directly with the numerical aperture of the fiber in question.

There are two types of splices now available, the mechanical splice and the fusion splice. With a mechanical splice an optical matching substance is used to reduce splicing losses. That substance must have a refractive index closely matching the index of the core fiber. A cement with similar properties is also used serving the dual purpose of refractive index matching and fiber bonding.

The fusion splice, also called the hot splice, is one where the fibers are fused together. First the fibers to be spliced are butted together and then heated with a flame or electric arc until softening and fusion occurs. Such splices show a 0.1–0.2-dB loss.

Splices are generally hard to handle in a field environment such as in a cable manhole. Connectors are much more amenable to field installation. However, connectors are lossier and can be expensive. Repeated mating of a connector may also be a problem, particularly if dirt or dust deposits occur in the area of the fiber.

14.5 LIGHT SOURCES

A light source, perhaps more properly called a photon source, has the fundamental function in a fiber optic communication system to efficiently convert electrical energy (current) into optical energy (light) in a manner that permits the light output to be effectively launched into the optical fiber. The light signal so generated must also accurately track the input electrical signal so that noise and distortion are minimized.

The two most widely used light sources for fiber optic communication systems are the etched-well surface LED and the injection laser diode (ILD). LEDs and ILDs are fabricated from the same basic semiconductor compounds and have similar heterojunction structures. They do differ considerably in their performance characteristics. LEDs are less efficient than ILDs, but are cheaper. The spatial intensity distribution of an LED is Lambertian (cosine), whereas a laser exhibits a relatively high degree of waveguiding and thus, for a given acceptance angle, can couple more power into a fiber than the LED. In other words the LED has a comparatively broad output spectrum and the ILD has a narrow spectrum, on the order of 1 or 2 nm wide.

With present technology the LED is capable of launching about 100 μW of optical power into the core of a fiber with a numerical aperture of 0.2 or greater. A laser diode with the same input power can couple up to 7 mW into the same cable. The coupling efficiency of an LED is on the order of 2%, whereas the coupling efficiency of an ILD is better than 50%, with one manufacturer reportedly achieving about 70%.

Methods of coupling a source into an optical fiber vary as do coupling efficiencies. To avoid ambiguous specifications on source output powers, such powers should be stated as out of the "pigtail." A pigtail is a short piece of optical fiber coupled to the source at the factory and, as such, is an integral part of the source. Of course the pigtail should be the same type of fiber as that specified for the link.

Component lifetimes for LEDs are on the order of 100,000 h, with up to a million hours reported in the literature. Many manufacturers guarantee an ILD for 10,000 h, and up to 100,000 h of lifetime are expected once more experience is gained in

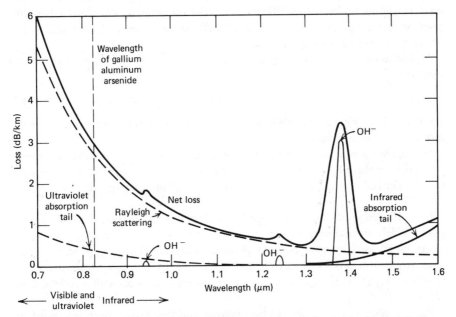

Figure 14.6 Net spectral loss curve for a glass core. Includes components from Rayleigh scattering, ultraviolet absorption, infrared absorption, and hydroxyl ion (OH^{-1}) absorption. (Ref. 18; Copyright © 1980 by Western Electric Company.)

manufacture and usage. The life expectancy of an ILD is reduced when it is over-driven to derive more coupled output power (i.e., more than 5–7 mW).

The ILD is a temperature-dependent device. Its threshold current increases non-linearly with temperature. Rather than attempt to control the device's temperature, a negative feedback circuit is used whereby a portion of the emitted light is sampled, detected, and fed back to control the drive current. Such circuits are similar to the familiar AGC circuits used in radio receivers.

Present fiber optic communication systems operate in the region (wavelength) of 820 nm. If we examine the attenuation versus wavelength curves in Figure 14.6, we see that at 820 nm the lowest attenuation per kilometer of glass fiber which can be expected is on the order of 3 dB. The net loss tends to drop to somewhere around 0.5 dB/km at just below 1300 nm, and there is another valley at about 1440 nm. Certainly it would be very advantageous to use the longer wavelengths. The fiber and connectors are available. It is the sources and detectors that are the bottlenecks. Development efforts are concentrated in the United Kingdom and Japan, and at least one vendor in Japan and one in the United Kingdom offer first-generation devices off the shelf.

14.6 LIGHT DETECTORS

The most commonly used detectors (receivers) for fiber optic communication systems are photodiodes, either PIN or APD. The terminology PIN derives from the semiconductor construction of the device where an intrinsic (I) material is used between the *p–n* junction of the diode.

A photodiode can be considered a photon counter. The photon energy E is a function of frequency and is given by:

$$E = h\nu \tag{14.3}$$

where h = Planck's constant (W/s^2) and ν = frequency (Hz). E is measured in watt-seconds or kilowatt-hours.

The receiver power in the optical domain can be measured by counting, in quantum steps, the number of photons received by a detector per second. The power in watts may be derived by multiplying this count by the photon energy, as given in equation (14.3).

The efficiency of the optical-to-electrical power conversion is defined by a photodiode's *quantum efficiency* η, which is a measure of average number of electrons released by each incident photon. A highly efficient photodiode would have a quantum efficiency near 1, and decreasing from 1 indicates progressively poorer efficiencies. The quantum efficiency, in general, varies with wavelength and temperature.

For the fiber optic communication system engineer *responsivity* is a most important parameter when dealing with photodiode detectors. Responsivity is expressed in amperes per watt or volts per watt and is sometimes called sensitivity.

Responsivity is the ratio of the rms value of the output current or voltage of a photodetector to the rms value of the incident optical power.

In other words, responsivity is a measure of the amount of electrical power we can expect at the output of a photodiode, given a certain incident light power signal input. For a photodiode the responsivity R is related to the wavelength λ of the light flux and to the quantum efficiency η, the fraction of the incident photons that produce a hole–electron pair. Thus

$$R = \frac{\eta\lambda}{1234} \, (A/W) \tag{14.4}$$

with λ measured in nanometers.

Responsivity can also be related to electron change Q by the following:

$$R = \frac{\eta Q}{h\nu} \tag{14.5}$$

where $h\nu$ = photon energy [equation (14.3)] and Q = electron charge, 1.6×10^{-19} coulombs (C).

Figure 14.7 plots typical responsivities for four photodetectors. The two upper curves are for semiconductor photodiodes shown with material codes Ge and Si. Curves S-1 and S-20 are for the two photodiode materials AgOCs and Na_2KsbCs, respectively. The curves are plotted with quantum efficiency η as a parameter. The dashed lines are for comparative purposes, where η is assumed to be constant with wavelength.

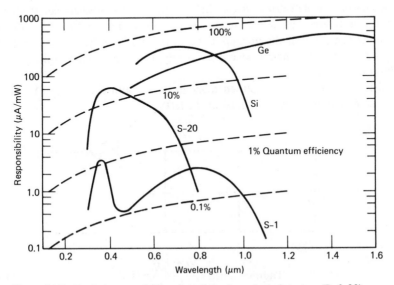

Figure 14.7 Typical responsivities plotted for four photodetectors (Ref. 20).

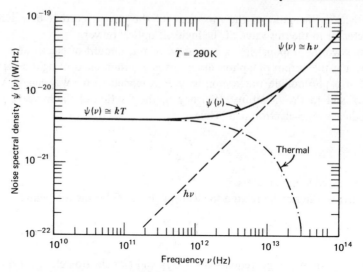

Figure 14.8 Noise in the frequency domain.

As in all transmission systems, noise is a most important consideration. Just as in the other systems treated in this text, the noise analysis of a fiber optic system is centered on the receiver. We must treat noise for these types of systems in both the optical and the electrical domain. In the optical domain, as shown in Figure 14.8, quantum noise in dominant. Quantum noise manifests itself as shot noise on the primary photocurrent from the detector. There is also shot noise on the dark current, and in the case of an APD there is excess noise from avalanche multiplication. Because the optical detector converts light energy into electrical energy, thermal noise must be treated on the electrical side of the detector. In most cases this will be the noise generated in the preamplifier that follows the light detector. In fact in PIN detectors, because their gain is very nearly 1, thermal noise is the principal contributor.

The drawing below is a simplified diagram of an optical receiver. The optical detector is the two intertwining circles at the left.

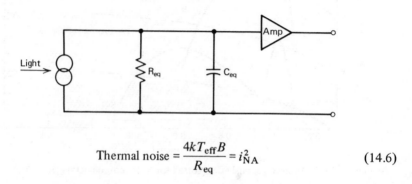

$$\text{Thermal noise} = \frac{4kT_{\text{eff}}B}{R_{\text{eq}}} = i_{\text{NA}}^2 \qquad (14.6)$$

where R_{eq} = equivalent resistance of driver amplifier

T_{eff} = effective noise temperature of load resistor (K)

B = bandwidth (Hz)

k = Boltzmann's constant

Noise is related, in this case, to the mean-square value of the current of the load resistor i_{NA}^2.

Shot noise can also be related to the mean-square value of current and consists of two parts, one from the fluctuations in the signal, i_{NS}^2,

$$i_{NS}^2 = 2q \left(2P_{opt} \frac{\eta q}{h\nu} \right) B \tag{14.7}$$

where P_{opt} = optical power

q = electron charge $(1.6 \times 10^{-19} \text{ C})$

η = quantum efficiency

$h\nu$ = photon energy [equation (14.3)]

B = system bandwidth

and the second from dark current in the detector, i_{ND}^2,

$$i_{ND}^2 = 2q i_D B \tag{14.8}$$

where i_D = photodetector dark current.

$$\text{Signal-to-noise ratio} = \frac{\text{signal power}}{\text{shot noise power + amplifier noise power}} \tag{14.9}$$

Signal power is a function of the mean-square value of the detector photocurrent, i_s^2,

$$i_s^2 = \frac{1}{2} \left(2P_{opt} \frac{\eta q}{h\nu} \right)^2 \tag{14.10}$$

Noise power, the two terms in the denominator in equation (14.9), was discussed above. The thermal noise power in this case is the preamplifier noise power. The signal-to-noise ratio equation (14.9) can now be written (Ref. 21):

$$\frac{S}{N} = \frac{2[P_{opt}(\eta q/h\nu)]^2}{[2q i_D + 4q P_{opt}(\eta q/h\nu) + 4kT_{eff}/R_L]B} \tag{14.11}$$

where R_L = load resistance and B has been factored out in the denominator. For a BER of 1×10^{-9}, a standard for fiber optic systems, requires a signal-to-noise ratio of 21.5 dB. The signal-to-noise ratio of the optical power incident on the detector is the square root of this figure, or 10.75 dB. For an APD the signal-to-noise ratio is given by (Ref. 21):

$$\frac{S}{N} = \frac{2[P_{opt}(\eta q/h\nu)]^2 M^2}{[[2qi_D + 4qP_{opt}(\eta q/h\nu)]M^2 F(M) + 4kT_{eff}/R_L]B} \tag{14.12}$$

where M = avalanche gain of photocurrent and $F(M)$ = excess noise introduced by avalanche gain.

An APD has what is called optimum gain M_{opt}. The gain M of an APD cannot be increased indefinitely; there is a point where as M is increased, the signal-to-noise ratio begins to degrade, and we have passed the point of M_{opt}. Reference 22 describes a group of 53 APDs produced by Bell Telephone Laboratories with a calculated theoretical gain of 140 (equivalent to a power gain of about 21.5 dB), whereas the practical M_{opt} for the group, as measured, turned out to be around 80 (19 dB). Other sources specify no more than 15 dB of practical gain M_{opt} for an APD.

In the noise analysis presented above we have seen that signal-to-noise ratios are a function of bandwidth B which, in turn, is a function of bit rate (see Section 14.8). Table 14.1 summarizes APD sensitivies in dBm for the standard BER of 1×10^{-9} for bit rates in common use for PCM transmission in the telephone industry (Ref. 17). Noise equivalent power (NEP) is commonly used as the figure of merit of a photodiode. NEP is defined as the rms value of optical power required to produce a unity signal-to-noise ratio (i.e., signal-to-noise ratio = 1) at the output of a light-detecting device. NEPs vary for specific diode detectors between 1×10^{-13} and 1×10^{-14} W/Hz$^{1/2}$.

Of the two types of photodiodes discussed here, the PIN is cheaper and requires less complex circuitry than its APD counterpart. The PIN diode has peak responsivity from about 800 to 900 nm for silicon devices. These responsivities range from about 300 to 600 μA/mW. The overall response time for the PIN is good for about 90% of the transient, but sluggish for the remaining 10%, which is a "tail." The poorer response of the tail portion of a pulse may limit the net bit rate on digital systems.

The PIN detector does not display gain, whereas the APD does. The response time of the APD is far better than that of the PIN, but the APD displays certain temperature instabilities where responsivity can change significantly with temperature.

Table 14.1   Receiver (APD)
             Sensitivities

Bit Rate (Mbps)	Sensitivity (dBm)
1.5	−67
6	−64
12	−61
45	−54
90	−51
170	−48

Compensation for temperature is usually required in APD detectors and is often accomplished by a feedback control of bias voltage. It should be noted that bias voltages for APDs are much higher than for PIN diodes, and some APDs require bias voltages as high as 200 V or more. Both the temperature problem and the high-voltage bias supply complicate repeater design.

14.7 MODULATION AND WAVEFORM

The most widely used type of modulation of a light carrier is a form of AM called *intensity modulation*. Both types of light sources, the LED and the ILD, can be conveniently modulated in the intensity mode. The detectors (PIN and APD) discussed above each respond directly to intensity modulation, producing a photo-current proportional to the incident light intensity. Today's optical fiber communication systems are more suitable to digital than to analog operation.

Considering the previous chapters, let us review some advantages and disadvantages of optical fiber communication systems, particularly for digital operation:

- We are dealing with much higher frequencies.
- Coherence bandwidths are much greater.
- Component response times are inherently faster than anything discussed previously.
- The linearity of optical fiber systems is comparatively poor.

As mentioned above, both the LED and the ILD light sources are semiconductor devices which, in most cases, are directly modulated. Biasing of the source and the adjustment of the quiescent operational point are most important considerations. In view of this, the following guidance should be followed.

1. The intensity of the driving source varied directly with the bias current either in the lasing region (for the ILD, see Figure 14.9) or in the spontaneous emission region (for the LED).

2. In the case of continuous analog modulation the quiescent bias current must be established at a point such that the modulating signal causes an equal plus and minus swing about the quiescent value. It should also be in the most linear range of the intensity characteristic.

3. For digital modulation of the several variants of PAM the quiescent bias current is adjusted, given a specific source, as follows. In the case of an LED, either near zero or at a quiescent point optimum for noise and/or response time. For the ILD case it should be adjusted at a point near or below threshold (Figure 14.9) providing that the transition noise is tolerable; slightly above threshold if transition noise must be reduced; or near zero if transition response time is adequate for the design modulation rate and if threshold transition noise is tolerable.

4. For continuous AM of an ILD consideration must be given to the laser's life when operating in this mode.

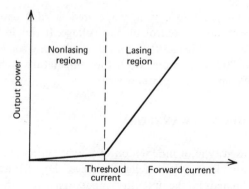

Figure 14.9   Power–current relationship for a laser diode (ILD) source.

An important digital system design consideration is the pulse format. For the discussion that follows only binary digital systems will be covered (see Chapters 8 and 9). By definition, then, information such as on-line data, PCM, or DM is transmitted serially as 1's and 0's. The manner (format) in which the 1's and 0's are presented to the modulator is important for a number of reasons.

First, amplifiers for fiber optic receivers are usually ac coupled. As a result each light pulse which impinges on the detector produces a linear electrical output response with a low-amplitude negative tail of comparatively long duration. At high bit rates tails from a sequence of pulses may tend to accumulate, giving rise to a condition known as baseline wander, and such tails cause intersymbol interference. If the number of "on" pulses and "off" pulses can be kept fairly balanced for periods that are short compared to the tail length, the effect of ac coupling is then merely to introduce a constant offset in the linear output of the receiver, which can be compensated for by adjusting the threshold of the regenerator. A line signaling format can be selected that will provide such a balance. The selection is also important on synchronous systems for self-clocking at the receiver.

Figure 14.10 illustrates five commonly used binary formats. Each is briefly discussed below.

1. *NRZ* (nonreturn to zero). This signal format was discussed in Chapter 8 where by convention a 1 represents the active state and a 0 the passive state. A change of state only occurs when there is a 1 to 0 or 0 to 1 transition. A string of 1's is a continuous pulse or "on" condition and a string of 0's is a continuous "off" condition. In NRZ information is extracted from transitions or lack of transitions in a synchronous format, and a single pulse completely occupies the designated bit interval.

2. *RZ* (return to zero). In this case there is a transition for every bit transmitted, whether a 1 or a 0, as shown in Figure 14.10, and, as a result, a pulse width is less than the bit interval to permit the return to zero condition.

3. *Bipolar NRZ.* This is similar to NRZ, except that binary 1's alternate in polarity.

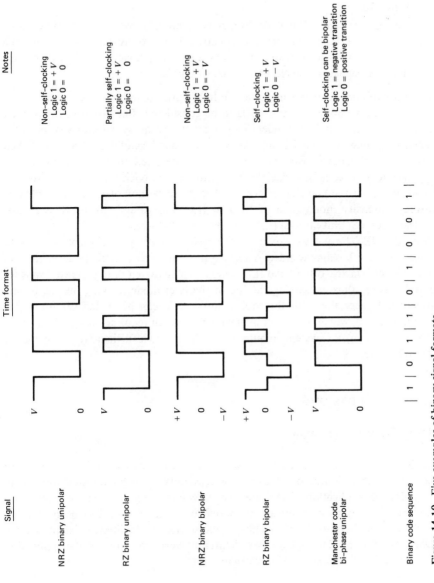

Figure 14.10 Five examples of binary signal formats.

583

4. *Bipolar RZ.* The same as bipolar NRZ, but in this case there is a return to zero condition for each signal element, and again the pulse width is always less than the bit interval.

5. *Manchester code.* This code format is commonly used in digital fiber optic systems. Here the binary information is carried in the transition. By convention a logic 0 is defined as a positive-going transition and a logic 1 as a negative-going transition. The signal can be either unipolar or bipolar.

The choice of code format is important in fiber optic communication system design, and there are a number of trade-offs to be considered. For instance, RZ formats assist in reducing baseline wander. To extract timing on synchronous systems the Manchester code and the RZ bipolar codes are good candidates because of their self-clocking capability. However, it will be appreciated from Figure 14.10 that they will require at least twice the bandwidth as the NRZ unipolar code format. An advantage of the Manchester code is that it can be unipolar, which adapts well to direct intensity modulation of LED or ILD sources and also provides at least one transition per unit interval (i.e., per bit) for self-clocking.

With the NRZ format we can attain the highest power per information bit if we wish to tolerate baseline wander. Achieving this power is particularly desirable with LED sources. On the other hand ILD sources can be driven to high power levels for short intervals, thus conserving the life of the laser diode, making the RZ format attractive. Longer life with the shorter duty cycle can be traded off for higher modulation rate, which will result in greater bandwidth requirements of the system. As we have seen, RZ systems require twice the bandwidth than NRZ systems for a given bit rate.

14.8 SYSTEM DESIGN

14.8.1 General Application

The design of a fiber optic communication system involves several steps. Certainly the first consideration is to determine the feasibility of such a transmission system for a desired application. There are two aspects to this decision, economic and technical. Analog applications are for wideband transmission of such information as video, particularly for CATV trunks, studio-to-transmitter links, and multichannel FDM mastergroups. Throughout this chapter, however, we stress digital application for fiber optic transmission and in particular for:

- On-premises data bus
- Higher level PCM or DM configurations for telephone trunks
- Radar data links
- Conventional data links well in excess of 9.6 kbps
- Digital video

The present trend is for the cost of fiber cable and components to go down, even with the impact of inflation. Fiber optic repeaters are more costly than their metallic PCM counterparts, but less repeaters are required for a given distance, and the powering of these repeaters is more involved, particularly if the power is to be taken off the cable itself. In this case a metallic pair (or pairs) must be included in the cable sheath if the power is to be provided from trunk terminal points. Another approach is to supply power locally at the repeater site with a floating battery backup. Consider the following features that make fiber optic transmission systems attractive to the telecommunication system designer:

1. *Low transmission loss* as compared to wire pairs or coaxial cable for broadband transmission (i.e., in excess of several megahertz (or Mbps), allowing a much greater distance between repeaters.

2. *Wide bandwidth* with present systems in the 820-nm region with bit rate distance products up to 2 Gbps/km. Future systems using the 1.2-μm band will permit 4–8 Gbps/km bandwidth products and with single-mode fiber at 1.2 μm, up to 20 Gbps or better.

3. *Small bending radius.* With proper cable design, a bending radius in the order of a few centimeters can have negligible effect on transmission.

4. *Nonradiative, noninductive, nonconductive, low crosstalk.* Complete isolation from nearby electrical systems, whether telecommunication or power, avoiding such problems as groundloops, radiative or inductive interference, and reducing lightning-induced interference.

5. *Lighter weight, smaller diameter* than its metallic equivalents.

6. *Growth potential.* Low-bit-rate systems can easily be expanded in the future on fiber to much higher bit-rate systems using the same cable. This is being done today with coaxial cable by increasing the number of repeaters on older cable.

14.8.2 Design Procedure

The first step in designing a fiber optic communication system is to establish the basic input system parameters. Among these we would wish to know:

- Signal to be transmitted
- Link length
- Growth requirements (i.e., additional circuits, increased bit rates)
- Tolerable signal impairment levels stated as signal-to-noise ratio or BER at the output of the terminal detector

Throughout the design procedure, when working with trade-offs, the designer must establish whether he is working in a power limited domain or in a bandwidth (bit rate) limited* domain. For instance, with low bit rates such as the T1 carrier (1.544

*Perhaps more properly called *dispersion limited.*

Mbps) we would expect to be power limited in almost all circumstances. Once we get into higher bit rates, say above 45 Mbps, the proposed system must be tested (by calculation) to determine if the system rise time is not bit rate limited. This test will be described in detail. The designer then selects the most economic alternatives and trade-offs among the following:

- Fiber parameters: (1) single mode or multimode; (2) step index or graded index; (3) number of fibers and cable makeup
- Source type: LED or ILD
- Wavelength (prior to 1982, probably in the 800–900-nm region)
- Detector type: PIN or APD
- Repeaters, if required, and their powering
- Modulation type and waveform (code format)

If the system is power limited, splicing rather than fiber connectors should be considered because splicing loss is smaller than connector loss. In general we will find that for systems in excess of 2 km, the fiber cable becomes the cost driver, whereas with shorter links the components (i.e., source and detector) are the cost drivers. As an example, in nearly every case of on-premises data bus systems, low-cost components and cable can be selected such as LED sources, step-index rather high-attenuation fiber, and PIN detectors.

When systems are extended over several kilometers, much more care has to go into system design to optimize cost versus system performance. For instance, the available power into a fiber from an LED runs about 10–15 dB below the available power from an injection laser. Furthermore, for a high-bit-rate link (i.e., becoming bandwidth limited), the narrower spectral width of the laser is necessary to avoid material dispersion and improve system rise time. The more expensive APDs have 15–20 dB gain over the less expensive PIN diode detectors.

14.8.3 General Approach

As a first step assume the system to be power limited. We then can say that the major constraint for systems over about 2 km in length is the distance from terminal to terminal (i.e., from source to detector). Another assumption is that systems over 10 km long will require intermediate repeaters. This assumes today's technology and operation in the 800–900-nm region. For systems of between 6 and 10 km between terminals a repeater may still be necessary. Without a repeater in this range considerable working and reworking of the design is required, adjusting and readjusting design parameters to define an optimal system that meets BER or signal-to-noise ratio objectives, allowing a proper system margin. For links over 5 km in length a 10-dB margin is acceptable for most applications, although some will argue that it is overly conservative. This margin is assigned to compensate for active component aging and for temperature variations.

Fiber cable is commonly available in 1-km lengths, and cable loss is specified in decibels per kilometer. We find that off-the-shelf cable attenuations are 3, 4, 5, 6, up to 15 or 20 dB/km. The price varies inversely with attenuation.

The system designer develops a power budget, and system loss is computed. Let us see what range of link losses is feasible. For instance, for a typical ILD the output power varies from - 3 to about +6 dBm, depending on the manufacturer and the driving signal type and format presented to the ILD. For digital systems with 50% density of 1's, the *average* source output is 3 dB less for NRZ and 6 dB less for RZ digital signals than the specified peak power. Care must be taken whether the power is measured at the source output before or after coupling to the fiber. This can be avoided if power is specified at the output of the pigtail.

A similar approach is suggested with the receiver (light detector) where the specified receiver threshold is given at the pigtail that inputs the PIN or APD detector. The threshold should be stated for the required signal-to-noise ratio or BER. The BER should be on the order of 10^{-8} or 10^{-9}.

Ranges of link power budgets in decibels can be roughly approximated from the following average power outputs: for an LED, - 12 dBm, for an ILD, +3 dBm. For detectors see the table below:

Detector Type	7 Mbps	45 Mbps
PIN (typical)	-47 dBm	-34 dBm
APD	-64 dBm	-54 dBm

Taking these in pairs (i.e., source and detector), first with an LED source and a PIN detector at 7 Mbps, the total link budget loss is -47 dBm - (-12 dBm) = 35 dB. The 35 dB is allotted to

- Fiber loss
- Splice or connector loss
- Margin

By observation we see that the link is short, and therefore the margin can be reduced to 8 dB, leaving 27 dB for real link losses. If we splice rather than use connectors, 5 dB/km cable possibly can be used for a 5-km link. For such a link, assuming cable in 1-km sections, six splices will be required. If, for some reason, connectors are called for (again six are required), then we must resort to the more expensive 4-dB/km cable, allowing 6 dB for total connector loss (1 dB per connector).

For a 45-Mbps system, such as the Bell System DS3, using an ILD source and an APD detector, the link budget is 57 dB [+3 dBm - (-54 dBm)]. Here we allow the 10-dB margin, leaving 47 dB for fiber and connector/splice losses. With 4-dB/km cable, a 10-km link can be installed, leaving 7 dB for connectors or splices. At 0.3 dB per splice and 11 splices required, 3.3 dB of splice loss would be incurred. With AMP connectors, at 1.5 dB per connector, with an 8-km link there would be 13.5

dB of connector loss (i.e., $9 \times 1.5 = 13.5$ dB). We could build an 8-km link with 5 dB/km cable using connectors, if these connectors had an insertion loss no greater than 0.7 dB, allowing 40 dB for cable loss and 9×0.7 or 6.3 dB for connector loss, totaling 46.3 dB, which is just within the 47 dB limit given above.

14.8.4 Dispersion Limited Domain and System Bandwidth

From the system rise time we can calculate the system bandwidth and the maximum data rate that the system can support. System rise time is a function of the root sum squares of the component rise times, namely,

- Source S
- Fiber due to multimode dispersion F_{mm}
- Fiber due to material dispersion F_{md}
- Detector d

For high-bit rate systems the rise time is measured in nanoseconds.

$$\text{System rise time (ns)} = 1.1 \sqrt{S^2 + F_{mm}^2 + F_{md}^2 + d^2}$$

$$\text{3-dB bandwidth (GHz)} = \frac{0.35}{\text{system rise time (ns)}}$$

Material dispersion can be neglected when laser diode sources are used because of the laser's narrow spectral emission characteristic. For LED sources, if no other data is available, 5.5 ns/km can be used for material dispersion.

Fiber rise times are stated in nanoseconds per kilometer and therefore must be extended for the total cable length. For instance, with an 8-km system and a fiber multimode dispersion rise time of 7 ns/km, the value to be used in the system rise time calculation is 56 ns (i.e., 8 km $\times$ 7 ns/km). This linear extrapolation may be used for conservative system design. Others (Ref. 17) suggest an rms law, and thus bandwidth (BW) is calculated:

$$BW = \left[\left(\frac{1}{BW_1} \right)^2 + \left(\frac{1}{BW_2} \right)^2 + \cdots + \left(\frac{1}{BW_N} \right)^2 \right]^{-M}$$

where M, for most cases, is between 0.5 and 0.6.

Suppose that we wished to verify whether we were in the dispersion limited domain or just where we entered that domain for a system transmitting 45 Mbps of NRZ data. What would be the system rise time limits and the required cable bandwidth per kilometer properties required?

First calculate the pulse width at 45 Mbps. This is $1/45 \times 10^6$, or 22.22 ns. The maximum tolerable rise time for NRZ data is 70% of this figure, or 15.55 ns. For RZ data it is half this value, or 7.77 ns. These are values of dispersion on a system basis which we must not exceed.

The problem is now to find the maximum-length system acceptable with these figures in mind, or the point where we enter the dispersion-limited domain. Of course to optimize system length with this high bit rate we would wish to use a laser diode source and an APD detector. With typical values used, calculate the system rise time with the rise time of the cable as the unknown X.

	Rise Time	Rise Time^2
Source (ILD)	1.5 ns	2.25 ns^2
Cable, graded index	X	X^2
Detector (APD)	2 ns	4 ns^2

$$15.55 \text{ ns} = 1.1 \sqrt{X^2 + 2.25 + 4.0}$$

$$X^2 = 193.75$$

$$X = 13.9 \text{ ns}$$

Thus we can allow 13.9 ns rise time for the fiber cable portion of the system. This total quantity can be allotted to intermodal dispersion because we use an ILD source rather than an LED. Cable, such as the ITT T-211, displays 2-ns/km rise time. Using this cable to determine the maximum system length without repeaters (or between repeaters) working in the dispersion-limited domain, we divide 13.9 ns by 2 ns/km, and the length limit is 6.95 km using the conservative linear extrapolation method.

Allowing that a fiber's step function response resembles that of a low-pass filter, its electrical bandwidth can be estimated from its rise time, or

$$\text{Bandwidth } (-3 \text{ dB}) \text{ (MHz)} = \frac{350}{\text{rise time (ns)}}$$

In this case with 13.9 ns for fiber rise time we have

$$\frac{350}{13.9} \text{ ns} = 25.2 \text{ MHz}$$

at the end of 6.95 km of fiber. When measured, we would probably find the bandwidth greater than this value and using the root mean square would give values in excess of the measured values.

14.9   THE ATT FT3 SYSTEM

The FT3 lightwave communication system is the major vehicle for introducing fiber optic technology into the Bell System of North America. This system is specifically designed for PCM transmission where F stands for fiber and T3 means the third level for the digital transmission hierarchy (i.e., DS3; see Chapter 11). The system, as shown in Figure 14.11, is made up of lightwave terminating multiplex assemblies

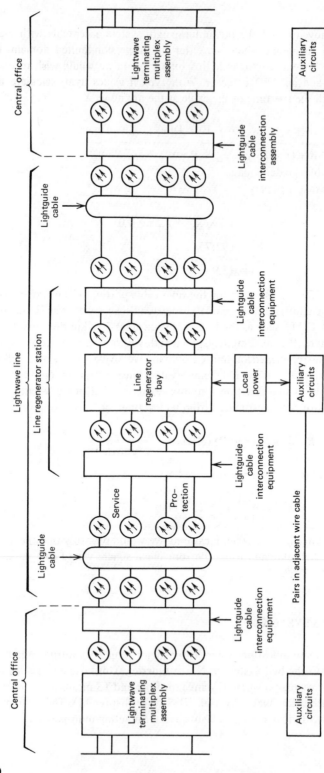

Figure 14.11 FT3 lightwave system. Regenerators are also located within each lightwave terminating multiplex assembly at each exchange (central office). (Copyright © 1980 by Western Electric Company.)

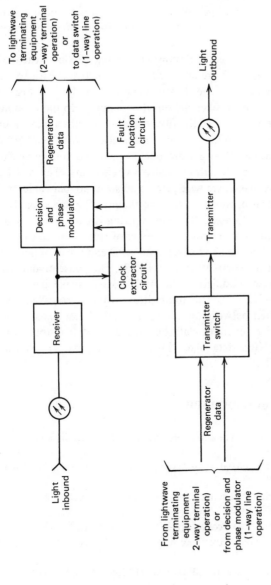

Figure 14.12 Simplified functional block diagram of regenerator shown in Figure 14.11. (Copyrite © 1980 by Western Electric Company.)

591

(LTMA), lightguide cable interconnection equipment (LCIE), and line regenerator bays. Figure 14.12 is a simplified block diagram of a lightwave regenerative repeater.

The FT3 system accommodates 28 DS1, 14 DS1C, or 7 DS2 at its input–output units. The line transmission rate on each fiber is 44.735 Mbps. A derivative system, the FT3LC, operates at 90 Mbps per fiber pair with a maximum equivalent 4-kHz VF channel capacity of 1344 circuits, and 80,000 circuits operate over one 144-fiber cable.

Individual fibers are packaged in ribbons containing 12 fibers each, and then up to a maximum of 12 ribbons are enclosed in a protective sheath to complete the cable. The outside diameter of the sheath is approximately $\frac{1}{2}$ in. (12.7 mm) and is independent of the number of fiber ribbons it contains up to the maximum.

Regenerators are provided along the fiber cable route with a maximum spacing of 4.2 mi (6.8 km). The system incorporates automatic 1 for 1 protection switching (i.e., for every operational fiber a spare fiber is provided, and upon failure of the operating fiber, the spare is automatically switched in). The auxiliary circuits shown in Figure 14.11, which are wire pairs in adjacent copper cable, include order-wires, telemetry, and gas pressure monitoring circuits. The line signal is binary NRZ data which has been randomized. The randomizing is carried out by a scrambler-descrambler. Randomizing increases the number of transitions to ease clock extraction in receivers. The basic FT3 receiver is an APD and preamplifier. Regeneration is carried on in the decision circuit (Figure 14.12). The phase modulator is used for fault location by phase modulating the 44.735-Mbps signal between two phase positions at the 470-kHz rate. The transmitter switch is so-called because it has been designed to switch between an externally applied test signal, a fault-locating signal, or the normal regenerator data line signal. Switch output is an electrical 44.735-Mbps binary NRZ data signal for all three inputs. The transmitter is an intensity-modulated ILD.

REFERENCES AND BIBLIOGRAPHY

1. S. E. Miller and A. G. Chynoweth, Eds., *Optical Fiber Telecommunications*, Academic Press, New York, 1979.
2. *Design Handbook for Optical Fiber Systems*, Vols. I–III, Information Gatekeepers, Inc., Brookline, MA, 1979.
3. R. L. Galawa, "Optical Communication via Glass Fibers," Institute of Telecommunication Sciences, Boulder, CO, 1978.
4. D. Gloge, Ed., *Optical Fiber Technology*, IEEE Press, New York, 1975.
5. "ITT Optical Fiber Characteristics and Application," ITT Tech. Note R-5, ITT, Roanoke, 1979.
6. "Optical Fiber Communications Link Design," ITT Tech. Note R-1, ITT, Roanoke, 1979.
7. R. B. Gill, "Injection Laser Sources for Optic Communications," Laser Diode Laboratories, Inc., 1977.
8. R. L. Galawa, "A User's Manual for Optical Waveguide Communications," NTIS PB-252-901, March 1976.

9. C. Kleekamp and B. Metcalf, "Designers Guide to Fiber Optics," Parts 1–4, Corning Glass Works, New York, 1977.

10. "Technical Manual Operation and Maintenance Instructions for Fiber Optic Communications System Offutt AFB," NOSC TM-258, 2 Jan 1979.

11. "Specifications for Fiber Optic Cables, General," DOD-C-85045, U.S. Department of Defense, Washington, DC, Feb. 16, 1979.

12. "Fiber Optic Cable Test Methods," DOD-STD-1678, U.S. Department of Defense, Washington, DC, Nov. 30, 1977.

13. M. Barnoski, Ed., *Fundamentals of Optical Fiber Communications*, Academic Press, New York, 1976.

14. *The Western Electric Engineer*, Special Issue: Lightwave, The Western Electric Co., New York, Winter 1980.

15. G. H. B. Yancy, "Fiber Optic Digital Telecommunication Systems," *Electronic Progress*, Vol. 22, Raytheon Co., Lexington, MA, Spring 1980.

16. C. P. Sandbank, Ed., *Optical Fibre Communication Sytems*, Wiley, New York.

17. E. E. Basch, H. A. Carnes, and R. F. Kearns, "Calculate Performance into Fiber-Optic Links," *Electronic Design*, Aug. 16, 1980.

18. J. A. Jefferies and R. J. Klaiber, "Lightguide Theory and Its Implications in Manufacturing," *The Western Electric Engineer*, Winter 1980.

19. E. Randall and R. Lavelle, "Optimize Optical Modem Cost/Performance through Emitter, Detector and Fiber Selection," *Electronic Design*, Apr. 12, 1980.

20. R. L. Galawa, Lecture Notes on Optical Communication Via Glass Fiber Waveguides, IEEE Course Series, Raytheon Equipment Division, Wayland, MA, April 1979.

21. M. J. Howes and D. V. Morgan, Eds., *Optical Fibre Communications*, Wiley, New York, 1980.

22. R. G. Smith, C. A. Brackett, and H. W. Reinbold, "Optical Detector Package," *Bell Sys. Tech. J.*, Vol. 57, 1809–1822, July–Aug. 1978.

23. H. Melchior *et al.*, "Planar Epitaxial Silicon Avalanche Photodiode," *Bell Sys. Tech. J.*, Vol. 57, 1791–1809, July–Aug. 1978.

15 | DIGITAL RADIO

15.1 DEFINITION AND SCOPE

A digital radio for this discussion is defined as a radio set in which one or more properties (amplitude, frequency, and phase) of the RF are quantized by a modulating signal. Turning back to Chapter 8, digital implies fixed sets of discrete values. A digital radio waveform, then, can assume one of a discrete set of amplitude levels, frequencies, or phases as a result of the modulating signal. Let us assume that the modulating signal is a serial synchronous bit stream.

For a given system the information in the serial bit stream may be PCM, DM (Chapter 11), or any other form of serial data information (Chapter 8). Although we defined data transmission in Chapter 8 as the digital transmission of record traffic, graphics, numbers, and so on, the industry has given *data* a wider meaning in many instances. Depending on the person one is talking to, data can mean any digital bit stream, no matter what its information content. For instance, *data rate* is a common term which is synonymous with bit rate. We shall also be using the term *symbol rate*, meaning the number of transitions or changes in state per second. In a two-state or two-level system the symbol rate measured in bauds is the same as the data rate in bits per second. If there are more than two states or levels, it is called an *M*-ary system, where *M* is the number of states. For example, 8-ary FSK is an eight-frequency system, and at any instant one of the eight frequencies is transmitted.

We must not lose sight of the system aspects of digital radio. A digital radio transmitter does not sit out there alone; it must work with a far-end receiver making up a link. There may be several links in tandem. These links may or may not be part of a larger network. Timing, bit count integrity, justification (slips and stuffing), BER, and special coding may all impact the design of a digital radio link.

Let us adopt CCIR practice (CCIR Rep. 378-3) to define digital capacity:

- Small-capacity radio-relay systems for the transmission of digital signals with gross bit rates up to and including 10 Mbps

- Medium-capacity radio-relay systems with gross bit rates ranging from 10 Mbps to about 100 Mbps
- Large-capacity radio-relay systems with gross bit rates greater than about 100 Mbps

Radio-relay systems for this discussion refer to radiolink (LOS microwave), tropospheric scatter, and satellite–earth station links in the point-to-point service.

15.2 APPLICATION

Digital radio-relay systems for the common carrier (telecommunication administration) or specialized common carrier find application in:

- Local digital trunks and, in particular, connecting digital exchanges.
- High-usage routes connecting digital exchanges in the local area.
- To expand a VF cable route where radio proves, under certain circumstances, more economic than expanding existing cable facilities either with PCM or by adding additional metallic pair cable.
- On toll (long-distance) routes connecting digital toll exchanges or with an analog space division exchange (switch) on one end and a digital exchange on the other. On shorter routes radiolinks and on longer routes satellite links may be more applicable.
- For all-digital networks where radio proves in on an economic basis over cable, whether metallic or fiber optic, for planning periods up to 15 years in the future.

Other applications of digital radio are:

- For military digital networks such as Tri-Tac and Ptarmigan. In the military environment digital transmission, especially radio, is more attractive than analog because it is much easier to make secure by on-line encryption. Digital radio tolerates fair quality links in tandem when compared to analog radiolinks as described in Chapter 5.
- Private data networks.
- Digital tropospheric scatter on tactical and strategic links 80–250 km long.
- Satellite systems for the western military forces such as the Multimission (MM) Satellite System, the Satellite Data System (SDS), FltSatCom, and the DSCS series of satellites.

The implementation of digital radio is being accelerated by the ever-increasing number of digital exchanges going into service. Much of this impetus derives from the very basic working of a digital exchange when compared to an analog exchange. Whereas an analog exchange (switch) requires a switching port for each and every voice channel to be switched, a digital exchange looks at a digital multiplexed group or other higher level multiplexed input at each port. Therefore for analog exchanges in tandem multiplexed groups carried over a radio or cable system must be demultiplexed before being presented to the switch. In this case each voice channel is switched in the space domain. At the output ports, voice channels are again multiplexed for the next particular switch in tandem. In the case of digital tandem switches, no demultiplexing and remultiplexing is required. Digital switches operate in the time domain being based on the technique of time slot switching, sometimes called *bit exchanging.* The result is a major savings by the elimination of much of the multiplex equipment in a digital network when compared to its analog counterpart. This fact has caused an acceleration of digital radio (and PCM metallic cable) implementation. However, it should be noted that the use of radio and conventional cable media for digital transmission is being tempered by digital fiber optic cable systems as a more economic alternative in many instances. (For a basic introduction to network design, analog and digital switching, consult Ref. 23.)

15.3 BASIC RADIO AND LINK DESIGN CONSIDERATIONS

15.3.1 The Primary Feature

Digital radio system design follows most of the basic procedures outlined in Chapters 5–7, but with one overriding issue which must be resolved by the system design engineer, namely, that of spectrum conservation without significant performance degradation. Consider that we can accommodate 1800 nominal 4-kHz VF channels in approximately 30 MHz of RF bandwidth using conventional FM, as discussed in Chapter 5. The same number of voice channels using PCM, allowing 1 bps/Hz of bandwidth, then would require 1800 × 64 kHz or 115 MHz of bandwidth. The RF spectrum is one of our natural resources, and we should not be wasteful of it. Thus something has to be done to make digital radio transmission more bandwidth conservative. It is the 1 bps/Hz assumption that we will be working with in this section. The problem of achieving more bits per hertz will be treated at length further on in the chapter.

The FCC has taken up the issue in Part 21.122 of *Rules and Regulations*, stating:

Microwave transmitters employing digital modulation techniques and operating below 15 GHz shall, with appropriate multiplex equipment, comply with the following additional requirements:

1. The bit rate in bits per second shall be equal to or greater than the bandwidth specified by the emission designator in hertz (e.g., to be acceptable, equipment transmitting at a 20-Mb/s rate must not require a bandwidth greater than 20 MHz).

2. Equipment to be used for voice transmission shall be capable of satisfactory operation within the authorized bandwidth to encode at least the following number of voice channels:

Frequency Range	Number of Encoded Voice Channels
2110–2130 MHz	96
2160–2180 MHz	96
3700–4200 MHz	1152
5925–6425 MHz	1152
10,700–11,700 MHz	1152

The FCC rules the following on emission limitation (Part 21.106):

When using transmissions employing digital modulation techniques

(i) For operating frequencies below 15 GHz, in any 4-kHz band, the center frequency of which is removed from the assigned frequency by more than 50% up to and including 250% of the authorized bandwidth: As specified by the following equation but in no event less than 50 dB:

$$A = 35 + 0.8(P - 50) + 10 \log_{10} B \qquad (15.1)$$

where A = attenuation (dB) below the mean output power level
   P = percent removed from carrier frequency
   B = authorized bandwidth (MHz)

(Attenuation greater than 80 dB is not required.)

(ii) For operating frequencies above 15 GHz, in any 1-MHz band, the center frequency of which is removed from the assigned frequency by more than 50% up to and including 250% of the authorized bandwidth: As specified by the following equation but in no event less than 11 dB:

$$A = 11 + 0.4(P - 50) + 10 \log_{10} B \qquad (15.2)$$

(Attenuation greater than 56 dB is not required.)

(iii) In any 4-kHz band, the center frequency of which is removed from the assigned frequency by more than 250% of the authorized bandwidth: At least $43 + 10 \log_{10}$ (mean power output in watts) dB or 80 dB, whichever is the lesser attenuation . . .

CCIR also provides guidance on spectral occupancy for digital radio systems. Reference should be made to CCIR Recs. 497-1 and 387-3 and CCIR Reps. 607-1, 608-1, and 609-1. However, CCIR is less specific on spectral occupancy versus bit rate and numbers of equivalent voice channels. It deals more with frequency allocations for specific operational bands.

15.3.2 Other Design Considerations in Contrast to Analog FM Systems

The measurement of performance of a digital radio system is the BER, often expressed in a time percentage. FM analog systems use signal-to-noise ratio in decibels or noise power weighted or unweighted (flat), which is often expressed in dBm or pW and referenced to the derived nominal 4-kHz VF channel. For radio systems this measure of performance should be related to a given time percentage of the year or worst month.

E_b/N_0 is an expression commonly used on digital systems, usually related to BER. E_b/N_0 expresses "energy per bit per hertz of noise spectral density." This is a fairly universal figure of merit for bit decisions. In particular it is independent of the waveform and bandwidth used to carry the information bits. It is valid in this simple form whenever the noise spectral density is flat across the necessary band, which is almost always the case in practical applications of digital radio, apart from spread spectrum systems and certain digital HF systems. Hence in link calculations bandwidth is disregarded and N_0 is calculated by:

$$N_0 = -228.6 \text{ dBW} + \text{NF}_{dB} + 10 \log_{10} \text{ (room temperature in K)*} \quad (15.3a)$$

where -228.6 dBW is derived from Boltzmann's constant (see Chapter 1) and NF = receiving system noise figure (dB). Alternatively, if dealing with noise temperature,

$$N_0 = -228.6 \text{ dBW} + 10 \log_{10} T_{sys} \quad (15.3b)$$

where T_{sys} = receiving system equivalent noise temperature.

Since we are dealing with synchronous serial data, synchronization is a very important system aspect. It would seem that this goes without saying. On radiolinks over, say, 7 km (4.3 mi) long, on all tropospheric scatter links and on some earth station links, particularly those above 10 GHz, we will encounter some sort of fading at least some of the time. Fading on earth station links will probably be attributed to rainfall, whereas on radiolinks (LOS microwave) and on tropospheric scatter fading is more probably caused by multipath, or at least we can say that it is the principal offender. When fades are deep enough, a dropout occurs and synchronization (i.e., the receiver is in synchronization with its companion far-end transmitter) may be lost. A digital radio system must be designed to withstand short periods of dropout without losing synchronization. This capability is expressed as maintenance of bit count integrity (BCI) and is often measured in milliseconds. The system should also have some rapid automatic means of resynchronizing itself, once BCI is exceeded.

One result of multipath on radiolinks and on tropospheric scatter is dispersion. Dispersion means that an arriving signal at the receiver has been dispersed or "elongated." In the case of a pulse, there would be a main signal component arriving in its proper time slot, this signal having followed the direct and expected path. During multipath conditions, reflected signal energy due to multipath conditions arrives somewhat later, spilling into the time slot of the next pulse. Dispersion is a measure of elongation or smearing, usually given in nanoseconds. Such dispersion causes intersymbol interference, deteriorating link error performance. Digital radio-

---

*Often it is better to use the highest temperature expected in the LNA enclosure, expressed in kelvins.

links often can operate up to 10 Mbps without concern for dispersion. Those operating at rates greater than 10 Mbps and especially those at rates higher than 20 Mbps should take dispersion into account in the link design and equipment selection. Dispersion impacts all digital tropospheric scatter links.

On digital radio systems there are basically three causes of deteriorated error performance. One is multipath dispersion just discussed. A second is due to Gaussian noise in the receiving system. Basic to any link design is to establish a value of E_b/N_0 to meet the required error performance criteria of the link expressed as BER. Deep fades may reduce this value for the period of the fade, causing an error burst. Another cause of errors is on systems designed to operate near the E_b/N_0 margin, resulting in bursts of errors and periods when the link is comparatively error-free. Thus the system designer at the outset must establish whether a digital radiolink is dominated by random errors or by bursts of errors. Earth stations* would probably fall into the random error category, whereas longer LOS radiolinks and tropospheric scatter links would fall into the bursty error category. If coding is used, and we mean channel coding here, to improve error performance, the code selected will much depend on whether a link will expect random errors or bursty errors.

15.3.3 Error Rate Performance Criteria

CCIR Rep. 378-3 describes the performance of a digital system in terms of two BERs and a third objective based upon error burst criteria. The two BERs are a high value and a low value. The higher BER and the error burst objective would in many systems be the most significant, and, for LOS systems, would determine repeater spacing. The lower BER objective would serve to control performance for the majority of the time (greater than 80%) when fading effects are negligible and when intersymbol and interchannel interference become significant.

Regarding the low BER value, CCITT has proposed an error rate design objective of 1 in 10^{10} per kilometer for the transmission system in a 25,000-km (15,500 mi) hypothetical digital path. For a 2500-km (1550 mi) digital radio-relay (LOS microwave) path this gives an error rate objective of 2.5×10^{-7}, which excludes contributions due to multiplexing equipment. For about 80% of the time a BER of 1×10^{-7} should be achieved for a 2500-km (1550 mi) hypothetical reference digital path.

The proportion of time for which the high BER may be exceeded has a very great influence on the design of a system. It would be desirable to establish the higher error rate criteria for a 2500-km (1550 mi) reference circuit for 0.01% of the time, but it is pointed out that this may place an impractical financial burden on a telecommunication company or administration to build such a system. Certainly a system designed where the BER should not be exceeded more than 0.1% of any month would result in a more economic system. On the other hand it then may not

*Operating below 116 Hz and with elevation angle $> 7°$.

compare favorably with existing FDM-FM microwave systems. Thus an error rate criterion between 0.01% and 0.1% trading off cost and quality of service would seem appropriate. The BER value would be somewhere between 1×10^{-3} and 1×10^{-6} for the higher rate on the 2500-km (1550-mi) path.

For digital systems carrying telephone traffic, error bursts can cause call dropout due to the loss of supervisory signaling data. A United Kingdom proposal to CCIR suggests that with an error rate threshold of 1×10^{-3}, any error burst of 0.07 s or less would not cause call dropout.

15.3.4 Modulation Techniques

As briefly reviewed in Chapter 8, there are three basic modulation techniques available: AM, FM, and PM. Terminology of the industry often appends the letters SK to the first letter of the modulation type such as ASK meaning amplitude shift keying, FSK meaning frequency shift keying (classified as digital FM), and PSK meaning phase shift keying. Any of the three basic modulation techniques may be two-level or multilevel. For a two-level system, one state represents a binary 1 and the other state a binary 0. For multilevel or M-ary systems there are more than two levels or states, usually a multiple of 2, with a few exceptions such as partial response systems. Duo-binary is an example of this latter. Remember from Chapter 8 that in a four-level system (four-state system) each level or state represents two bits of information. Therefore for each transition or change of state, two bits are transmitted. For an eight-level system, three bits are transmitted for each transition; for a 16-level system, four bits are transmitted per change of state or transition. Of course some form of coding or combining is required prior to modulation and decoding after demodulation to recover the original bit stream.

Let us consider that each of the three basic modulation techniques may be represented by a modulated sinusoid. At the receiver some sort of detection process must be carried out. Coherent detection requires a sinusoidal reference signal perfectly matched in both frequency and phase to the received carrier. This phase reference may be obtained from a transmitted pilot tone or from the modulated signal itself. Noncoherent detection, being based on waveform characteristics independent of phase (e.g., energy or frequency), does not require a phase reference.

After detection in the receiver there is usually some device that carries out a decision process, although in less sophisticated systems this process may be carried out in the detector itself. The decision process requires bit synchronization, which is usually extracted from the receive waveform (see Chapter 8). Some decision circuits make decisions on a bit-by-bit basis. Others obtain some advantage by examining the signal over several bit intervals prior to making each bit decision. The observation interval is the portion of the received waveform examined by the decision device.

Before proceeding further, a short discussion dealing with the selection of modulation technique may help clarify some points for the reader.

15.3.5 Selection of Modulation Technique

There are three aspects to be considered in selecting a particular modulation technique:

1. To meet the spectral efficiency requirements (i.e., a certain number of megabits for a given bandwidth, as described in Section 15.3.1)
2. To meet the performance requirements discussed above
3. As constrained by a complexity factor having economic impact on the system

Let us now follow an elementary mathematical analysis that will clarify and assist the design engineer in selecting the type of modulation that will meet the needs of a particular link or system. Assume that the only source of errors is thermal noise in the receiver. The normalized carrier-to-noise ratio, W may be expressed as (CCIR Rep. 378-3):

$$W = 10 \log_{10} \frac{W_{in}}{W_n f_n} \tag{15.4}$$

where W_{in} = received maximum steady-state signal power (i.e., the highest value of the mean power during any bit period) (valid for FSK, PSK, binary ASK)

W_n = noise power density at the receiver input

f_n = bandwidth numerically equal to the bit rate B of the binary signal before the modulation process

The bit rate B is the rate of the incoming serial bit stream, taking into account bits added for service channel(s), redundancy added for error control, FEC, and so forth.

The normalized carrier-to-noise ratio can be related to the more familiar carrier-to-noise ratio C/N by the following expression:

$$W = 10 \log_{10} \left(\frac{C}{N} \times \frac{B_{eq}}{B} \right) \tag{15.5}$$

where B_{eq} = equivalent noise bandwidth of the receiver.

The necessary bandwidth for a given class of emission is defined by CCIR as the width of the frequency band which is just sufficient to ensure the transmission of information at the rate and quality (error rate) required under specified conditions. In digital radio systems one desirable condition to be met is that the above-defined power requirement W, which is variable with the receiver bandwidth, should be at its minimum value. The necessary bandwidth B_n in megahertz may then be given by:

$$B_n = \begin{cases} FBR & \text{for DSB modulation} \\ 0.5FBR & \text{for SSB modulation} \\ 0.6FBR & \text{for VSB modulation} \end{cases}$$

Table 15.1 Comparison of Common Modulation Types

System	Variant	W (dB)[b]	Necessary Bandwidth B_n	Remarks[a]
Amplitude modulation	Full-carrier, binary DSB, with envelope detection	17	FB	Simple, wasteful of bandwidth, high signal power
	DSB, suppressed-carrier, two binary channels in quadrature with coherent detection	10.5	$FB/2$	Fairly complex, tolerant to distortion
	DSB, suppressed-carrier, two binary channels in quadrature with differentially coherent detection	12.8	$FB/2$	Fairly simple, fairly sensitive to distortion
	SSB binary, suppressed-carrier[a]	10.5	$0.5\,FB$	Complex, loss of low baseband frequencies
	VSB binary, suppressed-carrier, with coherent detection[a]	11.3	$0.6\,FB$	Fairly complex
	VSB binary, reduced-carrier, with coherent detection[a]	11.8	$0.6\,FB$	Fairly simple
	VSB binary, suppressed-carrier 50% AM with envelope detection[a]	17.8	$0.6\,FB$	Simple, subject to pulse distortion, high signal power
Phase modulation with coherent detection[c]	2-level	10.5	FB	Fairly simple, tolerant to distortion, wasteful of bandwidth
	4-level	10.5	$FB/2$	Fairly simple, tolerant to distortion
	8-level	13.8	$FB/3$	Complex, economic of bandwidth, sensitive to distortion
Phase	2-level	11.2	FB	Simple, fairly tolerant to

Method	Level		Bandwidth	Remarks
modulation, with differentially coherent detection [c]				distortion, wasteful of bandwidth
	4-level	12.8	$FB/2$	Fairly simple, fairly sensitive to distortion
	8-level	16.8	$FB/3$	Complex, high signal power, economic of bandwidth, sensitive to distortion
Frequency modulation, with discriminator detection [f]	2-level	13.4	FB	Simple, wasteful of bandwidth
	3-level (duo-binary)	15.9	FB [e]	Fairly simple
	4-level	20.1	$FB/2$	Fairly simple, high signal power
	8-level	25.5	$FB/3$	Complex, high signal power, economic of bandwidth
Other modulation methods with coherent detection	Two 3-level class 1 partial response channels in quadrature AM [a]	13.5	$FB/2$ [c]	Fairly simple, economic of bandwidth
	16-level quadrature AM [a]	17	$FB/4$	Fairly simple, economic of bandwidth, sensitive to distortion

Source: From CCIR Rep. 378-3 (Courtesy ITU–CCIR).

[a] The maximum steady-state signal power depends on the shape of the modulating pulses. These figures are therefore based on average power.

[b] $P_e = 10^{-6}$.

[c] All digital PM may be obtained directly by PM or indirectly by methods of AM or FM.

[d] Reconsideration of the validity of the remarks is desirable.

[e] The design factor F in this case can be close to the value of 1. This effective reduction in necessary bandwidth is achieved at the expense of a greater number of transmitted levels for a given number of input levels, or equivalently, at a greater value of W for a given error rate.

[f] The adaptation of analog FM radio-relay systems for the transmission of digital signals seems feasible at the present time for gross bit rates in the medium-capacity range. For FSK the bandwidth given by the relation $FD + BR$ is used by one administration and includes the peak frequency deviation D.

where F = a design factor depending on the implementation approach. For efficient
modulation methods F is generally between 1 and 2; values of F below 1
are possible, but only at the expense of increasing intersymbol interference

B = bit rate of the binary signal (Mbps) before the modulation process

R = symbol rate (Mbaud/B)

Table 15.1 compares various modulation types assuming no intersymbol or other
types of interference. For each modulation type it gives values of W, B_n, and includes comments on complexity. This table has been taken from CCIR Rep. 378-3
where it is stated that the value of F may differ for the different variants in the
table. The introduction of the factor F is an attempt to make an allowance for the
compromise that must be effected in any practical system.

As far as the three different type sizes of digital radio-relay systems are concerned
(i.e., small, medium, and large as defined previously), the modulation methods that
have reasonable characteristics from the aspects previously considered in Table 7.1
are (CCIR Rep. 378-3):

- AM, full carrier, DSB, with envelope detection, is wasteful of bandwidth and
  power and should not be considered for other than very-small-capacity requirements.
- FM, two-level, with discriminator detection, is also wasteful of bandwidth and
  should not be considered for other than very-small-capacity requirements.
- PM, two-level, with coherent or differentially coherent detection. These
  methods of modulation are fairly simple but comparatively wasteful of bandwidth and are considered most suitable for small-capacity systems (i.e., under
  10 Mbps).
- PM, four-level, with coherent or differentially coherent detection. These are
  the most suitable methods for medium- and large-capacity radio systems and
  may be used for small-capacity systems as well. For high-speed radio-relay
  systems the more desirable demodulation method may be coherent detection
  because it is more tolerant of interference. Four-level FSK or QPSK is the
  most commonly used form of modulation today for digital radio systems.
- FM, three-level (duo-binary), four-level, or eight-level discriminator detector,
  is preferred for the simple adaptation to digital transmission of FM (analog)
  radio-relay systems using FDM. In such cases a large carrier-to-noise ratio will
  be available and the method of modulation need not be changed.
- PM, eight-level, with coherent detection, is particularly suitable for medium-
  capacity systems operating below 12 GHz. The method is attractive when
  systems are to be mixed with existing analog channel arrangements because it
  offers good spectrum efficiency even when using a single polarization. Although
  it is theoretically more sensitive to distortion than four-level CPSK, comparable
  performance can be achieved by optimal equipment filtering and equalization.

- Partial response coding techniques may be attractive to reduce bandwidth occupancy by reducing the value of the design factor F with some increase in equipment complexity. This reduction in bandwidth is achieved at the expense of a greater number of transmitted levels for a given number of input levels, or equivalently, at a greater value of W for a given error rate.
- 16-level QAM is attractive for high-capacity digital radio-relay systems to increase spectral efficiency with only a small increase in equipment complexity.

15.3.6 Modified FM and PM Techniques

In its basic form FSK is a technique of FM where binary information is transmitted on two frequencies, one frequency representing the marking condition, and the other, the spacing condition. The two FSK frequencies are separated by Δf Hz, where the frequency deviation Δf is small compared to the carrier frequency f_c. The frequency separation can be expressed in terms of the modulation index d as follows:

$$d = \Delta f T \qquad (15.6)$$

where T = symbol period (duration). For binary systems we know that T is the inverse of the bit rate in bits per second.

In the past noncoherent detection was commonly used with FSK. Today more and more interest is being shown in certain modified versions of FSK, particularly those with coherent detection. A major constraint on limiting the bandwidth of FSK systems is the abrupt phase change when shifting from one frequency to the other. The modified FSK techniques are based on the idea of continuous phase, often denoted by CP-FSK. Such a modified version of FSK results in a rapid spectral roll-off and improved efficiency. Observation intervals greater than 1 bit allow narrower filter bandwidths than would otherwise be feasible. Such systems using coherent detection provide optimal performance. However, when the observation interval of 1-bit duration is used, the best performance is achieved for a value of $d = 0.715\ldots$, whereas with conventional FSK d must be equal to or greater than 1.

Minimum shift keying (MSK) is a subset of CP-FSK where $d = 0.5$ and coherent reception is utilized. It has the property of simple self-synchronization. Its performance is the same as that of coherent PSK, and it has superior spectral properties compared to CP-FSK (Ref. 12).

PSK requires some form of coherent detection. Without a phase reference symbol ambiguities can result. With coherent binary PSK (CBPSK) the carrier frequency is shifted 0 and 180° for the mark and space conditions. This requires a precise phase reference at the receiver, and such a reference is usually derived from the received waveform. One modified version of PSK is differentially encoded PSK, (DE-PSK). It is based on the state of the previous bit transmitted. If no transition occurs from the phase state of the previous bit, a spacing condition can be assumed and a 180°

transition would then correspond to a mark. If there is a "hit" (an error) on one bit, nearly always the subsequent bit will be in error. Thus its error performance is somewhat inferior to that of conventional BPSK.

Differential PSK (DPSK) is still another version of PSK transmission where again the transmitted information is differentially encoded. However, with DPSK there is no requirement to extract a coherent phase reference from the overall receive waveform. The phase reference in this case is the phase of the previous bit (or for multilevel systems, symbol). Because the phase reference is not smoothed over a long series of symbols or bit intervals, DPSK performance is not as good as BPSK performance with its coherent reference, but is within a decibel as good for high E_b/N_0. At low E_b/N_0 the advantage of BPSK is several decibels, but the practical difficulties of achieving a coherent reference are very great.

QPSK is simply four-level PSK and consists of two-level PSK at each of the in-phase and quadrature components (I carrier and Q carrier). From Chapter 8 we know that each transition in QPSK represents 2 bits. During each 2-bit time interval of T seconds, the I carrier is BPSK modulated by 1 bit and the Q carrier by the

Table 15.2 Comparison of Representative Modulation Techniques

Basic Type	Technique Common Identifier	bps/Hz	E_b/N_0 (dB) (for BER = 1×10^{-4})
AM	OOK, coherent detection	0.8	12.5
	QAM	1.7	9.5
	QPR	2.25	11.7
FM	FSK, noncoherent detection ($d = 1$) (discrim. detection)	0.8	11.8
	CP-FSK, noncoherent detection ($d = 0.7$)	1.0	10.7
	MSK ($d = 0.5$)	1.9	9.4
	MSK, differential encoding	1.9	10.4
PM	BPSK, coherent detection	0.8	9.4
	DE-BPSK	0.8	9.9
	DPSK	0.8	10.6
	QPSK	1.9	9.9
	DQPSK	1.8	11.8
	8-ary PSK, coherent detection	2.6	12.8
	16-ary PSK, coherent detection	2.9	17.2
Hybrid	16-ary APK	3.1	13.4

Source: (Ref. 12). d = FM modulation index; QPR = quadrature partial response.
Note. The information in Table 15.2 has been drawn from a number of differing calculations so that the entries are not precisely comparable, but it still is a useful and valuable compilation.

other bit. The resulting signal can take on any one of four states separated by 0, 90, 180, and 270° in phase shift. As one might imagine, other four-level forms are used. There is DQPSK using differential encoding and detection, as does DBPSK. There is also offset-keyed QPSK, sometimes called *staggered QPSK* (OK-QPSK or SQPSK) which is more spectrally efficient than QPSK and easier to synchronize. Whereas phase shifts occur every T seconds on both I and Q for QPSK, for OK-QPSK they occur every $T/2$ seconds on either I or Q, and when the I and Q channels are combined, the resulting signal can abruptly shift by 90° at most.

Various hybrid modulation schemes are now being implemented for additional spectral utilization (i.e., more bits per hertz of bandwidth). A hybrid scheme in this context is one where a carrier is modulated in both amplitude and phase, amplitude and frequency, or perhaps phase and frequency. One such hybrid technique showing promise is a combination of ASK and PSK, known as amplitude-phase shift keying (APK) which uses four amplitude states and four phase states, or an equivalent 16-level system achieving 4 bps/Hz theoretically.

Table 15.2 compares a number of modulation techniques for relative spectral utilization, measured in bps/Hz, and the required practical E_b/N_0 for BER of 1×10^{-4}.

15.4 BASIC DIGITAL LINK DESIGN PROCEDURE

Chapters 5–7 dealt with analog radio system design procedures for radiolinks (LOS microwave), tropospheric scatter and earth stations, respectively. The link design of digital radio systems does not differ significantly in the basic approach. Common to all three chapters was the calculation of the carrier-to-noise ratio C/N and the derived signal-to-noise ratio S/N, weighted or unweighted in the derived voice channel. Time availability of a link or system, given a certain noise criterion, was also an important characteristic or performance goal.

As covered in previous sections of this chapter, we will derive a BER from the receive signal level (RSL), use C/N_0 rather than C/N, and BER will be expressed for a time availability on fading links.

RSL, as defined in this text, is the signal level in dBW at any given time at the input to the first active stage of a receiver chain, whether an LNA or a mixer. RSL can be calculated from the following formula for a system with no fading:

$$\text{RSL (dBW)} = \text{EIRP (dBW)} - P_L - A_e + G_r - L_{LR} \qquad (15.7)$$

where EIRP = equivalent isotropically radiated power from transmitting antenna

P_L = familiar path loss (free space) or transmission loss

A_e = excess attenuation (over free space) due to water vapor, O_2, rainfall, and so on

G_r = gain of receiving antenna

L_{LR} = sum of receiving system line losses before the first active stage; these include waveguide loss, and the insertion losses of circulators, band-pass filters, directional couplers, and so on

$$\frac{C}{N_0} = \text{RSL} - (-204 \text{ dBW})^* - \text{NF}_{dB} \qquad (15.8)$$

where NF = receiver noise figure (dB).

E_b/N_0 may not be derived from C/N_0:

$$\frac{E_b}{N_0} = \frac{C}{N_0} - 10 \log_{10} (\text{bit rate}) \qquad (15.9)$$

where the bit rate is the transmitted bit rate (often called data rate) in bits per second, including all overhead bits, service channel bits, stuffing bits, and so on, and *not* the symbol rate.

For earth station link analysis (the space budget, uplink and downlink) it is often essential to work in the noise temperature domain and move the reference point from the receiver front end to the output of the feed or base of the antenna pedestal. Then

$$\text{RSL (dBW)} = \text{EIRP (dBW)} - P_L - A_e + G_r - L_{no} - L_{LR} \qquad (15.10)$$

where L_{no} = sum of nonohmic losses such as pointing loss, L_{LR} = sum of ohmic losses such as waveguide loss.

$$\frac{C}{N_0} = \text{RSL (dBW)} - (-228.6 \text{ dBW}) - 10 \log_{10} T_{sys} \qquad (15.11)$$

where T_{sys} = receiving system equivalent noise temperature expressed in kelvins.

$$T_{sys} = T_a + T_r \qquad (15.12)$$

where T_a = antenna noise, including the sky noise component (see Chapter 7), and T_r = receiving system noise temperature derived in Chapter 7, including the noise contributions of all ohmic losses from the antenna feed or other reference point and the noise contributions of the active receiving stages.

What, then, will be the BER for a link? Figure 15.1 gives theoretical BER values for coherent BPSK (CBPSK) and coherent QPSK (CQPSK) modulation plotted against E_b/N_0. We can also turn to Table 15.2 for practical values of E_b/N_0 for a BER of 1×10^{-4} for several modulation types. If we wish values of BER improved over a BER of 1×10^{-4}, then these can be roughly estimated by adding 1 dB for each power of 10 improvement in BER. For more accurate values, Section 15.6 may be used. Also the 1-dB value is only valid as an approximation for BER values between 1×10^{-4} and 1×10^{-7}. Above 1×10^{-7}, or as BER improves above this value, in many cases 0.5 dB or less is a more accurate estimate. The 1-dB estimate is evident in Figure 15.1 where a BER of 1×10^{-4} requires an E_b/N_0 of 8.4 dB

---

*−204 dBW is Boltzmann's constant adjusted for room temperature (17°C).

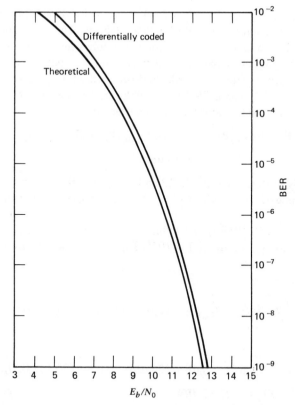

Figure 15.1 BER plotted against E_b/N_0 for CBPSK and CQPSK.

(theoretical curve), 1×10^{-5} requires about 9.6 dB, and 1×10^{-6}, 10.5 dB. Thus from Table 15.2, to derive the required E_b/N_0 of 8-ary PSK with coherent detection for a BER of 1×10^{-6}, add 2 dB to the value given in the table, or 14.8 dB. The discrepancy between the values in Table 15.2 and Figure 15.1 is that the table gives measured values (practical), whereas the figure gives theoretical values. The difference between theoretical and practical values is that the practical values take into account bandwidth limiting (filtering), whereas the theoretical ones do not.

15.5 EXAMPLE SYSTEMS

15.5.1 Digital Radiolink (LOS Microwave)

Given a 12-mi (19.3-km) single-hop system operating in the 11-GHz band, transmitting 1344 PCM channels with a 90.148-Mbps data rate (equivalent to two DS3 line rates, Chapter 11). The transmitter has an output at the flange of +10 dBW, and the receiver has a 6-dB NF. Allow 2 dB of line losses at each end, and assume a Rayleigh fading criterion and a 99.99% path availability. The modulation technique selected is 8-ary PSK with coherent detection. The following outlines one procedure or method of carrying out a path analysis to size the antennas required to meet the path availability given. The reader will find the methodology very similar to that used in Chapter 5 for analog FM-FDM radiolinks. The link error rate is given as 1×10^{-7}. Rainfall considerations have been neglected.

$$\text{Path loss (dB)} = 36.7 + 20 \log_{10} D_{mi} + 20 \log F_{MHz}$$

$$= 139.13 \text{ dB}$$

The next step is to determine the EIRP and the gain of the receiving antenna G_r in decibels, assuming $G_r = G_t$, where G_t is the gain of the transmitting antenna in decibels.

The fade margin is 40 dB (Rayleigh fading, 99.99% path availability). $E_b/N_0 = 15.8$ dB for a single-hop error rate of 1×10^{-7}. This value is derived from Table 15.2, adjusted for the error rate.

We know that

$$\frac{E_b}{N_0} = \text{RSL} - \text{NF}_{dB} + 204 \text{ dBW} - 10 \log_{10} \text{ (data rate)}$$

Then

$$15.8 \text{ dB} = \text{RSL} - 6 \text{ dB} + 204 \text{ dBW} - 10 \log (90,148,000)$$

$$\text{RSL} = -102.65 \text{ dBW}$$

Further,

$$\text{RSL (dBW)} = \text{EIRP (dBW)} - P_L - A_e + G_r - L_{LR} \qquad (15.7)$$

and

$$-102.65 \text{ dBW} = \text{EIRP} - 139.13 \text{ dB} - 0 + G_r - 2 \text{ dB}$$

$$\text{EIRP} + G_r = +38.48 \text{ dBW}$$

$$\text{EIRP} = G_t + P_t - L_{LT}$$

When $P_t = +10$ dBW (given), $L_{LT} = 2$ dB, and $G_t = G_r$,

$$2G_{t,r} = 30.48 \text{ dB (without fading)}$$

Adding to this value the required 40-dB fade margin,

$$2G_{t,r} = 70.48 \text{ dB}$$

$$G_t = G_r = 35.24 \text{ dB}$$

This, then, will require a 2.25-ft (0.68-m) antenna (parabolic reflector) at each end of the link.

Of course there are several methods of carrying out this same exercise. Another way is to compute E_b, then N_0, and to logarithmically subtract:

$$E_b = \text{RSL} - 10 \log_{10} \text{(bit rate)}$$

$$N_0 = -228.6 \text{ dBW} + 10 \log_{10} \text{(ambient temperature in K)} + \text{NF}_{dB}$$

Ambient temperature is often taken as 290 K. Many times this does not represent worst case conditions. For military systems, where the LNA enclosure is to be at outside (out-of-doors) ambient temperature, it would be wise to assume the LNA ambient at 55°C or 328 K.* This will deteriorate the performance from 290 K ambient temperature by 0.55 dB.

15.5.2 Tropospheric Scatter Link

Given a tropospheric scatter link where the transmission loss is 240 dB for 99.9% path availability, 95% service probability. The data rate on the link is 2.048 Mbps, and the modulation is DQPSK. The BER requirement is 1×10^{-5} for 99.9% of the time. Leaving aside diversity advantage, and using 10-kW transmitters and 3.1-dB NF receivers, what antenna sizes might serve this situation? Total transmission line losses are 2 dB at each end.

For a tropospheric scatter link, which could be dual or quadruple diversity, initial path analysis follows one transmit chain out along the path and, at the far end, one antenna down through one receive chain. We now look at the output of that receive chain.

We could use the same method as we did in Section 15.1 or turn to a substitution method. With the latter we assume an antenna size at the outset and adjust the size depending on the margin calculated, or margin shortfall. From Table 15.2, with the modulation method and BER given, the required E_b/N_0 is 12.8 dB.

In this case determine E_b/N_0 if 30-ft (9.1-m) antennas were used on each end of the link at an operating frequency of 4.6 GHz. Turning to the antenna gain nomogram, Figure 6.11 (Chapter 6), we see that the antenna gain is 50 dB. Then,

$$\text{EIRP} = +40 \text{ dBW} - 2 \text{ dB} + 50 \text{ dB}$$

(i.e., transmitter output power minus line losses plus antenna gain)

$$\text{EIRP} = +88 \text{ dBW}$$

$$\text{RSL(dBW)} = +88 \text{ dBW} - 240 \text{ dB} + 50 \text{ dB} - 2 \text{ dB}$$

$$= -104 \text{ dBW}$$

$$E_b = \text{RSL} - 10 \log_{10} \text{(data rate)}$$

---

*This assumes a room temperature or ambient temperature receiver and not one that is cryogenically cooled, as was discussed in Chapter 7.

$$= -167.1 \text{ dBW}$$

$$N_0 = -204 \text{ dBW*} + \text{NF}_{dB}$$

$$= -200.9 \text{ dB-K}$$

$$E_b/N_0 = 33.8 \text{ dB}$$

Since the required E_b/N_0 is

$$E_b/N_0 \text{ (required)} = 12.8 \text{ dB}$$

We have

$$\text{Margin} = 21.0 \text{ dB}$$

If we reduce the margin by 20 dB, the antenna size at each end can be reduced by the equivalent of 10 dB, leaving a 1-dB final margin so that 10-ft (3-m) antennas would serve.

Of course in this brief analysis no treatment was given of dispersion nor of explicit and implicit diversity. Depending on equipment design and the nature of dispersion expected on the path, an improvement in performance could be expected. For instance, one military equipment with 6.6-kW transmitters, 15-ft (4.5-m) antennas, quadruple diversity, 3.1-dB NF receivers with a data rate of 2304 kbps for a BER of 1×10^{-5} can tolerate a transmission loss of 256.9 dB for a nonfading nondispersive medium and 254.5 dB for a fading and dispersive medium for average dispersion.

It will be noted that aperture-to-medium coupling loss was not included for our treatment, and it must be included in any practical situation. For the example military equipment cited, the transmission losses included 9.4 dB aperture-to-medium coupling loss computed by the CCIR method.

15.5.3 Sample Satellite Links

For the first example we will treat a downlink from a geostationary satellite operating in the 4-GHz band. The satellite has an EIRP of +30 dBW, and we assume a free-space loss of 196 dB. The desired BER on the downlink is 1×10^{-6}, and the data rate is 1 Mbps. The modulation technique is DQPSK, and the terminal G/T is +20 dB/K.

Now turn to Table 15.2 to determine the minimum E_b/N_0 for the desired error performance. The table indicates an E_b/N_0 of 11.8 dB for a BER of 1×10^{-4}. We then estimate an E_b/N_0 of 13.8 dB for a BER of 1×10^{-6} by adding 1 dB for each order of magnitude (power of 10) improvement (i.e., add 2 dB). A link budget analysis is then carried out similar to the path analysis discussed in Chapter 5 (Figure 5.32). The link budget for this example is shown in Table 15.3.

---

*–228.6 dBW adjusted to room temperature of 290 K.

Table 15.3   Example Downlink Power Budget (4 GHz, 1 Mbps)*

1.	Satellite EIRP	+30 dBW
2.	Satellite pointing loss (when applicable)	-0.1 dB
3.	Atmospheric losses	-0.2 dB
4.	Rainfall loss	0.0 dB
5.	Free-space loss	-196.0 dB
6.	Terminal radome loss (when applicable)	-0.5 dB
7.	Terminal pointing (tracking) loss	-0.2 dB
8.	Signal level impinging on terminal antenna	-167.0 dBW
9.	Terminal G/T	+20.0 dB/K
10.	Boltzmann's constant expressed in dBW	-228.6 dBW
11.	C/N_0	81.6 dB-Hz
12.	$E_b/N_0 = C/N_0 - 10\log_{10}$ (data rate)	21.6 dB
13.	E_b/N_0, design criterion	13.8 dB
14.	Margin	7.8 dB

*Demodulation has been assumed on the uplink

DISCUSSION OF TABLE 15.3

1. Given. Of course the example could have many variations such as: With all items given but satellite EIRP, what satellite EIRP would be required?

2. Satellite pointing loss. If we dealt with an earth coverage antenna on the satellite for the downlink, this value would be 0.0 dB. If the antenna were for area coverage, two factors would have to be taken into consideration. The first factor is the location of the earth terminal in relationship to the satellite footprint. This information can be obtained from a map giving the satellite footprint contours expressed in decibels below the main beam value. The second factor is the ability of the satellite to keep its area coverage antenna pointed to exactly fit the contour. The 0.1 dB used in the example assumes the earth terminal to be located right on the main beam, and it then is a value expressing pointing loss alone.

3. Atmospheric losses. This represents excess path attenuation due to water vapor and oxygen. The value is a function of the frequency and elevation angle. The reader should consult CCIR Rep. 719 for a simple method of calculating a value to be used here.

4. Rainfall loss. This is a value of excess attenuation due to rainfall. At 4 GHz the value is nearly negligible and can be accommodated in the margin, item 14. For operating frequencies above 10 GHz excess attenuation due to rainfall takes on major importance, and, depending on the desired path availability and the rainfall statistics for the area of installation as well as the elevation angle, the link margins must be larger and/or we must resort to path diversity. One well accepted method to determine rainfall margin on satellite links is the Crane rainfall model, described in (Ref. 24). Also see Chapter 10.

5. Free-space loss. See Section 7.4.

6. Radome loss. Applicable when a radome is used to protect the earth terminal antenna from the environment. The value should be no greater than 1 dB, but special consideration must be given to wet radomes in salt or dusty environments.

7. Satellite earth terminal pointing or tracking loss. The pointing and/or tracking of a terminal antenna is not "perfect." This loss is a measure of that imperfection and is a function of the tracking (or pointing) error in degrees, the terminal antenna beamwidth, and the satellite antenna beamwidth.

8. Signal level impinging on the earth station antenna, sometimes called isotropic receive level. This should be stated in the same units as the satellite EIRP, in this case dBW. It is the satellite EIRP minus all the losses given in items 2–7.

9. Terminal G/T. Given in this example. See also Section 7.5.1.

10. Boltzmann's constant expressed in dBW, or -228.6 dBW/Hz/K.

11. C/N_0. Carrier-to-noise density ratio, where N_0 is the thermal noise in a 1-Hz bandwidth and C = RSL (dBW). $N_0 = -228.6$ dBW + $10 \log_{10} T_{sys}$.

12. E_b/N_0. Energy per bit-noise density ratio discussed at length at the beginning of this chapter. This is derived from C/N_0 by subtracting $10 \log_{10}$ (bit rate).

13. E_b/N_0 required was given or is specified by the designer at the outset. It is derived from the specified BER for the link and the modulation technique used.

14. Margin. E_b/N_0 is excess of specified. In this case 21.6 dB - 13.8 dB = 7.8 dB. The margin accommodates excess attenuation due to rainfall and the deterioration of equipment components due to aging and between maintenance periods.

The second case is an example of an uplink operating at 30 GHz with a data rate of 500 Mbps. The ground or terminal antenna is a 12-m (40-ft) parabolic reflector with a Cassegrain feed with a 60% aperture efficiency which translates into an effective gain of 69.3 dB. The terminal transmitter output is +23 dBW, and line losses to the antenna feed are 3 dB. The free-space loss is given as 213.7 dB. Terminal pointing loss is 0.4 dB, satellite antenna gain is 54.5 dB, the satellite receiving system noise temperature T_{sys} is 603 K (a fairly typical value), satellite pointing loss is 4.0 dB, modulation on the link is QPSK, and the link BER objective is 1×10^{-6}, translating into a required E_b/N_0 of 11.9 dB.

Table 15.4 is the space budget sheet for this example.

Table 15.4   Example Uplink Space Budget (30 GHz, 500 Mbps)*

1.	Earth terminal EIRP	+89.3 dBW
2.	Earth terminal pointing (tracking) loss	-0.4 dB
3.	Excess attenuation, atmospheric losses	-1.1 dB
4.	Free-space loss	-213.7 dB
5.	Rainfall loss	0.0 dB
6.	Satellite pointing loss	-4.0 dB
7.	Signal level impinging on satellite antenna	-129.9 dBW
8.	Satellite G/T (T_{sys} = 603 K)	+26.7 dB/K
9.	Boltzmann's constant	-228.6 dBW/Hz/K
10.	C/N_0	125.4 dB
11.	$E_b/N_0 = C/N_0 - 10 \log_{10}$ (data rate)	38.41 dB
12.	Required E_b/N_0	11.90 dB
13.	Margin	26.51 dB

*Demodulation in the satellite is assumed

DISCUSSION OF TABLE 15.4

1. EIRP of earth terminal is calculated in the conventional manner.

2. Terminal pointing loss. Given for this example. See discussion of item 7, Table 15.3.

3. Excess attenuation due to atmospheric losses. Given for this example. See discussion of item 3, Table 15.3.

4. Free-space loss. Given for this example.

5. Rainfall loss. We chose to enter 0.0. For a 30-GHz path, rainfall loss is an extremely important consideration. For such an uplink it must be accommodated in the margin or, a form of path diversity (described in Chapter 10) should be used. On uplinks such diversity can be a rather complex matter, unless the path(s) have been made adaptive where the diversity transmitter takes over when the primary path becomes marginal. One way of doing this is to monitor and measure BER of one's own signal on the signal's companion downlink.

6. Satellite pointing loss. Compare this to item 2 in Table 15.3 and its subsequent discussion. This is an area coverage satellite antenna, and the earth terminal is pretty far out on the contour, probably beyond the 3-dB contour, allowing several tenths of a decibel for satellite pointing error.

7. Signal level impinging on satellite antenna. Derived by subtracting all losses up to this point from the EIRP (item 1).

8. Satellite G/T. Calculated $G/T = G - 10 \log T_{sys}$, where $G = 54.5$ dB and $T_{sys} = 603$ K.

9. Boltzmann's constant expressed in dBW.

10. C/N_0. See discussion of item 11, Table 15.3.

11. E_b/N_0. See discussion of item 12, Table 15.3.

12. Required E_b/N_0. Self-explanatory.

13. Margin. Most of this margin can be used for rainfall loss. See discussion of item 5 of this table.

15.6 CALCULATION OF THEORETICAL BIT ERROR RATE (BER)

For several commonly used modulation techniques the theoretical value of the BER can be calculated as a function of the signal-to-noise ratio using the following formulas (where P_e denotes BER).

For coherent FSK and coherent ASK,

$$P_e = \frac{1}{2} \operatorname{erfc} \sqrt{\frac{S}{4N}} \qquad (15.13)$$

For FSK, noncoherent,

$$P_e = \frac{1}{2} e^{-S/2N} \qquad (15.14)$$

For PSK

$$P_e = \frac{1}{2} \, \text{erfc} \, \sqrt{\frac{S}{N}} \tag{15.15}$$

For ASK, noncoherent

$$P_e = \frac{1}{2} e^{-S/2N} \left(1 + \frac{1}{\sqrt{2\pi S/N}} \right) \tag{15.16}$$

For DPSK

$$P_e = \frac{1}{2} e^{-S/N} \tag{15.17}$$

where S/N is the signal-to-noise power ratio expressed as a numerical ratio. Those formulas involving just the exponential can be calculated on any scientific calculator. The complementary error function (erfc) is another matter. We recommend turning to Ref. 25 (sec. 6.7, pp. 136–140) for an explanation and calculation.

15.7   CONVERSION OF S/N TO E_b/N_0

If we allow that

$$E_b = S - 10 \log_{10} R$$

where S = signal level (dBm or dBW) an R = bit rate (bps), and

$$N = N_0 + 10 \log_{10} \text{BW}$$

Where N = level of noise (dBm or dBW) and BW = bandwidth (Hz), then

$$N_0 = N - 10 \log_{10} \text{BW}$$

Thus

$$\frac{E_b}{N_0} = (S - 10 \log R) - (N - 10 \log \text{BW})$$

$$= (S - N) + 10 \log \text{BW} - 10 \log R$$

or

$$\frac{E_b}{N_0} = \frac{S}{N} + 10 \log \text{BW} - 10 \log R$$

and

$$\frac{E_b}{N_0} = \frac{S}{N} - 10 \log \frac{R}{\text{BW}} \tag{15.18}$$

15.8 CODING GAIN

15.8.1 General

FEC coding is another method of optimizing a digital radio link, particularly on systems that are power limited. For instance, many satellite downlinks are power limited either by international agreement because of band-sharing (see Chapter 7) or by economic and/or technical factors. Coding gain can be defined as the improvement realized when an error correction code is appended to a digital communication system.

To digress, consider a satellite downlink where, by coding, the satellite EIRP can be reduced by half without affecting performance. This could allow the use of a satellite transmitter with half the output power than would be required normally. Transmitter weight, including power supply, may then be reduced 75%. Battery weight may be reduced by perhaps 50% (batteries required to power the satellite communication equipment during eclipse), with a concurrent reduction in the solar cells required. It is not only a savings in the direct cost of these items, but also a savings in the lifting weight of the satellite to place it in orbit. The economic impact on the system may be on the order of many millions of dollars. One price we have to pay is increased bandwidth due to an increased bit rate. Another is more complex processing at both ends of the link, but principally at the receiver.

Numerous coding schemes are available. The implementation of coding is shown in Figure 15.2 for the conventional RF repeater satellite and in Figure 15.3 for a processing satellite.

FEC was dealt with briefly in Chapter 8. Let's review that information and then provide some more depth. We are dealing with two broad classes of codes, namely, block codes and convolutional codes.

With block-coding techniques each group of K consecutive information bits is encoded into a group of N symbols for transmission over the channel (Figures 15.2 and 15.3). Normally the K information bits are located at the beginning of the N-symbol block code, and the last $N - K$ symbols correspond to the parity check bits formed by taking the modulo-2 sum of certain sets of K information bits. Block codes containing this property are referred to as systematic block codes. The *encoded symbols* for the $(K + 1)$th bit and beyond are completely independent of the symbols generated for the first K information bits and hence cannot be used to help decode the first group of K information bits at the far-end receiver. This essentially says that blocks are independent entities, and one block has no enhancement capability on another.

Because N symbols are used to represent K bits, the code rate R of such a block code is K/N bits per symbol, or

$$R = \frac{K}{N} \tag{15.19}$$

For instance, an encoder structure could be (7, 4) meaning $N = 7$ and $K = 4$. The

Near-end earth station transmitter

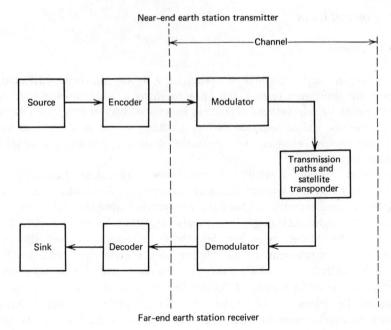

Far-end earth station receiver

Figure 15.2 Implementation of channel coding for the conventional RF repeater satellite.

information bits are stored in $K = 4$ storage devices, and the device is made to shift $N = 7$ times. The first K symbols of the block output are the information symbols and the last $N - K$ symbols are a set of check symbols that form the whole N-symbol word. A block code may be identified with the notation (N, K, t), where t corresponds to the number of errors in a block of N symbols that the code will correct.

Viterbi (Ref. 22) defines a *convolutional encoder* as

a linear finite state machine consisting of a K-stage shift register and n linear algebraic function generators. The input data which is usually, but not necessarily always, binary, is shifted along the register b bits at a time.

Figure 15.4 is an example of a convolutional encoder. If there were a five-stage shift register where the input data was shifted along 1 bit at a time and we had three modulo-2 adders (i.e., $n = 3$), using the Viterbi notation, the code would be described as a 5, 3, 1 convolutional code (i.e., $K = 5, n = 3, b = 1$).

In Figure 15.4 information bits are shifted to the right 1 bit at a time ($b = 1$) through the K-stage shift register as new information bits enter from the left. Bits out of the last stage are discarded. The bits are shifted one position each T seconds, where $1/T$ is the information rate in bits per second. The modulo-2 adders are used to form the output coded symbols, each of which is a binary function of a

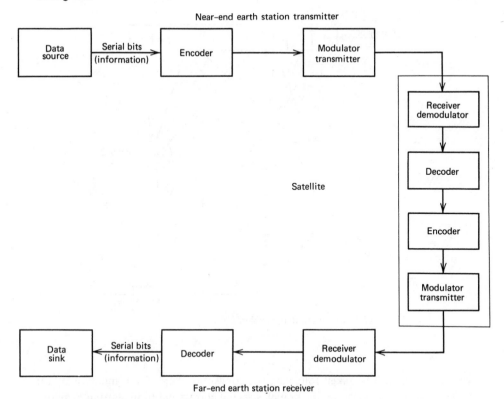

Figure 15.3 Implementation of channel coding for a digital processing satellite.

particular subset of the information bits in the shift register. The output coded symbols can be seen to depend on a sequence of K information bits, and thus we define the *constraint length* as K.

If we feed in 1 bit at a time to the encoded (i.e., $b = 1$), each coded symbol carries an average of $1/n$ information bits, and the code is said to have a rate R of $1/n$. For the more generalized case where $b = 1$, the rate R is expressed:

$$R = b/n \qquad (15.20)$$

where b = number of bits shifted into the register at a time.

In Figure 15.4 when the first modulo-2 adder is replaced by a direct connection to the first stage of the shift register, the first symbol becomes a replica of the information bit. Such an encoder is called a *systemic convolutional encoder*, as shown in Figure 15.5.

A convolutional code can be thought of as forming a tree structure (Figure 15.6). At each node (branch point) the information bit determines which direction (i.e.,

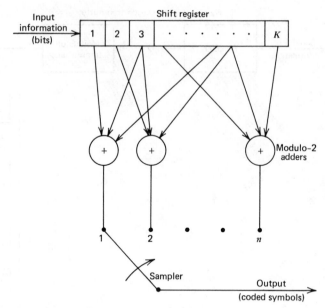

Figure 15.4 Convolutional encoder, $b = 1$.

which branch) will be taken following the convention "up" for 1 and "down" for 0. The n digits occurring on the branch selected correspond to the output symbols. A particular sequence of information bits then describes a particular path through the tree.

Consider the following example to illustrate the operation. Figure 15.5 is a rate $\frac{1}{2}$ encoder, and the shift register contains all 0s initially. We now assume that there is an input sequence of information bits of 1100. The path defined by the sequence is illustrated as the heavy line in the code tree in Figure 15.6. Since the code generated by this encoder (Figure 15.5) is a systematic code, the first digit of the output at each branch is the information bit. The second digit is the output of the modulo-2 adder, as shown in Figure 15.5. After the first bit is fed into the shift register, the contents are 1000000. . . so that the output of the adder is a 1, as shown as the second digit on the first branch of the code tree (Figure 15.6). After the second digit input to the shift register, the contents are 1100000. . . and the adder output is a 0, since the modulo-2 sum of an even number of 1s is a 0. Continuing in this fashion, the remaining portion of the tree can be constructed. The reader should note that unlike block codes, convolutional codes have no formal block structure in the code words. And unlike block codes, past information bits do have an influence on the symbols used to represent a present information bit. The code is constructed by taking the convolution of the shift register tap con-

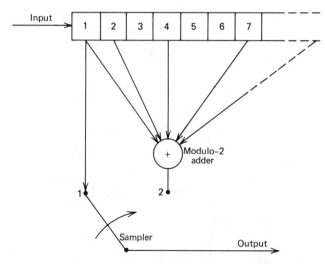

Figure 15.5 Convolutional encoder used to generate the code tree of Figure 15.6.

nection set (i.e., the code-generating polynomial) with the information bit pattern and hence a convolutional code.

Decoding algorithms for block and convolutional codes are quite different. Because a block code has a formal structure, advantage can be taken of the known structural properties of the words or the algebraic nature of the constraints among the symbols used to represent an information sequence. An example of a class of powerful block codes with well defined decoding algorithms is the Bose–Chaudhuri–Hocquenghem (BCH) codes.

The decoding of convolutional codes usually is carried out by means of the probabilistic techniques such as the sequential decoding algorithm by Wozencraft and the maximum likelihood technique by Viterbi (i.e., the Viterbi algorithm). Such techniques depend on the ability to home in on the correct sequence by designing efficient search procedures that discard unlikely sequences quickly. The sequential decoder differs from most other types of decoders in that when it finds itself on a wrong path in the tree, it has the ability to search back and forth, changing previously decoded information bits until it finds the correct tree path. The frequency with which the decoder has to search back and the depth of these backward searches is dependent on the value of the channel BER.

An important property of a sequential decoder is that, if the specified constraint length is large enough, the probability that the decoder will make an error approaches zero (i.e., a BER better than 1×10^{-9}). One cause of error is overflow, being defined as a situation in which the decoder is unable to perform the necessary number of computations in the performance of the tree search. If we define a

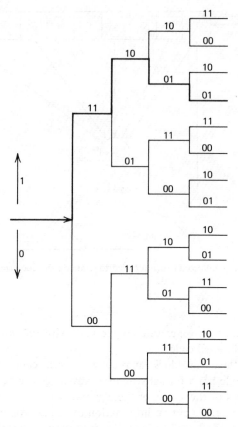

Figure 15.6   A code tree (related to the convolutional encoder shown in Figure 15.5).

computation as a complete examination of a path through the decoding tree, a decoder has a limit on the number of computations it can make per unit time. The number of searches and computations is a function of the number of errors arriving at the decoder input, and the number of computations that must be made to decode one information bit is a random variable. An important parameter for a decoder is the average number of computations per decoded information bit. As long as the probability of bit error is not too high, the chances of decoder overflow will be low, and satisfactory performance will result.

For the previous discussion it has been assumed that the output of a demodulator has been a hard decision, either a mark or a space, as an input to the decoder. If these decisions were soft decisions instead of hard decisions, additional improvement in error performance on the order of several decibels can be achieved. By a "soft" decision we mean that the output of a demodulator is quantized into four or eight levels (2- or 3-bit quantization, respectively), and then certain decoding algorithms can use this additional information to improve the output BER. Sequential and Viterbi decoding algorithms can use this soft decision information effectively, giving them an advantage over algebraic decoding techniques which are not designed to handle the additional information provided by the soft decision.

The soft decision level of quantization is indicated conventionally by the letter Q, which indicates the number of bits in the quantized decision sample. If $Q = 1$, we are dealing with hard decision demodulator output; $Q = 2$ indicates a quantization level of 4; $Q = 3$, a level of 8; and so on.

Table 15.5 is a summary of E_b/N_0 requirements of several coded communication systems using BPSK modulation, showing coding gain capabilities.

Table 15.5 Coding Gain for Several Coded Communication Systems, BPSK Modulation, BER 10^{-5}

Coding Type	Quantization Level (bits)	Coding gain[a] (dB)
Convolutional, $K = 7, R = \frac{1}{2}$, Viterbi decoded	1	3.1
Convolutional, $K = 7, R = \frac{1}{2}$, Viterbi decoded	3	5.2
Convolutional, $K = 7, R = \frac{1}{3}$, Viterbi decoded	1	3.6
Convolutional, $K = 7, R = \frac{1}{3}$, Viterbi decoded	3	5.5
Convolutional, $K = 9, R = \frac{3}{4}$, Viterbi decoded	1	2.4
Convolutional, $K = 9, R = \frac{3}{4}$, Viterbi decoded	3	4.3
Convolutional, $K = 24, R = \frac{1}{2}$, sequential decoded, 20 kbps, 1000-bit blocks	1	4.2
Convolutional, $K = 24, R = \frac{1}{2}$, sequential decoded, 20 kbps, 1000-bit blocks	3	6.2
Block Golay (24, 12)	3	4.0
Block Golay (24, 12)	1	2.1
Block BCH (127, 92)	1	3.3
Block BCH (127, 64)	1	3.5
Block BCH (127, 36)	1	2.3
Block Hamming (7, 4)	1	0.6
Block Hamming (15, 11)	1	1.3
Block Hamming (31, 26)	1	1.6

Source: From (Ref. 26).
[a] 9.6 dB required for the uncoded system, and only additive white Gaussian noise conditions assumed.

Table 15.6 Coding Gains for Several Modulation Types[a]

Modulation	Quantization Level (bits)	E_b/N_0 (dB) (for BER = 1×10^{-5})	Coding Gain (dB)
Coherent BPSK/QPSK	3	4.4	5.2
Coherent BPSK/QPSK	2	4.8	4.8
Coherent BPSK/QPSK	1	6.5	3.1
Octal PSK[b]	1	9.3	3.7
DBPSK[b]	3	6.7	3.6
DBPSK[b]	1	8.2	2.1
DE-QPSK[b]	1	9.0	3.0
Noncoherently demodulated binary FSK	1	11.2	2.1

Source: From (Ref. 26).
[a]In all cases, convolutional coding, $K = 7$, $R = \frac{1}{2}$, Viterbi decoded.
[b]Interleaving/deinterleaving assumed.

Coding gain is also a function of the modulation type selected for a particular digital radio link or system. Table 15.6 illustrates this point.

For a word of caution on coding gain, which is partially reflected in Table 15.6, we quote in part from Ref. 21:

> ... coding gain is strongly dependent on the system external to the code ... depending on the form of PSK (or other modulation type), interleaving and channel fading. Consequently one must be extremely cautious in using the term coding gain and in applying coding gain to systems other than that for which it was calculated or measured. The significant decrease in coding gain when DPSK is used rather than CPSK, and the extraordinary improvement when coding is applied to a rapidly fading channel are particularly noteworthy.

We should also not lose sight of the fact that we pay in increased bit rate for coding gain. For instance, for a rate $\frac{1}{2}$ code the symbols out of a coder are twice the bits coming in. For a rate $\frac{3}{4}$, the symbol rate is $\frac{4}{3}$ the input bit rate; for $R = \frac{1}{3}$, the symbol rate is three times the incoming bit rate to the coder. This eventually is paid for in equivalent bandwidth.

15.8.2 Guidelines on Code Specification

The following factors should be considered in selecting and specifying error control codes:

1. Performance required for a particular modulation/demodulation, channel,

and, if known, coding technique. For example, the error probability P_e for several E_b/N_0 ratios could be specified.

2. Modem interface requirements.

3. Synchronization requirements, that is, the method of determining the start of a block or other code grouping.

4. Data rate requirements.

5. Modem phase ambiguity requirements. Some decoders can internally compensate for the effects of a 90 or 180° phase ambiguity present in BPSK and QPSK modems which obtain a carrier phase referenced from the data modulation signals.

6. Encoder–decoder delay requirements, that is, the delay in bit times from the time an information bit is first put into the encoder to the time it is provided as an output from the decoder.

7. Decoder start-up delay.

8. Built-in test requirements.

9. Packaging requirements. The decoder could be on a card for insertion in an existing modem or a separate decoder could be provided. Power and thermal requirements should also be specified.

10. Complexity and cost of implementation of a particular coding scheme versus the coding gain achieved. Can the same gain be achieved by other means which are more economical?

REFERENCES AND BIBLIOGRAPHY

1. "Digital Radio," *Lenkurt Demodulator* Lenkurt Electric Corp. San Carlos, CA, Nov.–Dec. 1977.
2. W. J. Gill, "Transmission Considerations for Digital Microwave," *Telephony*, Oct. 11, 1976.
3. *Satellite Communications Reference Data Handbook,* Defense Communications Agency, Washington, DC, July 1972.
4. "Recommendations and Reports of the CCIR 1978," XIV Plenary Assembly, Kyoto, 1978, Vols. V and IX.
5. (U.S.) *Federal Communications Commission Rules and Regulations,* Part 21, Mar. 1974.
6. W. J. Garner, "Bit Error Probabilities Relate to Data-Link S/N," *Microwaves*, Nov. 1978.
7. P. Bylanski and D. G. W. Griffin, "Digital Transmission Systems," IEE Telecom Series 4, Peter Peregrinus Ltd., Sevenhenge, Herts (UK) 1976.
8. *Technical Publications on Digital Troposcatter*, Raytheon Co., Equipment Div., Sudbury, MA, Apr. 1979.
9. "Concepts for 18/30 GHz satellite Communication System Study," Vol. 1, Final Rep. NASA Contract NAS3-21362, Ford Aerospace and Communications Corp., Palo Alto, CA, Nov. 1979.

10. E. R. Berlekamp, "The Technology of Error-Correcting Codes," *Proc. IEEE,* Vol. 68, May 1980.

11. M. P. Ristenbatt, "Alternatives in Digital Communications," *Proc. IEEE*, Vol. 61, June 1973.

12. J. D. Oetting, "A Comparison of Modulation Techniques for Digital Radio," *IEEE Trans. Commun.,* Vol. COM-27, Dec. 1979.

13. J. W. Bayless, A. A. Collins, and R. D. Pedersen, "The Specification and Design of Band-limited Digital Radio Systems," *IEEE Trans. Commun.,* Vol. COM-27, Dec. 1979.

14. D. H. Marais and K. Feher, "Bandwidth Efficiency and Probability of Error Performance of MSK and Offset QPSK Systems," *IEEE Trans. Commun.,* Vol. COM-27, Dec. 1979.

15. W. T. Barnett, "Multipath Fading Effects on Digital Radio," *IEEE Trans Commun.,* Vol. COM-27, Dec. 1979.

16. C. W. Anderson, S. G. Barber, and R. N. Patel, "The Effect of Selective Fading on Digital Radio," *IEEE Trans. Commun.,* Vol. COM-12, Dec. 1979.

17. E. D. Sunde, "Digital Troposcatter Transmission and Modulation Theory," *Bell Sys. Tech. J.,* Jan 1964.

18. "Design Objectives for DCS Digital Radio Links," Eng. Publ., 27–77, Defense Communications Agency, Washington, DC, Dec. 1977.

19. W. C. Jakes, "An Approximate Method to Estimate an Upper Bound on the Effect of Multipath Delay Distortion on Digital Transmission," *IEEE ICC Proceedings,* 1978.

20. M. Engelson and L. Garrett, "Digital Radio Measurements Using the Spectrum Analyzer," *Microwave J.,* Apr. 1980.

21. P. D. Shaft and J. G. Isabeau, "Comments on Coding Gain," *IEEE ICC 1975 Conference Publ.,* Vol. II.

22. A. J. Viterbi, "Convolutional Codes and Their Performance in Communication Systems," *IEEE Trans. Commun.,* Vol COM-19, Oct. 1971.

23. R. L. Freeman, *Telecommunication System Engineering,* Wiley-Interscience, New York, 1980.

24. "Rain Attenuation over Earth–Satellite Paths," USN Contract N00039-79-C-0136, Science Applications, Inc., El Segundo, CA, May 1979.

25. J. F. Hart *et al.,* "Computer Approximations," *SIAM Series in Applied Mathematics,* J. Wiley, New York, 1968.

26. J. P. Odenwalder, *Error Control Coding Handbook,* Linkabit Corp., San Diego, CA, July 1976, under USAF contract.

GUIDE TO CCITT AND CCIR RECOMMENDATIONS*

CCITT and CCIR are two consultative committees operating under the auspices of the International Telecommunication Union (ITU) based in Geneva, Switzerland. More than 150 nations belong to the ITU and contribute to the consultative committees' work, as well as many of these countries' telecommunication administrations and other common carriers. Likewise a large number of the major producers of communication equipment also contribute as nonvoting members.

The work of CCITT (Comité Consultatif International de Téléphone et Télégraph) and CCIR (Comité Consultatif International de Radio) deals with providing standardized practices and system specifications so that the world's nations can intercommunicate, whether by telephone, telegraph, facsimile, data, program channel, or television. CCITT is concerned mainly with wire communications and CCIR with radio, although there is considerable interplay and overlap between the two consultative committees' work, which is embodied in a series of recommendations.

CCITT

The CCITT recommendations consist of ten volumes. The volumes are identified by color and the location in the world where the plenary assembly took place which authorized issuance of those recommendations, such as CCITT Green Books, Geneva, 1972.

CCITT uses the word "volume" with ambiguity. Some "volumes" contain several sections bound separately, identified by decimals.

Most recommendations are from two to four pages in length, although some are much longer. CCITT recommendations (Rec.) and reports (Rep.) are classified by letters A–Z as follows:

---

*The basic tabular information presented in Appendix A is by courtesy of ITU–CCITT, CCIR, Geneva.

Series Rec.	Contents*
A	Organization of the work in CCITT, Vol. I
B	Means of expression, Vol. I
C	General telecommunication statistics, Vol. I
D	General tariff principles, Vol. II.1
E	International telephone service—operation, Vol. II.2
	International telephone service—network management, Vol. II.3
F	Telegraph and facsimile, teletex, videotex—operations and tariffs, Vol. II.4
G	General characteristics of international telephone connections and circuits (G.101–G.171), Vol. III.1
	International analogue carrier systems (G.211–G.651), Vol. III.2
	Digital networks—transmission/multiplexing (G.701–G.939), Vol. III.3
H	Line transmission of nontelephone signals, Vol. III.4
I	Blank
J	Radio and television programme transmission, Vol. III.4
K	Protection against interference, Vol. IX
L	Protection of cable sheaths and poles, Vol. IX
M	Maintenance of telephone circuits and carrier systems, Vol. IV.1, 2
N	Maintenance of sound programme and television transmission, Vol. IV.3
O	Specification of measuring equipment, Vol. IV.4
P	Quality of telephone transmission, Vol. V
Q	Signaling and switching, Vol. VI.1
R	Telegraph transmission, Vol. VII.1
S	Telegraph terminal equipment, Vol. VII.2
T	Facsimile, videotex, teletex terminal equipment, Vol. II.2
U	Telegraph switching, Vol. VII.1
V	Data transmission over the telephone network, Vol. VIII.1
W	Blank
X	Data networks, Vol. VIII.2-3
Y	Blank
Z	Programming languages, Vol. VI.7-8
—	Terms and definitions, Vol. X.1
—	Index, Vol. X.2

*From CCITT Yellow Books, Geneva, 1980.

Contents of the CCITT Book Applicable after the Seventh Plenary Assembly (1980)

Volume	Yellow Book Contents
I	Minutes and reports of the VIIth Plenary Assembly of the CCITT—Resolutions and opinions issued by the CCITT—General table of study groups and working parties for the period 1981-1984—Summary table of abridged titles of questions under study in the period 1981-1984—Recommendations (Series A) on the organization of the work of the CCITT—Recommendations (Series B) relating to means of expression—Recommendations (Series C) relating to general telecommunication statistics.
II.1	General tariff principles—Charging and accounting in international telecommunications services. Series D recommendations (Study Group III).
II.2	International telephone service—Operation—Recommendations E.100-E.323 (Study Group II).
II.3	International telephone service—Network management—Traffic engineering—Recommendations E.401-E.543 (Study Group II).
II.4	Telegraph and "telematic services"* operations and tariffs—Series F recommendations (Study Group I).
III.1	General characteristics of international telephone connections and circuits—Recommendations G.101-G.171 (Study Group XVI, CMBD).
III.2	International analogue carrier systems—Transmission media characteristics—Recommendations G.211-G.651 (Study Group XV, CMBD).
III.3	Digital networks—Transmission systems and multiplexing equipments—Recommendations G.701-G.939 (Study Group XVIII).
III.4	Line transmission of nontelephone signals—Transmission of sound programme and television signals—Series H, J recommendations (Study Group XV).
IV.1	Maintenance; general principles, international carrier systems, international telephone circuits—Recommendations M.10-M.761 (Study Guide IV).
IV.2	Maintenance; international voice frequency telegraphy and facsimile, international leased circuits. Recommendations M.800-M.1235 (Study Group IV).
IV.3	Maintenance; international sound programme and television transmission circuits. Series N recommendations (Study Group IV).
IV.4	Specifications of measuring equipment. Series O recommendations (Study Group IV).

*"Telematic services" is used provisionally and includes such services as videotex, teletex, and facsimile.

Contents of the CCITT Book Applicable after the Seventh Plenary Assembly (1980)

Volume	Yellow Book Contents
V	Telephone transmission quality. Series P recommendations (Study Group XII).
VI.1	General recommendations on telephone switching and signaling—Interface with the maritime service—Recommendations Q.1-Q.119 (Study Group XI).
VI.2	Specifications of signaling systems Nos. 4 and 5—Recommendations Q.120-Q.180 (Study Group XI).
VI.3	Specifications of signaling system No. 6—Recommendations Q.251-Q.295 (Study Group XI).
VI.4	Specifications of signaling systems R1 and R2—Recommendations Q.310-Q.480 (Study Group XI).
VI.5	Digital transit exchanges for national and international applications. Interworking of signaling systems—Recommendations Q.501-Q.685 (Study Group XI).
VI.6	Specifications of signaling system No. 7—Recommendations Q.701-Q.741 (Study Group XI).
VI.7	Functional Specification and Description Language (SDL)—Man-machine language—Recommendations Z.101-Z.104, Z.311-Z.341 (Study Group XI).
VI.8	CCITT high level language (CHILL). Recommendation Z.200 (Study Group XI).
VII.1	Telegraph transmission and switching—Series R, U recommendations (Study Group IX).
VII.2	Telegraph and "telematic services"* terminal equipment—Series S, T recommendations (Study Group VIII).
VIII.1	Data communication over the telephone network—Series V recommendations (Study Group XVII).
VIII.2	Data communication networks: services and facilities, interfaces—Recommendations X.1-X.33 (Study Group VII).
VIII.3	Data communication networks: transmission, signaling and switching network aspects, maintenance, administrative arrangements—Recommendations X.40-X.180 (Study Group VII).
IX	Protection against interference—Series K recommendations (Study

*"Telematic services" is used provisionally and includes such services as videotex, teletex, and facsimile.

Contents of the CCITT Book Applicable after the Seventh Plenary Assembly (1980)

Volume	Yellow Book Contents
	Group V)–Protection of cable sheaths and poles–Series L recommendations (Study Group VI).
X.1	Terms and definitions.
X.2	Index of the Yellow Book.

The G. recommendations of CCITT have the most impact on transmission system design. Tables A.1–A.9 summarize those recommendations. The tables have been provided courtesy of the International Telecommunication Union (Orange Books, Geneva, 1976). These very condensed tables are not recommendations, and reference should be made to the complete recommendations.

Table A.1a Summary of Main Characteristics Specified by the CCITT for International Telephone Circuits[a] and International Connections

	For an International Circuit (1)	For a Complete Connection or for Its Parts (2)
Reference equivalent	G.111, B	For the connection and for the national systems G.111, G.121
Nominal four-wire equivalent (transmission plan, see G.101)	0.5 dB (G.141) Echo effects (G.131, B)	Four-wire chain national circuits G.101, B, b, G.121, G.122
Transmission stability	G.131, B	Balance return loss of national networks (G.122)
Band of frequencies effectively transmitted Limits (Hz)	At least 300–3400 (G.151, A)	From international center to local exchange: 300–3400 (G. 124) Four-wire chain of 6 circuits: 300–3400 (G.132)
Additional attenuation at limits of frequency	9 dB (G.151, A and G.132)	9 dB (G.151, A and G.132)
Attenuation/frequency distortion	G.151, A, Fig. 1/G.151	Graph No. 1, desirable objective for 12 circuits (Fig. 1/G.132). For data, see H.12
Group delay t	G.114	For the connection (G.114): $t \leq 150$ ms without reservation, $t \leq 400$ ms acceptable with conditions. For data, see H.12
Phase distortion (from group delay t)[b]	$t_m - t_{min} \leq 30$ ms[c] $t_M - t_{min} \leq 15$ ms[c] (G.133)	For the four-wire chain (G.133): $t_m - t_{min} \leq 60$ ms, $t_M - t_{min} \leq 30$ ms For each national 4-wire chain (G.133): $t_m - t_{min} \leq 15$ ms, $t_M - t_{min} \leq 7.5$ ms
Variation of overall loss with time	Mean deviation from nominal: $\leq \pm 0.5$ dB; Standard deviation ≤ 1 dB or 1.5 dB (G.151, C)	Extension circuits: as (1) (G.151). For data, see H.12
Linear crosstalk between different circuits (near- or far-end crosstalk ratio Δ)	$\Delta \geq 58$ dB (G.151, D, see Note 3)	Extension circuits: as (1) (G.151)
Near-end crosstalk ratio between the two directions of transmission	Ordinary circuits ≥ 43 dB (G.151, D); With speech concentrator: ≥ 58 dB With echo suppressors: ≥ 55 dB (G.151, D, see Note 4)	Extension circuits: as (1) (G.151)
Circuit noise	See Table 1 *bis*	
Impedance of the circuit		A single value for one trunk exchange (G.232, M)

632

Table A.1a (Continued)

	For an International Circuit (1)	For a Complete Connection or for Its Parts (2)
Frequency difference at two ends of carrier circuit	$\leq$ 2 Hz (G.135, G.225)	G.135, G.225
Power at zero relative level point Telephony, mean power in busy hour	Speech currents, etc.: 22 μW [d] (G.223); electric signals + tones: 10 μW [d] (G.223) (see G.224 for the power of signaling pulses)	
Voice-frequency telegraphy Maximum power per channel for VFT systems having: 24 channels 18 channels 12 channels or less	Amplitude modulation. Power when sending continuous mark (H.23, A, a): 9 μW 15 μW 35 μW	Frequency modulation mean power (H.23, A, b); 5.6 μW 7.5 μW 11.25 μW
Private wire telegraphy and telephony	Telegraphy level: $\leq$ -13 dBm0 (H.32) [e]	
Phototelegraphy	Amplitude modulation: 1 mW; frequency modulation: 0.1 mW (H.41)	
Maximum power for data transmission over leased circuits (H.51, A) [e]	1 mW on subscriber's line; frequencies: $\geq$ 2400 Hz (see G.224) Frequency modulation: -10 dBm0 or -20 dBm0; amplitude modulation: -6 dBm0 and 64 μW (mean for both directions in busy hour)	
Maximum power for data transmission over circuits in the switched network (H.51, B) [e]	1 mW on subscriber's line; frequencies: $\geq$ 2400 Hz (see G.224) Frequency or phase modulation: -10 dBm0 in simplex, Frequency or phase modulation: -10 dBm0 in simplex, -13 dBm0 in duplex; amplitude modulation: 64 μW (mean for both directions in busy hour)	

[a]Unless otherwise stated, circuits for voice-frequency telegraphy or phototelegraphy have the same characteristics.
[b] m = nominal minimum frequency effectively transmitted; M = nominal maximum frequency effectively transmitted; min = frequency corresponding to minimum group-delay time.
[c]These values apply to the chain of international circuits.
[d]Calculation target value or conventional value for a hypothetical reference circuit.
[e]This recommendation contains restrictions of use. See also Rec. H.34.

Table A.1b Summary of Noise Objectives Specified by the CCITT and the CCIR for Telephone Circuits[a]

Types of Systems	General Objectives			
	Cable[a] or Radio-Relay link	Single-Hop Satellite Link	Submarine Cable[a]	All Systems
Telephone circuits considered[b]	National 4-wire extension circuits and international circuits from 250 to 2500 km	Circuits[c] from 2500 to about 25,000 km	Circuits from 7500 to about 15,000 km	Circuits from 2500 to about 25,000 km	Chain of six international circuits
Recommendations of the CCITT	G.152 G.212[d] G.222 G.226	G.153		G.153	G.143 G.143
Recommendations of the CCIR	391, 392, 393-2, 395-1, 396-1, 397-2		352-2, 353-2		
Hypothetical reference circuit (HRC) or typical circuit considered	HRC of 2500 km[e] or similar real circuit	Circuit of 7500 km[e]	Basic HRC of at least 7500 km		Chain of about 25,000 km	Chain of more than 25,000 km

Recommended objectives						
Psophometric power						
Hourly mean						
Total power	10,000 pW				50,000 pW	
Terminal equipment	2500 pW				About 7000–9000 pW	
Line	7500 pW, i.e., 3 pW/km	15,000 pW^f 2 pW/km or better^g	10,000 pW^f	1 pW/kin^g	About 1.5 pW/km	1 pW/km for each section longer than 2500 km
For 1 minute exceeded during 20% of the month						
Line	7500 pW		10,000 pW^f			
% of a month during which the psophometric power for 1 minute due to the line indicated can be exceeded						
47,500 pW 50,000 pW 63,000 pW	0.1	0.3^f	0.3^f			

635

Table A.1b (*Continued*)

Types of Systems	General Objectives			
	Cable[a] or Radio-Relay link	Single-Hop Satellite Link	Submarine Cable[a]	All Systems
Unweighted power % of the month during which 10^6 pW (5 ms) can be exceeded	0.03^f	0.03^f		0.01

Special objectives

In National Networks	Radio-Relay Links						Tropospheric Radio-Relay Links in Special Conditions	Open-Wire Lines	
	Circuits not very different from HRC $280 < L$ < 2500 km	Composition of links very different from HRC					One or two circuits at most in one world connection	Up to 2500 km	More than 2500 km
		$50 \leq L$ < 280 km	$250 < L$ ≤ 840 km	$840 < L$ ≤ 1670 km	$1670 < L$ ≤ 2500 km				
Noise due to the national transmission systems									

636

General Objectives

Types of Systems	Cable[a] or Radio-Relay link	Single-Hop Satellite Link	Submarine Cable[a]	All Systems G.311	All Systems G.153
G.123	395-1	395-1	396-1,392		395-1
Total length L in km of the long-line FDM carrier systems in the national chain			HRC of 2500 km[e]	HRC of 2500 km[e]	Circuit of 10,000 km
(4000 + 4L) pW or (7000 + 2L) pW[h]				20,000 pW[i]	50,000 pW[i]
				2500 pW	
	3L pW	(3L + 400) pW		17,500 pW	
	(3L + 200) pW	(3L + 400) pW			
	3L pW	(3L + 600) pW	25,000 pW		
	(3L + 200) pW	(3L + 600) pW			
	$\dfrac{L}{2500} \times 0.1$ $\dfrac{280}{2500} \times 0.1$	$\dfrac{L}{2500} \times 0.1$	0.5		
	$\dfrac{L}{2500} \times 0.1$		0.05		

Table A.1b (*Continued*)

[a] For these systems it is sufficient to check that the objective for the hourly mean is attained.

[b] Special objectives for telegraphy are indicated in Recs. G.143, G.153, G.222, and G.442. Objectives for data transmission are shown in Recs. G.143 and G.153.

[c] For some very large countries, refer to Rec. G.222, c).

[d] See, in this recommendation, the details of the hypothetical reference circuits to be considered.

[e] The objectives for line noise, in the same column, are proportional to the length in the case of shorter lengths.

[f] Provisionally.

[g] Objective 3 pW/km for the worst circuits; if a real circuit has more than 40,000 pW, it should be equipped with a compandor.

[h] For planning purposes.

[i] Except for extremely unfavorable climatic conditions.

Note. All the values mentioned in this table refer to a point of zero relative level of a telephone circuit set up on the system under consideration (of the first circuit, for the chain). Furthermore (G.123), the psophometric EMF of noise induced by power lines should not exceed 1 mV at the "line" terminals of the subscriber's station. The mean value of the busy-hour noise power through a four-wire national exchange: ≤ 200 pWp. Limits of unweighted noise through exchange: 100 000 pW.

Table A.2 Summary of Main Characteristics Specified by the CCITT for Carrier Terminal Equipments

	Systems Wholly in Cable (G.232)	Systems on Open-Wire Lines	
		3-Channel (G.361)	12-Channel (G.232)
Level of carrier leak on the line			
a) within the 60–108-kHz band:			
Per channel	-26 dBm0	-17 dBm0	-26 dBm0
Per group[a]	-20 dBm0	-14.5 dBm0	-20 dBm0
b) Outside the 60–108-kHz band	-50 dBm0		-50 dBm0
Attenuation/frequency distortion	Figures 1/G.232 and 2/G.232		
Group delay	Table 1/G.232		
Nonlinear distortion	Figure 3/G.232		
Amplitude limiting	Definition (G.232, H)		
Crosstalk ratio	$\geq$ 65 dB for intelligible crosstalk (G.232, J) $\geq$ 60 dB for unintelligible crosstalk between adjacent channels (G.232, J)		
Near-end crosstalk ratio (A) between HF points	$\geq$ 47 dB without echo suppressors (G.232, J) $\geq$ 62 dB with echo suppressors (G.232, J)		
Near-end crosstalk ratio (X) between audio points	$\geq$ 53 dB without echo suppressors (G.232, J) $\geq$ 68 dB with echo suppressors (G.232, J)		
Relative levels	G.232, L		
Impedance	600 Ω (G.232, M)		
Protection and suppression of pilots	G.232, N		

[a]When part of the group is transmitted over open-wire lines (see G.232, E.1).
Note. See Recs. G.234 and G.235 for 8-channel and 16-channel equipments, respectively.

Table A.3a Summary of Main Characteristics Specified by the CCITT for Groups and Supergroups

	Group	Supergroup
Ratio between wanted component and the following components, defined in G.242, a, p. 2	at 84 kHz (G.242)	at 412 kHz (G.242)
Intelligible crosstalk (dB)[a]	70	70
Unintelligible crosstalk (dB)[a]	70	70
Possible crosstalk (dB)	35	35
Harmful out-of-band (dB)	40	40
Harmless out-of-band (dB)	17	17
Additional suppression to safeguard pilot frequencies (G.243)		At least 40 dB at 308 kHz ± 8 Hz At least 20 dB at 308 and 556 kHz ± 40 Hz (relative to 412-kHz value)
Additional suppression to safeguard additional measuring frequencies (G.243)		At least 20 dB at 308 and 556 kHz ± 20 Hz At least 15 dB at 308 and 556 kHz ± 50 Hz (relative to 412 kHz) (see also Figure 1/G.243)
Range of insertion loss over the passband for through-connection equipments	± 1 dB relative to 84 kHz (G.242)	± 1 dB relative to 412 kHz ≤ 3 dB for SG 1 and SG 3 (G.242)
Range of insertion loss over 10°C and 40°C for through-connection equipments	± 1 dB at 84 kHz relative to the insertion loss at 25°C (G.242)	± 1 dB at 412 kHz relative to the insertion loss at 25°C (G.242)

Pilot frequency for (G.241)	Frequency (kHz)[b]	Accuracy (Hz)	Absolute Power Level at 0 Relative Level Point (for tolerances, see G.241) (dBm0)
Basic group B[c]	84.080	± 1	-20
	84.140	± 3	-25
	104.080	± 1	-20
Basic supergroup	411.860	± 3	-25
	411.920	± 1	-20
	547.920	± 1	-20

[a] For telephony (G.242).
[b] See Rec. G.241 for use of these frequencies.
[c] Also applies to 8-channel groups (G.234).

Table A.3b Summary of Main Characteristics Specified by the CCITT for Mastergroups, Supergroups, and 15-Supergroup Assembly

	Mastergroup	Supermastergroup	15-supergroup assembly
Ratio between wanted component and the following components defined in G.242, a, p. 2	at 1552 kHz (G.242)	at 11 096 kHz (G.242)	at 1552 kHz (G.242)
Intelligible crosstalk (dB)[a]	70	70	70
Unintelligible crosstalk (dB)[a]	70	70	70
Possible crosstalk (dB)	35	35	35
Harmful out-of-band (dB)	40	40	40
Harmless out-of-band (dB)	17	17	17
Variation of insertion loss in pass-band of through-connection equipment	±1 dB with respect to value at 1552 kHz (G.242)	±1.5 dB with respect to value at 11 096 kHz ±1 dB in each master-group (G.242)	±1.5 dB with respect to value at 1552 kHz ±1 dB in each super-group (G.242)
Variation of insertion loss between 10°C and 40°C of through-connection equipment	±1 dB at 1552 kHz relative to insertion loss at 25°C (G.242)	±1 dB at 11 096 kHz relative to insertion loss at 25°C (G.242)	±1 dB at 1552 kHz relative to insertion loss at 25°C (G.242)
Relative levels at distribution frames (G.233) (dBr)			
Transmit	−36	−33	−33
Receive	−23	−25	−25
Return loss at modulator input (G.233) (dB)	≥20	≥20	≥20
Mastergroup, supermastergroup, or 15-supergroup assembly pilots (G.241) in:	Frequency (kHz)	Accuracy (Hz)	Level (for tolerances, see G.241) (dBm0)
Basic mastergroup	1,552	±2	−20
Basic supermastergroup	11,096	±10	−20
Basic 15-supergroup assembly	1,552	±2	−20

[a]For telephony (G.242).

Table A.4 Summary of Characteristics Specified by the CCITT for Carrier Systems on Open-Wire Lines

	Systems Acting on Each Pair		
	3-Circuit Systems	8-Circuit Systems	12-Circuit Systems
Line frequencies			
For a single system	Figure 1/G.361; (see also G.361, A, a; G.361, B, a; G.361, B, b; G.361, c)	Figure 1/G.314	Figure 1/G.311 or Figure 2/G.311
For several systems on the same route	Figure 1/G.361	(G.314, c)	See Figure 3/G.311, and Figure 4/G.311 for examples
Pilots			
Frequency	16.110 and 31.110 kHz or 17.800 kHz[a] (G.361, c)	(G.314, d)	(G.311, e)
Level	$^-$15 dBm0		$^-$20 dBm0[b]
Terminal equipment and intermediate repeater output.	≤17 dBr (G.361, b)	≤17 dBr (G.314, b)	≤17 dBr ± 1 dBr (terminal equipment)
Relative level per channel at 800-Hz equivalent frequency			≤17 dBr ± 2 dBr (intermediate repeater equipment) (G.311, c)
Frequency accuracy of pilot and carrier frequency generators	2.5×10^{-5} (G.361, c and h)	1×10^{-5} (G.314, d)	5×10^{-6} (G.311, f)

[a]Used only by agreement between administrations.
[b]Provisional recommendation.

Table A.5 Summary of Characteristics Specified by the CCITT for Carrier Systems on Symmetric-Pair Cables[a]

	System			
	1, 2, or 3 groups	4 groups	5 groups	2 supergroups
Line frequencies	Figure 2a/G.322	Figure 2b/G.322 Scheme 1 Scheme 1bis[b]	Figure 2c/G.322 Scheme 2 Scheme 2bis[b]	Figure 4/G.322 Schemes 3 and 4 Scheme 3bis[b]
Relative level at repeater output[c] (low-gain systems) (G.322, B.2, a)	-11 dBr	-14 dBr	-14 dBr	-14 dBr
Relative level at repeater output[c] (valve-type systems) (G.324, B, b):				
Nominal value	+4.5 dBr	+1.75 dBr	+1.75 dBr	+1.75 dBr
Tolerance	±2 dB	±2 dB	±2 dB	±2 dB
Return loss of repeater and line impedances (G.322, A, e)	$\leq 0.15 \sqrt{\dfrac{f_{max}}{f}}$ or	$\leq 0.08 \sqrt{\dfrac{f_{max}}{f}}$ or	$\leq 0.08 \sqrt{\dfrac{f_{max}}{f}}$ or ≤ 0.10 (paper-insulated cables)	
	≤ 0.25	≤ 0.10	$\leq 0.10 \sqrt{\dfrac{f_{max}}{f}}$ or ≤ 0.17 (cable types II bis and III bis[b], G.321)	
Relative level at repeater input[c]	≥ -56.5 dBr (G.324, B, b)			
Pilots	For alternative methods see Figure 5/G.322			60 kHz ± 1 Hz and 556 kHz ± 3 Hz (G.322, d, 2)
Monitoring frequencies (low-gain systems)	(G.322, B.2, b)			
Harmonic distortion (low-gain systems)	See Table 1/G.322			
Harmonic distortion (valve-type systems)	See Table (G.324)			

[a]For 12 + 12 systems, see Recs. G.325 and G.327.
[b]Used only by agreement between administrations.
[c]Note applicable to power-fed repeaters.

Table A.6 Summary of Characteristics Specified by the CCITT for Carrier Systems on 2.6/9.5-mm Coaxial Cables

	Systems			
	2.6 MHz[a] (1)	4 MHz (2)	12 MHz (3)	60 MHz (4)
Line frequencies	Figure 1/G.337 and Figure 1/G.338	Figure 1/G.338 and Figure 3/G.322	Figure 1/G.332 to Figure 4/G.332	Figure 1/G.333 and Figure 2/G.333
Pilot frequencies				
Line-regulating pilots	60 kHz ± 1 Hz or 308 kHz ± 3 Hz 2604 kHz ± 30 Hz (G.337, A, b)	60 kHz ± 1 Hz or 308 kHz ± 3 Hz 4092 kHz ± 40 Hz and see G.338, b, 1	4287 kHz ± 42.9 Hz for valve-type systems (G.339, b, 1) 12 435 kHz ± 124.3 Hz for transistorized systems (G.322, b, 1)	4287 kHz ± 42.9 Hz 12 435 kHz ± 124.3 Hz 22 372 kHz ± 223.7 Hz 40 920 kHz ± 409.2 Hz (G.333, b, 1)
Auxiliary line-regulating pilots	(G.337, A, b)	(G.338, b, 1)	308 kHz ± 3 Hz and 12 435 kHz ± 124.3 Hz for valve-type systems (G.339, b, 1) 308 kHz ± 3 Hz and 4287 kHz ± 42.9 Hz for transistorized systems (G.332, b, 1)	
Frequency comparison pilots				
National	As (2)	60 or 308 kHz 1800 kHz[b] (G.338, b, 2)	300 or 308 kHz (G.332, b, 2)	
International	As (2)	1800 kHz (G.338, b, 2)	308 and 1800 kHz 300 kHz[b] 808 kHz[b] and 1552 kHz[b] (G.332, b, 2)	4200 or 8316 kHz (G.333, b, 2)
Additional measuring frequencies	(G.337, A, c)	(G.338, b, 4)	(G.332, b, 3) and (G.339, b, 3)	(G.333, b, 3)
Level of line-regulating pilots and additional measuring frequencies				
Adjustment value	As (2)	−10 dBm0 ± 0.5 dB (G.338, b) −1.2 Nm0 for some systems (G.338, b)	−10 dBm0 ± 0.5 dB (G.332, b, 1) −1.2 Nm0 for valve-type systems (G.339, b)	As (2)
Error in the level	As (3)	As (3)	± 0.1 dB (G.332, b, 1)	As (3)
Variation with	As (3)	As (3)	± 0.3 dB (G.332, b, 1)	As (3)

Table A.6 *(Continued)*

	Systems			
	2.6 MHz[a] (1)	4 MHz (2)	12 MHz (3)	60 MHz (4)
Impedance match between repeaters and line N (as defined on G.332, c)	$N \geq 40$ dB for $f < 300$ kHz (G.338, e) $N \geq 45$ dB for $f > 300$ kHz (G.338, e)		$N \geq 48$ dB for $300 \leq f \leq 5564$ kHz [valve-type systems (G.339, e)] $N \geq 48$ dB for $f = 300$ kHz and $N \geq 55$ dB for $f \geq 800$ kHz [transistorized systems (G.332, e)]	$N = 65$ dB[c] (G.333, e)
Relative level on line			(G.332, f) and (G.339, f)	(G.333, f)

[a]Use of the 6-MHz system for telephony is specified otherwise (see G.337, B).
[b]Only used by agreement between administrations.
[c]The value of 65 dB is valid for telephone transmission.

Table A.7 Summary of Characteristics Specified by the CCITT for Carrier Systems on 1.2/4.4-mm Coaxial Cables

	Systems			
	1.3 MHz	4 MHz	6 MHz	12 MHz
Line frequencies	Figure 1/G.341	Schemes 1 and 2 of Figure 1/G.343	Schemes 1, 2, and 3 of Figure 1/G.344	(G.345)[a]
Pilot frequencies				
Line regulating pilots	1364 kHz ± 13.6 Hz (G.341, b, 1)	See G.343, b) 1. and Scheme 1 (G.338, b, 1); Scheme 2 (G.332, b, 1)	308 kHz ± 3 Hz (G.344)	The provisions of this recommendation are provisionally those appearing in Rec. G.332
Auxiliary line-regulating pilots	60 kHz ± 1 Hz or 308 kHz ± 3 Hz (G.341, b, 1)	4287 kHz ± 42.8 Hz[b] (G.343, b, 1)	4287 kHz ± 42.8 Hz[c] 6200 kHz ± 62 Hz (G.344, b, 1)	(see Table A.6), with the exception of the matching
Frequency comparison pilots	60 kHz or 308 kHz (G.341, b, 2)	Scheme 1 (G.338, b, 2) and Scheme 2 (G.332, b, 2)	Schemes 1 and 2 (G.338, b, 2) and Scheme 3 (G.332, b, 2)	
Additional measuring frequencies	(G.341, b, 3)	(G.343, b, 3)	(G.344, b, 3)	
Level of line-regulating pilots and additional measuring frequencies				
Adjustment value	−10 dBm0 or 1.2 Nm0 for some systems (G.341, b)	−10 dBm0 (G.343, b)	−10 dBm0 (G.344, b)	
Tolerances		(G.343, b)	(G.344, b)	
Impedance match between repeaters and line	$N \geq 54$ dB for 6-km repeater section $N \geq 52$ dB for 8-km repeater section (G.341, e)	$N \geq 50$ dB for $f = 60$ kHz $N \geq 57$ dB for $f \geq 300$ kHz [4-km repeater section (G.343, e)]	$N \geq 60$ dB for $f \geq 300$ kHz $N = 50$ dB for $f = 60$ kHz [3-km repeater section (G.344, e)]	$N = 63$ dB for a 2-km repeater section (G.345)
Relative levels on line and interconnection	(G.341, f)	−9 dBr at 4028 kHz or −8.5 dBr at 4287 kHz (G.343, f)	−17 dBr[a] (G.344, f)	(G.332, f)

[a]Provisional recommendation.
[b]Only used by agreement between administrations.
[c]Cannot be used with television transmissions.

Table A.8 Summary of Main Characteristics Specified by the CCITT for International Circuits for Program Transmissions

	Type of Circuits[a, b]		
	15 kHz[c]	10 kHz	6.4 kHz
Frequency band effectively transmitted by the complete link	0.04–15 kHz	0.05–10 kHz	0.05–6.4 kHz
Additional attenuation at these limits	2 dB at 0.04 kHz 3 dB at 15 kHz (J.21)	4.3 dB (J.22, a)	4.3 dB (J.23, a)
Attenuation/frequency distortion	± 0.5 dB; 0.125–10 kHz (J.21)	Figure 1/J.22	Figure 1/J.23
Group delay at frequency $f(\tau f)$ relative to the minimum value of group delay	$\tau_{15000} \leq 12$ ms $\tau_{14000} \leq 8$ ms $\tau_{75} \leq 24$ ms $\tau_{40} \leq 55$ ms (J.21)	$\tau_{10000} \leq 8$ ms $\tau_{100} \leq 20$ ms $\tau_{50} \leq 80$ ms (J.22, d)	$\tau_{6400} \leq 8$ ms $\tau_{100} \leq 20$ ms $\tau_{50} \leq 80$ ms (J.23, d)
Maximum absolute voltage level at a sound-program zero relative level point	+9 dB (J.14) – peak voltage 3.1 V (Figure 3/J.13)		
Definition of zero relative level at a point in a carrier circuit	Level to give no greater load than that for the telephone channels replaced (J.31, A)	As for telephony, is within ± 3 dB (J.14)	
Nominal relative voltage level at the input and output of the circuit defined in J.13⁻	6 dB (J.14)	6 dB (J.14)	6 dB (J.14)
Variation of relative level with time	$\leq \pm 0.5$ dB (daily variation) (J.21)	$\leq \pm 2$ dB (during a program transmission) (J.22, g)	$\leq \pm 2$ dB (during a program transmission) (J.23, g)
Intelligible crosstalk attenuation (near-end or far-end) ratio[d]	0.04 kHz $\geq$ 50 dB 0.5 kHz $\geq$ 74 dB 5 kHz $\geq$ 74 dB 15 kHz $\geq$ 60 dB (J.21)	Between 2 program transmissions circuits or telephony into sound programme: $\geq$ 74 dB (cables) $\geq$ 61 dB (open wire) Sound program into telephony: $\geq$ 58 dB (cables) $\geq$ 47 dB (open wire) (J.22, f and J.23, f, respectively)[d, e]	
Circuit noise including nonlinear crosstalk[f]	Level $\leq$ –47 dBm0ps (new weighting network according to J.16)	Psophometric voltage at the end of 1) cable circuit $\leq$ 6.2 mV 2) open-wire circuit $\leq$ 15.6 mV	

[a]Characteristics applicable to the hypothetical reference circuits, defined in Rec. J.11.
[b]Types of circuits described in Rec. J.12.
[c]For the additional characteristics specified by the CCITT for 15-kHz stereophonic sound-program circuits (see Rec. J.21).
[d]Provisional recommendation.
[e]Special precautions needed for crosstalk between the two directions of transmission (see Recs. J.18 and J.22).
[f]Measures taken to reduce the effects of noise in a group link (see Rec. J.17).

Table A.9 Summary of Main Characteristics of Analog Signals at Audio Frequencies at Terminals of PCM Equipments

Analog Characteristics Measured at Input and Output Parts[a, b]	Test Signal			
	Signal	Frequency Range	Power Level x (dBm0)	
Attenuation/frequency distortion			0	Figure 1/G.712
Envelope delay distortion			0	Figure 2/G.712
Idle channel noise				
Weighted				-65 dBm0p
Single frequency				-50 dBm0
Due to receiving equipment				-75 dBm0p
Image frequency	Sine wave	4.6–72 kHz	x	$< x - 25$ dBm0
Level of out-of-band image signals	Sine wave	300–3400 Hz	0	< -25 dBm0
Intermodulation products				
$2f, -f_2$	Two sine-wave	f_1 and f_2 (Hz)	$-21 < x < -4$	$< x - 35$ dBm0
Any intermodulation product	Sine wave	300–3400 Hz	-9	< -49 dBm0
	Sine wave	50 Hz	-23	< -49 dBm0
Variation of gain				
With input level	White noise		$-60 < x < -10$	Figure 6a/G.712
(reference = gain at input level	Sine wave	700–1100 Hz	$-10 < x < 3$	Figure 6b/G.712
of -10 dBm0)	Sine wave	700–1100 Hz	$-55 < x < 3$	Figure 6c/G.712
With time (stability)				$\pm$ 0.2 dB in 10 min
				$\pm$ 0.5 dB in 30 days
				$\pm$ 1.0 dB in 1 year
Crosstalk				
Interchannel	Sine wave	700–1100 Hz	0	< -65 dBm0
	White noise		0	< -60 dBm0
Go–return	Sine wave	300–3400 Hz		> 60 dB
Distortion	Gaussian noise		$-55 < x < 3$	Figure 4/G.712
	Sine wave	700–1100 Hz	$-45 < x < 0$	Figure 5/G.712

[a]Parameters of input and output ports: 600 Ω balanced, 4-wire ports; return loss better than 20 dB over frequency range 300–3400 Hz (provisional recommendation).
[b]For correct application to the equipments, see p. 1 of Rec. G.712.

CCIR

The CCIR recommendations (and reports, resolutions, and opinions) are numbered according to a system in force since the Xth Plenary Assembly. When a text is modified, it retains its original number, to which is added a dash and a figure indicating how many revisions have been made. For example, 396-1 indicates that the current recommendations now in force has been modified once from its original version. If Rec. 396-1 becomes 396-2, it is apparent that Rec. 396 has had two changes since the original. Table A.10 shows only original numbering; subsequent changes are not shown in the Table.

CCIR, Kyoto, 1978, has 13 volumes as follows:

Volume	Contents
I	Spectrum utilization and monitoring.
II	Space research and radioastronomy.
III	Fixed service at frequencies below about 30 MHz.
IV	Fixed service using communication satellites.
V	Propagation in nonionized media.
VI	Propagation in ionized media.
VII	Standard frequencies and time signals.
VIII	Mobile services.
IX	Fixed service using radio-relay systems. Frequency sharing and coordination between systems in the fixed satellite service and radio-relay systems.
X	Broadcasting service (sound).
XI	Broadcasting service (television).
XII	Transmission of sound broadcasting and television signals over long distances (CMTT). Vocabulary (CMV).
XIII	Information concerned the XIVth Plenary Assembly: Minutes of the Plenary Sessions. Texts of general interest. Structure of the CCIR. Complete list of CCIR texts. Alphabetical index of technical terms appearing in Volumes I to XII.

Table A.10 shows the most commonly referenced CCIT recommendations in the text, those referring to radiolink systems.

Table A.10 Index of CCIR Recommendations, Reports, and Decisions

Number	Volume	Number	Volume	Number	Volume
Recommendations					
45	VIII	341	I	457, 458	VII
48, 49	X	342–349	III	460	VII
77	VIII	352–354	IV	461	XII (CMV)
80	X	355–359	IX	463	IX
100	I	361	VIII	464–466	IV
106	III	362–365	II	467, 468	X
139, 140	X	367	II	469–472	XI
162	III	368–370	V	473, 474	XII (CMTT)
182	I	371–373	VI	475, 476	VIII
205	X	374–376	VII	478	VIII
214–216	X	377–379	I	479	II
218, 219	VIII	380–393	IX	480	III
239	I	395–406	IX	481–484	IV
240	III	407–412	X	485, 486	VII
246	III	414–416	X	487–496	VIII
257	VIII	417, 418	XI	497	IX
265, 266	XI	419	XI	498, 499	X
268	IX	422, 423	VIII	500, 501	XI
270	IX	427, 428	VIII	502–505	XII (CMTT)
275, 276	IX	430, 431	XII (CMV)	506–508	I
283	IX	433	I	509–517	II
290	IX	434, 435	VI	518–520	III
302	IX	436	III	521–524	IV
305, 306	IX	439	VIII	525–530	V
310, 311	V	441	VIII	531–534	VI
313	VI	442, 443	I	535–538	VII
314	II	444	IX	539–554	VIII
325–329	I	445	I	555–558	IX
331, 332	I	446	IV	559–564	X
334	I	447	X	565, 566	XI
335, 336	III	450	X	567–572	XII (CMTT)
337	I	452, 453	V	573, 574	XII (CMV)
338, 339	III	454–456	III		
Reports					
19	III	322	VI[a]	493	XII (CMTT)
32	X	324–326	I	496–498	XII (CMTT)
93	VIII	327	III	499–502	VIII
106, 107	III	329	III	504–507	VIII
109	III	336	V	509–512	VIII
111	III	338	V	516	X
112	I	340	VI (1)	518	VII

Number	Volume	Number	Volume	Number	Volume
122	XI	342	VI	519–528	I
		345	III	530–534	I
137	IX	347	III	535–546	II
176, 177	III	349	III	548	II
179	I	352–357	III	549–551	III
181	I	358	VIII	552–561	IV
183	III	362–364	VII	562–565	V
184	I	367–373	I	567	V
186	I	374–380	IX	569	V
195	III	382	IX	571	VI
196	I	383–385	IV	574, 575	VI
197	III	386–388	IX	576–580	VII
200, 201	III	390, 391	IV	581, 582	VIII
203	III	393	IX	584–591	VIII
204–208	IV	394	VIII	594–596	VIII
209	IX	395, 396	II	598, 599	VIII
212	IV	400, 401	X	602	VIII
214	IV	404, 405	XI	604	IX
215	XI	409	XI	607–610	IX
222–224	II	411, 412	XII (CMTT)	612–615	IX
226	II	413–415	I[a]	616, 617	X
227–229	V	418–420	I	619, 620	X
236	V	422, 423	I	622	X
238, 239	V	426	V	623	XII (CMTT)
249–251	VI	430–432	VI	624–634	XI
252	VI[a]	434–437	III	635–637	XII (CMTT)
253–255	VI	439	VII	639, 640	XII (CMTT)
258–260	VI	443–446	IX	642, 643	XII (CMTT)
262, 263	VI	448, 449	IX	646–649	XII (CMTT)
265, 266	VI	451	IV	651–671	I
267	VII	453–455	IV	672–700	II
270, 271	VII	456	II	701–705	III
272, 273	I	457, 458	X	706–713	IV
275–282	I	461	X	714–724	V
283–289	IX	463–465	X	725–729	VI
292, 293	X	468	X	730–738	VII
294	XI	469	XI	739–778	VIII
299–305	X	472	X	779–793	IX
306	XI	473	XI	794–800	X
311–313	XI	476–478	XI	801–814	XI
314	XII (CMTT)	481–485	XI	815–823	XII (CMTT)
315	XI	487, 488	XII (CMTT)		
319	VIII	491	XII (CMTT)		

Table A.10 (*Continued*)

Number	Volume	Number	Volume	Number	Volume
Decisions					
2	IV	18	XII (CMTT)	28, 29	VII
3–5	V	19	XII (CMV)	30–32	VIII
6–11	VI	21–24	VI	33	XI
17	XI	27	I		

[a]Published separately.

Table A.11 Characteristics of Radiolink Systems Specified in CCIR Recommendations

	Maximum Number of Telephone Channels														
Frequency Band	1	6	12	24	60	120	300	600	900/ 960ᵃ	1200/ 1260ᵃ	1800	2700	Television	Trans-horizon	Digital Systems
Occupied bandwidth	Bands 8 and 9 ←——————————————————————→													388	
Number of radio-frequency channels 2 GHz: Bandwidth															
Center frequencies and radio-frequency channel arrangements 200 MHz Bandwidth						←———— 283-3 ————→									283-3
Polarization arrangements 400 MHz									←—— 382-2 ——→				382-2		
4 GHz									←—— 382-2 ——→				382-2		
6 GHz								383-1		384-2	383-1	384-2	383-1, 384-2		
7 GHz					←—— 385-1 ——→										
8 GHz							386-1		386-1				386-1		
11 GHz									←—— 387-3 ——→				387-3		387-3
13 GHz									497-1				497-1		497-1
Interconnection at:															
Audio frequencies						←———— 268-1 ————→									
Baseband frequencies							←———— 380-3 ————→								
Intermediate frequencies									←———— 403-2 ————→						
Video frequencies													←— 270-2 —→		
Hypothetical reference circuit				←— 391 —×—					392					396-1	
Allowable noise power in the hypothetical reference circuit								←——— 393-3 ———→							
Noise in the radio portions of real circuits								←——— 395-2 ———→							
Hypothetical reference digital path															556
Availability objectives for a hypothetical reference circuit and a hypothetical reference digital path													555	397-3	557

653

Table A.11 (Continued)

Frequency Band	13	6	12	24	60	120	300	600	900/960ᵃ	1200/1260ᵃ	1800	2700	Television	Trans-horizon	Digital Systems
Maximum Number of Telephone Channels															
Frequency deviation			←				404-2	→					276-2	404-2	
Preemphasis and deemphasis characteristics			←					275-2	→				405-1	275-2	
Line regulating and other pilots					←					401-2	→				→
Signaling and service channels					←					400-2	→				→
Stand-by arrangements									305	→				→	
Auxiliary radio-relay systems					←			389-2	→					→	
Residues of signals outside the baseband					←			381-2	→				.463-1	381-2	
Maintenance measurement in actual traffic				←				398-3	→						
Measurements of noise using a continuous uniform spectrum								399-3	→						

ᵃOr the equivalent.

B GLOSSARY

ac	Alternating current
ADP	Automatic data processing
AF	Audio frequency, as distinguished from RF, radio frequency
AFC	Automatic frequency control
AGC	Automatic gain control
A-law	A PCM companding law commonly used with European (CEPT) PCM systems see also μ-law)
ALC	Automatic load control, automatic level control
AM	Amplitude modulation
AMI	Alternate mark inversion
AMP	˙ A connector manufacturer
ANSI	American National Standards Institute
APC	Automatic phase control
APD	Avalanche photodiode
APK	Amplitude phase shift keying
APL	Average picture level
ARQ	˙ Automatic repeat request. A feature in data links that allows for requesting the retransmission of data blocks, segments, or packets in which errors may have been detected
ASCII	American standard code for information interchange
ASK	Amplitude shift keying
ATB	All trunks busy
ATT	American Telephone and Telegraph Company.
Autodin	Automatic digital network (U.S. Department of Defense). A digital data and telegraph network with message and line switching
Autovon	Automatic voice network. A switched telephone, network under the auspices of the U.S. Department of Defense

AWG	American wire gauge
AZ	Azimuth. Refers to orienting earth station antenna systems, or any antenna system, the orientation being in the horizontal plane
baud	The unit of digital modulation rate
BCC	Block check count, block check character
BCD	Binary coded decimal. A binary code where the 10 decimal digits 0–9 are conventionally binary coded from 0001 through 1010 and all other numbers are built up from these basic digits
BCH	Bose, Chaudhuri, Hocquenqhem. A class of block codes with well-defined decoding algorithms
BCI	Bit count integrity
BER	Bit error rate
B factor	The penalty factor used in the VNL concept
BH	Busy hour. A term used in telephone traffic engineering
B_{IF}	IF bandwidth
BINR	Baseband intrinsic noise ratio
bit	Binary digit
BNZS	Bipolar with N zero substitution (PCM term)
BPO	British Post Office, the telephone administration of the United Kingdom
bps	Bits per second
BPSK	Binary phase shift keying
BR	Bit rate
B_{RF}	RF bandwidth
BSS	Broadcast satellite service
BTM	Bell Telephone Manufacturing Company, ITT's large Belgian affiliate
busy hour (BH)	A traffic engineering term. CCITT definition: "The uninterrupted period of 60 minutes during which the average traffic flow is maximum"
BW	Bandwidth
BWR	Bandwidth ratio =

$$10 \log_{10} \frac{\text{occupied baseband bandwidth}}{\text{voice channel bandwidth}}$$

	It is used to determining S/N from NPR
B3ZS	Bipolar with 3 zero substitution (used in PCM)

B6ZS	Bipolar with 6 zero substitution (used in PCM)
CARS	Community antenna radio service
CATV	Community antenna television, sometimes called cable television or "pay TV." It is a private line method of bringing broadcast (and other TV/communication) into the home or hotel room
CCD	Charge-coupled device
CCIR	International Consultive Committee for Radio. It is organized under the auspices of the ITU (International Telecommunication Union), Geneva
CCITT	International Consultive Committee for Telephone and Telegraph. It is organized under the auspices of the ITU.
CCS	Cent-call-second (100 call-seconds traffic intensity)
CDMA	Code division multiple access
CEPT	Conférence Européenne des Postes et Télécommunication, the regional European telecommunication conference committee
CFSK	Coherent frequency shift keying
cm	Centimeter
cm region	As designated by the ITU, that band of frequencies from 3 to 30 GHz
C/N	Carrier-to-noise ratio
C/N_0	Carrier-to-noise spectral density ratio
CNT	Canadian National Telephone Company now called Canadian National Telecommunications, based in Toronto
codec	Coder–decoder, an acronym used in PCM
Compandor	Acronym for compressor–expander
CPES	Customer premises earth station
CP-FSK	Continuous-phase FSK
cpm	Counts per minute.
CPU	Central processing unit
CR/BTE	Carrier recovery/bit timing recovery
CRC	(1) Communication relay center (HF terminology). (2) Cyclic redundancy check
CRPL	Central Radio Propagation Laboratory, operated by the U.S. Department of Commerce, Boulder, CO
CRT	Cathode ray tube
CSC	Common signaling channel
C/T	The ratio of carrier-to-thermal noise power, usually expressed in dBW per kelvin

CT   Centre de transit, an international routing term for the international hierarchical telephone network. CT is usually followed by a number (i.e., CT1, CT2, or CT3); the number indicates the hierarchical rank, 1 being the highest rank

CU   Crosstalk unit

CVSD   Continuous variable slope delta modulation

CW   Continuous wave, sometimes used to mean "Morse operation"

DA   Dependent amplifier

DAMA   Demand assignment multiple access

DASS   Demand assignment signaling and switching unit

dB   Decibel

dBa   Decibels adjusted, a noise measurement unit using F1A weighting (now obsolete)

dBi   Decibels above an isotropic, referring to antennas

dBm   Decibels referenced to 1 mW

dBm0   An absolute power unit referenced to the 0 TLP (test level point)

dBm0p   dBm0 psophometrically weighted

dBmp   A noise measurement unit based on the dBm using psophometric weighting

dBmV   An absolute voltage level measurement unit based on the dB and referenced to 1 mV across 75 Ω

dBr   Decibels referenced, the number of decibels of level above or below a specified reference point in a network. A minus sign indicates the level to be below or less than at the reference, the plus sign, above or more than

dBrn   dB reference noise (now obsolete)

dBrnC   A noise measurement unit using C-message weighting, usage generally limited to North America

dBrnC0   dBrnC referenced to the 0 TLP

dBV   Decibels referenced to 1 V

dBW   Decibels referenced to 1 W

dBx   The crosstalk coupling in dB above reference coupling, which is a crosstalk coupling loss of -90 dB

dc   Direct current

DCA   Defense Communications Agency, an agency under the U.S. Department of Defense

DCE   Data communications equipment

DCS   Defense Communications System (U.S.)

DEC	Digital Equipment Corporation
DE-PSK	Differentially encoded phase shift keying
dibit coding	Coding such that each transition of the modulated data signal represents 2 bits of data
DM	Delta modulation
dNp	Decineper or one-tenth of a neper (see neper)
DQPSK	Differential quaternary phase shift keying
DSB	Double sideband. A form of AM where the same information is sent on each sideband. Conventional AM broadcast is a form of DSB. If the term DSB is used, it should be stated whether the carrier is suppressed or not; on conventional AM it is not
DSBEC	Double-sideband emitted carrier
DSI	Digital speech interpolation
DTE	Data terminal equipment
DTU	Direct to user
duobinary	A three-level digital modulation scheme
EBCDIC	Extended binary coded decimal interchange code
EC	(1) Earth curvature. (2) Earth coverage
EDD	Envelope delay distortion
EDP	Electronic data processing
EHF	Extremely high frequency
EIA	Electronic Industries Association (U.S.)
EIRP	Effective isotropically radiated power
EL	Elevation. Refers to the orientation in the vertical plane of an antenna and, in particular, an earth station antenna (see AZ)
E_b/N_0	In digital transmission systems, signal energy per bit per hertz of thermal noise
erfc	The complementary error function
ERL	Echo return loss. In the VNL concept, a single weighted figure for return losses in the band of 500–2500 Hz
erlang	The international dimensionless unit of traffic intensity. One erlang is the intensity in a traffic path continuously occupied, or in one or more paths carrying an aggregate traffic of 1 call-hour per hour, 1 call-minute per minute, and so on. When based on the hour, 36 CCS = 1 erlang
ERP	Effective radiated power
exhaust, exhaustion	Term used when all wire pairs in a cable are considered, for planning purposes, to be used up (assigned), or when any installed system is considered to have reached its full capacity.

	Many are considered exhausted for planning purposes at 70 or 80% assignment. The remaining unassigned pairs are bad pairs, emergency spares, and so on
exhaust date	The date for planning purposes when exhaustion is expected to take place
F1A	An obsolete noise weighting used in the United States. Noise units used with this weighting network are dBa
FCC	Federal Communications Commission, the U.S. federal tele-communications (and radio) regulatory authority
FDM	Frequency division multiplex
FDMA	Frequency division multiple access
FEC	Forward error correction
FET	Field-effect transistor
FM	Frequency modulation
FOT	"Fréquence optimum de travail," an HF propagation term from the French. We more often call it OWF (optimum working frequency)
frame	A set of consecutive digit time slots in which each digit time slot can be identified by reference to a framing signal, in data transmission often synonymous with block or packet. For TV frame is defined as the total area, occupied by the picture, which is scanned while the picture signal is not blanked
FSK	Frequency shift keying
FSS	Fixed satellite service
ft	Foot
GaAs	Gallium arsenide
GBLC	Gaussian band-limited channel
GHz	Gigahertz, Hz $\times 10^9$
G/T	The figure of merit of an earth station receiving system. It is expressed in dB-K, $G/T = G_{dB} - 10 \log T_{sys}$, where G is the gain of the antenna at the receiving frequency and T = effective noise temperature in kelvins of the receiving system
HDB3	High-density binary 3. A PCM line signal term
HDLC	High-level data link control. An ISO datalink protocol
HE	Mode of propagation in waveguide
HF	High frequency. ITU definition: "The band of frequencies between 3 and 30 MHz."
highway	A common path over which signals from a plurality of channels pass with separation achieved by time division
HPA	High-power amplifier

HRC	Hypothetical reference circuit
Hz	Hertz, the unit of frequency measurement
IBM	International Business Machine (Company)
ICL	Inserted connection loss (VNL terminology)
IEEE	Institute of Electrical and Electronics Engineers, U.S. engineering society
IF	Intermediate frequency
ILD	Injection laser diode
IM	Intermodulation
inside plant	See outside plant
INTELSAT	International Communication Satellite, a series of satellites under an international consortium with the same name
I/O	Input-output (device). A generic group of equipment that serves as input and/or output for a telecommunication system. Examples are teleprinters, card readers, and telephone handsets
IR	Infrared.
IR	The product of the resistance R in ohms and the current I in amperes
ISB	Independent sideband. A form of AM where the information sent on the upper sideband is different from that sent on the lower sideband
ISO	International Standards Organization
ITA	International telegraph alphabet
ITS	Institute of Telecommunication Sciences (U.S.)
ITT	International Telephone and Telegraph (Co.)
ITU	International Telecommunication Union. This organization issues the Radio Regulations. CCIR and CCITT are subsidiary organizations to the ITU
junction	See trunk
kbps	Kilobits per second
K factor	A constant used in tropospheric scatter and radiolink (LOS microwave) path profiling which is indicative of the amount of bending a radio beam may undergo, given a particular set of circumstances. If K is greater than 1, the ray beam is bent toward the earth; if less than 1, it is bent away from the earth
kHz	Kilohertz, $Hz \times 10^3$
km	Kilometer
kW	Kilowatt

L-carrier	The FDM carrier series developed by the U.S. Bell System for long-haul transmission over broadband media. We may expect to encounter L1, L2, L3, L4, and L5 systems
lay-up	Assembly of tubes inside a single sheath (coaxial cable systems)
LBO	Line build-out. A method of extending the length of a line electrically, usually by means of capacitors
LCIE	Lightguide cable interconnection equipment
LED	Light-emitting diode
LHC	Left-hand circular (polarization)
LNA	Low-noise amplifier
LOS	Line-of-sight, generally in reference to line-of-sight microwave, which in this text is called radiolink systems
LP	Log periodic
LPA	Linear power amplifier
LPI	Lines per inch (facsimile)
LPM	Lines per minute
LPSS	Line protection switching system
LRC	Longitudinal redundancy check (data transmission)
LSA mode	Limited space charge accumulation mode
LSB	Lower sideband
LTMA	Lightwave termination multiplex assembly
LUF	Lowest usable frequency, an HF propagation term
mA	Milliampere(s)
M-ary	Multilevel digital modulation. A 2-level digital modulation system is *binary*
Mbps	Megabits per second
mesh connection	A telecommunication routing term describing the situations where telephone exchanges are interconnected directly by trunks. *Full mesh* is a term which is more explicit, meaning that all exchanges in an area are each and every one connected to all other exchanges in an area. It implies that tandem operation is not used at all (see star connection)
MHz	Megahertz, Hz $\times$ 10^6
mi	mile(s). In this work, unless otherwise specified, we refer to the statute mile (i.e., 5280 ft)
MIL, Mil	Military, U.S. military, often used in the nomenclature of U.S. military standards
mm	Millimeter(s)
mm region	As designated by the ITU, the frequency band from 30 to 300 GHz. In this text, from 13 to 100 GHz

modem	Acronym for modulator–demodulator
ms	Millisecond(s)
MSK	Maximum shift keying
MSL	Mean sea level
MTBF	Mean time between failures
MTTR	Mean time to repair
MUF	Maximum usable frequency, an HF propagation term
N_0	(1) Sea level refractivity; (2) Spectral noise density
NA	Numerical aperture (fiber optics)
NARS	North Atlantic Radio System. A broadband radio system, principally tropospheric scatter, connecting the U.K. to the U.S. via the Faeroes, Iceland, Greenland, Baffin Island, Canada and is principally used for defense
NBS	National Bureau of Standards (U.S.)
NEP	Noise equivalent power
neper (Np)	A logarithmic measurement unit expressing a ratio similar to the dB. However, the logarithm is to the base e. 1 dB = 0.1151 Np; 1 Np = 8.686 dB
NF	Noise figure
NL	Nonloaded (in reference to VF telephone cables)
NLR	Noise load ratio (or noise load factor)
nm	Nanometer(s)
NOSFER	Laboratory standard for reference equivalent
NPR	Noise power ratio
NRZ	Nonreturn to zero (digital modulation)
ns	Nanosecond(s)
NTSC	National Television Systems Committee (U.S.)
O & M	Operation and maintenance
OCL	Overall connection loss (used in VNL concept)
OCR	Optical character reader
off-hook	A telephone handset that has been taken off-hook closes the loop and "busys" the line
OK-QPSK	Offset keyed QPSK
on-hook	A telephone handset that is on-hook is not in use. Placing a handset on-hook opens the loop, making the line "idle"
OOK	On–off keying
ORE	Overall reference equivalent. ORE is the sum of the TRE (transmit reference equivalent) and the RRE (receive reference equivalent) of a telephone connection

outside plant	A telephone operating company term or derived from telephone operating company usage. It has various meanings. The total telephone plant may be considered the sum of "inside plant" and "outside plant" facilities. One may consider outside plant as all telephone facilities that are out of doors. In this text outside plant is that part of the telephone plant that takes the signal from the local switch to the subscriber, including the subscriber equipment, as well as the local trunk (junction) plant. Another connotation of outside plant is any telephone plant activity that involves civil (construction) engineering
OW	Orderwire, service channel
OWF	Optimum working frequency (see FOT)
PABX	Private automatic branch exchange
PAL	Phase alternation line. One of the two European color television systems; the other system is SECAM
PAM	Pulse amplitude modulation
PCM	Pulse code modulation
peak-to-peak	A voltage measurement from positive voltage peak to negative voltage peak
pel, pixel	Picture elements per unit area
PEP	Peak envelope power
PIN	Terminology derives from the semiconductor construction of a device where an intrinsic (I) material is used between the p–n junction of a diode
PM	Phase modulation
PPM	Pulse position modulation
ppm, pps	Pulses per minute, pulses per second
PRF	Pulse repetition frequency
PSK	Phase shift keying
psophometric	A noise weighting used in Europe with a noise (or level) meter called a psophometer. See CCITT Rec. G.223
PSV	Pair shield video
pW	Picowatts, $W \times 10^{-12}$; 1 pW = -90 dBm
pWp	Picowatts psophometrically weighted
QAM	Quadrature amplitude modulation
Q-carrier	Quadrature carrier (digital modulation)
QPR	Quadrature partial response (keying)
QPSK	Quaternary phase shift keying, 4-level PSK

QSY	International Q signal indicating that a change in frequency is required or is being carried out
RC	Resistance–capacitance product, determines the time constant of a circuit
Rec.	Recommendation, especially of CCITT and CCIR
reference equivalent	A measurement in dB of telephone subscriber satisfaction regarding "quality" of transmission. In broad terms reference equivalent considers only the level of a telephone connection (i.e., transmission level)
Rep.	Report, and in particular, a CCIR or CCITT report
RF	Radio frequency
RFI	Radio-frequency interference
RHC	Right-hand circular (polarization)
rms	Root mean square (mathematics)
RRE	Receive reference equivalent. The reference equivalent of the receiving portion of the telephone connection
RSL	Receive signal level
RZ	Return-to-zero. Refers to digital modulation line signal format
SBS	Satellite Business Systems, a U.S. domestic satellite communication system
scintillation	A random fluctuation of a received signal about its mean value, the deviations being relatively small in a benign (nonnuclear) environment. It should be noted that the word is an extension of the astronomical term for the twinkling of the stars, and the underlying explanation of the phenomenon may be similar (see IEEE dictionary)
SCPC	Single channel per carrier
S + D	Speech plus derived
S/D	Signal-to-distortion ratio (term used in PCM)
SDLC	Synchronous data link control (IBM protocol)
SECAM	Sequential color with memory. One of the two European color TV transmission systems (see PAL)
SHF	Super-high frequency, that band of frequencies from 3000 to 30,000 MHz
SIAM	Society of Industrial and Applied Mathematics
SIC	Station identification code
SID	Sudden ionospheric disturbance (term used in HF propagation)
S/N	Signal-to-noise ratio
SPADE	Single channel per carrier multiple access demand assignment equipment

SQPSK	Staggered QPSK
SSB	Single sideband
SSBSC	Single-sideband suppressed carrier
star connection	A telephone/telecommunications network routing term implying tandem operation to interconnect exchanges (switches) which have a comparatively low level of traffic flow, providing a savings of the additional trunk that would be required for a full mesh connection.
STC	Standard Telephone and Cable, ITT's large British affiliate
STL	(1) Studio-to-transmitter link. A radiolink connecting a broadcast studio (usually TV) to its associated transmitter site. (2) Standard Telephone Laboratories (see STC)
SSS/STRATSAT	Strategic Satellite System (U.S. military)
SWL	Shortwave listener. One who listens to HF signals as a hobby
sync	Synchronization, synchronizing
T1, T2	ATT-Bell System series of PCM line arrangements
TASI	Time assigned speech interpolation
TASO	Television Allocation Study Organization (U.S.)
TCI	Technology for Communications, Inc.
TDA	Tunnel diode amplifier
TDM	Time division multiplex
TDMA	Time division multiple access
TE (i.e., TE_{11}) mode	Tranverse electric. In circular waveguide the TE_{11} mode is the dominant wave
TELESAT	A Canadian domestic satellite system
Throughput	The amount of useful information that a data transmission system can deliver end to end. Parity bits, start and stop elements, repeats due to errors on ARQ systems, and so on reduce throughput
TLP	Test level point, such as the 0 TLP or zero test level point
TO	Technical order. The U.S. Air Force calls their technical manuals technical orders or TO
TOA	Take-off angle (refers to antennas)
TRE	Transmit reference equivalent. The reference equivalent of the transmit portion of a telephone connection
TRI-TAC	Tri-service tactical communication system. An advanced three-service military tactical communication system developed for the U.S. forces
trunk	In the local area those circuits connecting one exchange (switch) to another. This is called a junction in the U.K. In

	the long distance area (toll area), any circuit connecting one long-distance exchange to another
T_{sys}	System noise temperature
TT & C	Telemetry, tracking, and command (control)
TV	Television
TVRO	Television receive only (satellite systems)
TWT	Traveling wave tube
TWTA	Traveling wave tube amplifier
UHF	Ultra-high frequency. The band of frequencies encompassing 300–3000 MHz
USB	Upper sideband
USITA	U.S. Independent Telephone Association
UVR	Useful volume range
UW	Unique word
VDU	Visual display unit
VF	Voice frequency, encompassing the band of frequencies from nearly dc to about 30,000 Hz. In this text we consider the nominal VF channel to be 0–4000 Hz; CCITT, 300–3400 Hz
VFCT, VFTG	Voice frequency carrier telegraph
VHF	Very high frequency. The band of frequencies encompassing 30–300 MHz
VITS	Vertical interval test signal (TV)
VNL	Via net loss. The method used in North America of assigning minimum loss in a telephone network to control echo and singing
VNLF	Via net loss factor. Term used in computing VNL; the unit is dB/mi
VOGAD	Voice-operated gain adjust device
VRC	Vertical redundancy check(ing)
VSB	Vestigial sideband
VSWR	Voltage standing wave ratio
VU	Volume unit. A measure of level, usually used for complex audio signals such as voice or program channel traffic. The unit is logarithmic. For a continuous sine wave across 600 Ω, 0 dBM = 0 VU. For a complex signal in the VF range, Power$_{(dBm)}$ = VU − 1.4 dB
WARC	World Administrative Radio Conference (ITU)
WMO	World Meteorological Organization
wpm	Words per minute

Z_0	Characteristic impedance
0 TLP	Zero test level point. A single reference point in a circuit or system at which we can expect to find the level to be 0 dBm (test level). From the 0 TLP other points in the circuit or network can be referenced using the unit dBr or dBm0, such that dBm = dBm0 + dBr
ϵ	Relative dielectric constant
λ	Notation for wavelength, usually given in meters
μ	10^{-6}
μ-law	The companding law used with North American PCM systems. (See also A-law)
Σ, σ	Sum, summation; standard deviation
Ω	Ohms
η	Antenna efficiency (in %)

INDEX